AF863500

INTERFACING TECHNOLOGIES IN SOLUTION MINING

PROCEEDINGS OF THE SECOND
SME—SPE INTERNATIONAL SOLUTION
MINING SYMPOSIUM
DENVER, COLORADO — NOVEMBER 18-20, 1981

Editor

W. J. Schlitt

Associate Editor
J. B. Hiskey

SPONSORS

SOCIETY OF
MINING ENGINEERS
OF A.I.M.E.

SOCIETY OF
PETROLEUM ENGINEERS
OF A.I.M.E.

Printed in the United States of America
by Lucas-Guinn Co., Hoboken, New Jersey

Library of Congress Catalog Card Number 82-71423
ISBN 0-89520-295-6

PREFACE

The emergence of solution mining as an interdisciplinary technology has prompted both the Society of Mining Engineers (SME) and the Society of Petroleum Engineers (SPE) of AIME to actively address different aspects of this broad field. In order to promote an exchange of information between researchers and operators working in the various areas, the two societies have initiated periodic joint programming on solution mining. SPE had primary responsibility for the first symposium. This was held in 1980 as part of the SPE Fall Meeting in Dallas, Texas. Due to the short time available for developing the program, that meeting included only three joint sessions focussing on uranium solution mining. Although many of these papers found their way into SPE publications, a separate symposium proceedings volume was not issued.

Before the first joint meeting even took place, SME initiated efforts to organize the second symposium. This was scheduled as part of the 1981 SME Fall Meeting in Denver, Colorado. Since more time was available for this undertaking, a substantial effort was made to broaden the scope of the symposium and to encourage extensive foreign participation. The organizing committee took this charge seriously. Representatives of the Industrial Minerals, Mineral Processing, and Mining and Exploration Divisions of SME all joined with their SPE counterparts to help develop a comprehensive program on solution mining. The committee also adopted the theme, "Interfacing Technologies in Solution Mining" to reflect the interdisciplinary nature of the symposium.

The response to the committee's call for papers was overwhelming, with more than 60 abstracts submitted for possible inclusion in the program. In order to accommodate as many of these papers as possible, five technical sessions were scheduled. These covered trona, brine, and salt; copper, gold, and silver operations; uranium operations; advances in technology and the chemistry of solution-ore systems. In addition to the United States, papers were accepted from Australia, Canada, China, France, Netherlands, and Papua New Guinea. The foreign portion made up about a third of the program.

Ninety percent of the contributions were available in preprint form at the meeting. These manuscripts have been subjected to a peer review, careful editing, and indexing. They are included in this volume along with the papers which were not available at the meeting. In addition, the two papers on operations at Ortiz and Bulldog Mountain have been added to the proceedings volume. These could not be accommodated in the symposium due to limitations on each session. However, the papers were judged to be substantive in value. They were presented at the general SME meeting and warranted inclusion in the book.

The joint SME-SPE fall meetings compliment other AIME programming and publications on solution mining. Particularly noteworthy have been the annual uranium seminars sponsored by SME local sections. The series of minisymposiums and proceedings developed by the SME Solution Mining Unit Committee has also been significant. These symposiums have been held as part of each AIME Annual Meeting and have included the 1979 New Orleans sessions on in situ uranium mining and ground water restoration, the 1980 Las Vegas symposium on leaching and recovering copper, the 1981 Chicago sessions on gold and silver, and the 1982 Dallas symposium on solution treatment. The latter was cosponsored with the Mineral Processing Division's Chemical Processing Committee.

W. J. Schlitt
March 1, 1982

SYMPOSIUM ORGANIZING COMMITTEE

W. Joseph Schlitt, Chairman

J. Brent Hiskey, SME-Mineral Processing Division

William C. Larson, SME-Mining and Exploration Division

Bruce H. Mason, SME-Industrial Minerals Division

Erik I. Sandvik, SPE

Myles W. Scoggins, SPE

* * * * * * * * * *

SOCIETY STAFF LIAISON

Ruth M. Orologio, SME

Janelle Stewart, SPE

* * * * * * * * * *

The organizing committee and the editors are most appreciative of the time and effort contributed by the authors in preparing their presentations and manuscripts. The editors would like to thank Myrna R. Anderson, Fred C. Dunford and Frances Hinkins of Kennecott Minerals Company and Marianne Snedeker of the Society of Mining Engineers for their efforts and support in preparing this volume. A special thanks also goes to Milton W. Wadsworth of the University of Utah who suggested the term "Interfacing Technologies in Solution Mining".

TABLE OF CONTENTS

Table of Contents (Continued)

Table of Contents (Continued)

SYMPOSIUM SESSIONS AND CHAIRMEN

KEYNOTE SESSION
W. J. Schlitt, Kennecott Minerals Co., Salt Lake City, Utah

TRONA, BRINE AND SALT
B. H. Mason, The France Stone Co., Waterville, Ohio

ADVANCES IN TECHNOLOGY
L. W. Lake, The University of Texas at Austin, Austin, Texas
J. L. Shafer, Exxon Minerals Co., Houston, Texas

CHEMISTRY OF SOLUTION - ORE SYSTEMS
J. B. Hiskey, Kennecott Minerals Company, Salt Lake City, Utah
E. I. Sandvik, Exxon Production Research Co., Houston, Texas

COPPER, GOLD AND SILVER LEACHING OPERATIONS
R. H. Heinen, Dravo Engineers and Constructors, Denver, Colorado
D. A. Shock, Ponca City, Oklahoma

URANIUM LEACHING OPERATIONS
M. W. Scoggins, Nufuels Corp., Denver, Colorado
W. C. Larson, U.S. Department of the Interior,
Bureau of Mines, Minneapolis, Minnesota

Chapter 1

PAST, PRESENT AND FUTURE OF SOLUTION MINING

R. B. Bhappu
Mountain States Mineral Enterprise Inc.
Tucson, Arizona 85731

ABSTRACT

Over the last decade, solution mining has immerged from its infancy into a technically and economically mature process for in situ extraction of selected commodities such as copper, uranium, gold, silver, and evaporites. Moreover, because of increasing labor and energy costs, environmental considerations, and decreasing ore grades, solution mining is destined to play an increasingly important role in the extraction of the above commodities as well as for others which are currently extracted by hydrometallurgical means. This paper traces the development of this versatile extraction technique, and as the keynote address, sets the stage for the technical papers on solution mining which make up the balance of the symposium program.

INTRODUCTION

In situ, chemical or solution mining is the in-place extraction of metals from ores located within the confines of a deposit or a mine (unfractured or fractured ore, stope fill, caved material, and ores in permeable zones), low-grade dumps, slag piles, and tailing ponds. These materials represent an enormous but untapped source of all types of metals.

It is inevitable that solution mining and in situ extraction concepts will be increasingly utilized in production of metals to satisfy the insatiable demands of our minerals-based society. This premise is based on reasonings such as ever increasing demand for today's metals, the necessity for treating complex and lower grade ores, increasing capital and operating costs, the need for conserving our dwindling mineral resources, and the public awareness of environmental problems associated with conventional mining practices. These factors necessitate the development of new techniques for extracting metals.

The overall scope of this potential mining method embraces interdisciplinary science and technology requiring application of principles of basic sciences, geo-sciences, bio-engineering, petroleum production, mining engineering, minerals extraction technology, and economics. The field of in situ mining encompasses the preparation of ore for subsequent in-place leaching, the flow of solutions and ionic species through rock masses and within rock pores, the leaching of minerals with inexpensive and regenerable leaching reagents under conditions prevailing in-place, the generation and regeneration of such solutions, and the recovery of metals or metal compounds from the metal-bearing solutions.

It is not inconceivable that eventually our ore reserves will consist largely of higher grade but deep seated deposits, low-grade zones near previously worked deposits, caved ore and stopes filled with submarginal ore, waste dumps, tailing ponds, and slag heaps. In situ mining promises economic recovery of metals from such sources.

Heretofore, this kind of mining has been more or less limited to the extraction of evaporites (trona, brine and salt), copper and uranium from lower-grade ores; however, it has a much greater potential. Practically all metals are susceptible to leaching in the in situ environment. Techniques will soon be developed for the in-place extraction and recovery of metals such as lead, zinc, nickel, manganese, silver, gold, molybdenum, and aluminum.

Historical

In situ leaching, though considered a modern technique, has been practiced since the last hundred years. Earliest patent coverage dates back to at least 1896. Copper dump leaching at the Rio Tinto Mines in Spain has been reported since the early part of this century. Cananea started leaching in place in the 1920's from mined out and caved underground stopes to extract residual copper from chalcocite ores using weak sulfuric acid solution containing ferric ions. Ohio Copper Company of Utah similarly started in situ leaching of mined out areas in 1942. In this case, stope pillars left between mined out veins were fragmented by blasting to obtain a rubblized area 1400 x 600 feet at the surface.

On the other hand, in situ extraction of uranium was initially conceived by Professor Livingston under U.S. Patent 2,818,240 titled "Method of Mining Ores In Situ by Leaching" in 1955. Utah Construction and Mining Company started in situ leaching of uranium in the Shirley Basin area of Wyoming around 1961 using the bore hole mining technique. Millike Mine of Rio Algom ceased underground mining operation in 1964 and converted to in situ leaching of uranium by washing down stopes with dilute acid solutions. The Stanrock Mine and Dennison Mines, located in the same district, also recovered uranium using a similar technique around 1965.

Twin Cities Mining Research Center of the United States Bureau of Mines [1] has compiled a detailed bibliography on in situ leaching technology.

TECHNOLOGICAL CONSIDERATIONS

Solution Mining Techniques

There are basically three distinct applications and associated leaching systems encountered in the field of solution mining, depending mainly in the physical location of the ore deposit. These situations are illustrated in Figure 1.

Typical examples of Type 1 are the numerous dump and heap leaching operations practiced by the open pit copper mines in the Southwestern United States. [2] Dump leaching, in general, is a convenience for open pit mining operations to dispose of lower grade (submarginal) stripped material in a remote area with the intention of recovering the valuable metals contained therein. Because of limited available area and easy placement, such dumps tend to be very high and the leaching of such piles, either through spraying or ponding of solution, is relatively slow and inefficient. Nevertheless, the production of copper from such dump and heap leaching operations at the present amounts to about 200,000 tons of copper per year in the United States or approximately 15% of the total production.

Heap leaching, on the other hand, is a system in which low grade or mixed oxide-sulfide ores are leached under controlled conditions in relatively small heaps on prepared pads or surfaces with positive drainage over a shorter leaching period. In some cases the ore is brought to permanent pads for treatment and removal, followed by piling of fresh ore on the same pads for treatment. Several successful heap leaching operations for extraction of copper, gold, and uranium are carried out in the U.S.A. It should be noted that the effectiveness of copper extraction in these dump and heap leaching operations has been partially attributed to bacterial oxidation of sulfide minerals.

Other viable in-place leaching situations belonging to Type 1 are the truly in situ leaching systems in which the ore is broken in-place by some type of blasting technique followed by percolation leaching of the fragmented ore mass. Appropriate collection galleries are provided to collect the leach solution for subsequent recovery of metal and recirculation of the barren solution for additional leaching. Since the ore is broken and left in place, such leaching systems are economically attractive.

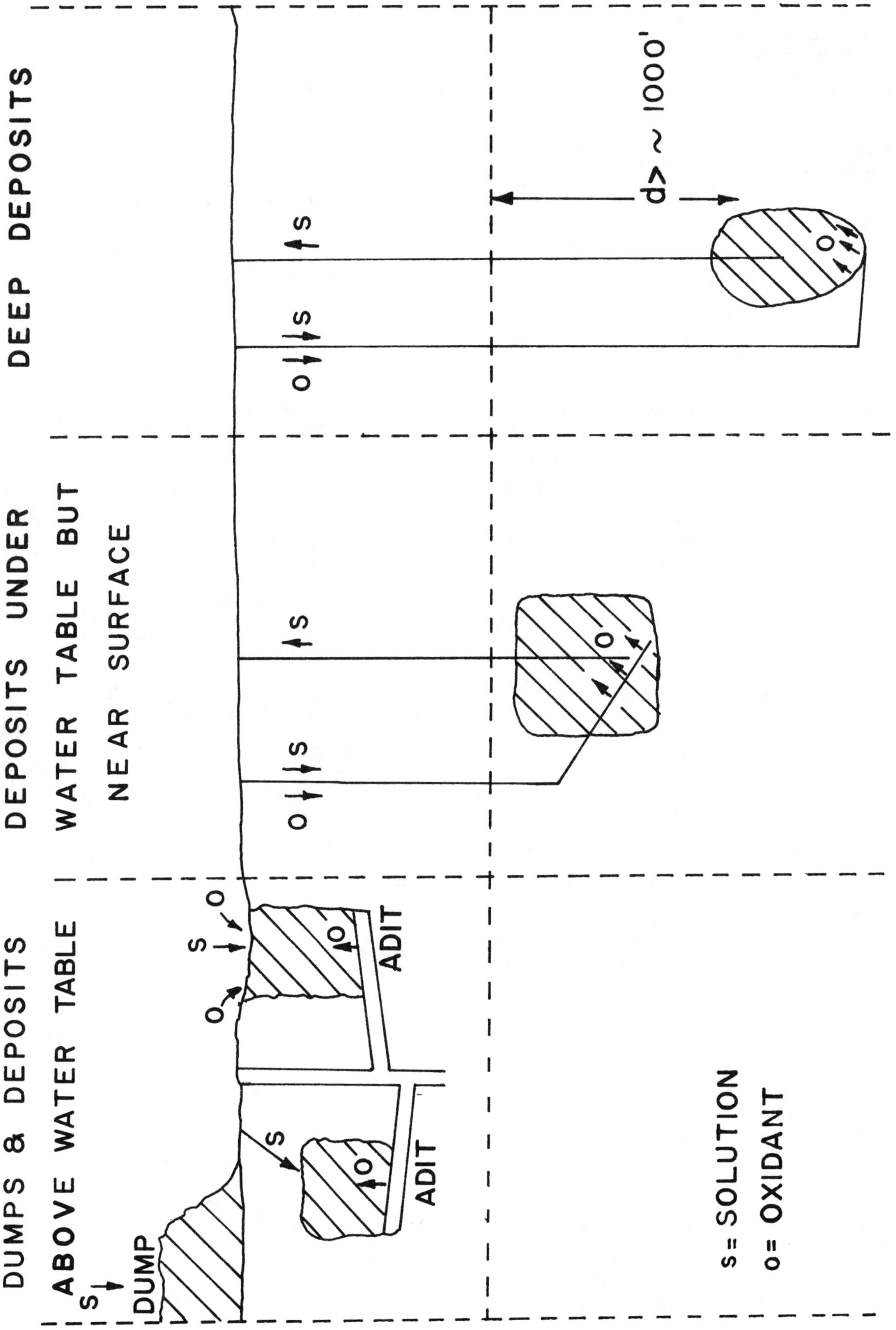

Figure 1 – Three types of in situ leaching situations. (According to Dr. Milton Wardsworth)

Ranchers Exploration and Development Corporation's Old Reliable Mine operation in Arizona and the Big Mike Mine in Nevada are fine examples of this in situ leaching method. [3-4]

Also, belonging to Type 1 are the classic in situ leaching operations of the Miami and Ray Mines, Arizona, U.S.A., in which the leaching has been carried out on fractured submarginal ore zones resulting from block caving operations. In the case of the Miami Mine operation, it has been estimated that 85,000 tons of copper metal have been produced over a 15 year period since completion of the underground mining operation in 1959, with an estimated recovery of about 60-70% of the metal left in place. The leaching operation continues with no decrease in recovery rate and no immediate end in sight. [5]

The second (Type 2) situation reflects the leaching of relatively shallow deposits located less than 300 meters from the surface and which are under the water table. Such deposits may need to be fractured in-place and the solution drained in order to commence the alternative leaching and drying cycles. Alternately, metal values from such deposits may be extracted by utilizing the bore hole mining technology as is currently practiced by uranium mining operations in the United States. The bore hole mining technique is also utilized for extraction of brines (salt and trona) and the modification of this system termed "slurry mining" may be used for mining phosphate rock, coal, uranium and copper ores.

The bore hole mining technique usually consists of a properly designed pattern of wells suitable for injection of the leaching solution and recovery of metal-bearing liquors. The wells are like water wells except that the casings are perforated for optimum leaching conditions. [6-8]

The final (Type 3) situations are characterized by relatively deep-seated deposits in excess of 300 meters and under the water table. Leaching of such copper deposits through chemical mining has been proposed by Lawrence Livermore Laboratory, University of California, U.S.A., after fracturing the ore body by nuclear devices. Leaching at such depths is enhanced by direct oxidation of sulfide minerals due to increased solubility of oxygen resulting from hydrostatic head. [9]

An alternate method proposed by University of Utah and Mountain States Research and Development for in situ extraction of deep seated copper deposits consists of effective rubblization through a modified block caving technique. In this case the required rubblization with about 8% porosity is achieved by employing conventional explosive to shatter the ore into a pattern of free space generated by mining appropriate lamellae every 100 feet of depth. Leaching would be carried out using modified bore hole mining technology with continuous oxidation provided by oxygen injection. [10]

A listing of various U.S.A. commercial and pilot plant operations employing the above three types of in-place leaching techniques in the extraction of copper, gold, and uranium are given in Table 1. These examples of practical and economic operations clearly indicate that the in situ extraction technology has developed into a matured and a viable metal extract process.

TABLE I

IN SITU LEACHING PRACTICES IN THE U.S.A.

Dump Leaching

1. Several copper operations in Southwest U.S.A.

Heap Leaching

1. Bluebird Mine, Arizona (copper)
2. Inspiration Mine, Arizona (copper)
3. Johnson Camp Mine, Arizona (copper)
4. San Juan Mine, Arizona (copper)
5. Carlin and Cortez Gold Mines, Nevada (gold)
6. Round Mountain Mine, Nevada (gold)

In Situ Leaching (copper)

1. Miami, Arizoana (block caving operation)
2. Ray, Arizona (block caving operation)
3. Mountain City, Nevada (block caving operation)
4. Old Reliable, Arizona (coyote blast)
5. Big Mike, Nevada (pit wall blast)
6. Zonia, Arizona (coyote blast)
7. Nuclear Blast (AEC - Kennecott concept for deep-seated deposits)
8. Kennecott's Experiment (near Safford, Arizona)
9. Occidental Mineral Co. (experiment at Miami, Arizona)

In Situ Leaching (uranium)

1. Utah Mining and Construction Company, Shirley Basin, Wyoming
2. Mine Water Leaching at Grants, New Mexico
3. Bacterial Leaching at Elliot Lake, Canada
4. Wyoming Minerals, Bruni, Texas
5. Wyoming Minerals, Ray Point, Texas
6. Arco-U.S. Steel-Dalco, George West, Texas
7. Union Carbide, Duval County, Texas
8. Exxon, Powder River Basin, Wyoming
9. Mobil Oil Corporation, Webb County, Texas

Advantages

1. In situ mining is environmentally more attractive, since it creates less surface disturbance and results in less water and air pollution.

2. In situ mining usually requires less capital and lower operating cost than a conventional mine and milling operation. Moreover, the total energy required in solution mining is much lower than that required to obtain the same quantity of metal by conventional processing methods.

3. In situ mining can often be used to recover metals economically from materials that could not be so treated by more conventional mining, milling, and smelting processes.

4. In situ mining increases a mine's ore sources and reserves. Low-grade or inaccessible ore zones, caved ore, and dumps and tailings may become ores, thereby

leading to more effective mineral utilization and conservation.

5. The leach solutions obtained through in situ mining usually lend themselves to a variety of metal recovery processes. The pure metal or metal compounds so obtained may be of greater value to a mine owner than the sulfide or oxide products normally obtained by conventional milling processes.

6. In situ mining can often be used in conjunction with a conventional mining or milling process to boost metal recoveries and increase ore reserves.

Disadvantages

1. Both physical and chemical restrictions may limit the usefulness of solution mining. The effectiveness of contacting ore with solutions and the recovery of leach solutions from the system without appreciable loss are two important physical factors. Dissolution or dissolution rates, metal precipitation, and solution regeneration chemistry are major factors.

2. Testing an in situ mining situation short of actual field operation sometimes proves difficult.

3. Ground water contamination may result from in situ mining operations.

4. Detailed information on the physical, chemical, and bacteriological factors involved in in situ extraction of metals and minerals has not been established commercially.

CONTROLLING PARAMETERS IN SOLUTION MINING

From a practical application viewpoint, the successful leaching of metals and minerals under in situ leaching environments is governed by several factors. These include elements of the leaching phase, geological and mineralogical considerations, solubility of minerals in various leaching reagents, effect of particle size, effective fragmentation, solution application and regeneration, and metal recovery methods.

Elements of the Leaching Phase

Accessibility, physicochemical interaction, and transport constitute the elements of the leaching phase involved in in-place extractions. Limitations imposed on any of these factors restrict the leaching process.

Accessibility is essential because interaction between the desired mineral and the lixiviant cannot take place in the absence of contacts, which depend on exposure and penetrability. The factors to consider are locations of the values, their volume and shape distribution, exposure area, specific surface, particle size, porosity, capillary pressure, viscosity, solubility of gases in leach solution, and surface roughness. Accessibility in most dump, heap and in situ situations also involves diffusion of lixiviant into ore fragments. This diffusion is governed by concentration gradient and diffusivity, which in turn are influenced by particle size, micropore radius, temperature, and molecular mass.

Physicochemical interaction converts the desired constituents from a fixed to a mobile condition and is governed by the solubility of the solid in leach solutions, vapor pressure in gases, and the presence of bacteria. Knowledge of free energies of reactants and products helps to determine whether a reaction is possible. The kinetic factors involved include time, concentration, diffusivity, specific rate constants, wettability, and oxidizing conditions, especially due to bacteria activity.

The first two elements by themselves do not ensure successful leaching without transport of products away from and reactants to the zone of interaction through diffusion and convection. As discussed earlier, diffusion is controlled by concentration gradient and diffusivity. On the other hand, convective flow concerns interparticle penetration and is restricted by pressure gradient, permeability, viscosity, and surface roughness. In the case of oxide copper and uranium leaching, "sweep" zones (i.e., moving reaction boundaries) are important due to rapid chemical kinetics.

Role of Mineralogy

In the past, little attention has been given by both the geologists and mineral processing engineers to the geological and mineralogical occurrences which may be potential sources of leachable ores under in situ environments. On the one hand, geologists have devoted less attention to lower grade, complex mineral deposits, and therefore, may have inadvertently excluded such deposits as potential sources. On the other hand, the metallurgists have neglected to study in detail the geological environments and mineralization of potential ores with the view of determining the critical leaching that could assist in effective processing of these complex, sub-marginal deposits.

In the final analysis, a chemical processing technique converts the valuable metals to a mobile state under conditions similar to those which deposited them originally in the host rock. Thus, the relationship between ore-genesis and chemical processing is close when the process of mineral or metal deposition in a particular ore deposit is clearly understood. Phase equilibria studies, Eh-pH diagrams, solubility products, mineral chemistry, surface chemistry and thermodynamics are utilized more and more by processing engineers to selectively separate minerals, dissolve minerals, and recover valuable metals and by-products. Doubtless, many of the future advances in processing technology will be derived from improved understanding of ore-genesis and related geological sciences.

It is interesting to note that the effectiveness of in solution mining of copper, uranium, and gold by appropriate reagents is economically feasible because of favorable mineralization of the metal values in the host rock. This is especially true for successful leaching of oxidized gold ores at coarser sizes since these ores contain submicron size gold particles in fracture fillings and cavities in the host rock consisting of either limey or silicified siltstone, sandy dolomite, or schist. These host rocks also tend to be porous and remain permeable throughout the leaching cycle. In the case of copper ores, the mineralization also occurs along macro and micro fractures in the host rock, while in the case of uranium, the values are in the form of secondary-enriched precipitates coating the sand grains in the sandstone host.

Solubility of Minerals

In general, nearly all the minerals are more or less soluble in either acids, alkalies, or other specific reagents such as cyanide. Most of the oxide-copper minerals such as azurite, malachite and chrysocolla are quite soluble in sulfuric acid. For this reason, sulfuric acid is the preferred leaching reagent for in situ extraction of oxide copper ores. However, in cases when the oxide copper ore contains considerable amounts of calcite ($CaCO_3$), the acid consumption is very high and it may not be economical to use sulfuric acid for leaching. Thus, the solubility of gangue contained in the ore may have profound influence in the selection of the appropriate reagent for leaching of valuable minerals.

As far as the sulfide copper minerals are concerned, simple sulfuric acid leaching is not very effective and the presence of $Fe_2\ (SO_4)_3$ is necessary to dissolve minerals such as chalcocite and chalcopyrite under in situ leaching conditions. The strong oxidizing conditions required for their dissolution are satisfied by the presence of ferric iron, and oxygen (air). Many commercial operations have failed because the required oxidizing conditions were not maintained during the leaching process. This is also the reason why the presence of oxidizing bacteria of the Thiobacillus thiooxidans is necessary in dump, heap, and in situ leaching operations. It should be noted that the required iron is either available in the natural ore or that it is added to the leach system through the cementation step in which copper from the leach solution is precipitated as elemental copper when the pregnant solution is contacted with scrap iron.

In the case of in-place leaching of uranium, both dilute sulfuric acid and carbonate-bicarbonate solutions are effective reagents for dissolving uranium minerals. However, both the reagents require a chemical oxidant to oxidize uranium from the insoluble +4 oxidation state to the soluble +6 state. Typical acid leach oxidants are $NaClO_3$ or MnO_2 and typical carbonate leach oxidants are hydrogen peroxide or NaOCl.

For gold and silver leaching, cyanide has been found to be the most selective reagent and is the universal solvent used. Here again, oxygen is required in the reaction and efforts must be made to maintain oxidizing conditions during the in-place leaching of gold and silver. [11]

Effect of Particle Size

Since leaching of minerals by various chemical reagents involves effective contact between the two, it is obvious that the mineral must be liberated or freed from the host rock. The only other way the reagent can contact the valuable mineral is by effective penetration or diffusion through fractures, cracks, capillary action, and other physical factors considered under the "elements of the leaching phase". Regardless of the process of contact, it is quite evident that the reduction in particle size would be of considerable benefit in in situ leaching of metals and minerals. With finer ore size it would be possible to obtain higher extraction rates in shorter leaching periods.

Effective Fragmentation

The desired particle size in in situ leaching situations can be obtained through proper blasting and fracturing techniques. The blast design for in situ leaching closely follows the principles of blast design used in more conventional mining practices. However, in case of in situ blasting unusually large dimensions are involved and these pose inherent problems associated with large blasting situations. To date, several unusually large blasts have been made for leaching with results that are very encouraging. [12-15]

The important factors that need to be considered in engineering a blast for in situ leaching are (a) shape of deposit, (b) characteristics of ore body, (c) seismic activity encountered, (d) missile consideration, (e) type of explosive, (f) powder factor and distribution, (g) initiating and delay system, (h) loading logistics, (i) economics, and (j) safety.

Besides conventional fracturing by blasting compounds, efforts have been made to explore the possibility of using liquid explosives, hydrofracing and nuclear explosives. No doubt these and many other new concepts for fracturing will be considered as in situ leaching becomes a more conventional method in the future. Because of environmental considerations, nuclear fracturing although proved to be quite effective, may not be applicable in the foreseeable future. Hydrofracing, on the other hand, has shown considerable promise in limited applications, [16] and with additional effort may prove to be a versatile fragmentation technique.

Solution Application and Regeneration

The primary goals in solution mining are (1) to obtain maximum metal concentration in the minimum solution volume and (2) to recover a major portion of the leach solution fed to the system with an economic metal concentration over the life of the operation. This economic concentration should be such that the value of recovered metal will offset the cost of operation, amortization, overhead and profit. It is also important from environmental considerations that the leach solution be confined in the leaching system and not appear as a pollutant in ground water. This is especially important in gold and silver leaching using cyanide as the specific reagent. Both the toxic nature of cyanide and preciousness of leach solution containing gold and silver necessitates in-place leaching conditions approaching a "zero discharge" objective.

Inasmuch as the chemical reagents used in in-place leaching generally constitute a major cost item and greatly influence the economics of leaching, reagent generation and regeneration play a very important part in any solution mining process. Some reagents can be generated and regenerated by natural processes in the leaching cycle, whereas others are totally consumed and need to be replenished.

The only known example of auto-regeneration of reagent is the natural reactions involved in the production of sulfuric acid and ferric sulfate from pyrite and spent ferrous sulfate solution under natural air or biological oxidation conditions. In some cases, part of the reagent is regenerated by the metal extraction process incorporated in the flowsheet, such as the regeneration of H_2SO_4 through solvent extraction and electro-winning as practiced in the extraction of copper from leach solutions. In other cases, though the regeneration of reagent from spent solutions is possible, it is not economically feasible or environmentally attractive such as the regeneration of cyanide from spent solution in gold leaching.

The Recovery of Metals from Leach Solutions

The final phase of any hydrometallurgical process, including solution mining, is the recovery of metals from leach liquors. Conventional purification of a metal-containing solution followed by recovery of metals or compounds from the solution either by chemical or electrolytic precipitation is employed to obtain the marketable product. These recovery techniques are adequately covered in the literature and their effectiveness is clearly demonstrated in several successful plant practices.

In connection with in situ leaching application, however, the recovery phase poses certain technical problems that may influence the overall effectiveness of the process. One such difficulty concerns treating a large volume of very dilute metal-bearing solution. This may require recirculation of the leach solution to build up the metal concentration followed by bleeding off a small part of the concentrated leach stream for metal recovery.

Newer techniques of ion exchange, solvent extraction, and charcoal adsorption are being used for effective concentration of metals from leach solution. Moreover, these procedures have proved very effective for processing large volumes of leach solutions containing more than one valuable metal. In recent years, the recovery of copper from leach solutions by solvent extraction-electrowinning has become more attractive than the copper cementation process used conventionally all over the world. Similarly, ion exchange has been exclusively used for recovery of low concentrations of uranium from in situ leaching operations. Finally, the use of activated carbon for the recovery of gold and silver values (as low as one part per million or less) from in-place leaching operations has become a favorite process. In this case, the precious metals are subsequently extracted by desorbing them from carbon by hot caustic-cyanide solution followed by electrolysis to obtain the dore.

ECONOMIC CONSIDERATIONS

From the economics viewpoint, the net return from a solution mining venture will depend upon the quantity and quality of metal or commodity extracted from the ore. The overall economics will also be influenced by the capital cost of unit processes such as cementation, solvent extraction, electrowinning and the like; the cost of utilities, supplies and reagents; and the prevailing prices of metals and by-products. Accordingly, it is imperative that realistic feasibility studies be carried out for any new solution mining venture in light of all these and many more influential factors.

It should be noted that feasibility studies play a very important role in evaluating solution mining projects since the percentage recovery of a metal under in situ leaching environment is very difficult to predict. The recoveries attained under actual operating conditions may not be as high as those indicated by laboratory or pilot plant tests. For this reason, it is customary in such feasibility studies to carry out cash flow analyses for a wide range of recovery levels (60 to 20%). If the cash flow does not appear to be favorable for 50 to 40 percent recovery levels, then it would be preferable to drop the project or undertake additional laboratory and field studies in order to increase the confidence level. On the other hand, if the cash flow looks favorable for 30 to 20 percent recovery levels, then the success of the venture is assured. Besides recovery levels, it is mandatory that cash flow analyses be made for varying metal prices since the economics of the venture is highly influenced by the price of metal or commodity.

In general, the net return from a mining venture such as in situ leaching depends on the quantity and quality of metal or compound extracted from the deposit. The overall economics is also dictated by the cost of exploration, capital costs of the unit operations involved in mining and extraction, the operating costs for labor, supervision, utilities and supplies; and on the prevailing price of metal and by-products. Accordingly, the economic feasibility of a solution mining venture must be accessed in light of all these and several other influencing factors.

The estimated capital costs of various unit processes encountered in typical solution mining operations for extraction of specified amounts of copper, gold and uranium are shown in Table II. On the other hand, Table III shows the estimated costs for different unit operations involved in extraction of the same commodities under in situ leaching extractions.

It is worth noting that in the case of copper, the cost of producing the metal by in situ leach-solvent extraction-electrowinning is considerably less than for cementation due to additional cost of smelting and refining (currently ranging from \$0.26-0.36 per pound) to produce high purity copper. On the other hand, for extraction of gold using the heap leaching-carbon adsorption-electrowinning concept, the direct cost of mining and processing is estimated to be less than \$5.00 per ton or less than \$7.50 including indirect, depreciation, etc. Accordingly, using this technique, it is currently feasible to process gold ores assaying 0.03 to 0.05 oz/ton if the gold price is \$400 per ounce.

Finally, for the extraction of uranium using the bore hole mining technology at a cost of \$17.50 to \$25.00 per pound of U_3O_8 extracted, it is feasible to work a uranium deposit averaging 0.05% U_3O_8 when the price of yellow cake is above \$25 per pound.

The effectiveness of in situ extraction of copper in comparison to conventional mining-milling-smelting-refining is clearly illustrated by comparing the operating costs as shown in Table IV. As can be seen, the overall cost per pound of copper produced by solution mining is in the range of \$0.45 to \$0.55 as compared to \$0.69 to \$0.88 for conventional method.

TABLE II

ESTIMATED CAPITAL COSTS FOR UNIT PROCESSES

Unit Processes	Basis	Estimated Cost
Copper (48,000 lbs./D;4,000 gpm at 1.0 gpl Cu)		
Cementation	\$100,000-115,000/T	\$2,000,000-2,300,000
Solvent Extraction	\$215-285/ft^2 settling area	\$4,300,000-5,700,000
Electrowinning	\$85-100/lb. cathode Cu	\$3,400,000-4,000,000
Total SX-EW	\$195-245/lb. Cu/D	\$7,700,000-9,700,000
Gold (200 oz/D; 2,000 gpm; 0.016 oz Au/T solution)		
Carbon Adsorption	\$350-430/gpm	\$ 700,000- 860,000
Desorption-Electrowinning	\$2500-2600/oz Au	\$ 500,000- 520,000
Total ADS-DES-EW	\$6000-7000/oz Au	\$1,200,000-1,400,000
Uranium (2,750 lb/D; 2,000 gpm at 0.125 gpl U_3O_8)		
Ion Exchange	\$2500-2850/gpm	\$5,000,000-5,700,000
Precipitation	\$655-725/lb U_3O_8	\$1,800,000-2,000,000
Total IX-PPT	\$2470-2800/lb U_3O_8	\$6,800,000-7,700,000

Note: (1) Tons (T) shown in the tables refer to short tons.

(2) Conversions: 1 foot (ft) = 0.3048 meters (m)

1 $foot^2$ (ft^2) = 0.0928 $meters^2$ (m^2)

1 pound (lb) = 453.6 grams (g)

TABLE III

ESTIMATED DIRECT OPERATING COSTS FOR IN-PLACE EXTRACTION

Unit Operation	Estimated Cost
Fracturing in place (with development)	$0.40-0.52/T
Mining (drilling, blasting, hauling and dumping)	$0.52-0.65/T
Crushing (primary, secondary, tertiary)	$0.14-0.28*-0.42*
Copper ore (averaging 0.5-1.0% Cu)	
Leaching (heap; pads; 4.0 lb H_2SO_4/lb Cu)	$0.45-0.58/T
Leaching (in situ, 4.0 lb H_2SO_4/lb Cu)	$0.29-0.42/T
Cementation	$0.13-0.16/lb
Smelting, refining, and transportation	$0.26-0.36/lb
Solvent extraction-electrowinning	$0.09-0.12/lb
Cost/lb Cu (in situ-cementation)	$0.51-0.60
Cost/lb Cu (in situ-SX-EW)	$0.20-0.29
Cost/lb Cu (heap-cementation)	$0.65-0.85
Cost/lb (heap-SX-EW)	$0.34-0.55
Gold ore (averaging 0.03-0.06 oz/T)	
Leaching (in situ; 1.0 lb NaCN; 3.0 lb CaO)	$0.58-0.72/T
Leaching (heap with pads; 1.5 lb NaCN; 3.0 lb CaO)	$0.72-0.85/T
Carbon adsorption, desorption & electrowinning	$0.21-0.26/T
Cost/T (in situ-carbon adsorption-EW)	$1.92-2.05
Cost/T (heap with pads; carbon adsorption-EW)	$2.33-2.46
Uranium ore (averaging 0.05-1.10% U_3O_8)	
Bore hole drilling (with casing)	$17.00-23.00/ft
Injection-leaching-recovery	$ 2.60- 3.90/lb U_3O_8
IX-U_3O_8 precipitation	$ 6.50- 9.00/lb U_3O_8
Total Cost (in situ leaching-IX-PPT	$17.50-25.00/lb U_3O_8

*Cumulative at each stage.

TABLE IV

COST COMPARISON (0.66% Cu = 8.0 LB/TON RECOVERED)
OPERATING COSTS PER TON ORE

Unit Operations	Conventional (strip ratio = 3.0:1)		In Situ
Drilling, Blasting, Hauling, Dumping	$1.90-2.30	Drilling, Blasting	$0.50-0.70
Crushing	$0.20-0.40		
Ore Handling	$0.13-0.20		
Leaching	$0.60-0.70		$0.60-0.70
Metal Recovery (SX-EW)	$2.00-2.40		$2.00-2.40
Supervision and Administration	$0.70-0.80		$0.50-0.60
Total	$5.53-6.80		$3.60-4.40
Cost/lb	$0.69-0.85		$0.45-0.55

FUTURE POTENTIAL

The successful application of solution mining to saline deposits in the past, and in situ extraction of uranium and copper during the last decade have clearly indicated the emergence of this versatile technique from its infancy into a technically and economically mature process. It is also evident that this newly developed extraction process may be applicable to practically all metals and minerals which are amenable to conventional hydrometallurgical techniques and which could be made susceptible to dissolution under in situ environment. Accordingly, potential candidates for the extension of the in situ mining concept include metals such as gold, silver, zinc, nickel, cobalt, manganese and aluminum.

Recent USBM-sponsored studies by Mountain States Research and Development have vividly shown the feasibility of extracting gold, silver, manganese, nickel, cobalt, and aluminum by solution mining. [17] The selection of these potential candidates for in situ extraction was based upon strategic and economic importance to the U.S. economy, marketability, probable amenability to in situ solution mining and recovery processes, environmental consideration, and economic feasibility of individual deposits.

Details regarding the above studies are provided in the paper titled "Feasibility of In Situ Leaching of Metallic Ores Other Than Copper and Uranium" to be presented under the Advances in Technology Session of this Symposium.

Although the mining industry in the future will be attempting to recover as many metals as practically and economically feasible by solution mining techniques, the greatest impact of in situ extraction will be centered around copper, especially in the United States. Existence of several deep-seated primary copper deposits in the southwestern United States has been well documented. Such deposits located in excess of 1000-meter depth and averaging about 1.0% copper cannot be mined at a profit now or in the future by conventional mining and processing techniques. These same deposits, however, may be worked economically by in situ mining technology and considerable efforts will be made in the future to develop the required technology for successful application of solution mining to recovery of copper from these deposits.

Finally, the future will see increasing use of solution mining techniques for recovery of residual coper and uranium left in the abandoned or recently worked out mines. USBM, in cooperation with Mountain States Research and Development, has identified several abandoned copper and uranium mines which could be subjected to in situ extraction techniques to recover residual copper and uranium values. [18]

RESEARCH NEEDS

Although in situ leaching in recent years has matured into a viable metal extraction process, considerable additional research is needed to bring it in competition with other high-confidence and universally acceptable processes.

The principle areas of research required to accomplish the above goal are:

1. Improved techniques for measuring physical properties of ores in-place.
2. Improved techniques for drilling and blasting to produce desired fragmentation.
3. Improvement in injection and recovery systems.
4. Improved understanding of hydrology in broken ore masses.
5. Design of leaching systems for maximum oxidation.
6. Basic studies on leaching of dispersed mineral particles.
7. Improved methods of scaling up from laboratory and field tests.

8. Determining optimum parameters for leaching of massive deep-seated deposits.

9. Improved environmental controls.

REFERENCES

1. Larson, W.C. and D'Andrea, D.V. "In Situ Leaching Bibliography", Parts I and II, U.S.B.M., February, 1976.

2. Sheffer, H.W. and Evans, L.G. "Copper Leaching Practices in the Western United States", U.S.B.M. IC 8341, 1968, 57 pp.

3. Rudershausen, C.G. "Copper Solution Mining at Old Reliable", Pres. at National Meeting of American Inst. of Chemical Eng., Salt Lake City, Nev, August 18-21, 1974, 11 pp.

4. Ward, M.H. "Surface Blasting Followed by In Situ Leaching the Big Mike Mine", Proc. Solution Mining Symp., AIME Annual Meeting, Dallas, Texas, February 25-27, 1974 pp. 243-251.

5. Fletcher, J.B. "In-Place Leaching - Miami Mine, Miami, Arizona", Pres. at AIME Centennial Annual Meeting, Soc. of Min. Eng., AIME, New York, N.Y., March, 1971, AIME Preprint 71-AS-40; Trans. Soc. of Min. Eng. AIME, 250, 324-337, December, 1971.

6. Crawford, E. "Developers Eye Texas Potential for In Situ Uranium Leaching", Engng. Min. J. 176, 81-82, 1975.

7. Shock, D.A. and Conley, F.R. "Solution Mining - Its Promises and Its Problems", Proc. Solution Mining Symp., AIME Annual Meeting, Dallas, Texas, February 25-27, p. 79-97, 1974.

8. Steckley, R.C., Larson, W.C. and D'Andrea, D.V. "Blasting Tests in a Porphyry Copper Deposit - In Preparation for In Situ Extraction, U.S.B.M. RI 8070, 1975, 57 pp.

9. Lewis, A.E. "Chemical Mining of Primary Copper Ores by Use of Nuclear Technology", Proc. Symp. of Engineering with Nuclear Explosives, Las Vegas, Nev., January 14-16, 1970, Conf. 700101, 21, 907-917, 1970.

10. Wadsworth, M.E. and Pitt, C.H. "An Assessment of Energy Requirements in Proven and New Copper Processes", USDOE Contract No. EM-78-S-07-1743, December, 1980.

11. Heinen, H. and Porter, B. "Experimental Leaching of Gold from Mine Waste", U.S.B.M. RI 7250, April, 1969, 5 pp.

12. D'Andrea, D.V., Dick, R.A., Steckley, R.C. and Larson, W.C. "A Fragmentation Experiment for In Situ Extraction", Proc. Solution Mining Symp., AIME Annual Meeting, Dallas, Texas, February 25-27, 1974, pp. 148-161.

13. Dick, R.A. "In Situ Fragmentation for Solution Mining - A Research Need", Pres. at the Second Internat. Symp. on Drilling and Blasting, Phoenix, Arizona, February 12-16, 1973; Available for consultation at U.S.B.M., Twin Cities Mining Research Center, Minneapolis, Minnesota.

14. Porter, D.D. "Blast Design for In Situ Leaching", Pres. at Southwest Mineral Industry Conf., Phoenix, Arizona, April 27, 1973, 7 pp; Available from E.I. duPont de Nemours and Co., Inc., Wilmington, Del. 19898.

15. Porter, D.D. and Carlevato, H.G. "In Situ Leaching: A New Blasting Challenge", Proc. Solution Mining Symp., AIME Annual Meeting, Dallas, Texas, February 25-27, 1974, pp 33-43.

16. Private communication and classified data in Mountain States Mineral Enterprises files.

17. Mountain States Research & Development, "Technological Investigation to Determine the Feasibility of In Situ Leaching Metallic Ores Other Than Copper and Uranium", USBM Contract No. J0295032, April, 1981.

18. Mountain States Research & Development, "Feasibility of In Situ Leaching of Depleted Underground Copper and Uranium Mines", USBM Contract No. J0295045, August, 1981.

Chapter 2

THE HISTORY AND PRESENT STATUS OF PRACTICE AND RESEARCH WORK ON SOLUTION MINING IN CHINA

Yao Dun Pu
Changsha Mining Research Institute
Changsha, Hunan, China

ABSTRACT

In this special chapter the history and current practices of solution mining in China are presented. Historically copper ore leaching has been practiced for about 1,000 years and was well-developed as early as the Soong Dynasty (960 - 1279 A. D.). Unfortunately, the practice remained unaltered for centuries and the relevant technical literature was also lost as time went on.

Research work and practical applications have only resumed in the last two decades. As a result, the state-of-the-art of solution mining in China is still lagging behind that of other highly industrialized countries. Current solution mining procedures at three non-ferrous metal mines are described, showing the present level of technology in China. These are followed by an account of recent research activities and of investigative reports published to date. The conclusion is that solution mining will certainly open a new and promising approach for the development and exploitation of mineral resources in China.

INTRODUCTION

Solution mining, sometimes referred to as chemical mining or bacterial mining, is an innovative technology which makes use of proper chemical reagents and takes advantage of biochemical action by some microorganisms to effectively dissolve most desirable metallic values from ores. Then the resultant metal-bearing solution is treated by a suitable means to form the metal product. Solution mining is preferable to conventional mining methods not only because of its capability to improve working conditions and minimize environmental problems either underground or on surface, but also for its application in exploiting low-grade or submarginal ore bodies or those deposits which are too deep-seated to be mined out economically. Furthermore, it also provides a potential approach to win metals directly from their ores or ore deposits in situ by simplified methods and at a lower cost than the conventional mining-milling-smelting process. Hence, in developing countries such as China, solution mining will most probably become a new mining tool of great value (1).

HISTORICAL DESCRIPTION

China is an ancient country with a long history of scientific development. The history of solution mining in China may date back to medieval age. The earliest record was found in a historical volume named the "Huainancius," a book written by Liu-An (177-122 B.C.) who was King of Huainan Kingdom in the West Han Dynasty. It stated, "Being attacked by iron, the blue vitriol turns into copper." A similar statement, "The bluestone can convert iron into copper," was also found in "Shennong's Classics of Materia Medica," a well-known pharmacopoeia of traditional Chinese medicine written by some pharmacologists in East Han Dynasty (25-220 A.D.) in the name of Shennongshi, the God of Agriculture and Medicine in China's prehistoric fairytales. The so-called blue vitriol, or blue-stone, or gall-stone, as it was denoted in other classical literature, was essentially chalcanthite or copper sulfate ($CuSO_4 \cdot 5H_2O$) in its natural state. This mineral formed by progressive oxidation of sulfides such as chalcocite, chalcopyrite, or bornite in contact with moist atmosphere. When contacted by surface or underground water, the sulfate dissolved, trickled down through the ground, and emerged as a gall-spring, i.e., a spring containing natural copper sulfate solution. It was so named simply because it tasted as bitter as gall.

Guo-Hoong (284-364 A.D.), a great chemist in the East Jin Dyansty, reported in his famous work, "Baopucius" that the color of iron could be changed to the red of copper if its surface were covered with a layer of blue vitriol, but this change would occur only superficially, leaving its interior unaltered.

These historical notes implied that the ancient Chinese had discovered not only the natural copper sulfate mineral, but also the significant phenomenon that this soluble copper salt could react with iron and yield metallic copper by replacement. This laid a sound basis for the invention and development of the gall-copper process in the subsequent years.

The gall-copper or leached-copper process was defined as winning metallic copper from gall-spring water by means of cementation with iron. The copper thus produced was termed gall-copper, leached-copper, or iron-copper. It had been identified as one of those ten types of normal copper production during the Five-Dynasties period (907-960 A.D.), as indicated by Shiangyuanshu in his paper entitled "On Precious Resources."

The gall-copper process was one of the most significant inventions in China's history of mining and metallurgy and dates back to the Tang Dynasty. It must also be considered as another important contribution made by the Chinese people to human culture and technology (2,3,4).

The gall-copper process was initially put into practice in the Tang Dynasty and became rather widespread in the Soong Dynasty.

"A Sketchbook of Study on History and Geography," written by Gu Zhu Yu (1631-1692 A.D.) noted: "Tungkwangshan is situated 5 kilometers south from the county town of Tungling, Chizhou Prefecture. Many springs are widespread there, which never exhaust either in summer or winter. The spring water may be used to produce copper by immersing iron into it. Copper-producing workshops run by the government had been set up there since the Tang Dynasty." This reveals that the gall-copper process had been put into practice on a state-run scale as early as the seventh or eighth century (5). Interestingly, Tungkwangshan is just the place where the Tungling Non-Ferrous Metals Co. is now situated.

Prior to and during the North Soong Dynasty, the gall-copper process was developed further. A special chapter, "On Foods and Goods" in the "Compendium of Soong Dynasty" had noted that there were 11 gall-copper workshops to the south in the lower reaches of Yangtse River. This was based on a statistical survey by an official who was responsible for copper production affairs in that area (6). Another book, "A General Investigation on Historical Literature" written by Ma Duang Lin (1254-1323 A.D.) in the Yuan Dynasty, noted: "Among all the prefectures and counties, those to the south in the lower reaches of the Yangtse River were the most prosperous in leached-copper production, where the total number of workshops exceeded 50" (7).

But the "Sketchbook of Study on History and Geography" said there were 77 workshops erected in the Soong Dynasty. No matter how diverse these figures are, they furnish evidence that the gall-copper process had been well developed and widely adopted in the Soong Dynasty. The latter Sketchbook also reported that this process got its momentum in 1135 (in the reign of Shaoxing), flourished further in 1182 (in the reign of Chunshi), and generally began to fade away after 1241 (in the reign of Chunyu). That is to say, quite a number of leached copper workshops, within the period of 1131-1241 when the gall-copper process was at its full bloom, were built and widely scattered over the vast area to the south of the Yangtze River, including Jiangxi, Fujian, Hunan, Zhejiang, and the southern part of Guangdong Province. Among them, three workshops, i.e., Chiangshan workshop in Xinzhou, Xingli workshop in Dexin, and Chenshui workshop in Shouzhou, were reportedly the largest and best known ones in the whole country. The Chiangshan workshop was founded in 1096 and had a rated capacity of nearly 190,000 kg Cu/yr. The Chenshui workshop had been built up earlier and was much larger in scale. In 1117 it produced more than 500,000 kg of leached copper. The labor force employed in each workshop peaked at 100,000. The sight of these in situ operations must have been really magnificent at that time.

Regarding productivity, "Miscellanea of Governmental and Nongovernmental Events Since the Reign of

Jiangyen (1127-1131 A.D.) in the Soong Dynasty" reported: "In the reign of Daguan (1107-1111 A.D.), total production of copper exceeded 3,300,000 kg per year, over 500,000 kg of which were gall-copper." By the end of the N. Soong Dynasty the annual output of gall-copper amounted to approximately 15 to 25% of the total national copper production. During the S. Soong Dynasty the overall production dropped sharply and the gall-copper output also decreased to only 100,000 kg annually, but its proportion of the total production increased to 85% and, hence, gall-copper became more critical to the national economy.

In the meantime, the practice of gall-copper production underwent improvement as well. There were two major procedures alternatively adopted, namely, the thermal method and the wet method, depending upon the occurrence of the mineral resources.

The Thermal Method

Shen-Kuao (1030-1094 A.D.), the outstanding scientist in the N. Soong Dynasty, in his famous academic writing "Sketches and Notes at Nengshi" described this method as follows: (8) "There are many bitter springs in Chiangshan County, Xinzhou Prefecture, which flow out continually like streams. By fetching some water from such streams and evaporating it, copperas alum will result. Further heating of the copperas alum in an iron vessel gives metallic copper and the vessel used for heating will also turn into copper in the long run." This indicated that the thermal method was carried out by heating successive amounts of gall-spring water in iron vessels to first produce concentrated copper-sulphate and then to convert it to metallic copper.

The Wet Method

It was noted in "The History of Soong Dynasty" that: "The process for leached-copper production involves the following steps. Thin plates made of forged iron are placed in a row of troughs wherein they are covered with gall-spring water. After immersion for a few days, these forged iron plates are corroded by the gall-water and become thinner, resulting in a reddish layer of soot on their surface. Scrape these soots down and melt and refine them three times, then you will obtain copper. Roughly, 2.25 kg of iron plate should be consumed for 1 kg of copper produced."

A detailed description of the process adopted by the Chiangshan workshop was also given in the "Sketchbook" cited earlier. It stated" "Each workshop stored the gall-spring water separately in its own ponds. Channels or troughs were built with timber slabbing to follow the topographical elevation and undulation of the surface. Straw mats were then laid on the bottom of the trough and a layer of cast iron which had been broken into pieces beforehand was placed upon the mats and arranged as required. The gall-spring water was then allowed to pass through and cover the cast iron until the latter had drastically changed in color. The cast iron was eventually taken out and forged to get copper."

Each method had its advantages and disadvantages. For the wet method, less labor was required and the profit was greater, but the source of gall-spring water might be limited. For the thermal method, the source of supply might be plentiful, but profit was less, as reported in "The History of Soong Dynasty."

As a matter of fact, the thermal method may be regarded as pyrometallurgical and the wet method, hydrometallurgical. It is noteworthy that the reaction mechanism, the operating procedure and production facilities of the latter were generally similar to the current leach-cementation process used in copper hydrometallurgy. So the accomplishment of the Chinese people in this field early in the 11th century was so splendid that it is of world wide significance to the history of metallurgy.

Furthermore, from a modern scientific-technological point of view, the gall-copper process was essentially an in-situ solution-mining process treating natural copper-bearing water. Historically, the central-southern area of China was rich in copper resources, a majority of which were porphyry deposits with a jointed and fractured struture. The outcrops and near-surface portions had been partially oxidized, forming readily soluble secondary minerals. Additionally, the climate there was rather warm and moist with plenty of precipitation all year round. Natural waterbodies, such as rivers, lakes, ponds and reservoirs, were widespread. All these made conditions favorable for the natural water to reach the subsurface ore deposit by seeping down along the fissures and cracks of the overlying strata, then the solution could percolate through the ore, dissolving the soluble metallic values. By mixing with undergournd water and being under hydrostatic pressure, the water containing soluble copper salt was forced to emerge at lower levels and form a gall-spring. This afforded the native inhabitants an inexhaustible source of raw material to make copper by the gall-copper process. Even today some peasants in Chiangshan, Dongshian or other counties in Jiangxi province are still using this method to make copper from mountain spring water as a sideline.

Hence, there is no doubt that solution mining originated in ancient China and was put into practice as the gall-copper process with a more or less advanced technique and on a considerable scale early in the 11th century. This dates back hundreds of years earlier than both the Hungarians and Spaniards who practiced their copper extraction from mine water and dump leaching in the 16th and 17th century respectively.

Furthermore, in the reign of Shaoshen (1094-1098 A.D.), Zhangchian and his son Zhangjiao from Dexin County, compiled a book entitled "The Principles and Guidelines of Copper Leaching", which summarized all techniques and experiences relating to copper leaching known at that time. Most probably, it is the earliest technical literature on solution mining produced anywhere in the world. Unfortunately, this book had disappeared by the Ming Dynasty and its broad outline could only be visualized from "The Preface to the Principles and Guidelines of Copper-Leaching" written by Wei-Shu at the end of Yuan Dynasty. (2)

Owing to social and economic conditions the production of leached copper declined gradually and no important modification was made after the Yuan Dynasty.

CURRENT PRACTICES

The metal mining industry in China has been vigorously expanding since the founding of the People's Republic. As metal mine production increases, solution mining practice makes progress as well. Tests and experiments were initiated in the 1960's and many mines began to adopt solution mining in the past decade. Initially, it was utilized as an adjunct to the conventional mining methods, at recovering metallic values from tailings, wastes, marginal or submarginal ores and remnants in worked-out areas. However, China is now a developing country with plenty of metalliferous mineral resources but temporary lack of energy and funds for industrial construction. Concurrently, some of her mineral resources, though extensive, are generally rather poor in grade, complex in composition and consequently more or less refractory for processing and smelting. It is therefore reasonable that solution mining, emerging as a new mining tool with low captial investment, rapid profit return, low energy consumption, less pollution and a savings in operating cost, is now attracting more and more public attention.

Current practices in three non-ferrous metal mines are described below to illustrate the present status of the technology in China.

In-Situ Solution Mining Practice at T-mine. (9) (10)

T-mine is a typical porphyry copper deposit with an apparent bedded structure. The main mineralized zone is 500 m long and averages 10 m wide, dipping 23.5°-46° to a depth of 410 m. The hanging wall rock is composed of hydrosiderite, griotte and loessial fillings and the footwall is mainly jointed quartzite. It has been in operation for many years and ore production is still continuing on the -25 m level and below. Above the 5 m level, there is a worked-out area which had been previously mined by the caving method. The overburden has an average thickness of 15-45 m and consists mainly of gossan, laterite and loess in successive layers. The copper content of the gossan assayed 0.2-0.5% Cu and that of the clayish layer 0.2-0.25% Cu. Below these, there was a large tonnage of remnant ore and pillars left in the worked-out stope on the 5 m level. This ore contained 1.2% Cu and was diverse in particle size. The overburden and the remnant ore as a whole were porous and amenable to leaching and the productive drifts could be utilized for solution collection. In 1972, in-situ percolation leaching was started from the surface and a cementation plant also was built and completed in time to treat the solutions. The proces flowsheet is shown in Figure 1.

The main features of this process are:

(1) The leach liquor is first inoculated with Thiobacillus ferroxidians in the bacteria cultivation ponds. Compressed air or oxygen-rich air is blown from the bottom to promote bacterial growth. NH_4OH, $(NH_4)_2SO_4$ and H_2SO_4 are added as nutrients and for pH adjustment, if necessary. The indigenous bacteria effective oxidize Fe^{2+} to Fe^{3+} in the copper-depleted solution. The inoculated solution is considered to be regenerated when it reaches a total Fe-content of 4-6 gpl, with a Fe^{2+}/Fe^{3+} ratio of 1/1 and a pH of 2.2-2.4; at this point it is made available for leaching.

(2) Before application on the surface, the pH value is readjusted to 2.0 ± 0.02 so that the clogging effect due to the formation and precipitation of iron and aluminum hydroxides can be minimized. The barren solution, plus some make-up water to balance evaporation and leakage, is recirculated until its copper content raises to 2.5-3 gpl. Then, the rich pregnant solution is pumped into the cementation column filled with scrap iron to precipitate copper as cement copper. This product contains 60% Cu on average and is sent to the copper smelter for further treatment. The copper

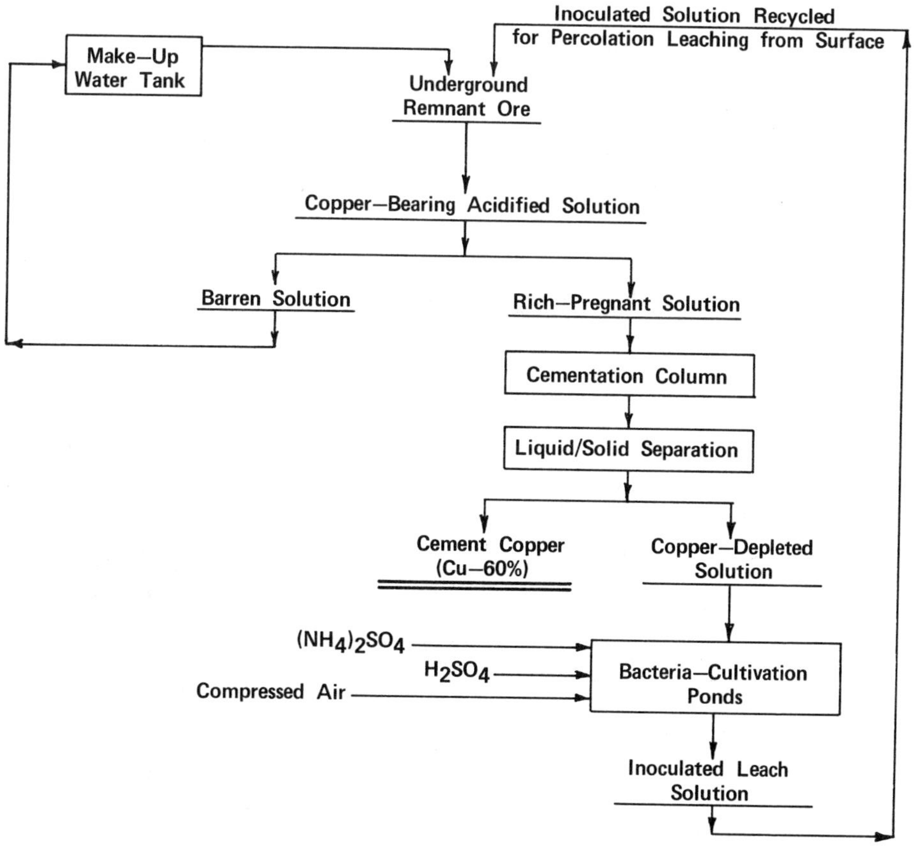

FIGURE 1. PROCESS FLOWSHEET OF IN-SITU SOLUTION MINING AT T-MINE

content of the depleted solution is <0.067 gpl and the copper recovery in cementation is 93%.

Heap Leaching Practice at N-Mine. (11)

N-Mine is a small underground copper mine which has applied heap leaching to its crude ore for 5 years. The run-of-mine ore, with an average grade of 2.49% Cu, is composed mainly of free oxides and secondary sulfide minerals. The sum of these two amounts to 54.5% while the primary sulfides represent only 4.6%. The ore size distribution is: 1-3 mm 20%; 4-7 mm 25%; 8-10 mm 35%; 10 mm 20%. The mine is situated in a remote mountainous district where modern traffic facilities and cheap energy are not available. As the ore could not be shipped anywhere, it was formerly treated pyrometallurgically on site and this caused the mine suffer a great deficit. A turning point was reached when the heap-leaching operation started. Subsequently, based upon the "Bacfox Treatment" theory, a leach dump of residual ore from 1st-stage heap leaching was used instead of cultivation ponds to supply the inoculated lixiviant. This has brought about more improvements. The flowsheet of this process is shown in Figure 2.

The main features of the process are as follows:

(1) The barren effluent from cementation is first passed through the residual ore dump for "Bacfox Treatment". By "Bacfox Treatment" we mean passing the Fe^{2+} salt solution through a vessel containing a bacterial supporting structure. This structure may be in in the form of honey-comb like structure, or may involve a series of plates. In this case, the bacterial oxidizing agents are present as a film and can speed up the oxidization of Fe^{2+} tremendously. This method was invented and patented by Eric Liversey and Gold Blatt in August, 1977 (GB Pat. No. 1542600, Pat. Off. of London). Generally, the barren solution contains 10-15 gpl Fe^{2+} and has a pH of 1.5-2.0. After being treated, these are changed to 5-6 gpl Fe^{2+}, 6-7 gpl Fe^{3+}, pH 2.0. The pH of the solution is further adjusted to 1.5-2.0 before it is pumped to a storage pond for use in the 1st-stage leaching.

(2) The regenerated lixiviant is evenly sprayed onto the heap surface at a rate of 30-50 l per sq m per hr. The lixiviant application is continued until the copper content of the ore decreases to 0.5% and that of the pregnant solution rises to 5-12 gpl. The rich pregnant solution is pumped to the cementation column to produce cement copper and the residual ore is transferred to the dump for Bacfox treatment. With this procedure, the time required for preparation of lixiviant has been reduced from 5 days to 5 hours; the rate of oxidization has been accelerated from 0.05 to 2.10 gpl per hr; the production cycle has been shortened from 60 to 25 days; copper-extraction in leaching has risen from 71% to more than 80% and the unit cost of copper is only 41% of that produced previously.

The practice at N-mine has shown that this flowsheet is fairly attractive for small scale heap leaching operations. However, scaling this up to a large operation still requires investigation.

Vat Leaching Practice of Complex Ore at P-Mine (12)

A great volume of tailings has accumulated on the surface of the P-mine in the past. The characteristics involve: (1) intimate association of copper and uranium minerals with copper and uranium contents of 0.11-1.5% and 0.011-0.34%, respectively; (2) an acidic ore with gangue rock high in carbonates; (3) finely grained material that is viscous and has a low permeability.

In order to take full advantage of these mineral resources and to prevent pollution of the environment, a decision was made to treat the tailings by vat leaching. Laboratory work was completed with an encouraging result in 1970, and the plant was put into production in 1972. The flowsheet is shown in Figure 3.

The main features of the process are:

(1) Both copper and uranium are leached and dissolved with acidified ferric sulphate solution. Before leaching, dilute H_2SO_4 solution should be added to minimize the adverse effect of carbonates in the

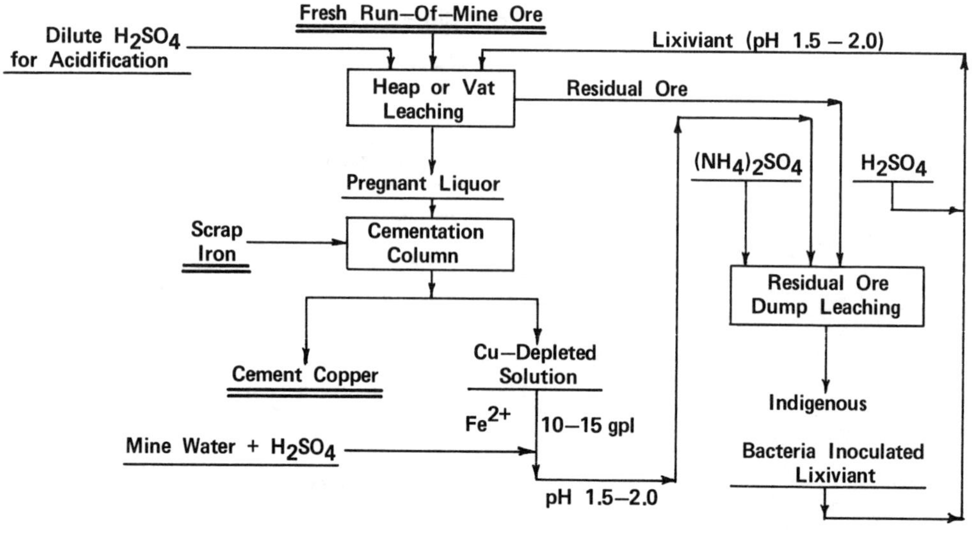

FIGURE 2. FLOWSHEET OF HEAP LEACHING AT N-MINE

gangue material. Leach extraction for copper and uranium may exceed 90% and 80% respectively.

(2) The rich pregnant solution, containing < 5 gpl Fe^{3+}, > 100 mgpl U, ∿10 gpl Cu^{2+} and with pH=2, is first forced to pass through the ion-exchange resin tower for U-extraction. The resin used in Resin 717 which is a homemade, strong-alkaline anion-exchange resin. The flowrate through the resin column is 80 l per min per ton resin. At the end point 33-35 kg of uranium may be absorbed per ton resin. Total time for contact is 5 min. Na_2CO_3+$NaHCO_3$ solution is used for stripping. Washing continues with a flowrate of 13 l per min per ton resin either concurrently or counter-currently until the uranium concentration of the effluent drops to 0.04 gpl. The pregnant liquor containing more than 2 gpl uranium is pumped to the yellow cake plant for further treatment to recover uranium.

(3) The uranium-depleted liquor with a copper concentration of 2-15 gpl is then treated in a cementation column to recover copper using scrap iron as a precipitant. Copper content of the cement product ranges from 60 to 80% and recovery in cementation is 93-98%. Iron consumption is 1.5-2.5 kg per kg of copper produced.

After cementation, the pH of the depleted solution is adjusted to 1.8-2.0 by adding H_2SO_4, and then this solution is stored in the bacteria cultivation pond in order to rapidly oxidize Fe^{2+} to Fe^{3+} by the biochemical action of Thiobacillus ferrooxidians. The inoculated $Fe_2(SO_4)_3$ solution is recirculated as an effective leaching agent for vat leaching. The overall recoveries of copper and uranium, taking the process as a whole, are 70-75% and 75-80% respectively.

RESEARCH WORK

Research work on the modern technology of solution mining started rather late in China. From the late 1960's to the early 1970's, tests involving vat leaching, dump leaching and underground leaching were conducted in a number of mines such as Shuikoushan Mine in Hunan, Tungkwangshan Mine in Anhui, Baohwashan Mine in Jiangsu, etc., by the Central-Southern Institute of Mining and Metallurgy, the Research Institute of Microbiology, Academica Sinica, Beijing and other universities and research centers. The preliminary result was promising. In 1972, the first symposium on solution mining of copper was held at Shuikoushan Mine. There was an extensive exchange of valuable information and technical know-how on this subject by specialists and engineers who attended the meeting. Part of the papers presented were published in special issues of "Non-Ferrous Metals" and "Metallurgy of Heavy Metals" in 1972. Another meeting, the symposium on leaching-solvent extraction-electrowinning technology. took place in December 1979 at Dongshian Mine, Jiangxi Province, to introduce and discuss this process.

These two symposia certainly provided an impulse to ensure that research work on solution mining made headway during the past decade. Up to now, quite a number of national or local research centers, institutes and colleges, as well as nearly twenty non-ferrous metal mines ranging from small to large in scale and including copper, uranium, manganese and cobalt-nickel mines, have been working in this field. Reportedly, significant results have been achieved in the following

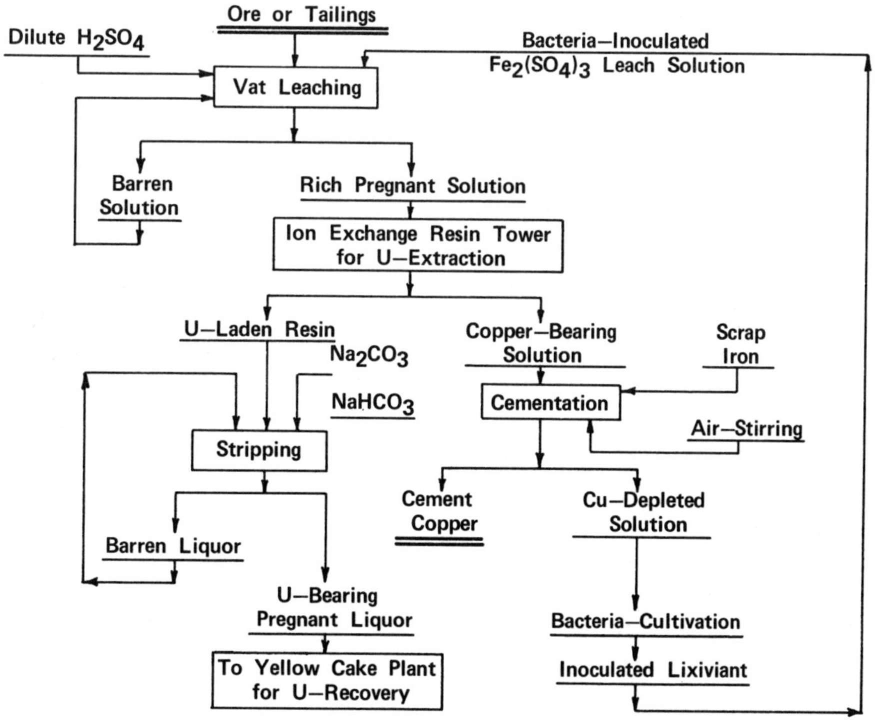

FIGURE 3. FLOWSHEET OF VAT LEACHING PRACTICE AT P-MINE

technologies: (1) the leaching of complex ores with closely associated copper and uranium; (2) heap-leaching of moderate to rich copper ore on a small scale; (3) underground in-situ leaching or dump leaching of low-grade secondary uranium ores; (4) leaching of high-sulphur manganese sulfide and carbonate ores to recover MnO_2 as an end product; (5) leaching of arsenical cobalt-sulfide ores to extract cobalt; (6) leaching of chalcopyrite concentrates to recover copper; (7) leaching of pyritic nickel-cobalt ores to recover Ni and Co. In addition, thorough studies on the mechanisms involved in the leaching processes and extensive investigations into the feasibility of applying solution mining techniques to various types of ores under different conditions are also underway in many institutes. Recent reports indicate that several research organizations are conducting detailed exploration of geological and hydrological conditions of two ore deposits, one in Jiangxi and the other in north-eastern China. The objectives of these studies are to assess the technical feasibility and economic justification for applying in-situ solution mining techniques to these two ore deposits. If warranted, the studies will include planning for the first commercial scale in-situ solution mining projects in China. (1) (3)

It should be pointed out that all the research work and experiments mentioned above were rather limited in scale and had been carried out independently by various research organizations associated with different trades and professions. The focus of their work varied widely depending on the tasks they were assigned. Under such circumstances, the research work on solution mining was restricted to identification of individual procedures or improvement of unit operations. However, solution mining as a whole is an interfacing technology which needs intimate cooperation among mining engineers, geologists, metallurgists, biologists and chemical engineers with centralized guidance to work out an overall plan and later to implement this plan through a common effort. But now this is not the case. Maybe this is why the research work on solution mining in China is lagging far behind that of highly industrialized countries.

CONCLUSION

Solution mining in China has demonstrated a long history of success. It was put into practice on a considerable scale and with comparatively advanced technology early in the 11th century. The gall-copper process developed in the Soong Dynasty was essentially a prototype of in-situ solution mining using natural water as a leaching agent. Unfortunately, owing to social, political and economical conditions, this technique ceased to develop in subsequent centuries.

Quite a number of significant achievements in scientific research work on solution mining have been realized since the founding of the People's Republic of China. Many experiments and field tests, particularly for copper, uranium and manganese ores, have been performed with many of their results applied in production. However, the scale of these operations is generally small, their application limited and the state-of-the-art still lagging far behind that of industrialized countries which accelerated their development of solution mining in the last decade.

China is now a developing country with extensive metalliferous mineral resources. Some of these, though abundant in reserve, are generally poor in grade, complex in composition and consequently more or less refractory to processing or smelting. However, since solution mining techniques feature low capital investment, rapid return on investment, low energy consumption, less environmental degradation and more savings on operating costs, it has certainly aroused tremendous interest within the mining industry. No doubt, the research work and practical application of solution mining in China will make still greater progress in the near future to meet the requirements for the four modernizations.

REFERENCES

1. Yao Dun Pu, Shiao Tung Yao, "On Solution Mining--Its History, Present Status and Prospect", An Informational Investigation Report (Unpublished), Changsha Mining Research Institute, Changsha, China, Oct. 1980.

2. Chiang Wei Chang, "Scientific Inventions in China's History", 1955.

3. Chang Zhi Gou, "History on Chemistry of China--Ancient Times (Draft)", Science Publish Co. of China, 1964.

4. Wu Zhi Zhen, "The Achievements in Copper Extraction by Leaching Process in Ancient China", Transactions of the Institute of Iron and Steel Metallurgy, Beijing, Vol. 5, 1958.

5. Gu Zhu Yu (Chin Dynasty), "A Sketchbook of Study on History and Geography".

6. Tu Ke Tu (Yuan Dynasty), "History of Soong Dynasty"; Zu Shen (Chin Dynasty), "Conpendium of Soong Dynasty"

7. Ma Duang Lin (Yuan Dynasty), "A General Investigation on Historical Literature".

8. Shen Kuao (Soong Dynasty), "Sketches and Notes at Menshi".

9. "Underground Bacterial Leaching Test for Remnant Ores", Bureau of Science and Technology, Tunglin Municipal Revolutionary Committee, Anhui Province, 1972.

10. "Bacterial Leaching of Copper", The Beijing Institute of Mining and Metallurgy, "Non-Ferrous Metals", Vol. 2, 1972.

11. "Practice of Bacterial Leaching of Copper", Science and Technology in Nanjian", Vol. 2, 1979.

12. "The Technology of Bacterial Leaching for Copper and Uranium Extraction from Low-Grade Ores", Shuikuoshan Bureau of Mines, Beijing Research Institute of Microbiology, Academica Sinica, "Non-Ferrous Metals", Vol. 3, 1972.

13. "The Research Work and Practice of Bacterial Leaching for Treating Low-Grade and Refractory Concentrates", Mineral Bacteria Group, Beijing Research Institute of Microbiology, Academica Sinica, Oct. 1980.

Chapter 3

THE INFLUENCE OF SOLUTION MINING ON THE WORLD SODA ASH MARKET

Dennis S. Kostick
U. S. Department of the Interior
Bureau of Mines
Washington, D. C. 20241

ABSTRACT

Solution mining is an established technique used to extract certain soluble minerals buried deep underground where conventional mining methods are impractical. This technology is being tested in the Green River basin of Wyoming to utilize the deeper and thicker trona deposits currently uneconimic because of rising conventional mining costs. If successful, the commercial application of solution mining could be operational by 1987. The domestic end uses of soda ash have reached maturity; however, the export market for soda ash has excellent growth potential. Approximately 75% of the soda ash production in the world is made synthetically by the Solvay process. Rising energy costs and environmental pollution are forcing many synthetic soda ash producers to close or limit the amount of production. Solution mining of natural sodium carbonate could be the technique used to provide a significant portion of future requirements. Three scenarios are presented that forecast the impact of solution mining on the Western European market in 1987. The conclusion is that, based on certain assumptions, there will be a need for increased natural soda ash production to meet anticipated increases in world demand. Present or future companies producing soda ash by conventional mining or by solution mining techniques will compete in the export market; however, solution mining could have the advantage because of reduced mining costs.

STATUS OF THE WORLD SODA ASH INDUSTRY

Soda ash is produced in over 39 countries by the Solvay ammonia-soda process or by mining natural sodium carbonate deposits. The total production of soda ash in 1980 is estimated at slightly over 29 million metric tons. Synthetic production accounted for over 75% of this amount with the balance of production originating from natural sources (either trona or sodium carbonate-rich brines) found primarily in the United States, Mexico, Kenya, and Chad.

United States

Most of the U.S. soda ash industry is situated in the Green River basin of Wyoming. Buried deep within the basin lies the richest deposit of trona in the world. Four companies - FMC Corporation, Stauffer Chemical Company of Wyoming, Allied Chemical Company, and Texasgulf Chemical Company (a fifth company, Tenneco Oil Company, will begin production in 1982) comprise the nucleus of the soda ash industry with a combined annual nameplate capacity of 7,267,000 metric tons. Including Kerr-McGee Chemical Corporation in California and Allied Chemical Company's synthetic soda ash plant in Syracuse, New York, the total U.S. soda ash nameplate capacity is 9,398,000 metric tons.

The United States produces about 26% of the world soda ash with the majority of the output produced in Wyoming. With additional natural soda ash production in California, the United States produces over 94% of the natural soda ash in the world.

<u>Conventional Mining Versus Solution Mining</u>. At the present time, all of the trona in Wyoming is mined by the room and pillar method. The method uses a conventional, continuous, or longwall mining technique or a combination thereof. The extraction efficiency ratio for the room and pillar method ranges from 50% for conventional and continuous mining to 75% for longwall mining. These established techniques have been used safely and successfully since 1947, when the Intermountain Chemical Company of FMC Corporation was the first company to mine and process trona ore into soda ash.

A new era in trona mining emerged in 1980 with the announcement by FMC Corporation that it would bring onstream in the mid-1980's additional soda ash capacity through the use of solution mining. Although the concept was not new, the practical application was. The theory was simple: inject a solvent under pressure down a well, establish contact with an adjacent recovery well, and pump the dissolved sodium carbonate to the refinery. The time spent in planning and experimenting to apply this principle, however, was considerable. Figure 1 shows a typical solution mining operation. The trona beds in the southern part of the basin are deeper than those in the North because all the trona, shale, and sandstone beds dip toward the Southwest.

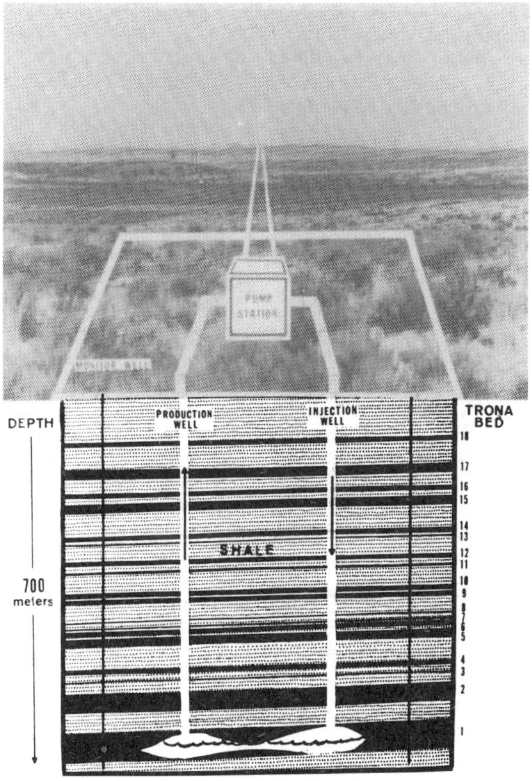

Figure 1. Cross-Section of a Wyoming Solution Mining Operation.

Vulcan Materials Company and Allied Chemical Company have also announced their intentions of using solution mining technology. By early 1981, the permits needed to conduct solution mining tests in the southern part of the Green River basin were granted by the Wyoming State Department of Environmental Quality. Presently all three companies are actively engaged in different stages of development and testing. Other companies have expressed interest in solution mining trona and are awaiting the results of the tests.

Why has the topic of solution mining become so popular within the soda ash community? Although Kerr-McGee uses solution mining to recover sodium carbonate-rich brines at Searles Lake, the only other domestic trona deposit that offers long-term mining potential is in Wyoming. With the high initial investment cost of between $600 million to $700 million for a 1-million-ton per-year conventional soda ash facility, solution mining affords advantages over conventional mining operations. Table 1 lists some of the apparent advantages using this new technique.

The five factors in Table 1 list the benefits derived from solution mining. The end result is a cheaper cost price to mine trona. The long-term net effect should be to keep the price of soda ash produced by solution mining lower with respect to the cost of conventionally mined or synthetic soda ash. synthetic soda ash process requires more energy to produce a ton of product compared with that produced from natural sources. Also the synthetic process discharges calcium chloride and sodium chloride as byproduct effluents into the local environment.

Table 1.--Advantages of Solution Mining

FACTORS	RESULT/BENEFIT
MINING COSTS	
o Sinking of mine shaft(s) eliminated	Time saved - month vs. year Cost saved - well vs. shaft
o Purchase of expensive mining machinery eliminated	Costs savings
MANPOWER REQUIREMENT	
o Mining staff	Labor cost reduced Socioeconomic impact reduced Safety costs reduced (training) Lives saved in mining accident
PROCESSING	
o No insoluble material	Eliminate need for large tailing ponds
o Ore is in solution	Surface processing step eliminated
ENERGY	
o Fuel used in mining	Energy and cost savings
o Fuel used in processing step	Energy and cost savings
RESOURCE	
o Deeper beds recoverable	Previous uneconomic beds now economic
o Deeper beds thicker	More trona can be mined

Overall benefit: initial investment cost and operating costs are lower.

Figure 2 shows a historical price comparison of natural and synthetic soda ash. By extrapolating the trend of historical soda ash prices to 1987, the future price of natural soda ash could be $174 per ton vs $226 per ton for synthetic soda ash. The split in soda ash pricing occurred in the mid-1970's as a result of the higher energy costs and the costs to implement the controls imposed by anti-pollution legislation. The

There are disadvantages of solution mining that should be mentioned. These include the problems of extracting soluble organic matter and halite from certain trona beds, possible excursions of the solvent into adjacent leased territory or neighboring aquifers, possible surface subsidence, and the cost of constructing miles of pipeline.

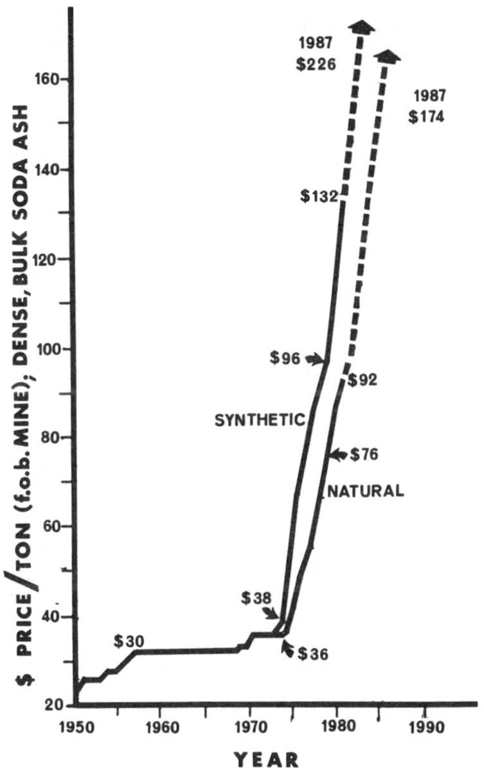

Figure 2. Historical and Projected Price Relationship of Natural and Synthetic Soda Ash

The future domestic and international demand for soda ash will determine the course of the U.S. soda ash industry. Other factors, such as the present status of foreign producers that manufacture soda ash and the impact of soda ash dumping by the members of the Council for Mutual Economic Assistance (Comecon), will influence decisions to expand domestic production capacity by the existing producers or to allow new entrants into the soda ash industry.

Rest of the World

Most of the soda ash in the rest of the world is manufactured by the Solvay process, which uses salt and limestone as raw materials. There are many natural sodium carbonate deposits in the world; however, most of them are too small or too remote to be developed. Sodium carbonate brines in Kenya and Mexico presently produce a total of about one-half million metric tons of soda ash per year. These deposits differ from Wyoming trona. The only known trona deposit in the world that is similar to the type found in Wyoming is reported to occur in mainland China in the Honan Province. Reports indicate that the Chinese are experimenting with a solution mining technique to recover the deep trona.

THE WORLD DEMAND FOR SODA ASH

Influence of Solution Mining on Domestic Demand.

The present end uses of soda ash have reached maturity and are not expected to grow significantly. These include the glass, chemical, pulp and paper, water treatment, and soap and detergent industries. The present structure of the soda ash industry, including expansions and new additions like Tenneco, can accommodate the anticipated domestic demand without the addition of new capacity from solution mining operations.

Figure 3 shows the historic and projected trends of production for the world, rest of world, and United States. The annual growth rates for production and U.S. demand were derived using annual data for the period 1960-1980. Because the U.S. demand has reached maturity, the associated production growth rate appears low in comparison with those for the world and the rest of the world.

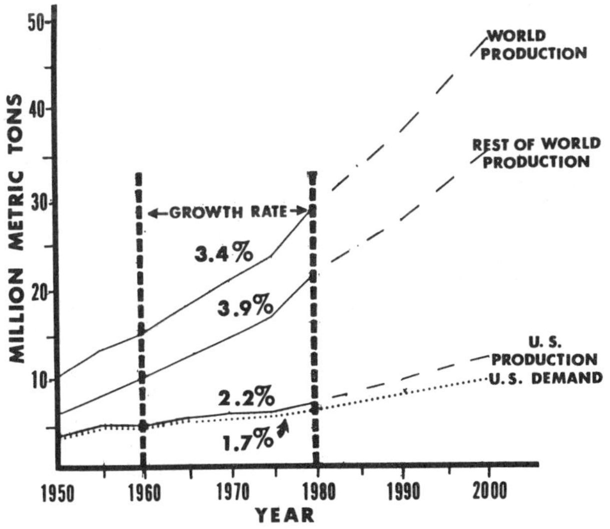

Figure 3. Present and Projected Growth Rates of Soda Ash

New end uses for soda ash are presently being evaluated. The replacement of sodium hydroxide by soda ash represents a large potential market. Presently sodium hydroxide is in short supply while the demand is strong. A parallel potential new end use for soda ash is in secondary oil recovery. Soda ash either by itself or after conversion into caustic soda, could be used as a chemical purge to recover additional quantities of petroleum trapped in oil fields. Soda ash is also being mentioned as a possible scrubber to remove excess sulfur from coal-fired plant emissions. The total demand for this end use is not expected to be strong, and certainly not high enough to warrant the need for solution mining. Until the size of any new market is established, it appears that the most favorable market for soda ash, whether it is from conventional mining or solution mining, is the export market.

Rest of the World Demand

The world demand for soda ash is assumed to be equal to the world production of soda ash. Figure 3 shows the annual growth rate of soda ash for the world and the rest of the world (excluding the United States). Because the United States produces about one-forth of the world's soda ash, it has a strong influence on the growth rate for the world. The rest of world growth rate is higher because many developing nations have growing industries and the demand for soda ash has not reached maturity. The demand for soda ash in some of these countries often exceeds the production capacity of their soda ash industries. Imports from foreign sources, such as the United States, can provide the needed supply. In 1980, the United States exported to 51 of the free world developing countries. Some developed nations have established soda ash industries that are old and in need of renovation. Some of the plants in Western Europe are good examples.

Western Europe - A Case Study

Western Europe offers an excellent opportunity and challenge for the U.S. soda ash industry. Its level of apparent consumption of soda ash is similar to that of the United States - approximately 7,300,000 metric tons per year. Table 2 lists the soda ash trade statistics for Western Europe from 1978-80. The apparent consumption of soda ash for each country has also been calculated and appears in the table. In 1980, about 3.6% of Western European imports came from the United States. Because most European synthetic soda ash plants are old, and have energy and pollution problems, the U.S. soda ash producers are optimistic about exporting increased quantities of soda ash in the future.

To forecast what influence solution mining may have on the export market and how much soda ash is exported to Western Europe, several basic assumptions must be introduced. The timing of new plant startups, changes in foreign economic and political conditions, and the degree of energy consumption can alter these assumptions and either restrict or encourage U.S. soda ash trade.

No New Construction Of Large-Scale Soda Ash Plants. Funding and construction of the Canadian Gaspe Peninsula soda ash plant is questionable at this time. The planned capacity was about 850,000 metric tons. The 400,000 metric-ton-per-year capacity Thailand-ASEAN project also has been postponed because of environmental and political conflicts. If U.S. soda ash producers can successfully negotiate contracts with foreign soda ash consumers, it will reduce the possibility of the Canadian and Thailand projects coming onstream.

New small-scale soda ash plants, 200,000 tons or less, will probably be built near sources of energy and raw materials - mainly salt and limestone. Some plants, such as the ones proposed in Greece, India, and Indonesia, are scheduled to be built only to satisfy the demand requirements of the particular country without

Table 2.--SODA ASH TRADE STATISTICS IN WESTERN EUROPE, 1978-1980
(Thousand Metric Tons)

Country	1978				1979				1980			
	P	I	E	AC	P	I	E	AC	P	I	E	AC
Austria	172	3	-	175	190	2	8	184	190	1	36	155
Belgium	427	56	44	439	480	49	32	497	480	49e/	32e/	497e/
Denmark	1	68	-	69	3	64	0	67	3	68	0	71
Finland	-	73	-	73	-	75e/	-	75e/	-	88	-	88
France	1354	29	54	1329	1708	21	397	1332	1700	29	389	1340
W.Germany	1230	178	87	1321	1544	197	218	1523	1550	191	216	1525
Greece	1	40	-	41	1	34	0	35	1	30e/	0	31e/
Ireland	-	15	5	20	-	16	0	16	-	18e/	0	18e/
Italy	726	73	15	784	800	76	20	856	800	194	14	980
Netherlands	280	40	129	191	460	50	150	360	460	55	153	362
Norway	26	34	1	59	30	47	0	77	30	57	1	86
Portugal	132	5	-	137	145	2	4	143	145	2e/	4e/	143e/
Spain	499	-	4	495	550	23	5	568	555	3	72	486
Sweden	1	108	-	109	1	120	-	121	1	119	-	120
Switzerland	45	1	-	46	50	4	34	20	52	4	65	9
U.K.	1597	24	158	1463	1550	59	146	1463	1500	34	150	1384
Total	6491	747	492	7389	7512	839	1014	7337	7467	942e/	1132e/	7295e/

P = Production I = Imports E = Exports AC = Apparent Consumption e/ = estimated

Apparent consumption = production + imports - exports

1978 - 3.4% of U.S. exports went to Western Europe; 1.7% of Western Europe imports came from U.S.
1979 - 1.2% of U.S. exports went to Western Europe; 1.3% of Western Europe imports came from U.S.
1980 - 4.1% of U.S. exports went to Western Europe; 3.6% of Western Europe imports came from U.S.

the intent of the country becoming a net exporter. Aside from providing a developing nation additional soda ash for its growing glass and chemical industries, the construction of small soda ash plants will provide employment and increase the general welfare of the economy of the country.

Tradeoffs. A few foreign projects, if successful, will result in the U.S. producers exporting less soda ash to those particular countries, thereby requiring new areas to export to in order to keep the level of exports constant. For example, U.S. soda ash exports to Argentina were 12,600 metric tons in 1980 and represented over 1% of the U.S. export market. In March 1980, construction began on a 225,000 metric ton-per-year soda ash complex at Punta Delgado. When completed in 1983, exports to Argentina will probably terminate. Likewise in Brazil, proposed soda ash projects will reduce soda ash imports from the United States. In 1980, exports to Brazil accounted for over 124,000 metric tons, or nearly 13% of the total U.S. export market. New markets, possibly in the Caribbean or Central American countries, can probably absorb these displaced tonnages.

Dumping of Comecon Soda Ash Reduced or Halted. Since 1978, the dumping of inexpensive light and dense soda ash from the Eastern bloc nations into Western Europe has been a problem for the Western European soda ash producers and has hindered the efforts of the U.S. soda ash producers to negotiate favorable export agreements with European consumers. Between 250,000 metric tons and 320,000 metric tons of soda ash are imported annually by Western Europe from the Comecon countries; namely, Bulgaria, the German Democratic Republic, Poland, Romania, and the U.S.S.R.

In 1979, the Federal Republic of Germany, Sweden, and Italy represented nearly one-half of the total soda ash imports to the 16 countries comprising Western Europe. The Federal Republic of Germany imported about 24% of the European total with approximately 32% of the nation's imports coming from Comecon sources. Sweden comprised 14% of the total with 23% of its imports originating from Comecon sources, and Italy represented 9% of the total with over half (56%) of its imports of Comecon origin. The three countries imported a total of 145,016 metric tons of Comecon soda ash during 1979, amounting to 17% of the total Western European imports for the year. Table 2 lists the quantity of imports of soda ash into Western Europe during 1978-80.

Comecon soda ash was inexpensive and available. However, several recent events occurred that may reduce the future quantity of Comecon soda ash available for export. In 1980, the U.S.S.R. encountered soda ash shortages and depended on supplies from local Comecon nations. The surplus Comecon soda ash normally available for export was reduced, and this reduction may continue in the future. Poland suffered labor strikes and coal shortages that impacted on the export goals of the Polish soda ash industry.

The Western European soda ash producers won an anti-dumping action against the Eastern bloc countries that should result in a reduction in Comecon imports. The problems in the Comecon community could allow the U.S. soda ash producers an opportunity to export soda ash to countries in Western Europe; however, the major factor is the delivered price. If U.S. soda ash is to be competitive, the price, including transportation costs, must be lower than the foreign producers' price. Assuming rail and ocean freight rates are fixed (with cheaper rates for larger bulk quantities), the price of soda ash is the variable factor. The question is: Can solution mined soda ash, with its less expensive operating costs, compete with conventionally mined soda ash to lower the actual delivered price?

Energy Consumption and Pollution Will Remain Important Issues. Energy consumption and pollution control have been important issues in the United States since 1973. In 1969, there were 10 synthetic soda ash companies operating in the United States. By 1979, all but one had closed because of energy costs, pollution control, and the competition from natural soda ash.

Although the amount of energy needed to manufacture synthetic soda ash should remain constant, the cost of energy is increasing. The costs associated with plant maintenance, installation of pollution reduction controls, and major renovations are also increasing. Many of the Western European Solvay plants are 50 to 100 years old and are not as energy efficient as some of the new Solvay facilities in the world or the natural soda ash plants in the United States. Some European Solvay plants are reported to be effectively operating at about 75% of capacity while the U.S. soda ash industry is operating at more than 86% of capacity.

Pollution, particularly in the Rhine River and its tributaries, is increasing and causing concern. Commissions have been formed to examine the problem, but recommendations have yet to be announced.

The cost of energy and the implementation of pollution controls are reflected in the higher prices of Western European soda ash. Foreign consumers, primarily the glass manufacturers, have recently renegotiated contracts with the major Western European soda ash producers - Imperial Chemical Industries (ICI) and Solvay et Cie. The contracts allow the consumer more freedom to purchase additional quantities of soda ash from foreign sources (United States, for example). As Western European soda ash producers experience increasing energy costs and higher levels of pollution, U.S. soda ash producers will be ready to increase exports.

However, recent preliminary negotiations between several Western European nations and the U.S.S.R. to build a 4,800 kilometer natural gas pipeline could affect the energy outlook for Europe. If the line is constructed in the next few years, approximately 36.8 billion cubic meters of natural gas will be transported each year from Siberia. It is not known at the present time what impact this will have on the Western European soda ash industry and the future of the U.S. export market.

Impact of Solution Mining on Western Europe. The assumptions previously discussed conclude that the U.S. soda ash industry has an excellent opportunity to export increased quantities of soda ash in the future - how much depends on the timing and outcome of some of the assumptions.

Forecasting to 1987. In an attempt to analyze the influence solution mining may have on the soda ash industry, 1987 was chosen as the probable year that the commercial application of solution mining will be in effect. The influence can be predicted by comparing the projected U.S. soda ash production capacities in 1987 with the projected soda ash demand in Western Europe. Normally it takes approximately five years for a company that is interested in entering the soda ash industry to acquire the necessary land in Wyoming, obtain the permits, construct the plant, and begin soda ash production. The company must determine the proper time to start construction in order to compete in the world soda ash market five years later.

Figure 4 shows the historical relationship of soda ash nameplate capacity, natural and synthetic soda ash production, and domestic demand for soda ash. As indicated in the graph, the difference is increasing between the level of production and the level of domestic demand. This difference represents the amount of soda ash available for export and inventory. The present growth rates for production and demand are projected to 1990.

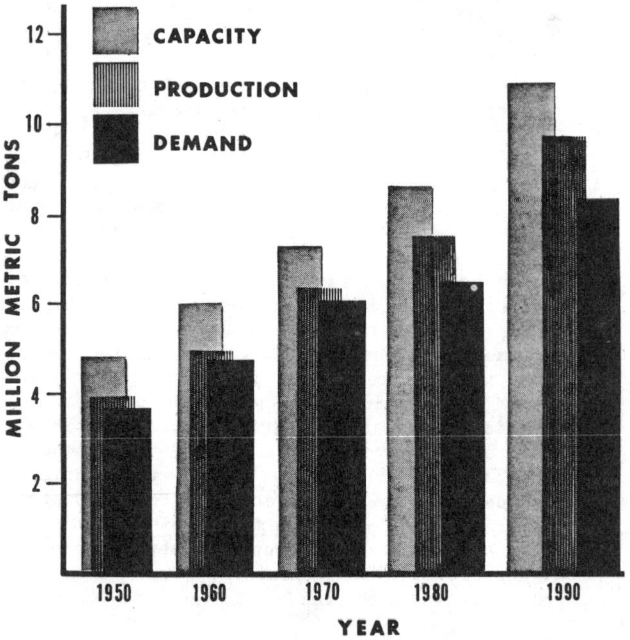

Figure 4. Historic Relationship Between U.S. Soda Ash Industry Nameplate Capacity and Domestic Production & Demand

Table 3 lists the present and projected nameplate capacities of the U.S. soda ash industry. Capacities through 1983 have been announced by the respective companies. Projections from 1985 through 1987 were based on the opinions of the author and do not necessarily reflect the opinions of the soda ash industry.

Projected production figures were derived using the historical annual growth rate of 2.2% as noted in figure 3, with 1978 as the base year, because it represented stronger economic growth than 1980. Domestic demand projections were based on a historical annual growth rate of 1.7% with 1978 as the base year. Projected 1990 and 2000 capacities were based on the projected production figures for those years, assuming the industry will be operating at an estimated 88% of capacity. The operating rates were obtained by comparing total production with total nameplate capacity.

Table 3 lists three possible structures of the soda ash industry that could exist in 1987 based on certain knowledge and assumptions. The three scenarios are hypothetical and represent capacities with a "high-range forecast" (I); a "medium-range forecast" (II); and a "low-range forecast" (III). Varying degrees of influence because of plant closures, plant expansions, new entrants, and solution mining are covered in each of the three scenarios. By forecasting the Western European soda ash demand in 1987 and also setting an optimistic goal for exports, a particular scenario will attempt to indicate whether the capacity forecast can meet the anticipated demand.

The present soda ash demand in Western Europe exceeds 7,300,000 metric tons. In 1987, the projected demand should be approximately 10.1 million metric tons. If the assumptions previously discussed concerning energy and pollution problems affecting foreign soda ash producers, Comecon import reduction, and no future construction of large-scale soda ash plants are correct, the U.S. soda ash industry could anticipate increasing exports to fulfill a portion of the demand. It is doubtful that by 1987 the Western European soda ash industry will collapse or reduce production sufficiently to warrant the export of 7 million tons of soda ash from the United States. However, a conservative estimate of the total amount of apparent consumption of soda ash in Western Europe that could be displaced would be about 20% of future demand, or about 2,020,000 metric tons. A very optimistic estimate would be 30%, or about 3,030,000 metric tons.

If no significant increases in U.S. soda ash demand occur in the next several years (caustic soda replacement, etc.), and using the domestic growth rate of 1.7% projected to 1987, the required U.S. production of soda ash would be about 9,150,000 metric tons. With a 2,020,000 metric ton (20%) demand replacement in

Table 3.--Present and Projected Nameplate capacities of the U.S. Soda Ash Industry (Thousand Metric Tons)

Company	1980	1981	1982	1983	1985	I 1987	II 1987	III 1987	1990	2000
SYNTHETIC										
Allied	820	820	820	820	820	--	--	--		
NATURAL										
Allied	2000	2000	2000	2180	2180	3080	2720	2270		
FMC	2310	2580	2580	2580	2720	3630	3630	3180		
Kerr-McGee (Argus)	1180	1180	1180	1180	1180	1180	1180	1180		
Kerr-McGee(Westend)	140	140	140	140	140	140	140	140		
Stauffer	1380	1780	1780	1780	1780	1780	1780	1780		
Texasgulf	900	900	900	900	1180	1810	1360	900		
Tenneco	--	--	270	680	900	900	900	900		
Company "X"						900	450	--		
Total	8,730	9,400	9,670	10,260	10,900	13,420	12,160	10,350	11,100	13,800
Production @2.2% 1978 base year	7,510	8,010	8,190	8,370	8.990	9,150	9,150	9,150	9,770	12,140
Demand @1.7% 1978 base year	6,470	6,810	6,920	7,040	7,620	7,970	7,970	7,970	8,380	9,960
Operating Efficiency Rate	86%	85%	85%	82%	83%	68%	75%	88%	88%	88%

Scenario I - High-Range Forecast

The assumptions are-

1. Allied closes Syracuse Solvay plant.
2. Allied expands capacity with major solution mining operation.
3. FMC expands capacity with major solution mining operation.
4. Texasgulf doubles capacity (effect of aquisition by Elf Aquitaine unknown).
5. Tenneco at full capacity.
6. New company enters soda ash industry - Company "X".
7. No planned expansions for Kerr-McGee and Stauffer.

High range forecast is 13,420,000 metric tons of total nameplate capacity

Scenario II - Medium-Range Forecast

The assumptions are-

1. Allied closes Syracuse Solvay plant.
2. Allied expands capacity by one-half million tons with solution mining operation.
3. FMC expands capacity with major solution mining operation.
4. Texasgulf has incremental expansion.
5. Tenneco at full capacity.
6. No planned expansions for Kerr-McGee and Stauffer.
7. Company "X" enters soda ash industry with small operation in Wyoming or California.

Medium range forecast is 12,160,000 metric tons of total nameplate capacity.

Scenario III - Low-Range Forecast

The assumption are-

1. Allied closes Syracuse Solvay plant.
2. Allied has small debottlenecking expansion.
3. FMC expands capacity by one-half million tons with solution mining operation.
4. Texasgulf does not expand (effect of aquisition by Elf Aquitaine unkown).
5. Tenneco at full capacity.
6. No planned expansions for Kerr-McGee or Stauffer.
7. No new company(s) enter soda ash industry.

Low range forecast is 10,350,000 metric tons of total nameplate capacity. Using the 1980 domestic growth rate of 1.7% to project estimated demand in 1987, the low range forecast would be the level of nameplate capacity to expect in 1987, assuming the existing companies will be operating at 88% of capacity, similar to the rate of the late 1970's.

Western Europe that represents additional U.S. exports, the total production needed by the United States would have to be about 11,170,000 metric tons. The forecasted industry capacities in Scenario I and II of Table 3 could meet the potential increase in demand; however, Scenario III, the low range forecast, could not. The assumption drawn from this are possibilities that could be used to expand domestic soda ash capacity as depicted in Scenarios I and II that include major industry capacity expansions from solution mining operations, from capacity expansions by existing companies, and from new entrants into the soda ash industry.

Supplying 30% of Western Europe's soda ash demand requires an estimated 3,030,000 metric tons of additional exports. Added to the projected 9,150,000 metric tons of domestic demand, the total would be 12,180,000 metric tons. Only Scenario I could meet this optimistic forecast, and if it did, the soda ash industry would be operating at about 91% of capacity. It is assumed that with this demand displacement, there could be opportunities for new companies to enter the soda ash industry and use either solution mining or conventional mining to increase the production capacity of the industry. Some companies may prefer to adopt solution mining technology because of the lower investment costs and the potential price advantage solution mined soda ash may have over conventionally mined soda ash. The conservative and optimistic estimates of demand displacements are summarized in Table 4.

Table 4.--Summary of Projected Western European Demand Displacements (Thousand Metric Tons)

Year	1978	1987
Western Europe demand	7,389	10,100
U.S. export goal:		
20% demand displacement	----	2,020
30% demand displacement	----	3,030
U.S. production @ 1.7% growth rate	7,521	9,150
Needed U.S. production:		
@ 20% displacement	----	11,170 results in 4.5% growth rate
@ 30% displacement	----	12,180 results in 5.5% growth rate

CONCLUSIONS

Hypothetical industry capacities forecast to 1987 were used in conjunction with the projected Western European demand in an exercise that illustrated the future prospects of domestic soda ash producers to capture an increased share of the export market. To accurately and realistically determine if solution mining will be used and which companies will use the technique is problematical at the present time. However, based on present knowledge and assumptions of the world soda ash supply-demand situation, the outlook is favorable that the export market will increase with the delivered price of soda ash to foreign ports being the most important factor. Foreign consumers will have the option of purchasing soda ash produced from synthetic, conventional, or solution mining operations. Realizing the companies that offer soda ash at the cheapest price will have the advantage, it appears that companies engaged in solution mining will have a competitive advantage provided the technical development of solution mining on a commercial scale is successful.

BIBLIOGRAPHY

Hou, T.P., Manufacture of Soda, Reinhold Publishing Corp., New York, 1942, pp. 590.

Kostick, D.S., "Soda Ash, Sodium Bicarbonate, and Sodium Sulfate," Bulletin 671, 1980, U.S. Bureau of Mines, pp. 16.

Nigbor, M.T., "State of The Art of Solution Mining For Salt, Potash, and Soda Ash, Preliminary Report, 1980, U.S. Bureau of Mines Twin Cities Research Center.

Parker, R.B., and Mannion, L.E., ed., "Trona Issue," Contributions to Geology, University of Wyoming, Vol. 10, No. 1, Spring 1971, pp. 72

Chapter 4

WASTE ISOLATION IN SALT CAVITIES

Larry Sevenker
Kenner, Louisiana, 70062

Michael Cox
Law Engineering Testing Company
Marietta, Georgia, 30060

ABSTRACT

On the basis of favorable salt characteristics, primarily compressive strength and impermeability, solution mined cavities can provide structurally stable isolation chambers for waste. Solution mining techniques are already established to provide storage capacity for hydrocarbons. The proposition of waste isolation in engineered salt cavities is similar to this practice relative to characterization, certification and safeguards. Principle differences include the materials being isolated and the way operations are conducted. Compatibility of wastes with other wastes must be considered, but hazards can be precluded by avoiding certain mixtures. Waste material characteristics and cavity considerations are discussed.

Considerable evaluation and characterization of salt formations has been conducted for storage and waste isolation. On the basis of favorable salt characteristics, it is believed that solution-mined cavities in salt can provide structurally stable isolation chambers for waste. Waste isolation in salt cavities is not the ultimate solution, but it may be a viable alternative to disposal by land fill and land treatment. Cavity utilization warrants careful thought and examination.

Isolation techniques and considerations will be described and illustrated as they apply to salt cavities. Waste materials will be addressed by broad physical characteristics rather than specific groups such as hazardous chemical wastes.

BACKGROUND

Rock salt deposits underlie millions of acres in several areas of the United States. Salt deposits exist as dome salt, bedded salt or playa lakes. There are in excess of 500 known salt domes in the Gulf Coast, of which about half are onshore and the others offshore. Bedded salt deposits in restricted marine basins responded to downbuilding of overlying sediments and density differences between salt and sediments to develop domes by horizontal and vertical plastic flow of the salt. Fig. 1 shows known U.S. salt deposits.

Depth to bedded salt deposits varies from near the surface to 12-18,000m (40-60,000') deep in the Coastal basin. Thickness also varies considerably along with the amount of interbedding shale and anhydrite layers.

Playa lakes exist on or near the surface and are located in the western United States. Desert environment has resulted in high evaporation rates and has formed salt lakes.

Salt has a number of properties which make it an attractive isolation medium. The insitu permeability is so low that it is difficult to measure. Porosity is also generally low although voids do exist which contain gases or natural brine. Compressive strength of salt is high and is equivalent to that of concrete. Generally above 1830m (6000') depth salt is stable. Below this depth, pressure and temperature increase and salt reacts in a plastic manner. Purity of rock salt in most areas is greater than 95% halite, with a small percentage of anhydrite and trace minerals. Discontinuity and lack of homogeneity are factors which affect development of solution-mined cavities. Other surrounding formations or features must be evaluated to predict the impact on cavity development.

Salt is easily mined because of its solubility in water. Chemical companies have been dissolving salt for a number of years by injecting fresh water and extracting brine for use as feedstock and have created a considerable number of solution-mined cavities. Many of these were converted to storage of hydrocarbons and other products. A variety of cavity shapes and sizes exist. Some were solution-mined for brine extraction and little attention was paid to cavity geometry. When LPG and other hydrocarbon storage became more popular and stability and pressure maintenance concerns were encountered in some existing cavities, companies began to pay more attention to cavity geometry and integrity. In association with Solution Mining Research Institute and the Strategic Petroleum Reserve, in separate efforts, many studies of cavity stability and maintenance have been undertaken recently to establish acceptable certification criteria and development procedures.

Geological characteristics have been the controlling factor in cavity configuration and depth limitations for solution mining cavity development, especially in bedded salt. A depth limitation of 1520-1830m (5-6000') has generally been accepted due to the plastic nature of salt below this depth.

Larger cavities are generally located in dome salt because of the relative size of salt domes. In order to insure that a cavity meets design criteria, it is imperative to accurately characterize adjacent features and surrounding sediments since these will affect the final size and stability of the cavity.

WASTE ISOLATION CONSIDERATIONS

As previously discussed, the storage of hydrocarbons in bulk fluid form is a proven concept. The proposition of waste isolation in engineered salt cavities is similar to this practice. The basic solution mining technology, characterization, certification, safeguards and maintenance are similar. What materials are isolated and how operations are conducted are the principle differences.

Although we see no technical reason that hazardous chemical wastes could not be disposed of individually or in compatible groups in separate cavities, economics and environmental conservation suggest that salt cavity isolation should be treated as long term storage. Regardless of the length of time, salt cavities lend themselves to this concept of utilization.

Waste materials have been classified as solids, dense liquids, light liquids and gases for the purpose of their placement and location in a solution-mined cavity. Operation and casing settings are modified to accommodate storage in the upper and lower section of the cavity. Cavities in bedded salt and dome salt are adaptable to various types of waste isolation. Dome salt cavities being larger and deeper are more suitable to light liquids, gases and larger volume wastes. Bedded salt cavities being smaller and shallower lend themselves to dense liquids and smaller volumes of wastes. (See Fig. 2-3).

To give an idea of relative capacity, a comparison has been made to approximate between a hypothetical municipal landfill facility for sludges from a large city and a large salt dome cavity. A large landfill facility could have a capacity for 3 million m^3 (4 million cubic yards) of sludge and could handle the city's waste for 20 years. One moderately large salt dome cavity could have a capacity for 0.7-1.5 million m^3 (1-2 million cubic yards) of sludge for a life span of 5-10 years. Although more than one cavity would be required at a dome site, one can see that the working volume would certainly be quite large.

The compatibility of wastes with the cavity walls of nearly pure salt is a relatively minor concern. Very few potential chemical reactions yield greater concern. Wastes incompatible with salt should be excluded or limited to safe levels. Some wastes do react with each other and produce greater hazards. Many mixtures of wastes can produce heat, violent reactions, phase changes, and toxic gases. Compatibilities must be considered, but hazards can be precluded by avoiding certain mixtures.

Physically, salt is insoluble in organic liquids in general and therefore continued dissolving of the cavity would not occur. Wastes in aqueous solution do have dissolving capibilities. This must be allowed for when designing the size and shape of the cavity and most importantly, the distance between adjacent cavities to assure that they do not coalesce. To safe-guard the facility, the design would allow for greater cavity separation and future development.

Looking ahead toward potential reclamation, it may be technically and economically wiser not to mix dissimilar wastes to facilitate the reclaiming process. More and more we are finding that today's wastes are tomorrow's resources.

Consideration must be given to the well construction and conversion of the solution-mined cavity to waste storage. Design of the well casing program must protect fresh water sources and the environment for a suitable life term. Since the wastes are to be injected into the cavity through the cased drillhole, these wastes must be fluidized and pumped under pressure. The pressure is required to counteract unbalanced forces due to density differences and friction and displace brine from the cavity.

It would be most desirable to fill a cavity with a single waste material to circumvent compatibility concerns. Small quantities of a large variety of wastes are best suited to individual small cavities for each type. Similar wastes of large volume are better suited to single large cavities.

Prior to converting a cavity to waste isolation, the structural integrity of the well and cavity must be tested. An acceptable certification procedure must be established to prevent endangering the formation integrity to the degree that hazardous wastes may be released to the environment.

Pressure maintenence within the cavity is necessary to reduce the differential with the lithostatic pressure. Operating pressure must be maintained between design limits to prevent structural damage to the cavity.

Operations monitoring is very important to assure cavity and well integrity and that design limitations have not been exceeded. Cavity surveys and well logging are used to determine fill predictions and changes in well conditions. Inventory of waste storage is a part of the data collection and record keeping. Also included should be operating data, roof protection and logging surveys.

A waste isolation facility is conceived of having multiple cavities for handling multiple wastes or expansion. The wastes would be received into temporary holding tanks, analyzed and then pumped through a piping system to the appropriate cavity. As initial cavities are being filled, additional cavities could be solution mined under a continuous expansion plan. The possibility exists that brine from the solution mining of new cavities and displaced from the waste storage cavities could be utilized as feedstock to a chemical plant. Depending on the nature of the waste, the discharge brine from the waste storage cavity may require brine treatment to meet process requirements.

To minimize the quantity of discharge brine from the waste storage cavity, wastes must be relatively concentrated, such as sludge slurries from an industrial waste water treatment plant. However, sludges need not be dewatered as required for solid landfill disposal. Dewatering is a costly process to increase solids content, and therefore a negative factor in the overall economics.

Although site characterization and cavity certification are the two most important considerations, there are others that need to be dealt with in the development of waste isolation cavities. EPA would be involved in the application and permitting process. Environmental impact assessment would be required for site development, but removal of wastes from the biosphere would naturally be beneficial to the environment. A water supply is necessary to dissolve cavity space. Brine disposal by injection into saline aquifers is regulated by states if the brine cannot be utilized as feedstock. Ownership of the mineral rights would be necessary to minimize legal complications. The method of waste transportation to the storage site must be considered.

Solution-mined salt cavities are attractive for waste isolation for all the reasons noted. Outstanding advantages are stability and impermeability of salt. Properly certified solution-mined cavities in salt can be used to contain and isolate large volumes of waste.

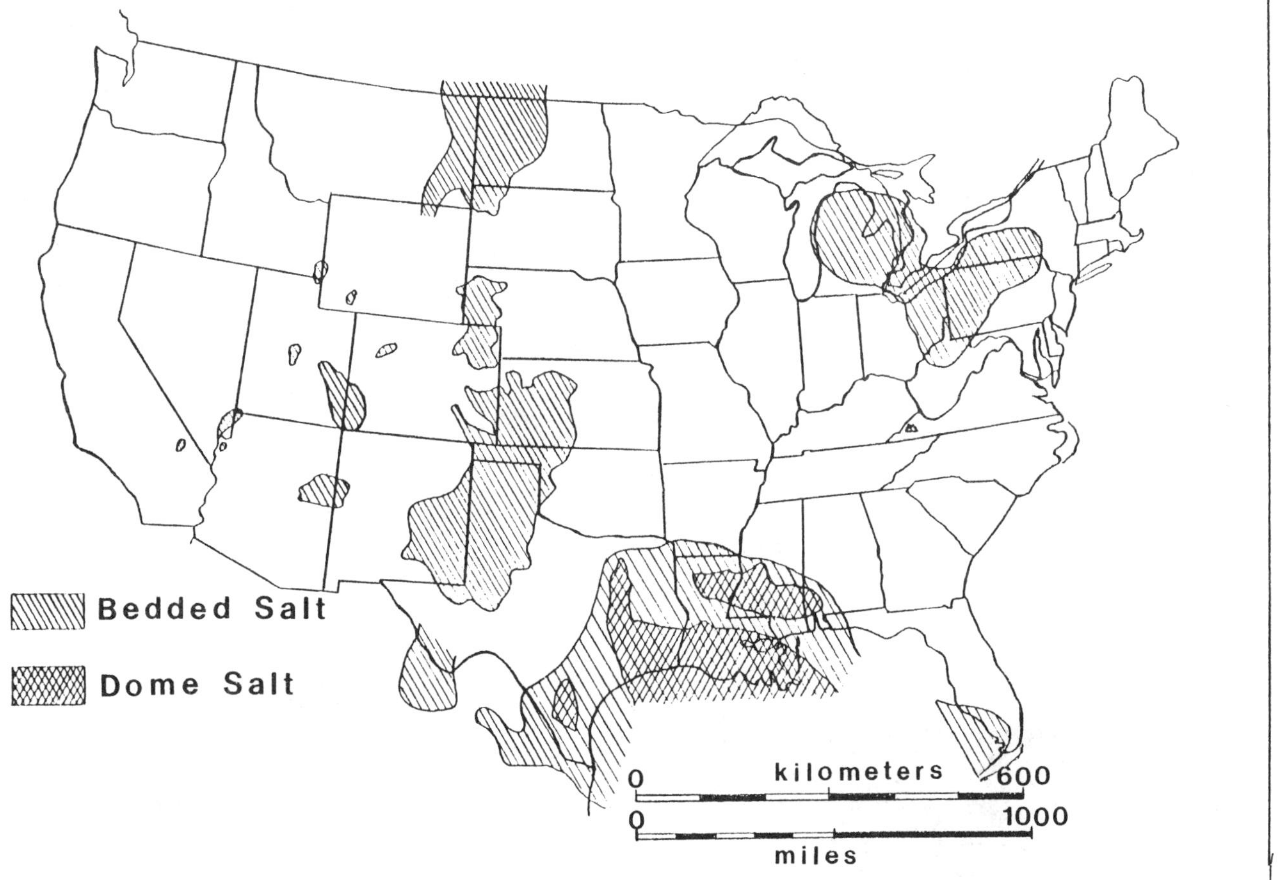

Figure 1. Salt Deposits in the United States

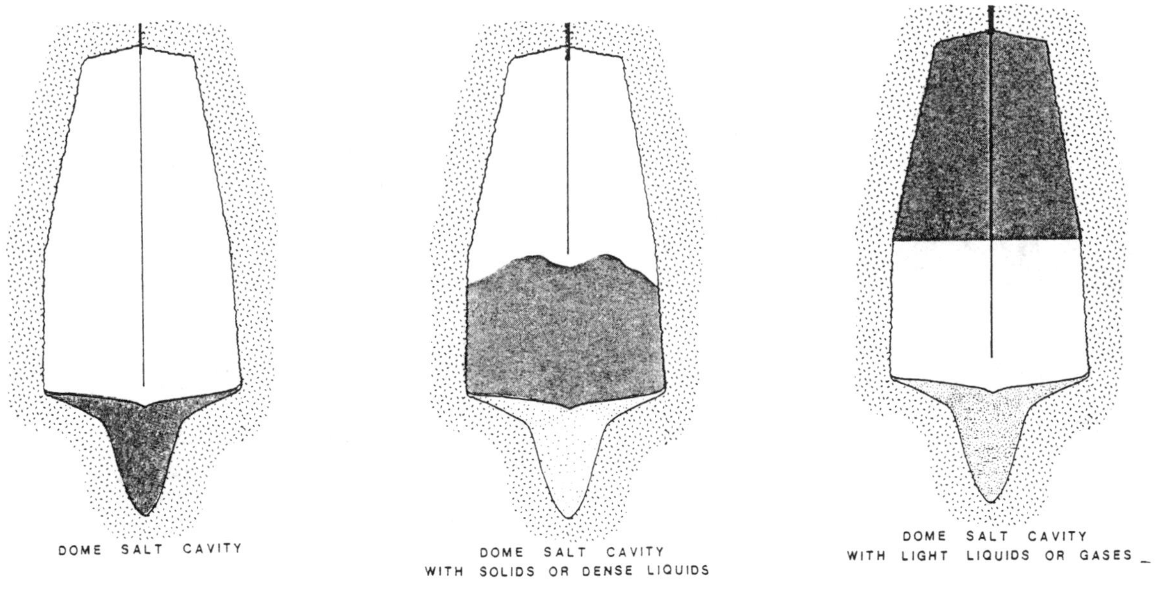

Figure 2. Dome Salt Cavity Storage for Different Waste Classifications

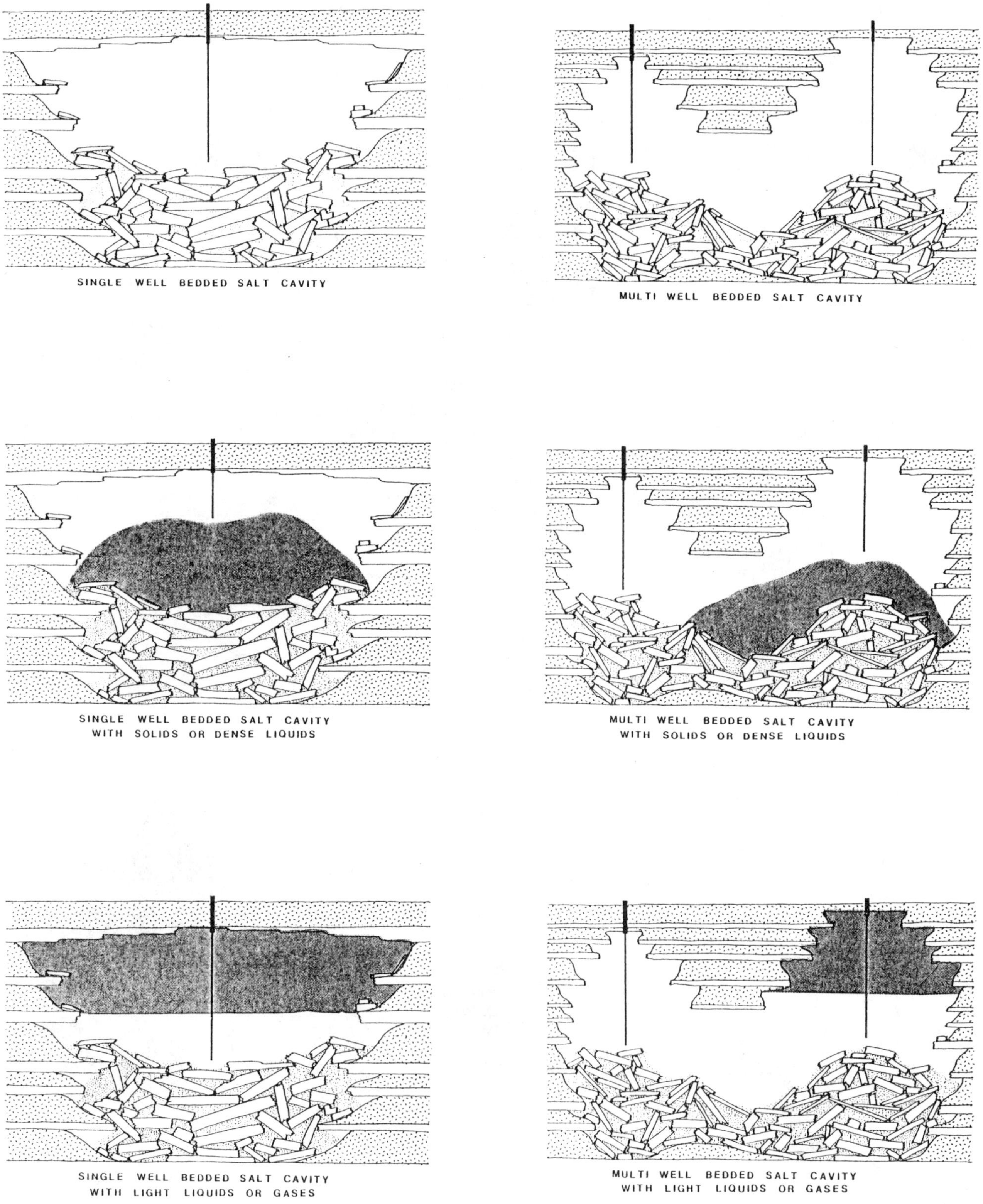

Figure 3. Bedded Salt Cavity Storage for Different Waste Classifications

Chapter 5

FEASIBILITY OF SOLUTION MINING FOR SODIUM CARBONATE AT SEARLES LAKE, CALIFORNIA*

James Giulianelli
Department of Chemistry/Geochemistry
Colorado School of Mines
Golden, Colorado 80401

Sam Carpenter
Department of Chemistry
Northwest Missouri State University
Maryville, Missouri 64468

Warren Dowler
Roger Carlson
Jet Propulsion Laboratory
Pasadena, California 91125

ABSTRACT

Evaporite salts have been processed commercially from brines at Searles Lake, California, for over sixty years. Research at the Jet Propulsion Laboratory on the feasibility of solution mining of trona was stimulated by the ideal conditions which the site offers for the proposed use of salinity gradient solar ponds as an optional means of providing inexpensive low-temperature heat for the injection brine.

The concept, which is based on phase relationships, proposes a novel in situ leaching operation initiated by injecting a warmer-than-ambient saturated sodium sulfate solution into the trona-bearing strata. This technique yields a brine that is richer in sodium carbonate than that which is presently available in the natural brine. The density of the sodium sulfate solution is controlled to match the density of the existing brines, thereby allowing penetration. This may also prevent subsurface collapse of the strata by its precipitation of solids that will partially replace the extracted sodium carbonate.

---

* The research described in this paper was carried out at the Jet Propulsion Laboratory, California Institute of Technology, and was sponsored by the U. S. Department of Energy through an agreement with NASA.

INTRODUCTION

Personnel at the Jet Propulsion Laboratory (JPL) are studying the technology and applications of salt-gradient solar ponds (see below) for both electrical and thermal energy generation. The brine extraction and processing site operated by the Kerr-McGee Corporation at Searles Lake, California, offers ideal conditions for economic construction of solar ponds. Furthermore, since rising conventional energy costs adversely affect the economic recovery of brine by the present process, this site was selected for a study on the feasibility of integrating the solution mining of trona ($Na_2CO_3 \cdot NaHCO_3 \cdot 2H_2O$) in the lower salt with solar ponds for heating.

The major emphasis of this study was to determine whether inexpensive thermal energy available at the brine extraction site would be useful in extracting increased mineral value from the resource. The present operation, which involves the simple extraction of the complex brine and its processing at the adjacent plant, is sometimes referred to as solution mining. Since the concept discussed in this paper involves injecting a heated sodium sulfate brine to dissolve carbonate underground, the proposed scheme is referred to as "enhanced" solution mining. The subject is discussed in greater detail in a recent JPL report.(1)

SALT-GRADIENT SOLAR PONDS

A salt-gradient solar pond (SGSP) is a body of still water that captures solar radiation and stores it as usable heat.(2) In an SGSP (Figure 1), thermal energy is absorbed and retained in the nearly saturated convecting bottom layer, which is typically about two meters thick. This heavy storage layer is prevented from rising and losing heat to the upper layers by a non-convecting salt-gradient layer above the dense storage layer.

Solar ponds are one of the least expensive and simplest solar collectors, have solar-to-thermal efficiencies of around 20 percent, and can store and deliver huge quantities of thermal energy, even throughout the winter months. Temperatures near 95°C have been attained in operating outdoor ponds.

For maximum effectiveness of solar ponds, inexpensive land, an inexpensive salt source, and water or brine must be available. The ponds must also be located in an environment of compatible meteorological conditions, such as insolation, wind, ambient temperature, and local topography. The Searles Lake site satisfies all these criteria.

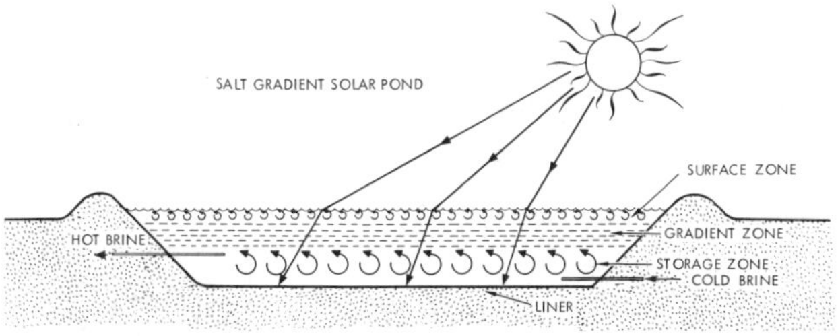

FIGURE 1. SCHEMATIC OF A SALT-GRADIENT SOLAR POND

SEARLES LAKE

Deposits

Searles Lake, located in the high desert in east central California between the Sierra Nevada Mountains and Death Valley, is a large evaporite deposit with mineral values available in the form of interstitial brines and layered solids beneath the surface. The lake still has water below the surface in the form of saturated interstitial brines. The brines and their enclosing salines are composed chiefly of sodium carbonate, bicarbonate, sulfate and chloride. There are also commercially useful amounts of potash and borate. Figure 2 details the structure of the Searles Lake deposits.

Most of the chemical production from Searles Lake has come from the pumping of the Upper Salt brines, which has occurred on a major scale since the mid-1920s. The processing plants are now operated by the Kerr-McGee Corporation. Pumping from these operations has led to increasing variation in brine composition.(3)

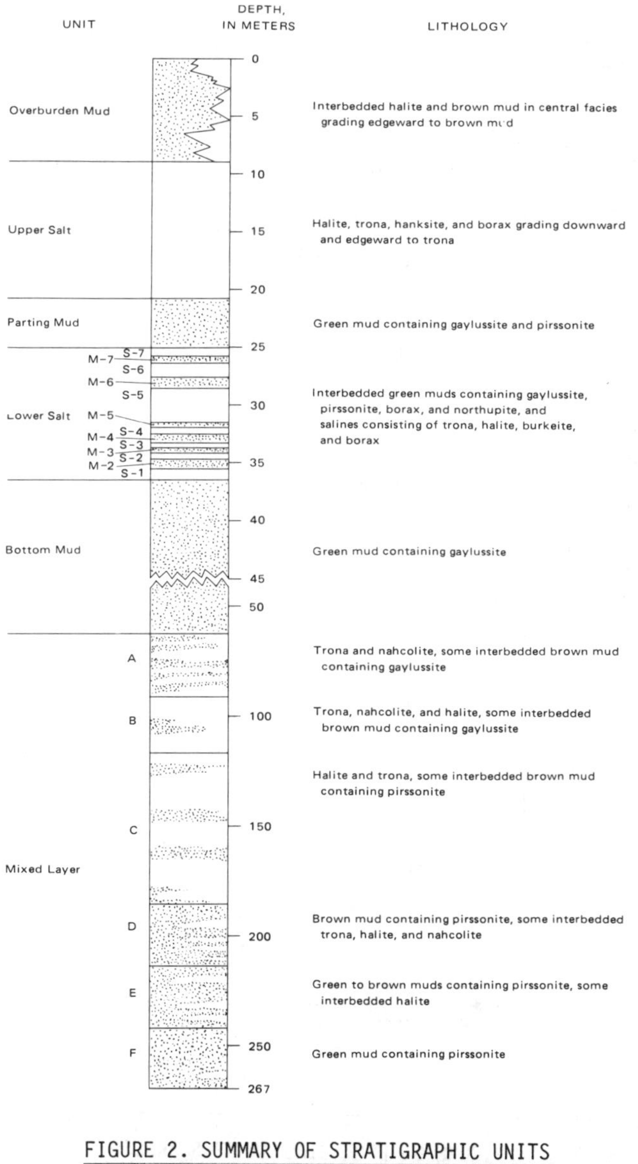

FIGURE 2. SUMMARY OF STRATIGRAPHIC UNITS IN SEARLES LAKE EVAPORITE SEQUENCE

The evaporite salt layers are thin, but cover a large area, so a number of wells are required to slowly pump brine from each location (excessively rapid pumping might damage the formation). A complex set of processes are used to extract the maximum amount of saleable product chemicals at the least cost.

Brine Processing

The brine is concentrated by solar evaporation in large ponds. Because solar concentration of brine principally occurs during the summer, concentrated brine is stored in wells in parts of the Upper Salt for use during the winter. Brackish water from wells several miles from Searles Lake is used in the plants and as replacement water for the salt layers. Also, the spent brines from the plants are returned to the lake surface.

Present plant operations are summarized in Figure 3 which shows the overall flow of brine into and out of each plant at the site and the final commercial product(s) of each plant. Roughly 360,000 metric tons (400,000 tons) of coal and petroleum coke are used yearly to heat and power the operations.

The mining operation at Searles Lake operates under the following two basic physical-chemical constraints:

1. Limited Brine Concentration. Concentration of mineral values in the brine is limited by their solution phase relationships. These underground brines tend to be in equilibrium, with the undesired chloride ion being dominant. Only four to ten percent by weight of the desired sodium carbonate and sodium bicarbonate is dissolved in the brine even though the brine originates from a strata consisting almost entirely of these two salts (i.e., S-1 and S-2 of the Lower Salt). Furthermore, although underground brines are replaced by waters that dissolve enough salts to reach local chemical equilibrium, the ratio of components in the replacement brines may be becoming less favorable for future commercial operations.(3)

2. Collapse and Mixing of Separated Strata. Since Searles Lake strata are unconsolidated muds and salines, removal of solids from the layers without at least partial replacement of their volumes will inevitably lead to local collapse and intermixing of layers. Attempts to inject water rather than a saturated brine fail because the less dense water is forced upwards by the more dense brine. Vertical mixing of layers via abandoned cores and wells is also a primary concern for maintaining the resources.

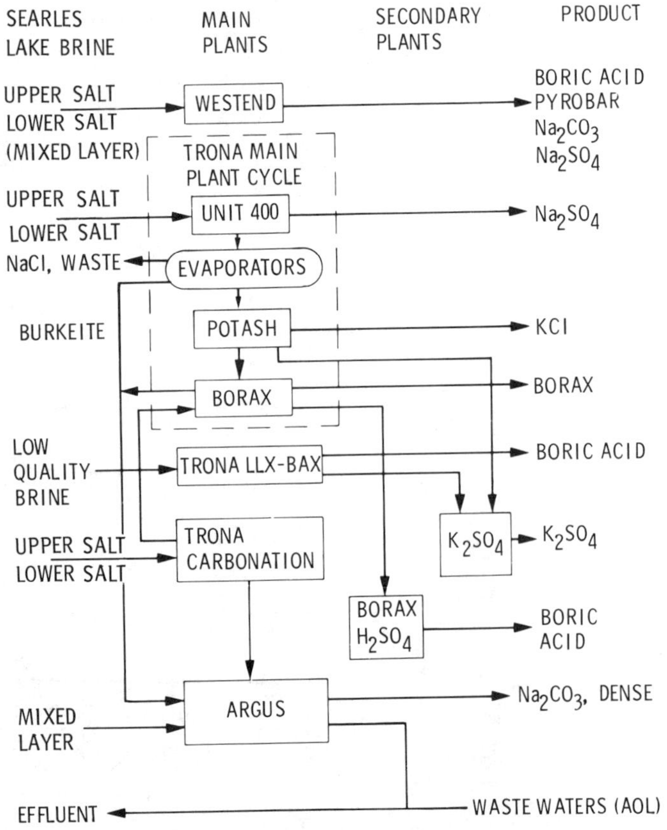

FIGURE 3. FLOW DIAGRAM OF SEARLES LAKE PLANT OPERATIONS

ENHANCED SOLUTION MINING

Concept

At present, solid minerals are extracted from the natural brine pumped from the three main saline strata. The proposed mining operation, discussed herein, concentrates upon the mining of a single mineral, trona, from the Lower Salt layer with specific reference to Levels S-1, S-2, and S-3 which are almost entirely trona. In the future, these studies could extend to multiple or sequential solution mining to obtain other mineral values such as borax from the strata.

As indicated previously, a dense make-up injection brine must be used since a less dense fluid would simply either return to the surface at the injection well or rise through the many old core or bore holes connecting and contaminating adjacent layers. In the proposed mining scheme (Figure 4), Glauber's salt ($Na_2SO_4 \cdot 10H_2O$) is investigated as the candidate salt for the make-up brine. Burkeite ($Na_2CO_3 \cdot Na_2SO_4$) could possibly replace Glauber's salt as the sulfate source.

A second reason for chosing a sulfate solution is concern with the future of the site being mined. Removal of the solids from below ground without at

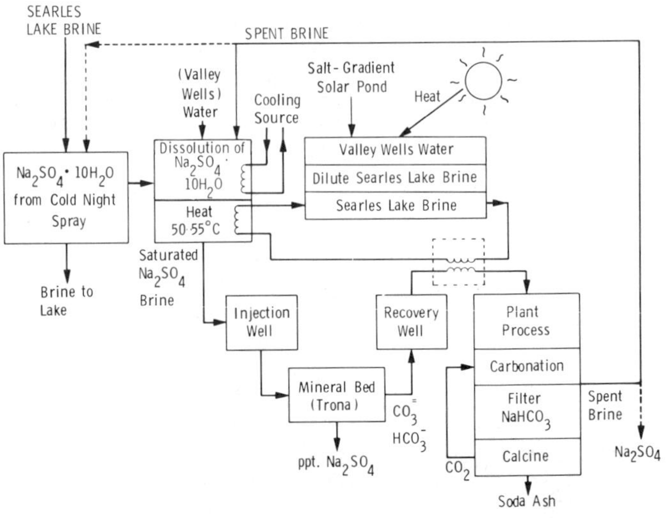

FIGURE 4. SCHEMATIC OF PROPOSED SOLUTION MINING USING SOLAR POND PROCESS HEAT

least partial replacement would inevitably lead to local collapse, subsidence, and intermixing of layers, undesirably upsetting the balance of today's mining operation. Examination of the phase relationships indicates that solids can be precipitated during and after the mining of a particular stratum, and these should aid in holding up the structure.

The source of sulfate is from evaporative winter night cooling of sprayed brine. Because the exiting sulfate brine will be chemically much different from the present brines, the entire operation proposed here is divorced from the present plant process since a separate plant recovery scheme may be required.

The baseline concept proposed for integrating the SGSP into the mining operation (as shown in Figure 4) is a closed-loop pond using Searles Lake brine. The dense bottom layer is circulated to supply the heat needed and then returned to the pond. Choices other than Searles Lake brine, such as sodium chloride, sodium sulfate, or trona itself, might also be used as pond salts.

A typical set of calculated operating characteristics is shown in Table 1.

TABLE 1

Possible Characteristics of Proposed Scheme

Heated Injection Brine	
Composition:	32% Na_2SO_4 by weight
Density:	1.3 mg/mm^3
Flow rate:	0.19 m^3/s
Heating above ambient:	30°C
Injection wells:	Array around exit well--going to S-1, S-2 layers
Solar Pond Option	
Insolation at site:	2100 kWh/m^2 per year
Assumed solar to thermal efficiency:	20%
Area of solar ponds required:	70 km^2
Sodium Carbonate Yield	
Present (approximate):	7-8%
Proposed (see text):	up to 14%

Thermodynamics

Phase Diagrams. Since the early 1900's, the struggle to profitably mine Searles Lake has depended on the exact definition of the phase relationships among the various salts.[4] These relationships were studied at the higher temperatures in our work in order to predict the equilibrium composition of the heat-injected brine within the strata.

The Searles Lake Lower Salt S-1 and S-2 saline layers are best represented by the five-component system of sodium carbonate, sodium bicarbonate, sodium sulfate, sodium chloride and water. By representing only the relative compositions of the salts within the saturated solutions with water as an understood member, a simpler four-component system is created. Although data are abundant at 20°C, nothing seems to have been been reported above 60°C.

The most complete studies at elevated temperatures [5,6] unfortunately contained some errors in the tabulated values, making it necessary to redo some of their measurements. A three-dimensional pyramid was constructed using data at 45°C, a reasonable temperature for the enhanced mining operation. This pyramid, looked down upon at the correct angle, projects the two-dimensional image shown in Figure 5. Points 73 and 74 are the so-called "hidden points" within the three-dimensional volume, all others being on or near the surface. The cross-hatched plane is significant in that it intercepts all points having solid trona in equilibrium with solution.

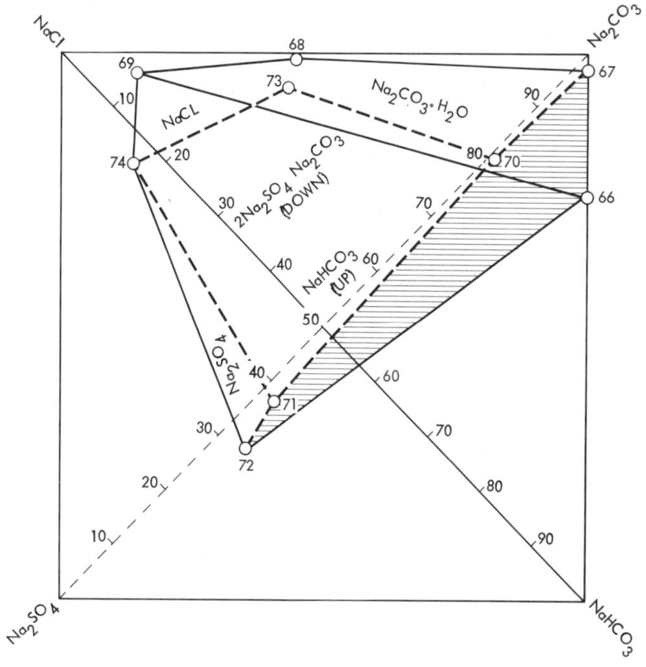

FIGURE 5. GRAPHIC PROJECTION OF INVARIANT POINTS AT 45°C FOR NaCl-Na_2CO_3-Na_2SO_4-$NaHCO_2$-H_2O SYSTEM AT 45°C (From Makarov and Sedelnikov, 1940) (See Table 2)

Table 2 lists the corresponding weight percentages of this system given at the so-called invariant points at which the solution will tend to stay as long as these solids are in equilibrium with the solution.

This data shows that the presence of chloride depresses the carbonate solubility and that this effect is significant in the lower salt layer in the natural brine. However, when the dense sodium sulfate leach solution is injected and displaces the original brine, chloride will no longer be present. This initial flushing away of chloride ions is a crucial function of the sulfate solution.

In the absence of chloride and with water as an understood member, data can be represented by the three-component system shown in Figure 6. The *in situ* equilibrium must always contain solid trona and

TABLE 2

INVARIANT POINTS AT 45°C
(Makarov, S. Z. and Sedelnikov, G. S., 1940)

Relative Weight Percent in Solution

Pt	Na_2CO_3	$NaHCO_3$	NaCl	Na_2SO_4	Solids
66	74.1	25.9	--	--	B + T
67	97.3	2.7	--	--	C_1 + T
68	45.6	0.9	53.5	--	C_1 + X + T
69	11.1	4.6	84.3	--	B + X + T
70	81.1	1.1	--	17.8	C_1 + γ + T
71	36.1	4.4	--	59.5	S_0 + γ + T
72	26.5	8.35	--	65.15	S_0 + B + T
73	43.5	1.0	50.1	5.4	C_1 + γ + X + T
74	9.8	5.1	69.8	15.3	S_0 + γ + X + T

Notes:

B = $NaHCO_3$

C_1 = $Na_2CO_3 \cdot H_2O$

S_0 = Na_2SO_4 (thenardite)

T = $Na_2CO_3 . NaHCO_3 . 2H_2O$ (trona)

X = NaCl

γ = $Na_2CO_3 \cdot 2Na_2SO_4$

thus have its composition along the boundaries of the "trona plane" also shown cross-hatched in this figure.

If trona (whose composition is shown by the arrow of Figure 6) is placed in distilled water, the incongruent decomposition into bicarbonate and carbonate

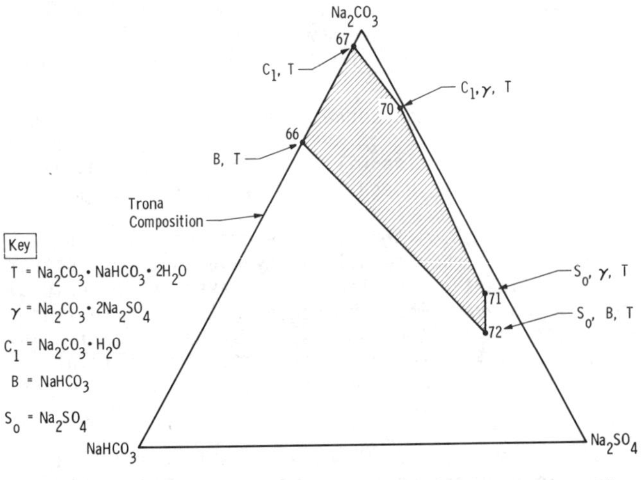

FIGURE 6. COMPOSITION OF SOLUTION (Excluding H_2O) IN EQUILIBRIUM WITH INDICATED SOLIDS AT 45°C (From Makarov and Sedelnikov, 1940) (See Table 2)

yields the solution shown at point 66. This point has a much higher carbonate value, and the concept of injecting hot water or carbonate brine was patented.[7] These methods are obviously not applicable for the unconsolidated sediments at Searles Lake. Water is not a viable injection fluid due to its density, and neither water nor carbonate brine leaves any material to prevent future strata collapse.

Dissolving trona in the presence of sulfate ion causes the composition to approach point 72 along the line connecting points 66 and 72 in Figure 6. If enough burkeite precipitates, however, point 71 is approached, which is undesirable because carbonate is precipitated as part of the burkeite and may not be recoverable.

Equilibrium Carbonate Yields. The yield of soluble sodium carbonate within a saturated sodium sulfate leach solution at various temperatures was calculated from the available data. The data, summarized in Table 3, show that increasing the temperature causes the phase equilibrium, and thus the carbonate yield, to shift in favor of a greater percentage of carbonate in solution. Selecting the optimum operating temperature will depend on the trade-off between the cost of additional heat to the injection brine versus the cost benefits of processing a richer carbonate brine.

TABLE 3

CALCULATED YIELDS OF Na_2CO_3 IN LEACH SOLUTIONS OF SATURATED Na_2SO_4, IN EQUILIBRIUM WITH TRONA

T (°C)	Na_2CO_3 (wt %)	Solid Phase (trona and $NaHCO_3$ always present)	Phase Diagram Point*
25	14.27	$Na_2SO_4 \cdot 10H_2O$	36
30	11.04	Na_2SO_4	39
35	12.43	Na_2SO_4	42
45	12.57	Na_2SO_4	72
60	13.96	$Na_2CO_3 \cdot 2Na_2SO_4$	81

*Numbering of phase diagram points conforms to the data presented in Reference 5.

There is some uncertainty concerning yields at 60°C and above due to a confusion about whether burkeite or sodium sulfate precipitates at point 81.[6] This is important because burkeite precipitation would decrease the amount of carbonate available for eventual solution.

Dissoluton of Trona. A review of published accounts led to concern that trona may be prevented from completely dissolving by the formation of a coat of sodium bicarbonate or other insolubles.[8] A few simple experiments were performed which suggest that this would not be a problem. Pellets made from Searles Lake trona were dissolved in water and sodium sulfate brines at different temperatures while their weights in these solutions were being recorded. The pellet weights decreased uniformly with no evidence of the occurrence of armoring.

CONCLUSIONS

The phase relationship correlations and static experiments performed support the feasibility of the proposed enhanced solution mining operation. An analysis of the phase relationships shows that by using sulfate input brines at 45°C or 60°C, the output brine would consist of about 12.6 or 14 percent total Na_2CO_3, respectively (compared to a maximum existing concentration of 9.8) and precipitates left in the cavity could be enough to support the resulting structure. The next phase of this study obviously should involve laboratory dynamic flow experiments or field studies to verify these conclusions.

The interest of the Jet Propulsion Laboratory was to demonstrate the use of SGSP for low-temperature industrial process heat. It is hoped that the proposed use of solar ponds, as well as the enhanced solution mining approach, will be of interest to the mining industry.

ACKNOWLEDGMENTS

The authors would like to thank the Kerr-McGee Corporation for their encouragement and for providing samples and data from their Searles Lake operation. Two of the authors, James Giulianelli and Sam Carpenter, acknowledge the support of the DOE-ASEE Summer Faculty Research Program for the stipends which resulted in this work.

REFERENCES

1. Giulianelli, J. L., et al, Solution Mining of Searles Lake Evaporites Using Solar Ponds, Jet Propulsion Laboratory, Internal Document 5105-58, July 1981.
2. Nielsen, C. E., "Nonconvective Salt Gradient Solar Ponds", Solar Energy Handbook, Dickinson, W. C., and Cheremisinoff, P. N., ed., Marcel Dekker, Inc., 1980.
3. Smith, G. I., "Subsurface Stratigraphy and Geochemistry of Late Quaternary Evaporites, Searles Lake, California", U. S. Geol. Surv. prof. paper 1043, 1979.
4. Teeple, J. E., "The Industrial Development of Searles Lake Brines with Equilibrium Data", Amer. Chem. Soc., Mon. Ser., No. 49, 1929.
5. Makarov, S.Z. and Sedelnikov, G. S., "A Study of the System Na_2CO_3 - $NaHCO_3$ - Na_2SO_4 - NaCl - H_2O in the Region of Trona Crystallization", Izv Akad. Nauk. SSR, Ser. Khim., No 6, 1940, pp. 835-863.
6. Sedelnikov, G. S. (in Russian), Zh. Prikl. Khim., Vol. 18, 1945, pp. 430-438.
7. Pike, R. D., "Solution Mining of Trona", U. S. Patent 2,388,009, 1945; Bays, Carl "Solution Mining of Trona", U. S. Patent 2,979,317, 1961; Caldwell, N. A. and Frint, W. R., "Method of Recovering Sodium Values by Solution Mining of Trona", U. S. Patent 3,050,290, 1962.
8. Gancy, A. B., "Process for the Production of Soda Ash from Underground Trona Deposits", U. S. Patent 3,184,287, 1965.

Chapter 6

APPLICATION OF 3-D REFLECTION SEISMIC IN SALT MINING: A CASE HISTORY

F. A. Molloy
Shell Internationale Petroleum Mij. B.V.
The Hague, Netherlands

G. M. W. Haug
Billiton International Metals B.V.
The Hague, Netherlands

ABSTRACT

As a result of feasibility studies made in 1975 a 3-D seismic reflection survey was shot over part of the Veendam salt pillow during 1978–1979. The Zechstein salt pillow, located in the northeast of the Netherlands near the town of Veendam, occurs at a depth of 1400 meters and contains three carnallite rich horizons over an interval of 300 meters below top salt. $MgCl_2$ brines are produced from these layers as feedstock for a magnesia plant, which starts up in 1981.

Prior to the acquisition of the seismic data four wells had been drilled. The thickness of the carnallite layers encountered in these wells varied between 50 and 125 meters. In view of these variations it was necessary to obtain a reliable isopach of the different layers. The 3-D seismic survey covered an area of ± 4x4 km with a sampling interval of 20 meters in all directions. The data was recorded in an areal array using a 120 channel instrument. Final presentation of the data was in the form of 3-D migrated acoustic impedance cross sections. The quality and resolution of the seismic data allowed reliable structural and isopach maps to be made. It was shown that the carnallite layers are continuous over the area surveyed, whilst local thickening is "fault" related.

Having this information brine production locations could be choosen in the thicker parts, allowing maximisation of production potential whilst avoiding faults and fulfilling mining regualtions.

A subsequent benefit of this approach is the possibility of using the seismic data to predict ore grade. Using the available wells as calibration points the acoustic impedance response of the seismic at the well was matched to the ore grade found. From these points a synthetic section was made with varying ore grades. Preliminary results are encouraging and the reliability of this predictive tool will be evaluated in the light of future drilling.

INTRODUCTION

Having determined the feasibility of solution mining techniques for the production of magnesium brine, three additional wells were drilled in 1976 on the Veendam salt pillow in order to establish sufficient magnesium reserves for the project. The wells show a wide range of thicknesses of the magnesium bearing layers penetrated. These variations in thickness precluded an accurate reserve estimate being made with the available information. As the economics of the project dictate a high magnesium brine yield per cavity, an accurate isopach map of the magnesium layers is necessary in order to plan the exploitation stage.

A study was done to ascertain whether seismic reflection acquisition and processing methods could be used to map the magnesium layers. To carry out this study seismic data covering the Veendam salt pillow was purchased from the Nederlandse Aardolie Maatschappij (NAM). Six additional seismic lines were shot in 1975 to improve the reliability of the study. The results of the study showed that within the constraints of the seismic system the magnesium bearing layers could be mapped. The next step was the planning and shooting of a 3-D data set, a noise spread was shot in the area and at the same time charge size, depth of shot-holes and geophone/shothole pattern were evaluated.

The survey was shot during November 1978 and January 1979 by Prakla-Seismos, a seismic contractor. Processing of the seismic data to a 3-D migrated stage was done by Geophysical Services Incorporated (G.S.I.) in Bedford, England. Acoustic impedance processing of the 3-D migrated data was done by Shell.

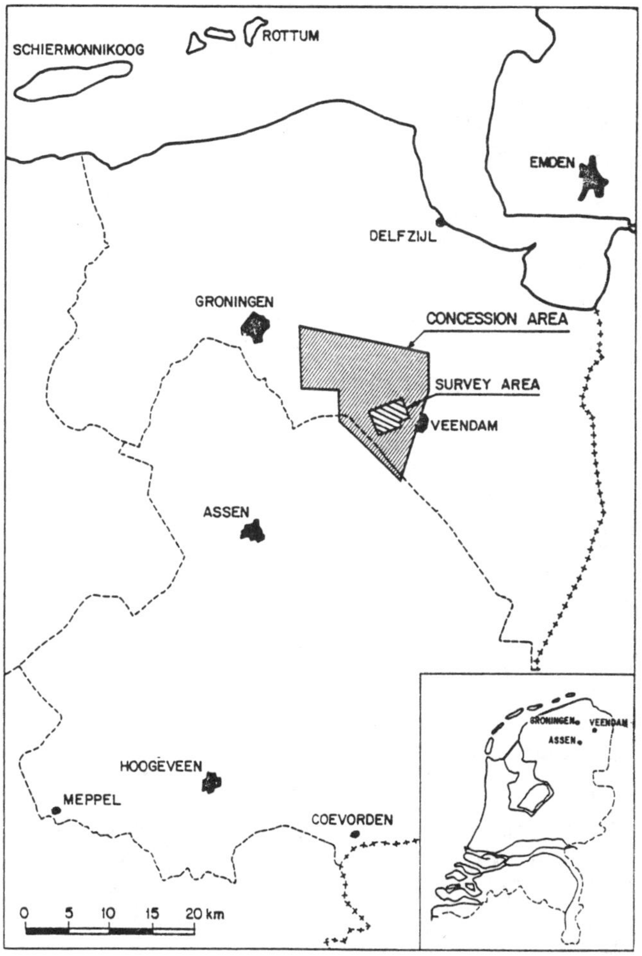

FIG. 1

GEOLOGICAL SETTING

A vast body of literature is available on the Zechstein (Permian) evaporites in Northern Europe, in particular on Germany, where salt mining dates back many years. Comprehensive studies have been written by Borchert and Muir, Braitsch, Richter-Bernburg, Sanneman, Stauble and Milius, Visser, and others; in general the published information on the Northern Netherlands is limited.

The concession area (see Fig. 1) is located south of the roughly ESE-WNW trending axis of the Zechstein basin, and was only moderately affected by the widespread salt tectonics. Here salt flow caused the formation of salt pillows only. The salt movements ceased in early Tertiary times, leaving the Quarternary sediments undisturbed. In this area the evaporation cycle reached completion only in the Zechstein III. Within this, three major subcycles are present, containing an aggregate thickness of magnesium/potassium salts up to 100 meters; only the lower subcycle has substantial amounts of pure magnesium chloride (bischofite). Figure 2 shows the general stratigraphy, as logged in one well, and is representative for the area. Results of earlier exploration, discussing the stratigraphy and depositional environment, as well as the use of conventional 2-D seismic, was reported before by Coelewij et.al.

The economic depth range as presently considered for this project is from 1500 to 1700 meters for the top of the magnesium bearing salt, and the bottom of the lowest layer from 1700 to 2000 meters. The average thickness of the barren halite "roof" measures 90 meters. The thickness of the underlying halite is on average 1400 meters. The overburden of the salt pillow is heavily faulted at Base Tertiary level. The intensity of faulting decreases with depth and only a minor number of faults effect the Zechstein layers. Whether the observed displacement on seismic of the magnesium bearing layers can be considered as faults "sensu stricto", or as the extremes of plastic flow is not known.

OBJECTIVE OF SEISMIC SURVEY

In view of the mining restrictions and economics pertaining to solution mining, it is imperative to obtain a high magnesium brine yield per cavity size to a diameter of 100 meters and the horizontal area covered by cavities to 10% of the total area. A further restriction of a minimum cavity separation of 200 meters also applies. This being the case, it is essential to exploit the thicker parts of the magnesium bearing layers where present.

Studies made showed that the technique of lateral prediction could be used in mapping the top and bottom of the magnesium layers in the Zechstein III (Fig. 2) evaporitic cycle present in the Veendam salt pillow. This study also indicated that the thickness of the Zechstein III 2b/3b and 1b layers varied greatly over short distances. Furthermore it was shown that the Zechstein III 2b/3b layers would have to be regarded as one. This is due to the inability of the seismic frequency band at that depth to resolve the thin (± 10 meters) halite layer separating the two magnesium bearing layers. The available seismic data and drill holes also indicated rapid thickness variations of the layers.

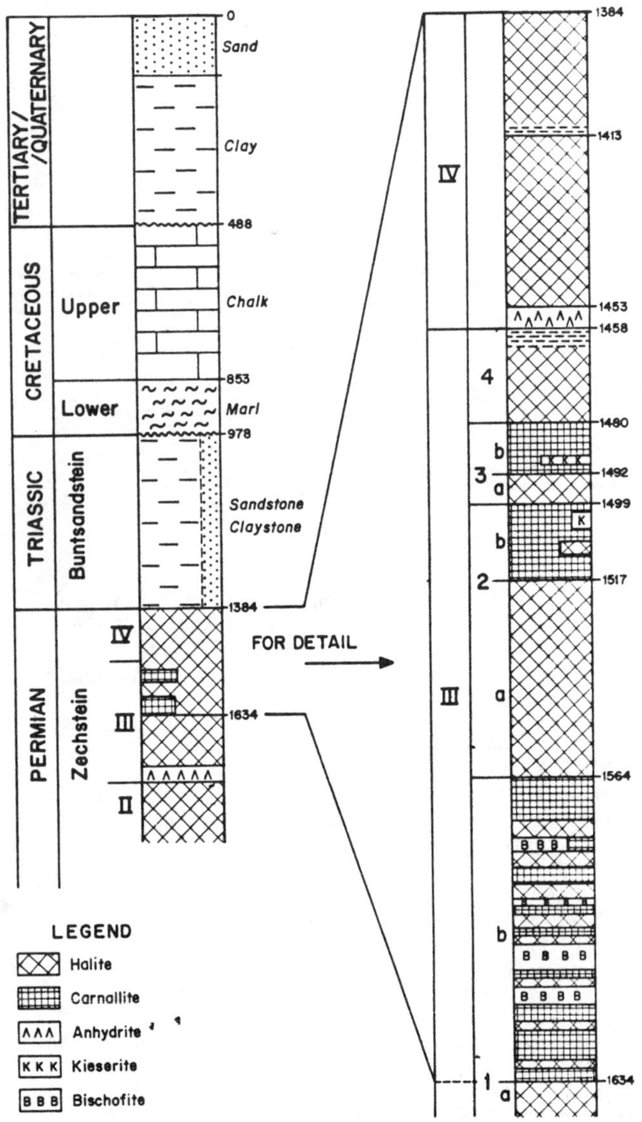

FIG. 2

REPRESENTATIVE STRATIGRAPHY IN AREA OF SURVEY
(Bore hole VEENDAM 4)

For the mining stage of the project it was necessary to construct maps which would accurately depict the thickness variations of the magnesium layers. To do this using reflection seismic required a very close grid of sample points in view of the size of the anomalies and the dips involved. This method is most cost effective; the total cost of the project equalled the cost of one drill hole (in 1976 money).

SEISMIC

General

The seismic field survey was carried out by Prakla-Seismos G.M.B.H. All instrumentation and field techniques to be used were defined by Shell. Topographical surveying and permitting started on the 8th November, 1978. Seismic data acquisition was carried out from the 28th November 1978 to the 20th January 1979. Weather conditions were extremely inclement during the survey period with recorded temperatures in excess of -20°C on a number of occasions.

Area of Survey

The area in which the survey was shot lies to the W.N.W. of the town of Veendam and covers about 4 x 4 kilometers. The S.E. corner of the survey had to be omitted due to the development of Veendam town. The area is intensely cultivated and crossed by many small ditches. As the area comprises of some 200 land holdings, obtaining access permission from all the land owners involved considerable effort. Permission was withheld on one occasion only. A rather dense network of gas lines, water mains, electric cables and drainage pipes resulted in many shot-points being moved. Out of a total of 1880 shot-points, 660 had to be displaced from their planned location.

Field Technique

It is essential, if all the anticipated benefits of a 3-D survey are to be achieved, that the proper advanced planning be performed. In general a seismic programme can be called 3-D if it is of such a sampling density that a minimum of one half wavelength samples exist in all directions. However one quarter or even one eighth wavelength would still be better. The fundamental considerations to meet the above space sampling requirements are as follows:-

a. Dominant frequency of the objective horizon at various levels or an estimate of the maximum possible bandwidth.

b. Dip rates in all directions, including ideas on major structural axis and fault trends.

c. Representative velocity function for the area.

d. Information regarding the shallowest and deepest data required.

A further critical factor in planning is calculating the migration aperture in any direction which is required to properly migrate the data. Once this aperture has been defined the programme boundaries can be established by adding the calculated migration aperture to all edges of the area of interest. However, care should be taken to ensure that no critical structural flank, important fault block segment or other features of interest, falls outside of the area to be surveyed. Furthermore, all the general geophysical problems that would be considered during the planning of a high quality 2-D survey need to be analysed as well, since they play a crucial part in the successful execution of a 3-D survey. Key features to consider in this respect are:-

a. An analysis of the noise problems of the area together with a study of all arrays used in the past.

b. A review of the methods for the derivation of near surface corrections and an estimation of the apparent wavelengths of any static problems.

c. If possible, a comparison of different energy sources to see if there are any source related problems.

d. An examination, if possible, of different vintages of data for field and offset effects to decide whether there is a multiple problem.

e. A critical study of the velocities for the area to determine the magnitude of the velocity gradients.

Only when the foregoing factors are taken into account is it possible to establish the practical and economic advantages of a 3-D survey in a given area. Field parameters for this survey were based on field work and studies made in 1978.

In order to obtain discrete information on the seismic properties of the area a number of seismic experiments were made in June 1978. As a result of these tests the optimum geometries with regard to geophone station interval and patterns were designed. Shothole depths and patterns were also evaluated at this time.

Having considered all parameters such as common depth point (C.D.P.) spacing, maximum and minimum offsets, azimuths from shotpoint to receiver groups, the areal layout was fitted to the prevailing infrastructure. This was done in order to obtain as complete as possible a 20 x 20 m C.D.P. spacing with six fold multiplicity over the whole area to be surveyed. The survey was executed by dividing the area into five strips, fig. 3. Within each strip the areal geophone station array was eight parallel rows containing fifteen stations each. The total 120 stations were then recorded from sixteen shotpoints, arranged along two traverses located 320 m and 44 m in front of the receiver field.

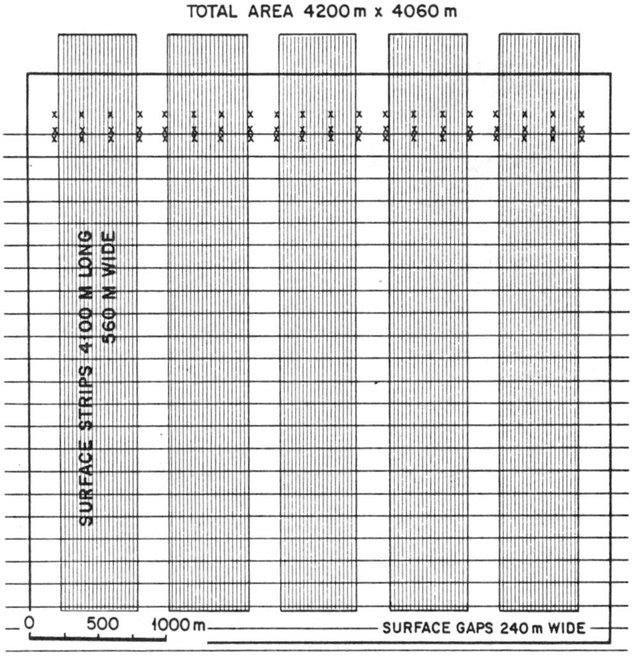

THIN LINES IN WSW-ENE MARK THE 26 SP-TRAVERSES
HEAVY LINES MARK THE OUTLINE OF SUBSURFACE COVERED

FIG. 3

STRIP CONFIGURATION USED
FOR DATA ACQUISITION

To obtain a optimum damping of the receiver groups a pattern of 19 geophones in the form of a 3-armed windmill was laid out at each station. As a result of experiments it was found that a single shothole, twenty meters deep, charged with 3 kg of dynamite, gave the best results.

The Sercel SN 348 telemetric instrument was used for recording. Data was recorded on 9 track ½ inch tape with 1600 B.P.I. The SN 348 system differs from conventional recording systems in that the analogue signal is digitized at the geophone station. The data is then transmitted by a single six conductor cable to the recording unit.

On hundred and twenty channels were recorded for each shot. Playbacks were made for all shots to ensure proper encoding of the data on tape. Two problems were experienced with the instruments. The cooling capacity of the recorder was insufficient. This gave rise to various faults through overheating. The second, and more serious problem, was failure of geophone station units. These units were affected by extremes of temperature from 5°C to -20°C during the survey. This fault was intermittant, making it difficult to trace.

Seismic Processing

The major benefits to be gained from a properly planned and executed 3-D survey are:-

1. Correct horizontal and vertical placement of reflection images and dip evaluation.

2. Elimination of diffraction events.

3. Restoration of signal strength lost through scattering problems.

4. More accurate mapping of events due to control point density.

5. Flexibility of display orientation.

6. Improved statistical base for the estimation of near surface corrections and velocities.

7. Data gathering methods can be adapted to fit the surface problems.

Individual processing steps are given in figure 4. The most critical steps in the processing sequence are those involving velocity determinations and the building of velocity models prior to entering the 3-D migration stage. Up to this stage the processing follows the same sequence as that used with 2-D data.

The final unfiltered stacks are used as input to G.S.I.'s 3-D wave equation migration programme. The algorithms used in this programme are based on an integral solution to the scalar wave equation in which seismic observations are the boundary values. The problem with this approach is knowing the velocity field with which the data should be migrated. As far as is known, all current migration algorithms which neither ray-trace nor pre-distort the input section tend to overmigrate the data in the presence of lateral velocity variations and are thus correct for horizontally layered velocity structures only.

In the case of Veendam at depth the velocities do not fit this model and the tendency to overmigrate the data has been compensated for by a "slowing down" of the stacking velocities. The problem of estimating these "slow down" factors is tackled in a very subjective and qualitative way. It is done by testing different factors and then choosing the ones which give a geologically acceptable picture for the migrated data.

The processing of 3-D data demands rigorous quality control. Prior to entering the 3-D migration stage it is important to combine all the geological and velocity information available in building the velocity model by which the data is to be migrated. Apart from the routine data validation steps, frequency spectra were generated for selected traces after each major processing stage. In this way, by considering the effect on the high frequencies, the effectiveness of the processing could be assessed. By generating seiscrops before and after migration it was possible to evaluate the validity of the velocity model used. This was done on the basis of the improved energy focusing obtained.

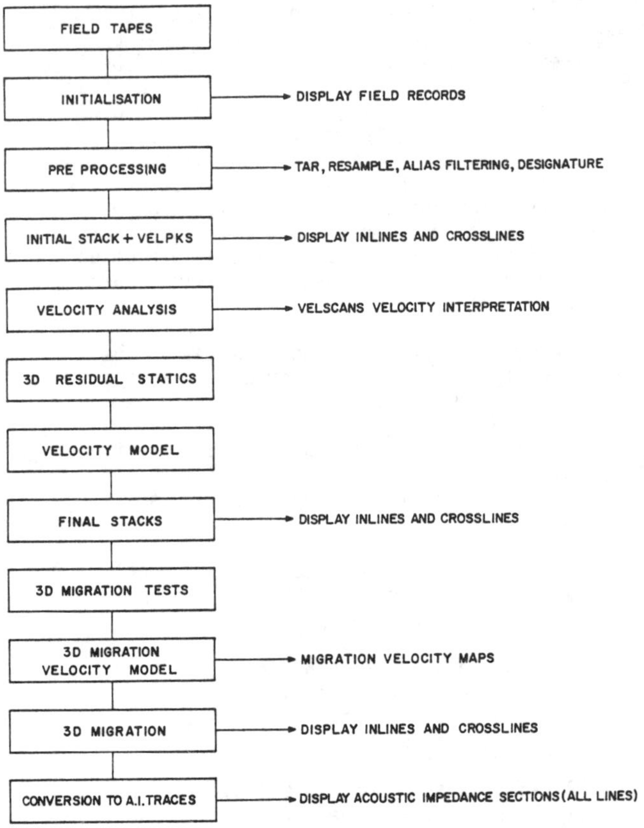

FIG. 4

SEISMIC PROCESSING DATA FLOW

INTERPRETATION

General

The objective of the survey was to resolve anomalies of a 100 m in areal extent, and to measure the thickness of the magnesium bearing intervals. Considering the size of the anomaly to be mapped, it was necessary to make observations on the seismic sections every 40 m. This necessitated the measuring of some 12,000 reflection times for each magnesium bearing layer in order to make detailed isopach maps. Additional measurements were made to construct structural contour maps. To handle such large data sets it was necessary to digitize the readings and have them plotted and contoured by computer. To do this the 3-D migrated acoustic impedance sections were interpreted, and the picked horizons digitized, using a Hewlett-Packard 9845A and digitizer. All reflection times were stored on digital tape and hard copies produced of the digitized times. The data base for each horizon was then checked prior to being input to the automatic contouring and data manipulation package SEIFI. All basic data are stored on "Flexibile Disc" and cassettes. This allows any future updates to be made with a minimum of effort.

Model Studies

The acoustic response for various salts is shown in figure 5. The separation between halite and carnallite is sufficient to be used. Model studies were made for wells Veendam 1 to 4 to check the correlation between the acoustic impedance response of the seismic data and the Zechstein III section penetrated in the well. An example of this correlation is given in figure 6. The correlation was found to be good in wells uncomplicated by faulting. When interpreting the Zechstein III 2b/3b layer it was seen that a knee or additional loop, which did not intersect the zerocrossing, was present. This distortion of the waveform was attributed to the Zechstein III 3a halite layer known to be present. Model studies were also made to evaluate the effect of changing thickness on acoustic response, an example of which is given in figure 7.

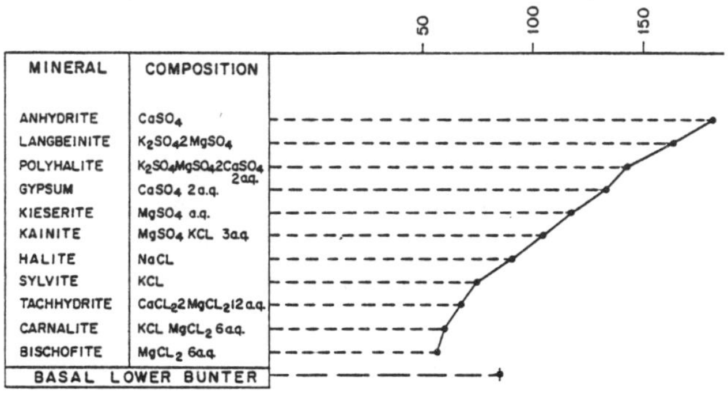

FIG. 5

ACOUSTIC IMPEDANCE

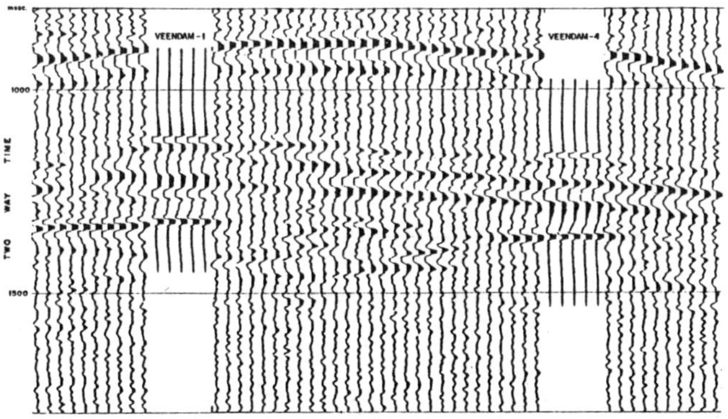

FIG. 6

ACOUSTIC IMPEDANCE SECTION LINE 72 N-S

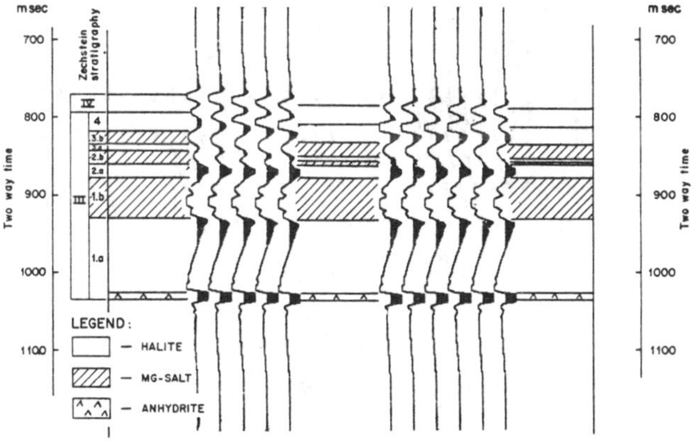

FIG. 7

ACOUSTIC RESPONSE
VARYING THICKNESS ZECHSTEIN III 2a/3b

The results of these studies implied a degree of uncertainty in the thickness measurements for the Zechstein III 2b/3b layer. To compensate for this a general discount of 25% (due to the presence of halite layer Zechstein III 3a) was made based on the well results and interpretation of the model studies.

Ore Grade

A qualitative method to predict magnesium ore grade has been made. This was done by modelling the well VE 2 and measuring the amplitude of the acoustice impedance response across the magnesium bearing layers. This amplitude measurement was then matched with A.I. amplitude of the seismic trace recorded at the well. Once the amplitudes corresponding to a known grade were matched the model was run again this time varying the grade of the layers from zero to 100 percent (figure 8).

The seismic A.I. sections can now be correlated with the model and a qualitative prediction of the ore grade made. Any estimate of ore grade derived in this way should be considered as a range and not used in any detailed production planning.

Studies made of the acoustic response for the Zechstein III 2b/3b and 1b layers showed that a minimum grade of 30% is necessary in order to distinguish the magnesium layer from the host rock halite. These studies also indicated that a quantitative estimate of grade for the proposed second location would be 60% for the Zechstein III 2a/3b layers and 70% for 1b.

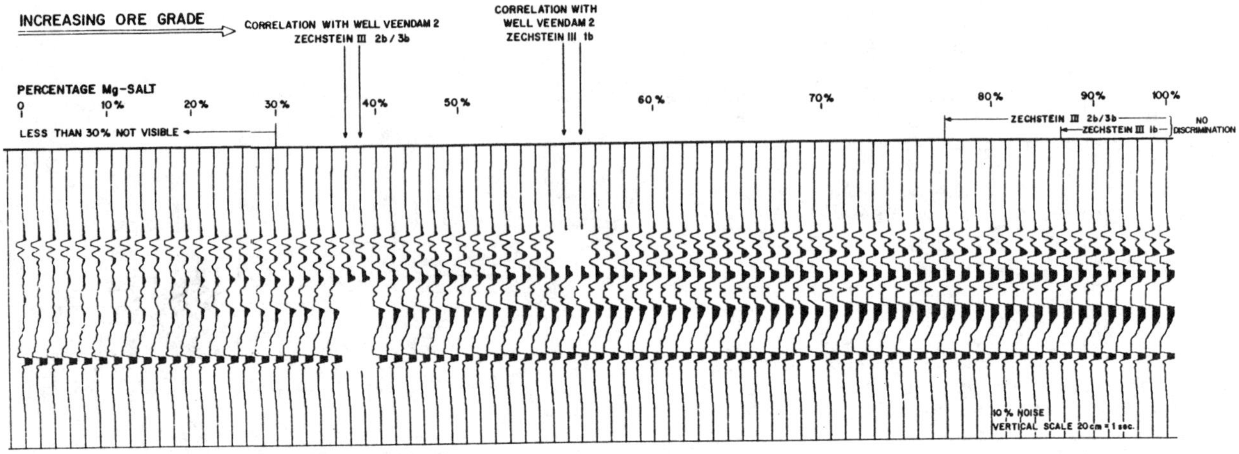

FIG. 8

CORRELATION OF ACOUSTIC RESPONSE
WITH
GRADE OF MAGNESIUM BEARING LAYERS

Results

The survey fulfilled its objective. The top and base of Zechstein III 2b/3b and 1b layers were mapped and isopach maps generated. Additional contour maps were made at Base Tertiary, Base Cretaceous and Base Zechstein levels to show the structural configuration of the area. As a result of the model studies previously described the Zechstein III 2b/3b layer was discounted by 25%. The model studies showed that the uncertainties in the observed thickness for these layers were due to the presence of the Zechstein III 3a halite.

The smoothing factors to be used in contouring were established by means of semi-variograms. Results of these anlyses showed a range of up to 200 m, which is well outside the maximum 13 point smoothing used. A typical semi-variogram for the area is given in figure 9. All contour maps are presented on a scale of 1:5,000 and 1:10,000, with five and thirteen point smoothing.

The isopach maps (figures 10 and 11) show the magnesium bearing layers to be continuous over the area surveyed. The expected short range thickness variations were not found. Localised thickness variations are associated with faulting. A fault is now shown to cut the Veendam 1 well explaining the excessive thickness found.

An area undisturbed by faulting was outlined in which both magnesium baring layers are thick. This area is proposed as the second cluster location. Though it is deemed advisable to avoid faulted areas when locating the second well cluster, the fact that the well Veendam 1 is located in a zone disturbed by faulting may be relevant.

No reliable map could be made of the Top Zechstein. The acoustic contrast between the Base Triassic and Top Zechstein is too small for this interface to be seen on seismic.

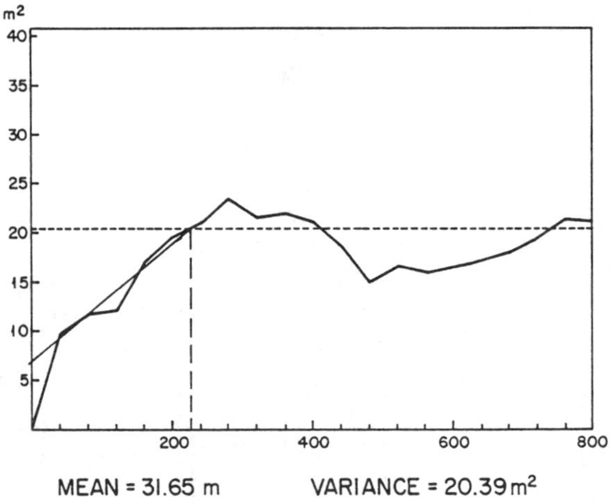

FIG. 9

SEMI-VARIOGRAM
LINE 99 EAST-WEST

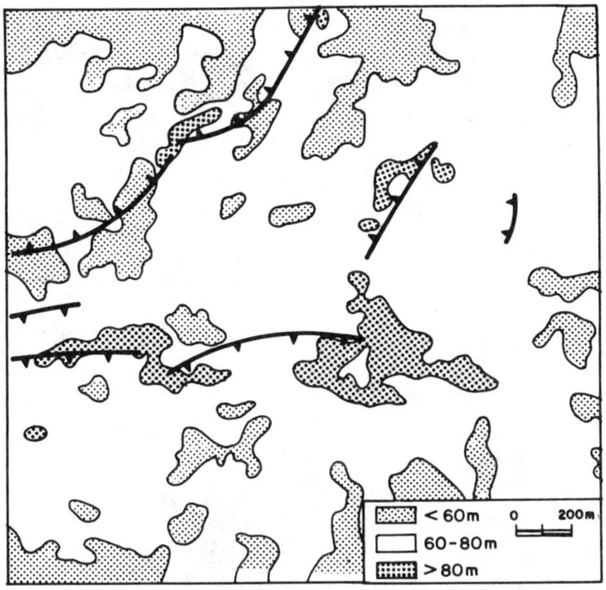

FIG. 10

ISOPACH ZECHSTEIN III 2b/3b
CONTOUR INTERVAL 20 metres

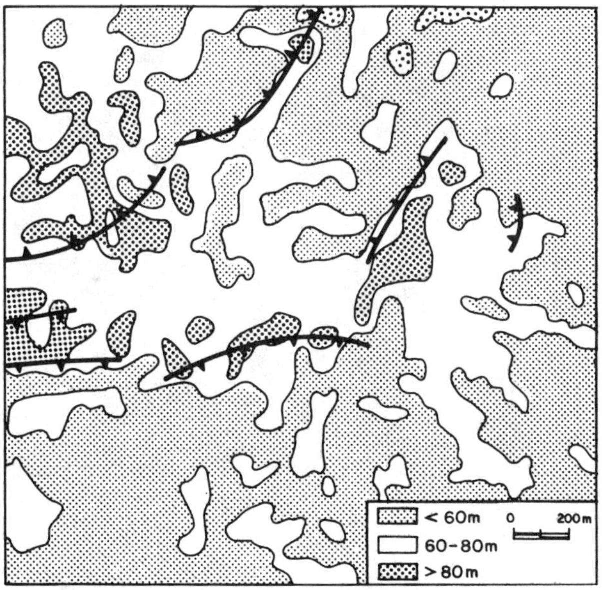

FIG. 11

ISOPACH ZECHSTEIN III 1b
CONTOUR INTERVAL 20 metres

Figure 12 gives a few examples of the final 3-D migrated seismic sections over the study area. As can be seen the high quality and resolution validate the use of seismic for exploration targets which are in principle too deep (economically) to evaluate using conventional drilling. An additional benefit is the high sampling density which allows a more accurate delineation of the objective.

REFERENCES

1. Borchert, H. & Muir, R., "Salt Deposits": Van Nostrand Co. Ltd., 337p., 1964.

2. Braitsch, O., "Salt Deposits, Their Origin and Composition": Springer Verlag, 297p., 1971.

3. Coelewij, P.A.J. & Haug, G.M.W. & Van Kuyk, H., "Magnesium - Salt Exploration in the Northeastern Netherlands": Geologie en Mijnbouw: Vol. 57(4), pp 487-502, 1978.

4. Kühn, R., "Geochemistry of the German Potash Deposits": Geol. Soc. Am. Spec. Paper 88, pp427-681 , 1968.

5. Richter-Bernburg, G., "Saline Deposits in Germany : a review and general introduction to the excursions": Unesco Proc. Hanover Symp. 1968, pp275-287, 1972.

6. Sanneman, D., "Salt Stock Families in Northwestern Germany": Am. Ass. Petroleum Geologists, Mem.8, pp216-270 , 1968.

7. Visser, W.A., "Upper Palaeozoic Evaporites": K.N.G.M.G. Verh. Geol. Serie 21-2, pp66-71, 1963.

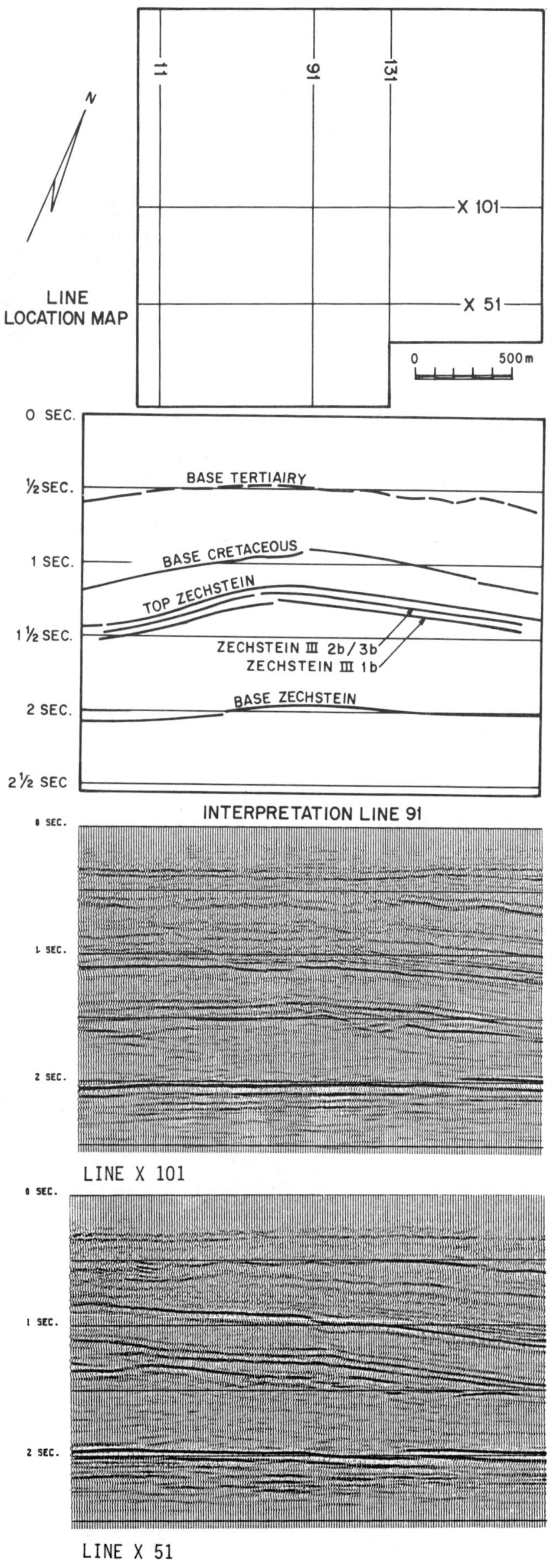

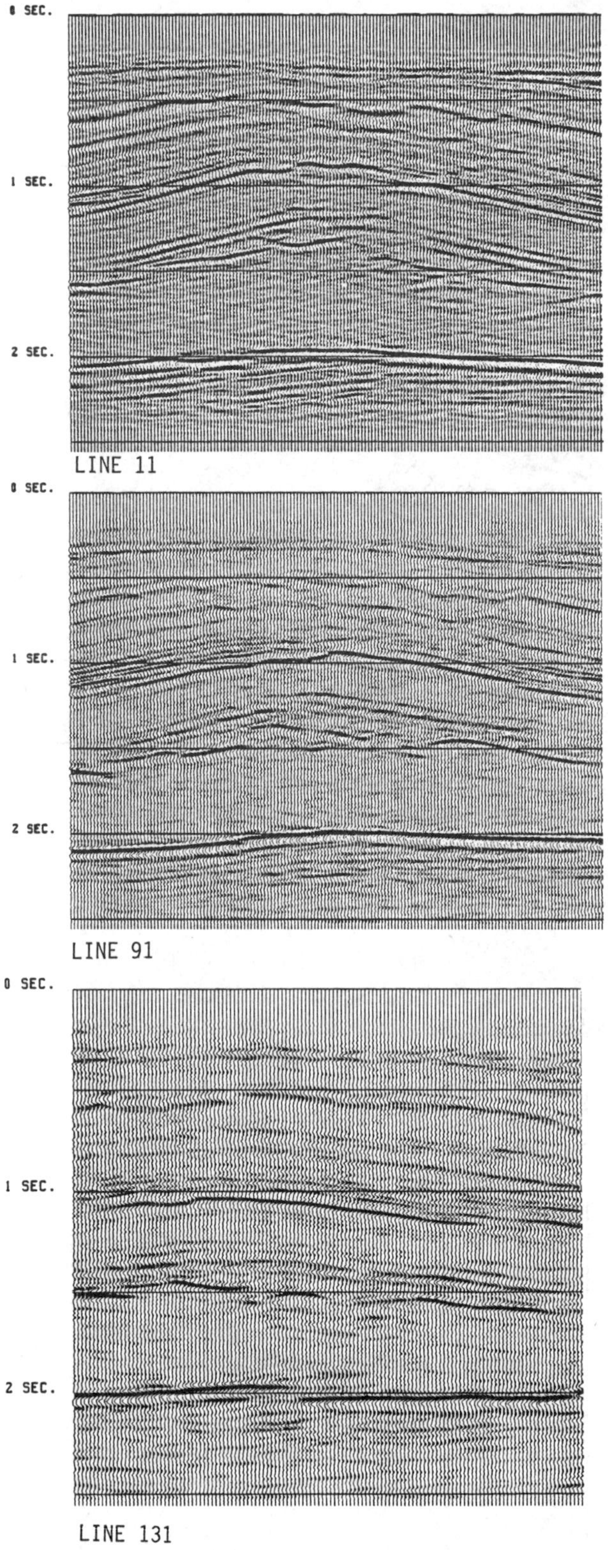

FIG. 12 : EXAMPLES OF 3-D MIGRATED ACOUSTIC IMPEDANCE CROSS SECTIONS

Chapter 7

AQUEOUS OXIDATION OF PYRITE

J. B. Hiskey
W. J. Schlitt
Process Technology
Kennecott Minerals Company
Salt Lake City, Utah 84147

ABSTRACT

Pyrite is by far the most widespread and abundant of all sulfide minerals. As a result, pyrite is observed in a variety of geological environments and often influences the recovery of associated metal values. However, the pyrite itself is rarely of economic importance and is often viewed as a gangue mineral or an undesirable source of sulfur. This paper reviews our present understanding of the aqueous oxidation of pyrite. This includes the thermodynamics, chemistry, kinetics and electrochemistry in both acidic and alkaline systems. Microbiological aspects are also considered. By providing a consistant interpretation of the available data, the extractive metallurgists and mineral processing engineers will have a basis for solving problems associated with the presence of pyrite in coal and metallic ores.

INTRODUCTION

Pyrite is the most widespread and abundant of all sulfide minerals.[1] It can occur in almost any geologic environment. Pyrite is particularly common in massive hydrothermal deposits formed at moderate to high temperatures. Other hosts include hydrothermal veins and replacements, igneous rocks and sedimentary beds. These include anaerobic sediments where pyrite was apparently precipitated with the aid of sulfate-reducing bacteria.[2]

In many of these settings pyrite is the only sulfide present, making it one of the so-called persistent minerals. This characteristic is not due entirely to thermodynamic stability. As shown in Table 1, many more reactive sulfides are theoretically more stable than pyrite because they exhibit larger negative free energies of formation.[3] Thus, other factors must contribute to pyrite's persistence. As discussed later, these include the existence of surface films that retard oxidation and an exceptionally high rest potential that prevents dissolution except at high oxidation potentials.

TABLE 1
FREE ENERGY OF SULFIDE FORMATION AT 298.15°K AND ONE ATMOSPHERE[3]

Mineral	Formula	ΔG^o, kJ/mole
Argentite	Ag_2S	- 40.2
Orpiment	As_2S_3	- 168.7
Oldhamite	CaS	- 473.4
Greennockite	CdS	- 145.7
Chalcocite	Cu_2S	- 86.7
Pyrite	FeS_2	- 160.3
Hydrogen Sulfide	H_2S	- 33.5
Cinnabar	HgS	- 50.7
Alabondite	MnS	- 218.1
Molybdenite	MoS_2	- 297.6
Galena	PbS	- 96.3
Stibnite	Sb_2S_3	- 173.7
Tungstenite	WS_2	- 298.0
Sphalerite	ZnS	- 203.4

As one of the most widely occurring minerals, pyrite is found in association with various ores. This makes pyrite an important factor to be considered by the extractive metallurgist or mineral processor. Since it usually lacks any economic value itself, pyrite generally acts as an undesirable gangue constituent which complicates processing. Some of the major problems associated with pyrite are noted below.

1. Acid drainage from coal mining operations.

2. Dilution of base metal values in sulfide concentrates where pyrite contributes to the sulfur burden and the subsequent need for iron removal.

3. High consumption of oxidant in the leaching of uranium, together with coextraction of iron during uranium recovery from acidic leach liquors.

4. Inability to use a cyanide leach to effectively recover gold and silver from auriferous or argentiferous pyrite.

Only in the heap or dump leaching of sulfidic copper does pyrite play a beneficial role.

Except in smelting, most metallurgical problems associated with pyrite involve oxidation in moist or aqueous environments. Therefore, the following sections bring together information on the thermodynamics, chemistry, kinetics and electrochemistry of the aqueous oxidation of pyrite. One section also deals with microbiological aspects of pyrite oxidation. These attempt to provide a clear overview that provides the first step in dealing with metallurgical problems related to pyrite.

THERMODYNAMICS

Thermodynamic relationships involving pyrite in an aqueous environment are conveniently illustrated by Eh (potential) - pH diagrams. Although developed for metal corrosion equilibria (Pourbaix[4]), these diagrams are equally appropriate for representing mineralogical and geological relationships (Garrels and Christ[5]). This is because the diagrams are actually graphical expressions of the fundamental reaction thermodynamics. For example, consider generalized oxidation-reduction reaction

$$\text{reduced state} \rightarrow \text{oxidized state} + ne \qquad (1)$$

where n denotes the number of electrons involved in the half-reaction. The free energy of reaction is given by the following expression

$$\Delta G_r = \Delta G_r^o + RT \ln Q \qquad (2)$$

where ΔG_r^o is the standard free energy change; R, the gas constant; T, absolute temperature; and Q, the reaction quotient. The standard free energy is related to the standard half-cell potential E^o by

$$\Delta G^o = n FE^o \qquad (3)$$

where F is the Faraday constant. Similarly, the relationship between the free energy change for the reaction and the half-cell potential is

$$\Delta G = nFE \qquad (4)$$

Then Eh=E and the oxidation potential for the half-cell is expressed as follows

$$Eh = E^o + \frac{RT}{nF} \ln Q \qquad (5)$$

This is the fundamental Nernst equation which relates Eh to the activities of species involved.

Figure 1 shows an Eh-pH diagram for the Fe-S-H_2O system under standard conditions (298°K, 1 atm total pressure, 10^{-1} M activity dissolved sulfur, and 10^{-6} M activity dissolved Fe). Equilibrium relationships for the various sulfur species are plotted in Fig. 2. The Eh-pH diagram for sulfur was constructed for the standard conditions stated above.

As shown by Fig. 1, oxidizing conditions are required to dissolve pyrite. At low pH values (i.e., pH 1 to 2), pyrite is oxidized to soluble iron and sulfur species at potentials between +0.19 and +0.25 volts. The reaction is

$$FeS_2 + 8H_2O \rightarrow Fe^{2+} + 2HSO_4^- + 14H^+ + 14e^- \qquad (6)$$

$$(Eh = 0.305 - 0.059\ pH)$$

In addition, Fig. 1 shows that at a Fe^{2+} activity of 10^{-6} M, pyrite cannot oxidize to yield elemental sulfur at pH values above 2.4. In other words, at potentials where pyrite oxidizes, sulfur is not a stable phase at pH's greater than 2.4. However, the stability relationships depicted in Fig. 1 do reveal that FeS_2 can be oxidized to produce elemental sulfur and Fe^{2+} according to the following reaction

$$FeS_2 \rightarrow Fe^{2+} + 2S^0 + 2e \qquad (7)$$

$$(Eh = 0.340 + 0.0295 \log a_{Fe^{2+}})$$

For a dissolved iron activity of 10^{-6} M, there is a constant potential for reaction (7) equal to +0.16 volts. Actually, slow chemical kinetics associated with the oxidation of elemental sulfur to bisulfate will yield Fe^{2+} and S^0 outside the stability regions predicted from thermodynamic considerations. This point will be discussed in more detail in the following section. The shaded portion of Fig. 1 highlights the region where FeS_2 coexists with elemental sulfur.

In the sulfate domain, FeS_2 undergoes oxidation through the following reaction

$$FeS_2 + 8\ H_2O \rightarrow Fe^{2+} + 2SO_4^{2-} + 16H^+ + 14e \qquad (8)$$

$$(Eh = 0.354 - 0.067\ pH + 0.0084 \log a\ SO_4^{2-} + 0.0042 \log a_{Fe^{2+}})$$

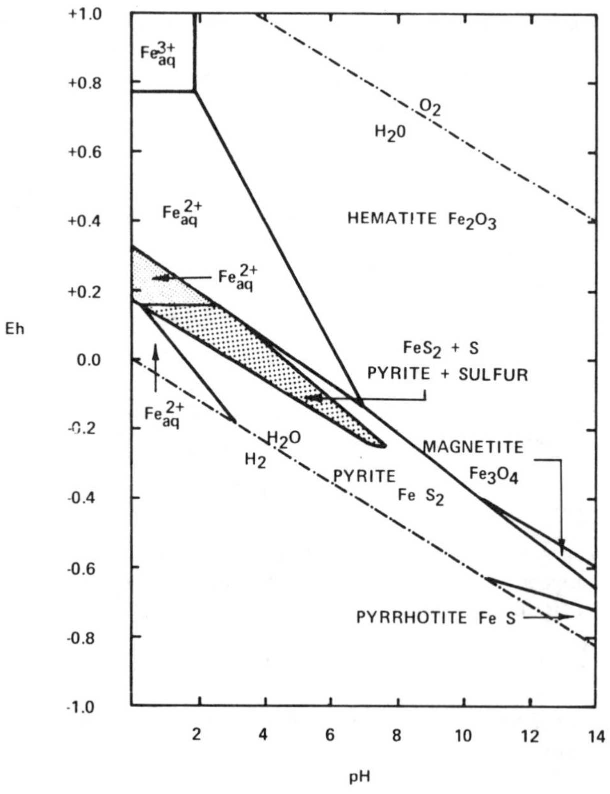

Fig. 1. Pourbaix diagram for the Fe-S-H_2O system at 298°K and 1 atm total pressure. Total dissolved sulfur species = 10^{-1} M. (After Ref. 5)

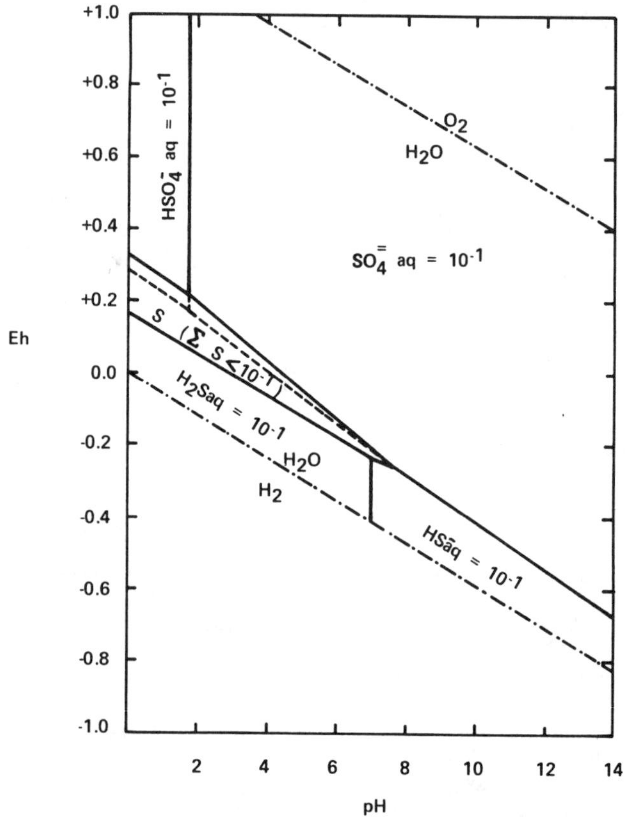

Fig. 2. Pourbaix diagram for the S-H_2O system at 298°K and 1 atm total pressure. Total dissolved sulfur species = 10^{-1} M. (After Ref. 5)

This relationship holds up to about pH 6.7 where the ferrous ion is no longer stable. At that point, pyrite is directly oxidized to insoluble iron hydroxide (hydrated iron oxide) type products and soluble sulfate. The oxidation of pyrite at these conditions can be represented by a reaction of the form

$$FeS_2 + 11H_2O \rightarrow Fe(OH)_3 + 2SO_4^{2-} + 19H^+ + 15e \quad (9)$$

$$(Eh = 0.394 - 0.075\ pH)$$

Actually, $Fe(OH)_3$ is a metastable phase as implied by the favorable thermodynamic driving force for the following transformation at 25°C

$$2Fe(OH)_3 \rightarrow Fe_2O_3 + 3H_2O \quad (10)$$

$$\Delta G^o_{298} = -63.5\ kJ/mole$$

When associated with pyrite oxidation Feitknecht and Schindler[6] have suggested the following aging scheme for freshly precipitated iron (III) hydroxide

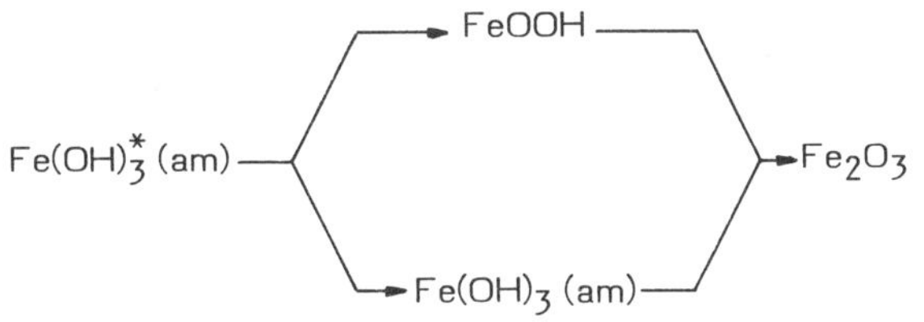

where $Fe(OH)_3^*$(am) and $Fe(OH)_3$(am) represent active and inactive amorphous ferric hydroxide, respectively. Active amorphous ferric hydroxide is unstable and slowly transforms to geothite (FeOOH) or to a more stable amorphous ferric hydroxide phase. The conversion requires approximately 6 to 8 days at 25°C; however at 100°C, less than one day is required. Goethite and inactive amorphous $Fe(OH)_3$ are stable at room temperature, but decompose to hematite (Fe_2O_3) at temperatures above 130°C. Therefore, Fe_2O_3 is the thermodynamically stable end product of the alkaline dissolution of pyrite at potentials above -0.30 volts.

A limited field of Fe_3O_4 (magnetite) stability exists at pH values above approximately 10. However, if conditions are sufficiently oxidizing Fe_3O_4 will be thermodynamically unstable and will exist only if the solid-state transformation to hematite is slower than the rate of magnetite formation. Again, Fe_2O_3 would be the stable end product.

Natural leaching systems have two oxidants available for pyrite dissolution, namely oxygen (air) and ferric ion. When oxygen is available, ferric ion, along with sulfuric acid, can actually be generated in situ as a consequence of pyrite oxidation itself. Such solutions have relatively high oxidation potentials associated with them. This Eh value depends on the ferric/ferrous couple ($Fe^{3+} = Fe^{2+} + e^-$) and is characterized by the following relationship

$$Eh = 0.771 + 0.059 \log \frac{a_{Fe^{3+}}}{a_{Fe^{2+}}} \quad (11)$$

Thus as pointed out by Dutrizac and MacDonald[7] in their review, acid ferric sulfate solutions provide a powerful lixiviant.

Even though acid solutions of Fe^{3+} have high (approximately +0.8 volts) oxidation potentials, the pH stability range is extremely limited. Ferric ion begins to hydrolyze at approximately pH 3, precipitating $Fe(OH)_3$. The hydrolysis is a complicated process and involves the formation of polymeric hydrolysis products. The existence of some of these complexes has been corroborated experimentally. Stumm and Morgan,[8] and Butler[9] have furnished examples illustrating ferric iron hydrolysis solution chemistry. The equilibrium concentrations of Fe^{3+} and various cationic hydroxo complexes are shown as a distribution of species diagram in Fig. 3. The shaded area represents oversaturation with respect to solid $Fe(OH)_3$, ($K_{so} = 10^{-38}$).

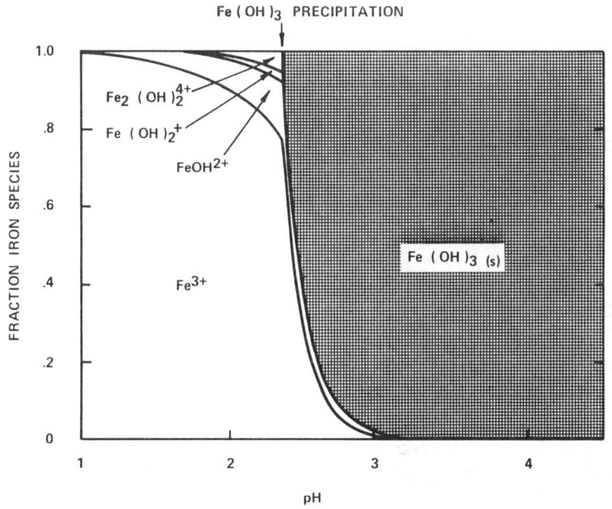

Fig. 3. Fraction of iron present as ferric hydroxide complexes as a function of pH.

DISSOLUTION CHEMISTRY

Acidic solutions

Considerable attention has been devoted to the natural oxidation of pyritic materials. Many early investigators recognized the simultaneous effects of oxygen and ferric iron on pyrite oxidation in solution. Natural pyrite/aqueous systems were characterized by Singer and Stumm[10] according to the following stoichiometric reactions

$$FeS_2 + 7/2\ O_2 + H_2O \rightarrow Fe^{2+} + 2SO_4^{2-} + 2H^+ \quad (12)$$

$$Fe^{2+} + 1/4\ O_2 + H^+ \rightarrow Fe^{3+} + 1/2\ H_2O \quad (13)$$

$$FeS_2 + 14Fe^{3+} + 8H_2O \rightarrow 15Fe^{2+} + 2SO_4^{2-} + 16\ H^+ \quad (14)$$

$$Fe^{3+} + 3H_2O \rightarrow Fe(OH)_{3(s)} + 3H^+ \quad (15)$$

The dissolution reactions stated are reasonable, but fail to include the formation of elemental sulfur as a stable product. McKay and Halpern[11] described the oxidation of pyrite by either oxygen or ferric ion according to reactions that yield elemental sulfur and sulfate. As discussed by Dutrizac and MacDonald,[7] both elemental sulfur and sulfate are produced concurrently according to separate but related oxidation processes.

Pyrite dissolution in acidic solution can produce elemental sulfur by the following principle reactions

$$FeS_2 + 1/2\ O_2 + 2H^+ \rightarrow Fe^{2+} + H_2O + 2S^o \quad (16)$$

$$FeS_2 + 2Fe^{3+} \rightarrow 3Fe^{2+} + 2S^o \quad (17)$$

Under certain conditions, the sulfur produced is thermodynamically unstable and should undergo oxidation to either bisulfate or sulfate. Except at low pH or oxidation potentials, sulfur can be oxidized to sulfate by oxygen or ferric ion

$$S^o + \frac{3}{2}O_2 + H_2O \rightarrow H_2SO_4 \quad (18)$$

or

$$S^o + 6Fe^{3+} + 4H_2O \rightarrow H_2SO_4 + 6Fe^{2+} + 6H^+ \quad (19)$$

However, like conversion to bisulfate, these reactions are limited by slow kinetics over a wide range of conditions. Correspondingly, elemental sulfur with its unusual stability is more than an intermediate species and should be considered in the overall reaction stoichiometry. As will be discussed in a later section, the formation of elemental sulfur layers on the pyrite surface can influence the relative rates of pyrite oxidation. Under certain conditions, the sulfur layer is protective and limits the rate of oxidation.

Habashi and Bauer[12] studied the aqueous oxidation of elemental sulfur and found that at temperatures below the melting point of sulfur (120°C) the rate of oxidation was extremely slow. At 4.1 atm oxygen pressure and 60°C, sulfur was not oxidized to any measurable extent. At the same overpressure of oxygen, the rate was observed to increase by an order of magnitude when the reaction temperature was increased from 90 to 130°C.

Ichikuni[13] examined the aqueous oxidation of pyrite with oxygen at 80°C. He was able to show that pyrite oxidizes in aqueous solution by different routes depending on pH. At lower pH levels, pyrite reacts to give

$$FeS_2 + \frac{1}{2}O_2 + 2H^+ \rightarrow Fe^{2+} + H_2O + 2S^o \quad (16)$$

At higher pH's, the reaction is

$$FeS_2 + \frac{7}{2}O_2 + H_2O \rightarrow Fe^{2+} + 2\,SO_4^{2-} + 2H^+ \quad (7)$$

If these two reactions are considered as extremes, then the molar ratio of SO_4^{2-}/Fe^{2+} will vary between 0 and 2. Figure 4 shows the effect of pH on the ratio SO_4^{2-}/Fe^{2+}. The ratio increases linearly from pH 0 to 2.5; in this range, the system moves from being predominantly elemental sulfur producing to predominantly sulfate forming. The sharp rise in the SO_4^{2-}/Fe^{2+} ratio above pH 3 is explained by the disappearance of iron from the system due to its hydrolysis.

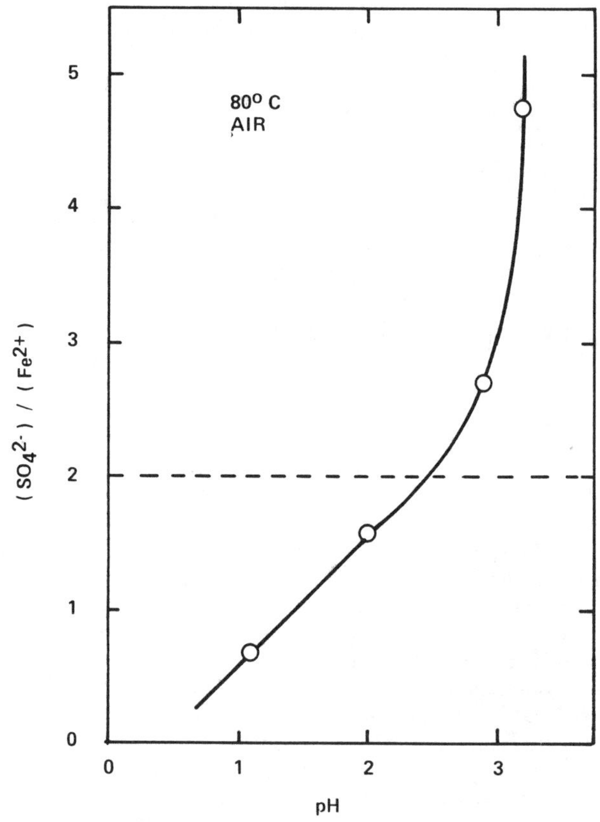

Fig. 4. Effect of pH on the SO_4^{2-}/Fe^{2+} ratio observed for the leaching of pyrite. (After Ref. 13)

Bailey and Peters[14] suggested that pyrite decomposition occurs by a generalized overall reaction of the form

$$FeS_2 + (0.5 + 1.5y + 0.25x)\,O_2 + (2 + x - 2y)\,H^+ \rightarrow (1 - x)\,Fe^{2+} + xFe^{3+} + (2 - y)S^o + y\,SO_4^{2-} + (1 - y + 0.5)H_2O \quad (20)$$

The ratio of Fe^{3+} to Fe^{2+} produced and the distribution of sulfur species (oxidation of S^o to SO_4^{2-}) depends upon the exact reaction conditions. In addition, it is recognized that reaction (20) proceeds by a complex dissolution mechanism and could entail the formation of intermediates like thiosulfate, sulfite, dithionate, and dithionite.

Singer and Stumm[10] proposed a model for the aqueous oxidation of pyrite in natural systems that utilizes both oxygen and ferric ion as oxidants in an initiation step and a propagation cycle. This mechanism can be represented as follows:

Initiation reaction:

$$FeS_2 \xrightarrow{(+\,O_2)} Fe^{2+} + \text{S-compound} \qquad (21a)$$

Propagation cycle:

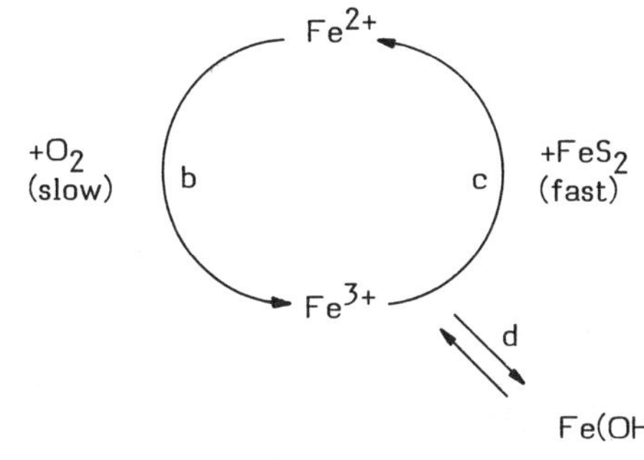

(21b,c,d)

Reaction (21a) initiates the overall dissolution process and provides a constant supply of ferrous iron to the system. In the absence of bacteria, ferrous ion is oxidized to ferric ion by oxygen (reaction 21b) at a relatively slow rate under ambient or near ambient conditions. Ferric ion reacts rapidly with FeS_2 in step (21c). Therefore, oxygen is not directly involved in the oxidation of pyrite, but maintains the propagation cycle by regenerating ferric ion. In side reaction (21d), ferric iron is precipitated as $Fe(OH)_3$. This allows for removal of iron from the system and balances the iron derived from pyrite oxidation. The specific rate of pyrite oxidation as implied by this mechanism is controlled by the rate of oxidation of ferrous ion to ferric ion with oxygen.

The rate of oxidation of Fe^{2+} by oxygen over a range of pH values (1 to 7) is shown in Fig. 5. At pH values below approximately 3.5, the rate of reaction is independent of pH and is explained by the following relationship

$$\frac{-d(Fe^{2+})}{dt} = k\,(Fe^{2+})(O_2) \qquad (22)$$

Above pH 4.5, the rate of ferrous ion oxidation is inversely proportional to the square of the hydrogen ion concentration. The rate for this pH region is given by the following equation

$$\frac{-d(Fe^{2+})}{dt} = \frac{k'(Fe^{2+})(O_2)}{(H^+)^2} \qquad (23)$$

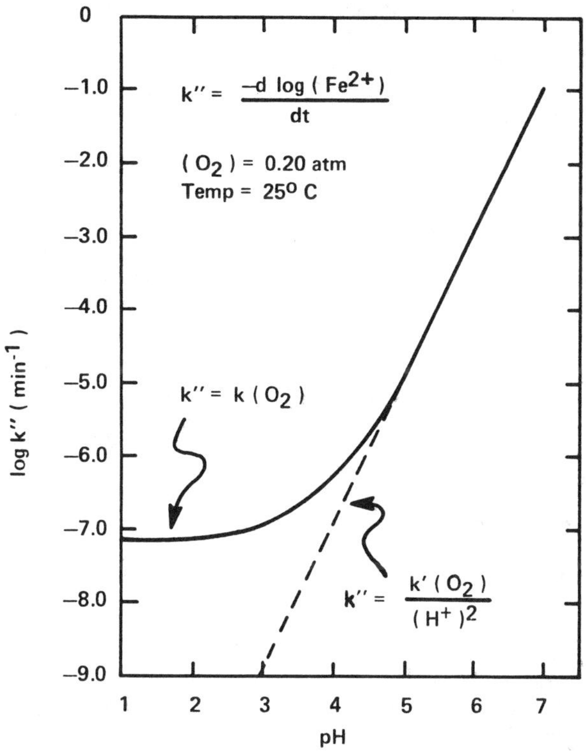

Fig. 5. Rate of oxidation of ferrous iron by oxygen as a function of pH. (After Ref. 10)

Alkaline solutions

The aqueous oxidation of pyrite in alkaline solutions has received only limited attention. Stenhouse and Armstrong[15] studied the dissolution of monosized pyrite particles in sodium hydroxide solutions. They concluded that the leaching involves direct oxidation to sulfate, intermediate sulfoxy species not being detected. Iron is oxidized and remains as an insoluble surface layer. This surface coating represents a diffusional barrier that limits the rate of pyrite oxidation. Although mainly amorphous, this product was found to contain both Fe_2O_3 and Fe_3O_4. In general, the overall reaction can be described as follows

$$2FeS_2 + 15/2\ O_2 + 8OH^- \rightarrow Fe_2O_3 + 4SO_4^{2-} + 4H_2O \qquad (24)$$

The actual reaction mechanism which describes this oxidation process is extremely complex. Stenhouse and Armstrong[15] suggested that the oxidation of sulfur and iron shows a certain degree of independence. Their reaction model was based on the assumption that oxidation must proceed across an impervious oxide layer. Simply, they proposed that sulfur diffuses through the oxide film as positive ions which are oxidized at the surface to sulfate ions. Iron is oxidized at the pyrite/oxide interface by substitutional diffusion of oxygen atoms through the oxide film.

Recently Burkin[16] reviewed the alkaline leaching of pyrite with special emphasis on the oxide coatings and their influence on the kinetics of the dissolution reaction. He reported that a thin coating of γ-Fe_2O_3 (maghemite) was observed on the surface of a large crystal of pure pyrite leached for one hour at 90°C with 0.5 M NaOH and 1 atm of air. In this test, there was some indication of preferred orientation of the oxide product. Another crystal of pure pyrite was leached for five minutes at the same conditions; in addition to maghemite, other unidentified phases were detected. Partially reacted pyrite samples also showed preferential attack along natural crystal defects and along grain boundaries. Furthermore, when the pyrite possessed a well defined shape, the solid product appeared to replicate the original shape. This is accepted as a characteristic of systems in which solid state reactions occur.

Under these relatively mild leaching conditions, both massive and finely divided porous oxide products were observed. The massive product coated the residual pyrite; however, the porous material was not in contact with the pyrite but appeared as suspended particles in the leaching solution. Electron probe microanalysis showed that the massive product contained 70% iron and no sulfur. This composition is in agreement with that of maghemite.

Figure 6 shows the sulfur and iron content at the edge of leached pyrite and in the adjacent oxide product layer. The composition ranges from 47% Fe and 53% S at point A (typical of pyrite) to 62% Fe and 2.5% S near the surface at point B. There is a concentration gradient across the pyrite/oxide interface instead of a sharply defined boundary. This region is depicted in Fig. 6 as the shaded area and may represent compositions for intermediate reaction products.

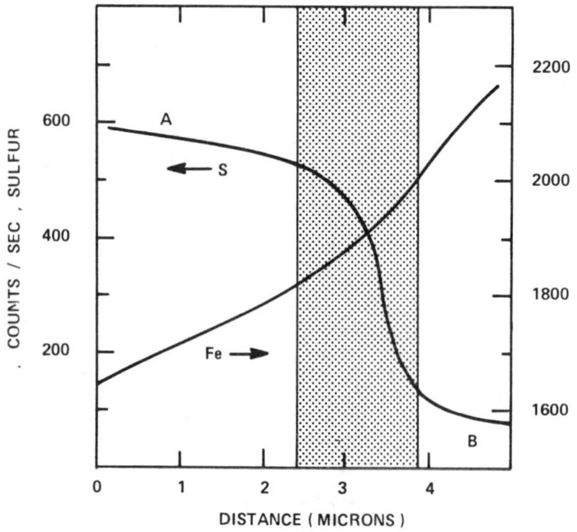

Fig. 6. Typical composition of edge of leached pyrite and adjacent iron oxide product coating. (After Ref. 16)

Burkin and Edwards[17] interpreted the oxidation of FeS_2 to occur by the exchange of S_2^{2-} groups with O^{2-} ions which assume the same lattice postion. Wustite (FeO) is produced initially; however, it is reactive at low temperature and will be quickly oxidized by oxygen to Fe_3O_4 and γ-Fe_2O_3 (the final oxidation product). The formation of wustite during the transition from sulfide to oxide is important because of the accompanying reduction in lattice dimensions. The molar volume of wustite is about 50% that of pyrite. Therefore, strains at the pyrite-oxide interface will cause cracks and fissures to form in the coating which then spalls from the pyrite surface when the layer is very thin. This was observed for large crystals of pure pyrite. However, solid reaction products do accumulate along reactive surface sites (i.e., faults or grain boundaries). The oxide coating can build up in thickness, the extent of which depends on the leaching conditions.

In summary, the dissolution of pyrite in alkaline solutions is characterized by the following general considerations:

1. At low temperatures, protective oxide layers form.

2. At intermediate temperatures, maghemite layers build up and eventually rupture, reducing diffusional resistance and causing the reaction to speed up.

3. At high temperatures (>120°C), initial kinetics are so fast that no massive protective oxide layer is formed.

CHEMICAL KINETICS

In reviewing the enormous volume of literature on the aqueous oxidation of pyrite acidic systems have clearly received greater attention. This interest is understandable, since natural pyrite oxidation is a source of in situ generated sulfuric acid. Therefore, the following discussion will focus on the kinetics of pyrite dissolution in acid solution and will review noteworthy literature in approximately chronological order.

The investigations of Nelson, et al.[18] and Burke and Downs[19] represent early research on the kinetics of pyrite oxidation in aqueous solution. Even though these studies addressed the oxidation of pyritic sulfur in coal, they provide valuable information about the effects of temperature, particle size, oxygen concentration, and ferric ion. Nelson, et al.[18] examined the influence of temperature in the range between 22 and 90°C for pyritic sulfur oxidation using low pressure air. An experimental activation energy was calculated from their data, its value being 33.8 kJ/mole for the initial oxidation stage. Oxygen, when substituted for air, did not affect the kinetics of pyrite leaching. Ferric ion was added to the system (the rate of oxidation represents the combined effects of oxygen and ferric ion) in the form of ferric sulfate. Tests at 22°C and 70°C using 0.01 M Fe^{3+} indicated that Fe^{3+} has only a slight effect on the rate of pyrite oxidation in comparison to simple air oxidation.

Burke and Downs[19] employed pure oxygen at atmospheric pressure during their experiments. At pressures below 1 atm, oxygen concentration had no detectable effect on the pyrite oxidation rate. Results of this work revealed an activation energy of 16.5 kJ/mole for temperatures between approximately 0 and 90°C. Above 100°C the rate was insensitive to temperature probably because of decreased oxygen solubility at these higher temperatures. Furthermore, there were no precautions taken to ensure a constant oxygen overpressure in the autoclave. Most of the experiments were conducted in hydrochloric acid solutions; however, when sulfuric acid was used, the rate of oxidation decreased. Using 0.5 M H_2SO_4 solutions, the reaction was extremely slow. The formation of an inhibiting elemental sulfur film at conditions of low oxidation potential and high acid strength can explain the slow kinetics.

The first detailed kinetic studies were reported by Warren[20] and McKay and Halpern.[11] Warren's experiments were conducted under pressure leaching conditions using high grade pyrite samples. In this work, temperatures were varied between 130 and 210°C, and oxygen partial pressures were applied in the range between 2.7 and 13.9 atm. At these conditions, the activation energy for pyrite oxidation was 83.7 kJ/mole suggesting that the rate controlling step was a heterogenous chemical process occurring at the pyrite surface. In addition, the rate of reaction was found to obey a 0.5 order dependence with respect to oxygen partial pressure. Because of this finding, Warren theorized that the reaction involved dissociation of molecular oxygen at specific surface sites. However, the observed 0.5 order dependence on oxygen partial pressure can be rationalized by an electrochemical surface reaction where oxygen is cathodically reduced at the pyrite surface. The reacted pyrite surface showed signs of oxidation at selected sites or active centers. This resulted in fissuring and disintegration of the FeS_2 particle as the reaction proceeded. It was suggested that the active sites for oxidation were a function of lattice geometry of the pyrite crystal and not of defects or impurities in the lattice.

McKay and Halpern[11] studied the kinetics of pyrite oxidation in sulfuric acid at temperatures between 100 and 130°C and oxygen pressures from 0 to 4 atm. Unlike previous investigations which claimed that pyritic sulfur oxidized only to sulfate, at the conditions of this work, elemental sulfur appeared as a major reaction product. An experimental activation energy equivalent to 55.7 kJ/mole was calculated for the temperature range investigated. The distribution of sulfur oxidation products exhibited a pronounced dependence on temperature. As the temperature increased from 120 to 130°C, the ratio of H_2SO_4 to S(total) at 40% FeS_2 oxidation increased from 0.11 to 0.20. For the same level of FeS_2 oxidation, the ratio of S^0 to S(total) decreased from 0.27 to 0.14 as the temperature increased from 120 to 130°C. The S^0/S(total) ratio was found to be equal to 0.5 at 100°C for the early stages (<10%) of pyrite oxidation.

In contrast to Warren's results which indicated a 0.5 order with respect to oxygen partial pressure, the work of McKay and Halpern[11] showed that the oxidation of pyrite was first order with respect to oxygen partial pressure. The proposed mechanism of leaching considered O_2 chemisorbing rapidly at the pyrite surface with one molecule of O_2 occupying each active FeS_2 site according to the following reaction

$$FeS_2 + O_{2(aq)} \xrightarrow{\text{fast}} FeS_2 \cdot O_2 \tag{25}$$

It was then proposed that the rate controlling step involved the attack of a second O_2 molecule on the oxygen covered site according to the reaction

$$FeS_2 \cdot O_2 + O_{2(aq)} \underset{k_{26}}{\xrightarrow{\text{slow}}} \left[FeS_2 \cdot 2O_2\right]^{\neq} \tag{26}$$

with the product decomposing as

$$\left[FeS_2 \cdot 2O_2\right]^{\neq} \rightarrow FeSO_4 + S^0 \tag{27}$$

The overall rate equation is given by the following relationship

$$\frac{-d(FeS_2)}{dt} = k\, A_{FeS_2} P_{O_2} \tag{28}$$

where k is the rate constant, A_{FeS_2} is the total surface area, and P_{O_2} is the partial pressure of oxygen. The rate of oxidation was directly proportional (first order dependence) to the total surface area. However, the rate should have decreased as the reaction progressed because of an accompanying decrease in surface area. This apparent anomaly was explained by an increase in surface roughness which compensates for the reduction in macroscopic surface area. Microscopic inspection of large pyrite crystals revealed a substantial amount of surface etching and pitting which provided support to this interpretation.

The dissolution of pyrite in acid chlorine solutions was examined by Sherman and Strickland.[21] At pH less than 2 and at temperatures between 10 and 40°C, the reaction proceeded with only the formation of ferric ions and sulfate ions. An activation energy of 20.9 kJ/mole was reported and they contended that the kinetics were diffusion controlled. There was some evidence of microscopic etching of the reacted pyrite surface.

As part of a study of the pressure leaching of a manganese ore using pyrite as a reductant source, Cornelius and Woodcock[22] considered the oxidation of an upgraded (95% FeS_2) concentrate from Mt. Lyell, Tasmania. Temperatures between 130 and 165°C were investigated, and oxygen pressures ranging from 6.1 to 23.8 atm were examined for various concentrations of FeS_2. The oxidation rate exhibited a half order dependence on oxygen pressure at 165°C. When the rate was expressed in terms of SO_4^{2-} production, an experimental activation energy of 77.4 kJ/mole was calculated. Based on the production of Fe^{2+}, the experimental activation energy was found to be 70.3 kJ/mole. In these experiments, iron oxides were the principle solid reaction products.

Garrels and Thompson[23] used ferric ion as the sole oxidant for their study of pyrite oxidation. Using oxygen and carbon dioxide free solutions, they observed as have others that, between 0 and 2, pH has little effect on oxidation rate. An important conclusion from this work is that kinetics of pyrite oxidation are directly proportional to the solution oxidation potential (Fe^{3+}/Fe^{2+} ratio), and are independent of the total iron concentration. Under near equilibrium conditions (low oxidation potentials), elemental sulfur was produced by the ferric ion oxidation of pyrite; however, with excess ferric ion (high oxidation potentials), sulfate was the dominant sulfur product.

As noted earlier, Singer and Stumm[10] proposed that the rate of ferrous oxidation is rate limiting in the overall leaching of pyrite. This is consistent with the interpretation advanced by Garrels and Thompson.[23] In an extensive study by Smith and Shumate[24] two basic oxidation modes proved to be important in natural systems: namely, oxygenation by dissolved oxygen, and ferric ion oxidation. In the absence of microbial activity, the oxygenation rate is dependent only on the oxygen concentration in the aqueous phase. Oxidation by ferric ion is established by the Fe^{3+}/Fe^{2+} ratio and free ferric ion concentration, but not by the dissolved oxygen concentration. Data at low oxygen concentrations ($6.3 \times 4 - 3.0 \times 10^{-3}$ M O_2) indicated approximately a 0.5 order rate dependence. The following mechanistic steps were proposed:

Step 1 O_2 adsorption on "reactive site" of pyrite

Step 2 O_2 dissociation to form [activated complex]$^{\neq}$

Step 3 [activated complex]$^{\neq}$ $\rightarrow$ oxidation products

Step 4 desorption of oxidation products

Decomposition of the activated complex was suggested as being the rate controlling step. Adsorption and desorption steps were considered as being in equilibrium. In the presence of dissolved nitrogen, the rate of oxidation was slightly less than for pure oxygen. Competitive adsorption reduces the surface concentration of oxygen, thus reducing the overall oxidation rate.

At high dissolved oxygen concentrations (1.3×10^{-3} - 2.0×10^{-2} M O_2), the rate of oxygenation remained 0.5 order with respect to $O_{2\,(aq)}$, and overpressures of nitrogen continued to have a detrimental effect on the oxidation rate. Figure 7 illustrates the effect of pH on the oxidation rate at 25°C and 1.2×10^{-3} M $O_{2\,(aq)}$. The reaction rate as shown by previous work and reaffirmed here is relatively independent of pH up to approximately pH 3. As was previously shown for Fe^{2+} oxidation, the rate of pyrite oxidation increases rapidly as pH exceeds 6.

Smith and Shumate[24] calculated an activation energy of 58.6 kJ/mole for the aqueous oxidation of pyrite with oxygen. This value is in agreement with those previously reported, and as stated before supports a chemically controlled rate-limiting step.

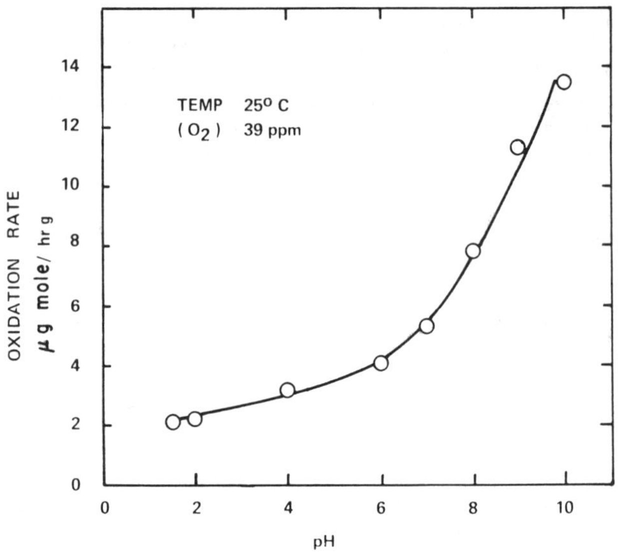

Fig. 7. Rate of pyrite oxidation as a function of pH. (After Ref. 24)

Ferric ion oxidation of pyrite was studied at low pH values (0.2 to 0.5) using different Fe^{3+}/Fe^{2+} ratios and different total iron concentrations. Oxidation potentials (volts vs SCE) ranged from 0.477 to 0.700 volts and total iron (at constant oxidation potential) varied through an extreme range of concentrations. A reaction model based on "dual-site" adsorption (ferric ion is adsorbed on two reactive sites) was found to explain the experimental results. The rate equation developed from this model was as follows

$$r = \frac{k\sqrt{K_1} - \frac{\sqrt{K_2}}{K}\sqrt{\frac{Fe^{2+}}{Fe^{3+}}}}{\frac{1}{\sqrt{Fe^{3+}}} + \sqrt{K_1} + \sqrt{K_2}\sqrt{\frac{Fe^{2+}}{Fe^{3+}}}} \qquad (29)$$

where

k = rate constant, electron transfer reaction

K = equilibrium constant, electron transfer reaction

K_1 = adsorption equilibrium constant for Fe^{3+}

K_2 = adsorption equilibrium constant for Fe^{2+}

Fe^{2+}, Fe^{3+} = activity of ferrous and ferric ions

At high oxidation potentials, the $\sqrt{\frac{Fe^{2+}}{F3^{3+}}}$ term is negligible and equation (29) reduces to

$$r = \frac{k'}{\frac{1}{\sqrt{Fe^{3+}}} + \sqrt{K_1}} \qquad (k' = k\sqrt{K_1}) \quad (30)$$

or

$$\frac{1}{r} = \frac{1}{k'\sqrt{Fe^{3+}}} + \frac{\sqrt{K_1}}{k'} \qquad (31)$$

Equation 31 predicts that 1/r plotted versus $1/\sqrt{Fe^{3+}}$ should yield a straight line, the slope of which equals 1/k' and the intercept equals K_1/k'. Figure 8 shows the application of equation (31) to the leaching of high purity pyrite at +0.7 volts. A straight line correlates well with the experimental data. Smith and Shumate,[24] after calculating all the constants for equation (29) demonstrated that predictions based on this reaction model compare closely with experimental rate values obtained for various pyrite specimens.

Mishra[25] studied the oxidation of pyrite by measuring the rate of oxygen uptake by wetted pure mineral at constant temperature. The tests were conducted at atmospheric pressure and were run for extended periods of time. In the temperature range of 40 to 90°C, the oxidation followed linear kinetics and was characterized by an activation energy of 51.9 kJ/mole. Most of the solubilized iron remained in the ferrous state; however, that which oxidized precipitated as Fe_2O_3 on the pyrite surface. Elemental sulfur formed during the reaction was amorphous and remained with the Fe_2O_3 as patches on the mineral surface. There was enough surface accessible to oxygen to sustain the reaction (linear kinetics).

King and Lewis[26] have examined the simultaneous effects of oxygen and ferric ion on the aqueous oxidation of pyrite (-325 mesh). The oxidation kinetics were found to be sensitive to temperature, oxygen pressure, and ferric ion concentration. At 0.5 M $Fe_2(SO_4)_3$, 9.2 atm O_2, and 100°C, the fraction of FeS_2 reacted in 4 hr was 0.90. In Fig. 9, the dissolution data are plotted according to a surface reaction model of the form

$$(32) \qquad 1 - (1 - \alpha)^{1/3} = k_o t$$

where α is the fraction of FeS_2 reacted, k_o is the rate constant, and t is time. The straight line fit indicated that the reaction boundary is moving at a constant velocity (i.e., linear kinetics), and the products of reaction are not limiting the reaction. The non-zero intercept is possibly the result of an initial release of iron from the dissolution of surface oxides. These may not have been removed prior to leaching with oxygenated acid ferric sulfate solutions.

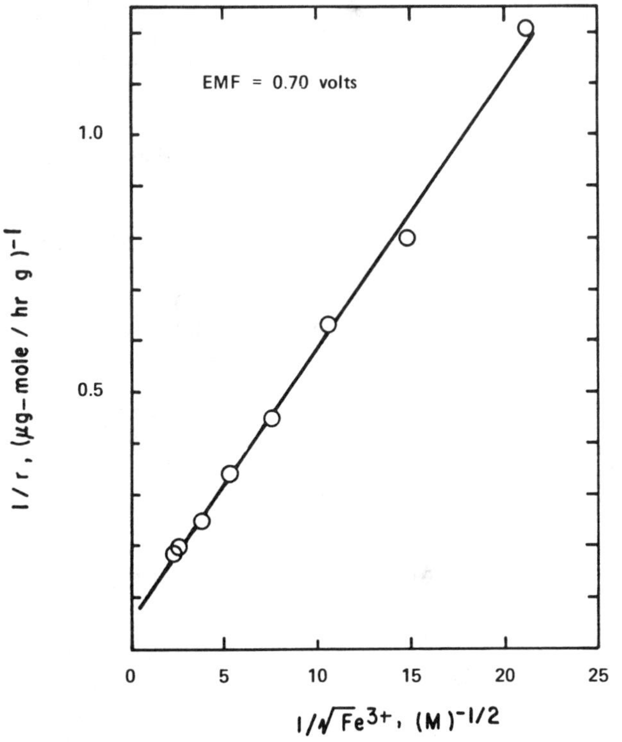

Fig. 8. Reciprocal rate vs reciprocal of the square root of free ferric ion concentration. (After Ref. 24)

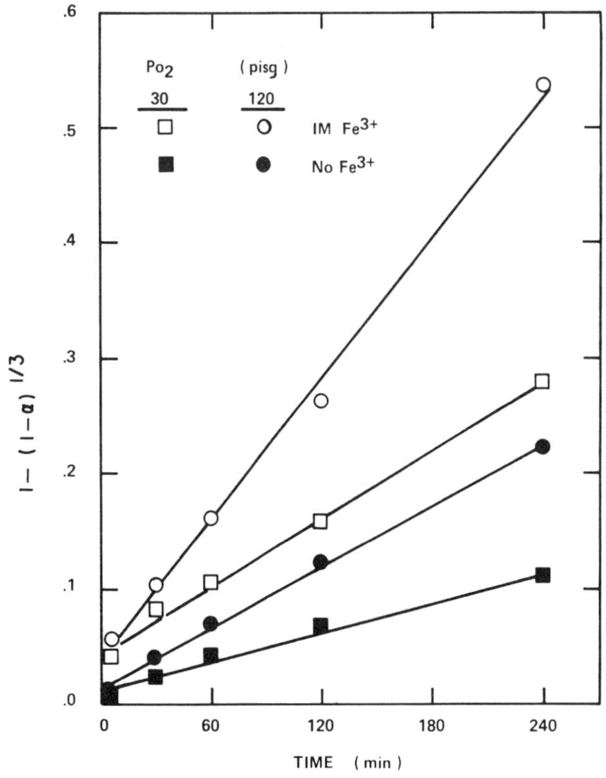

Fig. 9 Correlation of pyrite dissolution data with surface reaction rate equation.

The rate constants determined by this treatment of the data were used to calculate the experimental activation energy and reaction order with respect to oxygen partial pressure. In the absence of ferric ion initially, the oxidation of pyrite is characterized by a 0.5 order dependence on oxygen partial pressure at both 80 and $100^{o}C$. Activation energies calculated for oxygen pressures of 3.0, 5.1, and 9.2 atm equaled 44.8, 39.6, and 46.0 kJ/mole, respectively. These values are consistent with those previously reported by others. In the presence of ferric ion (0.5 M $Fe_2(SO_4)_3$) for a temperature of $100^{o}C$, the rate is also 0.5 order with respect to oxygen partial pressure. As shown in Fig. 9, the effect of ferric ion is obviously significant. The rate is approximately a factor of 2 greater in the presence of 0.5 M $Fe_2(SO_4)_3$. This work also indicated no significant difference in oxidation rate with ferric ion under nitrogen atmosphere versus ambient air pressure. This is because ferric ion is consumed and is not adequately regenerated under these conditions. However, significant regeneration of ferric was found at oxygen pressures above 3.0 atm.

Bailey and Peters[14] conducted extensive pressure leaching tests at oxygen pressures from 0 to 66.4 atm and temperatures between 85 and $130^{o}C$. At 1 M H_2SO_4 and 66.4 atm O_2, the activation energy for initial kinetics was equal to 51.1 kJ/mole. At temperatures above $110^{o}C$, the rate of reaction decreased substantially after 2 hr. This was attributed to the formation of a liquid sulfur coating on the pyrite particles. In the oxygen pressure range of 12.0 to 66.4 atm O_2, the reaction was found to exhibit a 0.5 order dependence on oxygen. The oxygen dependence observed for pyrite dissolution can be rationalized in terms of the electrochemical model described in the next section. The overall electrochemical interpretation of leaching must incorporate an oxygen adsorption step. This approach is similar to the electrochemical model developed for the ammoniacal sulfate leaching of chalcopyrite[27] and the ammoniacal carbonate leaching of uraninite[28] which included oxygen adsorption in the kinetic treatment.

Besides the literature cited above, there are numerous other studies of the aqueous oxidation of pyrite. A partial list of other fundamental investigations include: the work of King and Perlmutter[29] examining pyrite oxidation in aqueous ferric chloride; the study of Mathews and Robins[30] providing a rate expression for the aqueous oxidation of pyrite using molecular oxygen; and the investigation by Kim and Choi[31] describing the pressure leaching of a pyrite ore. In addition, the kinetics of pyrite oxidation have been examined as part of other metal recovery systems. Examples of this applied work include: a method developed by Thomas and Whalley[32] for leaching manganese from pyrolosite using pyrite as a source of Fe^{2+} for the reductive leach, a pressure leaching process described by Forward and Halpern[33] which utilized the sulfuric acid generated from pyrite oxidation to dissolve uranium, and a procedure reported by Johnson and Nackowski[34] where pyrite was added to a beryllium ore to produce sulfuric acid in situ during pressure leaching.

Table 2 is a summary of kinetic results for the aqueous oxidation of pyrite under a variety of conditions.

TABLE 2
SUMMARY OF THE KINETICS OF AQUEOUS OXIDATION OF PYRITE

Material	Conditions	Medium	O_2 Dependence	Activation Energy kJ/mole	Mechanism	Reference
Natural FeS_2	130-210 ^{o}C 2.7-13.9 atm P_{O_2}	H_2O	0.5	83.7	Surface Controlled Chemisorption	Warren (1956)
Upgraded FeS_2 Concentrate	$100-130^{o}C$ 0-4 atm P_{O_2}	H_2SO_4	1.0	55.7	Surface Controlled Chemisorption	McKay & Halpern (1958)
Upgraded FeS_2 Concentrate	$130-165^{o}C$ 6.1-23.8 atm P_{O_2}	H_2O	0.5	70.3-77.4	Chemical Control	Cornelius & Woodcock (1958)
Natural FeS_2	$33^{o}C$ $4x10^{-5}$ - $2x10^{-3}$ M Fe^{3+} Deoxygenated	Acid Ferric Sulfate	--	--	Controlled by Eh	Garrels & Christ (1960)
Natural FeS_2	See text	H_2O	0.5	58.6	Fe^{3+} Adsorption	Smith & Shumate (1970)
Natural FeS_2	$40-90^{o}C$ 1 atm P_{O_2}	H_2O	--	51.9	Linear Kinetics Chemical Control	Mishra (1973)
Natural FeS_2	$80-100^{o}C$ 0-9.2 atm 0-1.0 M Fe^{3+}	H_2SO_4	0.5	39.6-46.0	Linear Kinetics Chemical Control	King & Lewis (1980)
Natural FeS_2	$85-130^{o}C$ 0-66.4 atm P_{O_2}	H_2SO_4	0.5	51.1	Electrochemical/ Controlled by the Mixed Potential	Bailey & Peters (1975)

ELECTROCHEMISTRY

Electrochemical processes in the leaching of metal sulfides and oxides were recently reviewed by Hiskey and Wadsworth.[35] Koch[36] and Peters[37] have also surveyed the electrochemistry of sulfide minerals. In general, metal sulfides can participate in coupled charge transfer (electrochemical) processes because of their favorable electronic properties. The semiconducting properties of selected sulfide minerals have been documented by Shuey[38].

Pyrite is recognized as having a cubic structure with $Fe^{2+}(S_2)^{2-}$ representing its ionic bonding model. Pyrite has been identified as being both a n- and p-type semiconductor. At room temperature, electron mobility is approximately 100 times greater than hole mobility. Furthermore, n-type pyrite has a lower resistivity than p-type pyrite. The mean resistivity for n- and p-type pyrite is about 1×10^{-3} and 3×10^{-2} ohm-m, respectively.

An important property of semiconducting minerals is the characteristic "rest" potential. For interfacial electrode processes, the "rest" potential corresponds to the equilibrium (no net anodic or cathodic current) electrode potential. The thermodynamically predicted equilibrium (rest) potential for the FeS_2/Fe^{2+} couple (reaction (7)) is approximately 0.34 volts (vs SHE).* The electrochemical studies of Peters and Majima[39] have shown pyrite to exhibit an experimentally measured rest potential of about 0.62 volts. The 0.3 volt difference between the observed and predicted rest potentials is explained below.

Figure 10 shows the polarization curves for pyrite with the rest potential denoted. These results show that the anodic oxidation of pyrite only takes place at relatively high potentials (potentials more noble than the rest potential). As pointed out by Majima[40] and Majima and Peters,[41] the rest potential of pyrite is greater than that of other base metal sulfides. Therefore, in comparison to other sulfides, pyrite is quite refractory and does not oxidize at any appreciable rate until the oxidation potential exceeds 0.62 volts. Peters and Majima[39] demonstrated that normally passive natural pyrite with a rest potential of +0.62 volts is rendered "active" by a cathodic treatment which generates a clean pyrite surface. The active form of pyrite exhibits a rest potential 0.2 to 0.3 volts lower than the passive (normal) pyrite. The lower value is that predicted thermodynamically. Thus, the natural surface tarnish is an important factor in the persistence of pyrite in many geological settings.

Anodic oxidation of pyrite at 1.05 volts has been shown[39] to proceed by the following reaction

$$FeS_2 + 8\,H_2O \rightarrow Fe^{3+} + 2\,SO_4^{2-} + 16\,H^+ + 15e \qquad (33)$$

with sulfate ion as the only sulfur species. The cathodic reduction of pyrite proceeds at negative potentials (i.e.,

*All potentials will be reported as volts versus standard hydrogen electrode (SHE).

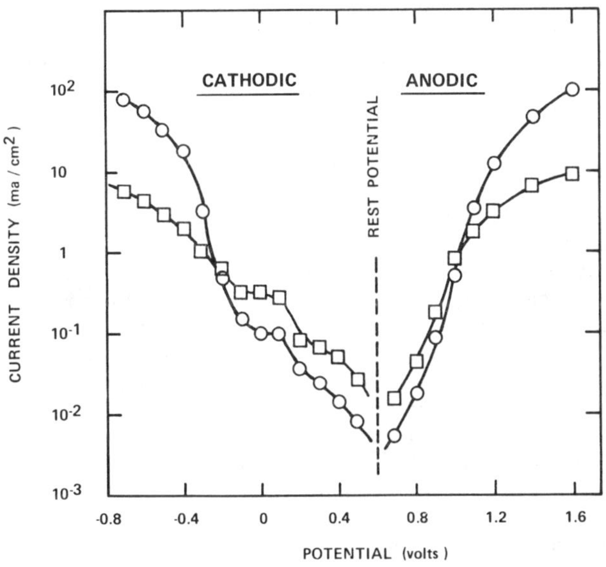

Fig. 10 Typical polarization curves for two pyrite samples in 1 M $HClO_4$. (After Ref. 39)

potentials more active than the SHE according to the following reaction

$$FeS_2 + 4H^+ + 2e \rightarrow Fe^{2+} + 2H_2S(g) \qquad (34)$$

The electrochemical nature of pyrite oxidation was clearly established by Bailey and Peters[14] as part of their investigation of the oxygen pressure leaching of pyrite. The sulfate forming and elemental sulfur forming anodic reactions are as follows

$$FeS_2 + 8H_2O \rightarrow Fe^{2+} + 2SO_4^{2-} + 16H^+ + 14e \qquad (8)$$

and

$$FeS_2 \rightarrow Fe^{2+} + 2S^0 + 2e \qquad (7)$$

The complementary cathodic reaction involves the reduction of oxygen

$$O_2 + 4H^+ + 4e \rightarrow 2H_2O \qquad (35)$$

This reaction is supplemented by the cathodic reduction of ferric ions if sufficient regeneration by oxygen is achieved.

A major finding resulting from this work was that sulfate formed from pyritic sulfur contains oxygen acquired from water instead of oxygen from the gas phase. Unequivocal evidence in support of this conclusion was obtained from O^{18} tracers experiments. The fact that sulfate oxygen is derived from water rather than molecular oxygen confirms the electrochemical dissolution mechanism. In addition, detailed stoichiometric determinations showed that sulfate yield varied between 60 and 90% and increased with increasing oxidation potential.

Table 3 summarizes the effect of potential on the dissolution of pyrite. The oxidation of pyrite at these potentials reveals that the actual dissolution proceeds by a combination of anodic dissolution routes. A dependence on charge of between 2 and 14 (or 15 if oxidation of Fe^{2+} to Fe^{3+} is considered) equivalents per mole of iron produced is indicated by reactions (7) and (8). Actually, the data indicated a dependence of between 8.02 and 13.4 for an increase of potential from 0.699 to 1.000 volts. Furthermore, the SO_4^{2-} / Fe^{2+} ratio should vary between 0 and 2 depending upon the dissolution reaction. Lower oxidation potentials favor a condition where both reactions (7 and 8) contribute to the dissolution of pyrite. However, as the potential increases, reaction (8) becomes the dominant route. Reaction (7) by itself would not account for significant leaching because the elemental sulfur insulates and protects the surface. When reaction (8) contributes to greater than 27% of the leaching, the sulfur formed is non-protective and higher oxidation rates are observed.

TABLE 3
THE EFFECT OF POTENTIAL ON PYRITE DISSOLUTION STOICHIOMETRY

Potential (SHE) volts	SO_4^{2-} / Fe^{2+}	Charge/Fe^{2+}	SO_4^{2-} Yield* %
0.699	1.19	8.02	59.6
0.800	1.26	9.72	63.1
0.900	1.35	10.50	67.6
1.000	1.80	13.4	90.2

*Measured analytically.

The cathodic reduction of dissolved molecular oxygen at the pyrite surface according to

$$O_2 + 4H^+ + 4e \rightarrow 2H_2O \tag{35}$$

is important to pyrite oxidation by oxygen. Biegler and co-workers[42] have investigated the electrode kinetics of oxygen reduction on a rotating pyrite electrode. Associated with this study, Biegler[43] examined the effect of electronic properties on the electrochemical behavior of pyrite from different origins. There was no apparent correlation between electrochemical response for oxygen reduction and semiconduction properties for various pyrite samples. The minor variations observed for the individual pyrite specimens were attributed to trace impurities or possibly crystallographic considerations (e.g., dislocations, defects). Finally, the electrochemistry of the FeS_2/O_2 system was essentially the same for both n- and p-type pyrite.

In concentrated acid solutions (1M), the cathodic reduction of oxygen (saturated solution) has a Tafel slope equal to -130 mV per decade and an exchange current density equal to 10^{-11} A cm^{-2}. The limiting current density (mass transfer control) is given by

$$i = 0.298\, n\, \omega^{1/2} \quad mA\ cm^{-2} \tag{36}$$

at $O_2 = 1.21 \times 10^{-3}$ M where n = number of electrons required for the half-cell reaction and ω = angular velocity of the electrode. The experimental value of n in the limiting current region was approximately 4; therefore, the cathodic reduction of O_2 as given in reaction (35) is supported by this data.

The mechanism of oxygen reduction on pyrite involves two distinct processes. First, at low pH values (no pH dependence up to about pH 7 and a Tafel slope of -130 mV per decade) the electrochemical data is consistent with an initial rate-controlling step involving the first electron transferred

$$O_2 + e \rightarrow O_2^- \tag{37}$$

As pH increases above 7, electrochemical and analytical measurements suggest that a second mechanism is rate controlling. It includes the formation of a soluble intermediate (i.e., hydrogen peroxide). The nature of the Tafel slope and the pH dependence supports the following mechanism

$$O_2 + e \underset{k_2}{\overset{k_1}{\rightleftharpoons}} O_2^- \tag{37a}$$

$$O_2^- + 2H^+ + e \xrightarrow{k_3} H_2O_2 \tag{38}$$

where step (37a) is in quasi-equilibrium and the addition of the second electron step (38) is rate-controlling. In actuality, step (37a) must be considered in the kinetic treatment to account for the electrochemical results in alkaline solutions.

Springer[44] subjected different types of pyrite to anodic and cathodic polarizations in acid solution. There was a significant difference in rest potential between n- and p-type pyrite specimens. However, there was essentially no difference in the polarization curves for the various kinds of pyrite. The Tafel slope is approximately 120 mV per decade of current, both for anodic and cathodic polarization. It was concluded that the type of semiconduction properties did not play a significant role in determining the rate of pyrite oxidation.

The electrochemistry of pyrite was investigated by Meyer[45] in both acid and alkaline solutions. Polarization data for the anodic oxidation of pyrite revealed that the Tafel slopes in acid solution varied from about 80 to 110 mV per decade for polished specimens. In Fig. 11, the effect of pH on the anodic oxidation is shown for 25°C, Ar atm, 0.5 mV/s scan rate. In the pH region between 0.32 and 3.89, current density (rate of oxidation) was dependent on the hydrogen ion concentration to the -0.33 order. Above pH 3.89, the electrochemical reaction was stirring sensitive (as shown by dashed line in Fig. 11). The activation energy averaged 59 kJ/mole for potentials between +550 and +650 mV, an observation, which indicates chemical (electrochemical) reaction rather than diffusion control. Meyer,[45] as did Bailey and Peters,[14] observed a higher percentage of sulfate produced with increasing oxidation potential.

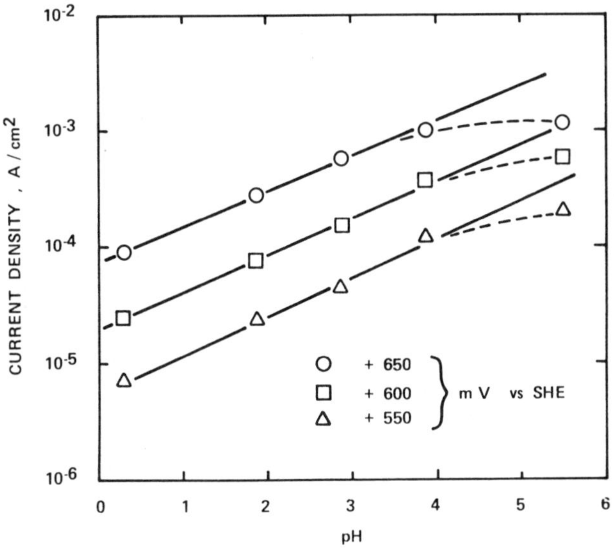

Fig. 11. Current density as a function of pH for various potentials on the pyrite electrode. (After Ref. 45)

Anodic polarization of pyrite in alkaline solutions proceeds at potentials noticeably less anodic than those required by acid solutions. The reaction was found to be approximately first order with respect to hydroxide ion concentration. The electrochemical dissolution mechanism was complicated by the formation of an insoluble iron hydroxide surface film. The film was identified as being mostly amorphous with some FeO also detected. At intermediate potentials, the reaction is limited by mass transfer of OH^- in the bulk solution. At more anodic potentials, the reaction reverts to charge transfer control and because the surface film is porous, the electrochemically active interface is that between pyrite and the oxide layer.

Biegler and Swift[46] examined the anodic oxidation of pyrite in acid solutions. The initial rest potential of freshly prepared pyrite surface measured about 0.55 to 0.65 volts. The Tafel slope for potentials extending from 0.9 to 1.2 volts was about 95±10 mV per decade. The reaction stoichiometry (expressed as percentage sulfate produced) for the anodic oxidation of pyrite is shown in Fig. 12. This figure consolidates the data of Meyer,[45] Bailey and Peters,[14] and Biegler and Swift.[46] There is an obvious correlation between sulfate yield and potential.

The electrochemical mechanism is viewed as being very complex, possibly involving two competing reactions. One pathway leads to the formation of sulfate

$$FeS_2 + 8H_2O \rightarrow Fe^{3+} + 2SO_4^{2-} + 16H^+ + 15e \quad (33)$$

and the other pathway yields elemental sulfur

$$FeS_2 \rightarrow Fe^{3+} + 2S^o + 3e \quad (39)$$

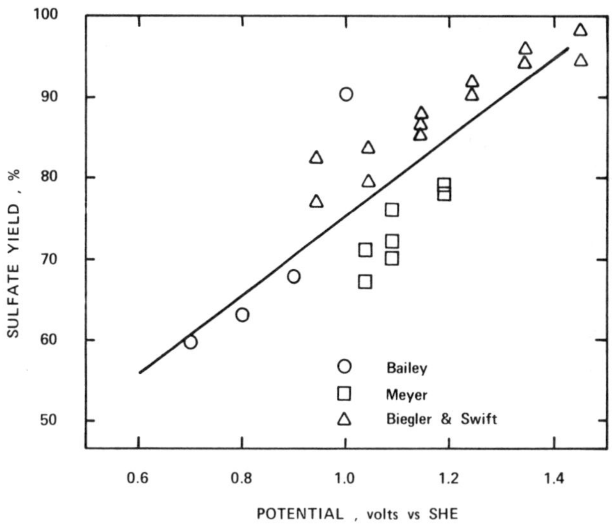

Fig. 12. Percentage of SO_4^{2-} produced during the oxidation of pyrite at different anodic potentials.

The degree of reaction for each varies linearly with anodic potential. Sulfur produced during the anodic dissolution of pyrite is a stable end product, even at potentials where sulfate formation from pyrite is favored. Therefore, sulfur is not an intermediate product in the sulfate reaction pathway. As supported by the work of Bailey and Peters,[14] Biegler and Swift[46] proposed a mechanism involving the electrochemical transfer of oxygen from water to the sulfur atoms in the pyrite lattice. The mechanistic scheme proposed is as follows

$$FeS_2 + H_2O \rightleftarrows FeS_2(OH)_{ads} + H^+ + e \quad (40)$$

$$FeS_2(OH)_{ads} \rightarrow FeS_2(O)_{abs} + H^+ + e \quad (41)$$

Step 40 is considered as being in quasi-equilibrium, and the second charge transfer step (41) is regarded as being rate-controlling. However, analysis of the current-voltage relationships (Butler-Volmer equations) indicated that the predicted Tafel slope is only 59 mV per decade which is considerably less than the experimentally determined Tafel dependence. Other mechanisms were advanced; however, they were less accurate in predicting the electrode kinetics.

Recently, Hamilton and Woods[47] investigated the electrochemistry of pyrite oxidation using linear potential sweep voltammetry. They examined the oxidation process in both acidic and basic solutions. Some important data was obtained regarding the products of anodic oxidation. In Fig. 13, the sulfur products (elemental sulfur and sulfate) are plotted versus potential for pH values of 4.6, 9.2, and 13.0. At potentials above about 0.4 volts, the iron product is in the ferric state. At pH's between 4.6 and 13.0, the initial oxidation product is elemental sulfur with only minor amount of sulfate being produced. At pH 9.2 and 13.0, the yield of sulfur remains nearly constant at 0.45×10^{-8} moles/cm^2. At the same time sulfate yield increases sharply with increasing potential. At pH 4.6, the amount of sulfur produced increases with increasing potential; however, so does the sulfate which predominates above about 0.8 volts. This work also shows anodic dissolution of pyrite occurring at lower potentials as pH increases.

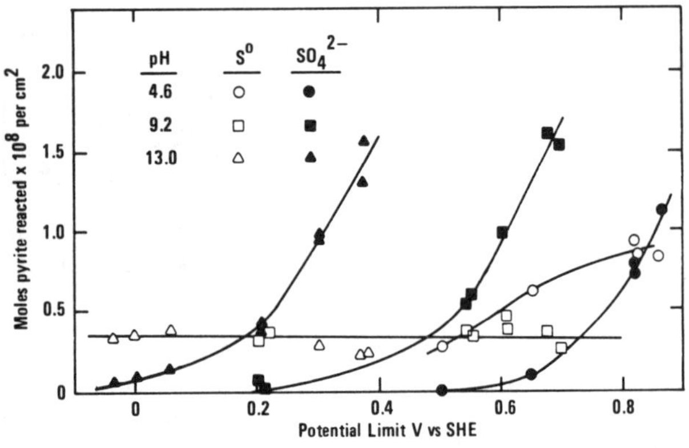

Fig. 13 Quantity of pyrite oxidized to sulfur and sulfate as a function of potential. (After Ref. 47)

BACTERIAL EFFECTS

Previous sections have dealt with a strictly chemical mode of oxidation. However, both industrial experience and laboratory studies have demonstrated that selected types of bacteria can play a significant role in the leaching or oxidation of pyrite. Thiobacillus ferrooxidans was the first bacterium to be isolated in acid mine drainage now attributed to pyrite oxidation.[48] Later this organism was found to be widely distributed in industrial leaching systems involving treatment of pyritic material.[49] Other early studies showed that the bacterial strain Thiobacillus thiooxidans also participates in sulfide oxidation.[50,51] Both strains are chemosynthetic, autotrophic, aerobic bacilli. T. ferrooxidans derives energy from oxidation of both ferrous iron and inorganic sulfur while T. thiooxidans utilizes only reduced forms of inorganic sulfur. Both microorganisms depend on carbon dioxide as their source of carbon and also require nitrogen and phosphorous for chemosynthesis and growth.

As living organisms, these bacteria are sensitive to their environment. Even though Sulfolobus-like microbes extend the temperature at which bacteria actively oxidize ferrous iron and metal sulfides,[52] conditions for bacterial oxidation of pyrite are more restricted than those in strictly chemical regimes. Particularly important are limitations on temperature, pH and ionic concentration. For Thiobacillus organisms, Murr[53] reports optimum conditions ranging from 25-45°C at a pH level of 2.3-2.5. As the pH drops below this, Atkins[54] showed that metal release from pyrite declines linearly, actually stopping at a pH of about 1.0. The upper end of the pH range for bacterial growth is reported to be 5-6.[53] Known thermophilic bacteria are apparently operative over the same pH range but can oxidize iron at 60°C and possibly as high as 80°C.[55] Even in the proper temperature and pH regimes Thiobacillus, like other living organisms, can be affected by heavy metal poisoning.[56] Uranium, cadmium, mercury, thallium, and molybdenum have been reported as inhibitory, but silver appears to be the most toxic.[57] Bacterial adaptation can reduce some of these toxic effects and can produce strains well suited for particular environments. For example, bacteria thrive in copper leach solution at Kennecott's Bingham Canyon mine where uranium levels are 4-8 ppm and the chloride content is 120 ppm. The latter will completely destroy Thiobacillus cultured on the usual laboratory substrates.

In spite of their limitations, bacteria can have a startling effect under ambient or near ambient conditions. As shown by previously unpublished Kennecott data in Fig. 14, the reaction rate for

$$Fe^{2+} + \frac{1}{4} O_2 + H^+ \rightarrow Fe^{3+} + \frac{1}{2} H_2O \qquad (13)$$

is several times greater in natural (bacteria-containing) mine water than in the same solution after addition of a chemical bacteriacide.

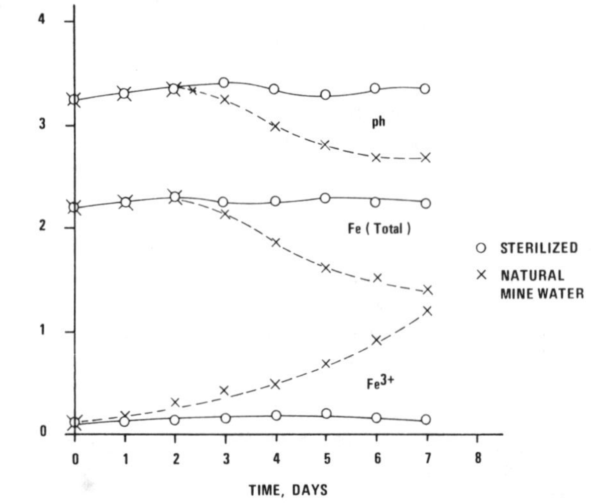

Fig. 14. Impact of bacteria on iron oxidation in mine water. The drop in pH and total iron in the natural (bacteria-containing) mine water is due to iron hydrolysis.

Two principle mechanisms have been suggested to explain the role of bacteria in enhancing pyrite oxidation.[24] One represents an indirect mechanism. In such a system, the bacteria catalyze the conversion of ferrous iron to the ferric state. Since the responsible organism is aerobic, oxygen is the only electron acceptor which supports the conversion

$$Fe^{2+} + \frac{1}{4} O_2 + H^+ \xrightarrow{\text{bacteria}} Fe^{3+} + \frac{1}{2} H_2O \quad (13a)$$

The elemental sulfur may be oxidized to sulfate by reaction with additional ferric ions or by a bacterially catalyzed reaction with dissolved oxygen. The oxidation mechanism is considered to be indirect since the bacteria need not be attached to any substrate. The total bacterial population may therefore be the rate determining factor.

In contrast to this, the second mechanism is direct oxygenation in which bacteria exert their catalytic effect only while attached to the pyrite surface. In such a system the oxidation rate is independent of the ferric ion concentration since the reaction is

$$FeS_2 + \frac{7}{2} O_2 + H_2O \xrightarrow{\text{bacteria}} Fe^{2+} + 2\,SO_4^{2-} + 2H^+ \quad (12a)$$

Here pyrite oxidation will be a function only of the number of bacteria actually attached to exposed pyrite surfaces.

There is no reason why either mechanism cannot be rate controlling under appropriately different conditions. Evidence for the direct mechanism is substantial. In addition to the selective bacterial attachment frequently observed on pyrite (Fig. 15)[53], Bruynestyn[58] cites the difficulty in detecting the elemental sulfur expected with ferric oxidation. He also points out that enzymes controlling iron and sulfide oxidation can be selectively inhibited so that T. ferrooxidans still oxidizes sulfide minerals without being able to oxidize ferrous iron. Finally, Bryner, et al.[59] have shown that at least copper sulfides (covellite and chalcocite) can be oxidized by bacteria in the absence of iron.

Likewise, there is considerable evidence supporting the indirect mechanism. For example, Berry and Murr[60] have detected sulfur crystals on freshly leached pyrite surfaces. Atkins[54] also noted conversion of a sulfur substrate to sulfate even after all pyritic iron was oxidized. Furthermore, pyrite grains within rock fragments can be oxidized without detectable bacteria present. Manometric studies by Sutton and Corrick[61] also support the indirect mechanism.

On balance, the extensive review by Murr[53] suggests that the indirect bacterial mechanism generally predominates in systems involving pyrite associated with copper leaching. Similarly, detailed studies by Smith and Shumate[24] have shown that the pyrite oxidation responsible for acid coal mine drainage is generally indirect. Under such conditions, ferric ion oxidation is the

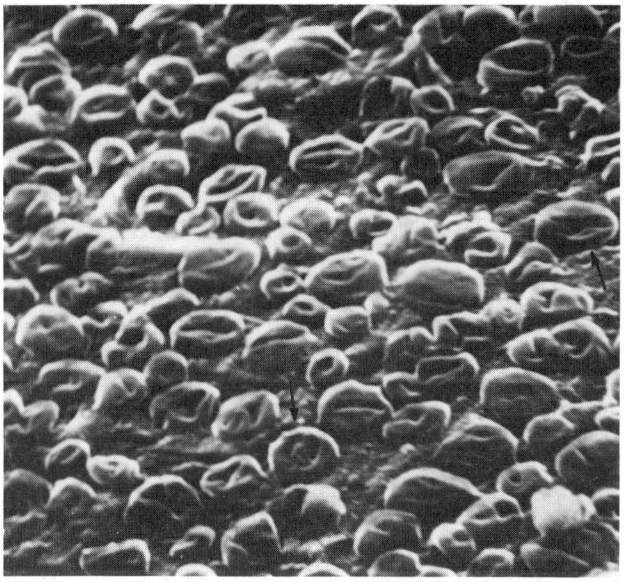

Fig. 15. Thermophilic microorganisms observed in the scanning electron microscope, sulfolobus-like bacteria on pyrite. (Courtesy of L. E. Murr)

chemical analogy of microbiol-enhanced oxidation. Thus, chemical and kinetic considerations for the ferric oxidation system apply to the bacterial one. In fact, Smith and Shumate found that both the indirect bacterial mechanism and the direct chemical one proceeded simultaneously but were independent of each other (see Fig. 16). The latter was much slower where bacterial conversion of ferrous ions provided excess ferric iron. Under these conditions, the specific oxidation rate is fixed by the surface sites on the pyrite where ferric reduction (pyrite oxidation) can occur. However, as the pyrite-to-water ratio increases, the bacterial population will not be able to provide sufficient ferric ions and chemical oxidation will gradually predominate.

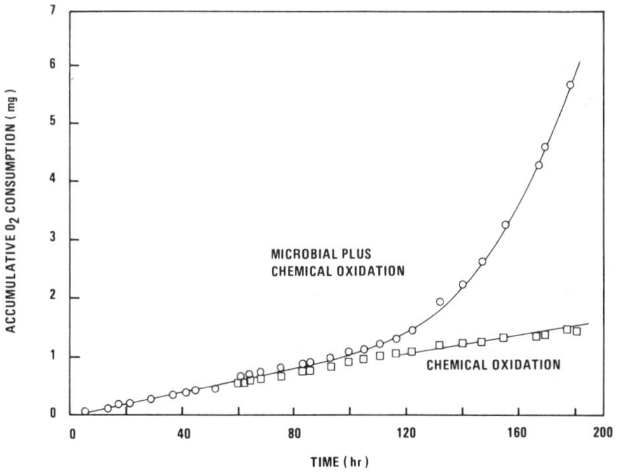

Fig. 16 Comparison of rate of O_2 uptake for chemical and microbiol/chemical oxidation of pyrite. (After Ref. 24)

When bacterial activity contributes to pyrite oxidation in natural systems, the effect is one of self-regulation. As the bacterial population increases to its steady state level of 10^8-10^9 cells/g[62] the ferrous-to-ferric oxidation rate will accelerate. This increases the ferric-ferrous ratio in solution, in turn raising the rate of ferric reduction through pyrite oxidation. The steady state ferric-ferrous content is then determined through pyrite oxidation by the relative rates for the homogenous ferrous oxidation reaction and the heterogeneous reduction of ferric ion. The latter is a function of the surface area of the exposed pyrite.

Further self regulation is derived from heat release and acid generation during pyrite oxidation. As noted by Schlitt and Jackson,[63] the net enthalpy for pyrite oxidation (including generation of the necessary ferric ion) is -1.4 MJ/mole. When iron is converted to the ferric state and sulfur is converted to sulfate, pyrite also becomes the only sulfide that oxidizes to provide a net source of acid. As the temperature increases, bacterial growth accelerates, ultimately leading to more rapid pyrite oxidation. However, if the temperature rises above some optimum, bacterial activity will decline and the ferrous oxidation rate will fall. The overall impact will be to reduce the pyrite oxidation rate as well. Increased acid levels derived from pyrite oxidation can also exceed the optimum for the bacteria. Again, the ultimate effect is to lower the pyrite oxidation rate until a steady state is achieved. Thus, when pyrite oxidation occurs in a natural system as a result of bacterial activity, the oxidation rate will tend toward a steady state that generates enough heat and acid to provide optimum conditions for bacterial growth.

Overall, the bacteria function to provide a high ferric-ferrous ratio and a high total iron content. As implied by the "dual site" reaction model, the relative adsorption of ferrous and ferric ions on pyrite (and not solution Eh) controls the pyrite oxidation rate. As determined by Smith and Shumate,[24] ferrous adsorption is much greater than ferric. They found that the ratio of the ferrous and ferric adsorption constants varied from 900 to 45,000 depending on the type of pyrite. Thus, by oxidizing the bulk of the iron, bacteria indirectly promote the high concentration of ferric ions on pyrite surfaces that is needed for rapid oxidation.

The bacteria also help offset the adverse effect of anion concentration. In an acidic pyrite system, increases in sulfate concentration decrease the oxidation rate. Evidence developed by Smith and Shumate[24] suggest that sulfate will complex ferric ions, reducing their ability to adsorb onto pyrite surfaces and to oxidize the pyrite itself. There is also evidence that sulfate ions are adsorbed on pyrite, reducing the number of sites available for ferric reduction. Again, by generating a high ferric-ferrous ratio and a high total iron content, the bacteria work to overcome the adverse effect of sulfate. A relatively high ferric content will tend to crowd sulfate ions off the pyrite and to increase the concentration of free (uncomplexed) ferric ions at any sulfate level.

ACKNOWLEDGEMENTS

The authors wish to acknowledge the efforts of Dr. J. S. Jackson for his efforts in helping collect the necessary literature. We also extend our thanks to Ms. M. R. Anderson for her important contribution in typing this paper.

REFERENCES

1. Palache, C., Berman, H., and Frondel, C.,, Dana's System of Mineralogy, 7th Ed. - 8th printing, John Wiley & Sons, Inc., New York, 1966, Vol. 1, p. 285.
2. Trudinger, P. A., Lambert, I. B., and Skyring, G. W., "Biogenic Sulfide Ores: A Feasibility Study," Econ. Geol., 1972, Vol. 67, pp. 1114-1127.
3. Robie, R. A. and Waldbaum, D. R., Thermodynamic Properties of Minerals and Related Substances, Geol. Survey Bull. 1259, U.S. Government Printing Office, Washington, 1968, pp. 14-15.
4. Pourbaix, M., Atlas of Electrochemical Equilibria in Aqueous Solutions, Pergammon Press, New York, 1966.
5. Garrels, R. M. and Christ, C. L., Solutions, Minerals, and Equilibria, Harper and Row Publishers, New York, 1965.
6. Feitknecht, W. and Schindler, P., "Solubility Product of Metal Oxides, Metal Hydroxides and Metal Hydroxide Salts in Aqueous Solution," Pure Appl. Chem., 1963, Vol. 6, pp. 130-199.
7. Dutrizac, J. E. and MacDonald, R. J. C., "Ferric Ion as a Leaching Medium," Minerals Sci. Engng., 1974, Vol. 6, No. 2, pp. 59-100.
8. Stumm, W. and Morgan, J. J., Aquatic Chemistry - An Introduction Emphasizing Chemical Equilibria in Natural Waters, John Wiley & Sons, Inc., New York, 1970.
9. Butler, J. N., Ionic Equilibria - A Mathematical Approach, Addison-Wesley Publishing Co., Inc., Palo Alto, 1964.
10. Singer, P. C. and Stumm, W., "The Rate-Determining Step in the Production of Acidic Mine Waters," Amer. Chem. Soc., Div. Fuel Chem., Prepr., 1969, Vol. 13, No. 2, pp. 80-87.
11. McKay, D. R. and Halpern, J., "A Kinetic Study of the Oxidation of Pyrite in Aqueous Suspension," Met. Trans., 1958, Vol. 212, pp. 301-309.
12. Habashi, F. and Bauer, E. L., "Aqueous Oxidation of Elemental Sulfur," Ind. Eng. Chem. Fundamentals, 1966, Vol. 5, No. 4, pp. 469-471.
13. Ichikuni, P. M., "Sur la Dissolution des Minerais Sulfures en Divers Milieux," Bull. Chem. Soc. Japan, 1960, Vol. 33, pp. 1052-1057.

14. Bailey, L. K. and Peters, E., "Decomposition of Pyrite in Acids by Pressure Leaching and Anodization: The Case for an Electrochemical Mechanism," Can. Met. Quart., 1976, Vol. 15, No. 4, pp. 333-344.

15. Stenhouse, J. F. and Armstrong, W. M., "The Aqueous Oxidation of Pyrite," Trans. Can. Inst. Min. Metall., 1952, Vol. 55, pp. 38-42.

16. Burkin, A. R., "Solid-State Transformations during Leaching," Min. Sci. Engng., 1969, Vol. 1, No. 1, pp. 4-14.

17. Burkin, A. R. and Edwards, A. M., "The Formation of Insoluble Iron Oxide Coatings during the Alkali Pressure Leaching of Pyrite," Proceedings VI International Congress on Min. Proc., Cannes, 1963, pp. 159-169.

18. Nelson, H. W., Snow, R. D., and Keyes, D. B., "Oxidation of Pyritic Sulfur in Bituminous Coal," Ind. Eng. Chem., 1933, Vol. 25, No. 12, pp. 1355-1358.

19. Burke, S. P. and Downs, R., "Oxidation of Pyritic Sulphur in Coal Mines," Trans. AIME, 1938, Vol. 130, pp. 425-444.

20. Warren, I. H., "The Generation of Sulphuric Acid from Pyrite by Pressure Leaching," Aust. J. Appl. Sci., 1956, Vol. 7, pp. 346-358.

21. Sherman, M. I. and Strickland, J. D. H., "Dissolution of Pyrite Ores in Acid Chlorine Solutions," Trans. AIME, 1957, Vol. 209, pp. 1386-1388.

22. Cornelius, R. J. and Woodcock, J. T., "Pressure Leaching of a Manganese Ore," Proc. Aust. Inst. Min. Met., 1958, No. 185, pp. 65-107.

23. Garrels, R. M. and Thompson, M. E., "Oxidation of Pyrite by Iron Sulfate Solutions," Aust. J. of Sci., Bradley Volume, 1960, Vol. 258-A, pp. 57-67.

24. Smith, E. E. and Shumate, K. S., "The Sulfide to Sulfate Reaction Mechanism, A Study of the Sulfide to Sulfate Reaction Mechanism as it Relates to the Formation of Acid Mine Waters," Water Pollution Control Research Series, Ohio State University Research Foundation, 1970.

25. Mishra, R. K., "Low Temperature Surface Reactions in Leaching of Chalcopyrite and Pyrite," Ph.D. Thesis, University of Utah, 1973.

26. King, W. E. and Lewis, J. A., "Simultaneous Effects of Oxygen and Ferric Iron on Pyrite Oxidation in an Aqueous Slurry," Ind. Eng. Chem. Process Des. Dev., 1980, Vol. 19, No. 4, pp. 719-722.

27. Beckstead, L. W. and Miller, J. D., "Ammonia Oxidation Leaching of Chalcopyrite-Reaction Kinetics," Met. Trans. B, 1977, Vol. 8B, pp. 19-38.

28. Hiskey, J. B., "Kinetics of Uranium Dioxide Dissolution in Ammonium Carbonate," IMM Trans., Sect. C, 1979, Vol. 88, pp. 145-152.

29. King, W. E. and Perlmutter, D. D., "Pyrite Oxidation in Aqueous Ferric Chloride," AIChE J., 1977, Vol. 23, No. 5, pp. 679-685.

30. Mathews, C. T. and Robins, R. G., "Aqueous Oxidation of Iron Disulfide by Molecular Oxygen," Aust. Chem. Eng., 1974, Nov./Dec., pp. 19-24.

31. Kim, J. W. and Choi, K. J., "A Study on Pressure Leaching of Pyrite Ores," J. Korean Inst. Metals, 1968, Vol. 19, No. (2), pp. 83-90.

32. Thomas, G. and Whalley, B. J. P., "The Leaching of Manganese from Pyrolusite Ore by Pyrite," Can. J. Chem. Eng., 1958, Vol. 36, pp. 37-43.

33. Forward, F. A. and Halpern, J., "Acid Pressure Leaching of Uranium Ores," Trans. AIME, 1955, Vol. 203, pp. 463-466.

34. Johnson, P. H. and Nackowski, M. P., "Pyrite Leaching of a Beryllium Ore in an Air-Agitated Autoclave," Paper presented at AIME Annual Meeting, Los Angeles, California, 1967.

35. Hiskey, J. B. and Wadsworth, M. E., "Electrochemical Processes in the Leaching of Metal Sulfides and Oxides," Process and Fundamental Considerations of Selected Hydrometallurgical Systems, M. C. Kuhn, Ed., SME-AIME, New York, 1981, pp. 303-338.

36. Koch, D.F.A., "Electrochemistry of Sulfide Minerals," Modern Aspects of Electrochemistry, No. 10, pp. 211-237, Plenum Press, New York, 1975.

37. Peters, E., "The Electrochemistry of Sulfide Minerals," Trends in Electrochemistry, J. O'M. Bockris, D. A. J. Rand, and B. J. Welch, Eds., pp. 267-290, Plenum Press, New York, 197.

38. Shuey, R. T., Semiconducting Ore Minerals, Elsevier Publishing Company, New York, 1975.

39. Peters, E. and Majima, H., "Electrochemical Reactions of Pyrite in Acid Perchlorate Solutions," Can. Met. Quart., 1968, Vol. 7, No. 3, pp. 111-117.

40. Majima, H., "How Oxidation Affects Selective Flotation of Complex Sulphide Ores," Can. Met. Quart., 1969, Vol. 8, No. 3, pp. 269-273.

41. Majima, H. and Peters, E., "Electrochemistry of Sulphide Dissolution in Hydrometallurgical Systems," Proceedings VIII International Min. Proc. Cong., Leningrad 1968, pp. 1-13.

42. Biegler, T., Rand, D. A. J., and Woods, R., "Oxygen Reduction on Sulphide Minerals - Part I. Kinetics and Mechanism of Rotated Pyrite Electrodes," J. Electroanal. Chem., 1975, Vol. 60, pp. 151-162.

43. Biegler, T., "Oxygen Reduction on Sulphide Minerals - Part II. Relation Between Activity and Semiconducting Properties of Pyrite Electrodes," J. Electroanal. Chem., 1975, Vol. 70, pp. 265-275.

44. Springer, G., "Observations on the Electrochemical Reactivity of Semiconducting Minerals," IMM Trans., Sect. C, 1970, Vol. 79, pp. 11-14.

45. Meyer, R. E., "Electrochemistry of FeS_2," J. Electroanal. Chem., 1979, Vol. 101, pp. 59-71.

46. Biegler, T. and Swift, D. A., "Anodic Behavior of Pyrite in Acid Solutions, "Electrochimica Acta., 1979, Vol. 24, pp. 415-420.

47. Hamilton, I. C. and Woods, R., "An Investigation of Surface Oxidation of Pyrite and Pyrrhotite by Linear Potential Sweep Voltammetry," J. Electroanal. Chem., 1981, Vol. 118, pp. 327-343.

48. Colmer, A. R. and Hinkle, M. E., "The Role of Microorganisms in Mine Drainage: A Preliminary Report," Science, 1947, Vol. 106, pp. 253-255.

49. Beck, J. V., "The Role of Bacteria in Copper Mining Operations," Biotech. & Bioengng., 1967, Vol. 9, pp. 487-497.

50. Bryner, L. C. and Anderson, R., "Microorganisms in Leaching Sulfide Minerals," Ind. Engng. Chem., 1957, Vol. 49, pp. 1721-1723.

51. Malouf, E. E. and Prater, J. D., "New Technology of Leaching Waste Dumps," Mining Congr. J., 1962, Vol. 48, pp. 82-107.

52. Brierly, C. L. and Murr, L. E., "Leaching: Use of a Thermophilic and Chemoautrophic Microbe," Science, 1973, Vol. 179, pp. 488-491.

53. Murr, L. E., "Theory and Practice of Copper Sulphide Leaching in Dumps and In Situ," Minerals Sci. Engng., 1980, Vol. 12, pp. 121-190.

54. Atkins, A. S., "Studies on the Oxidation of Sulphide Minerals (Pyrite) in the Presence of Bacteria," Metallurgical Applications of Bacterial Leaching and Related Microbiological Phenomena, Murr, L. E., Torma, A. E., and Brierly, J. A. (Eds.), New York, Academic Press, 1978, pp. 403-426.

55. Murr, L. E. and Berry, V. K., "Observations of a Natural Thermophilic Microorganism in the Leaching of a Large, Experimental Copper-Bearing Waste Body," Metall. Trans. B, 1979, Vol. 10B, pp. 523-531.

56. Saddler, W. R. and Trudinger, P. A., "The Inhibition of Microorganisms by Heavy Metals," Mineralia Deposita, 1967, Vol. 2, pp. 158-170.

57. Norris, P. R. and Kelly, D. P., "Toxic Metals in Leaching Systems," Metallurgical Applications of Bacterial Leaching and Related Microbiological Phenomena, Murr, L. E., Torma, A. E., and Brierly, J. A. (Eds.), New York, Academic Press, 1978, pp. 83-102.

58. Bruynesteyn, A., "Microbiological Leaching, "Research to Date and Future Applications," Annual AIME Meeting, Denver, CO, February 1970, Preprint 70-B-104.

59. Bryner, L. C., et al., "Microorganisms in Leaching Sulfide Minerals," Ind. Engng. Chem., 1954, Vol. 46, pp. 2587-2595.

60. Berry, V. K. and Murr, L. E., "An SEM Study of Bacterial Catalysis and Its Effect on Surface Reactions at Sulfide Phases in the Leaching of Low-Grade Ore," Scanning Electron Microscopy/1977, Johari, O. (Ed.), Chicago, IIT Research Inst., 1977, Vol. 1, pp. 137-146.

61. Sutton, J. A. and Corrick, J. D., "Microbial Leaching of Copper Minerals," Mining Eng., 1963, Vol. 15, pp. 37-40.

62. Murr, L. E., "Observations of Solution Transport, Permeability and Leaching Reactions in Large, Controlled Copper-Bearing Waste Bodies," Hydromet., 1979, Vol. 5, pp. 67-94.

63. Schlitt, W. J. and Jackson, J. S., "In Situ Generation of Acid During Dump Leach Production of Copper," In Situ, 1981, Vol. 5, pp. 103-131.

Chapter 8

A MINIFIELD TEST OF IN SITU URANIUM LEACHING—INTERPRETATION OF RESULTS

M. I. Kabir
L. W. Lake
R. S. Schechter
Departments of Petroleum and Chemical Engineering
University of Texas at Austin
Austin, Texas 78712

ABSTRACT

A new minitest of in situ uranium leach processes at reservoir conditions is proposed. The test consists of a single observation/production well and two observation wells. It is designed to permit the evaluation of both restoration and leaching parameters. It is shown that by using a straight forward history matching procedure, sometimes called a type curve analysis, both the rate of leaching and the ultimate uranium recovery can be evaluated based on the results of the test. Type curves are presented for the case in which the reaction rate is kinetically controlled and there are not other minerals competing for oxidant. Other situations can be approached using the same techniques as those presented here.

Introduction

In situ leaching offers the promise of efficient recovery of uranium with minimal environmental impact. These in situ processes must be carefully designed and executed, however, since ore bodies are almost always in groundwater aquifiers whose quality is becoming of increasing concern[1]. Many lixiviants, notably ammonium-based ones, are difficult to remove from an ore body after leaching because of the unfavorable exchange existing between sodium and ammonium cations on resident clays[2].

The University of Texas, sponsored in part by the U.S. Bureau of Mines, is preparing to conduct a small uranium leaching field test using a potassium carbonate lixiviant preceded by a potassium chloride preflush[3]. The purpose of the test is to establish that South Texas uranium ores can be leached with this process, to demonstrate the ease of restoration of potassium cations, and to determine kinetic and mass transfer-related constants. The latter quantities are necessary for a successful scale-up of the process to commercial scale[4,5].

In this paper we briefly describe the proposed test and discuss in detail the procedures to be used to interpret test results. Specifics about the actual results of the field test will follow in a publication after the test is completed. One of the approaches to interpreting the anticipated results is to use a form of "type-curve" matching to fit experimental results to the output of a numerical simulation model. The type-curve approach is similar to the procedure used in pressure transient analyses[6] which is designed to provide a rapid, easy, and unique determination of model parameters. If successfully applied, the parameters so determined are subject only to the limitations on the physics used in the numerical model.

The proposed field test is patterned after the minitest procedures used in micellar/polymer field testing[7]. Figure 1 shows an areal view of the well arrangement for the test now being conducted in conjunction with Everest Minerals Corporation at their Hobson, Texas site. The test consists of an injection/production test well at the center of a 10-ft. radius circle on whose perimeter are located two sampling observation wells. The very small well spacing is to hold the length of the test to a few weeks. The two observers are oriented at right angles to each other to detect fluid channeling or non-radial flow. If possible the observer test well axes should be parallel to the major and minor permeability directions of the ore body. The observers are to be sampled at a low rate (a few gallons per day) to not disturb the flow patterns caused by fluid injection in the test well.

The test procedure itself consists of an injection (push) cycle of a few pore volumes of preflush and leach solution, and a subsequent production (pull) cycle wherein the injected chemicals are to be removed from the ore body. During the push cycle the observers are to be sampled for chlorides and the cations calcium, potassium, and sodium, as well as for oxidant, and uranium. If competing minerals are present, other oxidation products may be assayed also. From the observer cation data we may derive cation exchange constants using a procedure similar to that presented by Hill and Lake[8]. Previous work has shown that such parameters are not extractable from the test well effluents alone[9]; this, in fact, is one of the major reasons for including the observers in the minitest. Kinetic data may be derived from the uranium and oxidant observer effluents using the procedures described below. This paper does not discuss the matching procedure on the cation data. During the pull cycle uranium and oxidants will be measured at the test well as will the cations and chlorides which will be a measure of the approach to a successful restoration. Kinetic data derived from the test well effluent are to serve as a verification of the quantities derived from the observer effluents. Such redundancy is necessary to adequately quantify the non-ideal flow that is almost certain to be present in an actual ore body. Between the push and the pull cycles there can be a "soak" cycle where no fluids are injected or produced. For the current test the soak cycle will be relatively short; we, therefore, neglect it in this paper.

In the following sections we discuss the mathematical model we propose as an analytical tool, sample results of this model, the effects of some typical non-ideal behavior, and the generation of the type curves.

Mathematical Formulation

Assumptions

The following assumptions have been made in the development of the simulation model:

1) The theory is presented in radial geometry. Permeability varies with depth only, and porosity is constant.

2) The fluid and ore are single phase and incompressible.

3) Mobilities and densities of the injected and formation fluid are equal and constant.

4) Mass transfer rates are assumed to be large enough such that reactions are kinetically controlled.

Push Cycle

The concentration balances for the various species in the fluid phase[10], subject to the above assumptions, are for the oxidant,

$$\phi \frac{\partial C_{ox}}{\partial t} + u \frac{\partial C_{ox}}{\partial r} - \frac{1}{r}\frac{\partial}{\partial r}\left(K_\ell \phi r \frac{\partial C_{ox}}{\partial r}\right) - \frac{\partial}{\partial z}\left(\phi K_t \frac{\partial C_{ox}}{\partial z}\right) = \sum_{i=1}^{N} \phi a_i R_i \qquad (1)$$

and for minerals i=1...N,

$$\phi \frac{\partial C_i}{\partial t} + u \frac{\partial C_i}{\partial r} - \frac{1}{r}\frac{\partial}{\partial r}\left(K_\ell \phi r \frac{\partial C_i}{\partial r}\right) - \frac{\partial}{\partial z}\left(\phi K_t \frac{\partial C_i}{\partial z}\right) = -\phi R_i \qquad (2)$$

From left to right in equations (1) and (2) the terms represent species accumulation, radial convection, radial and transverse dispersion, and source terms.

The longitudinal and transverse dispersion coefficients are given by[11]

$$K_\ell = \frac{D_o}{F\phi} + \frac{\alpha_\ell}{\phi}|u| \qquad (3)$$

$$K_t = \frac{D_o}{F\phi} + \frac{\alpha_t}{\phi}|u| \qquad (4)$$

where α_ℓ and α_t are longitudinal and transverse dispersivities. In this study the molecular diffusion term D_o has been neglected. For cases in which exceedingly low injection rates are used, this term must be included.

The initial and boundary condition for equations (1) and (2) in the push cycle are

$$C_{ox}(r,0) = 0 \ , \ \left(\frac{\partial C_{ox}}{\partial r}\right)_{r=R} = 0 \ ,$$

$$(C_{ox})_{r=r_w} = C_{ox}^{Inj} \qquad (5a)$$

for the oxidant, and for the minerals

$$C_i(r,0) = 0 \ , \ \left(\frac{\partial C_i}{\partial r}\right)_{r=R} = 0 \ ,$$

$$(C_i)_{r=r_w} = 0 \ ,$$

$$W_i(r,0) = W_{iI} \qquad (5b)$$

The second expression in these equations expresses the continuity of mass flux at some hypothetical boundary R. This boundary is much greater than R_o, the observer radius. Other symbols in equations (1) - (5) are given in the nomenclature.

The stationary phase material balance equation is given by the first-order reaction[12,13]

$$(1-\phi)\frac{\partial W_i}{\partial t} = \phi R_i = -k_i C_{ox} \Delta W_i \cdot \rho_o (1-\phi) \qquad (6)$$

where k_i is reaction rate constant of mineral i and $\Delta W_i = (W_i - W_{iR})$. W_{iR} is the residual, unleachable concentration of mineral i which will vary with each formulation. The W_i are ore concentrations expressed on a mass of i/mass ore basis. Because of the assumption of a kinetically-controlled reaction, the bulk oxidant concentration appears in equation (6), rather than a surface concentration[5]. More complete reaction rate expressions including mass transfer effects have been written and will perhaps be required to adequately represent a particular test. Equations (1) and (2) may be made dimensionless by introducing the following variables:

$$r_D = \frac{r}{R_o} \ ; \ z_D = \frac{z}{H} \ , \ t_D = \frac{tQ}{\pi H R_o^2 \phi} \qquad (7a)$$

$$C_{oxD} = \frac{C_{ox}}{C_{ox}^{Inj}} \ ; \ \Delta W_{iD} = \frac{\Delta W_i}{\Delta W_{iI}} \qquad (7b)$$

Here R_o is the distance from the test well to the observation well and t_D is a dimensionless time expressed as the cumulative volume of fluid injected or produced divided by the pore volume. If the well radius is small with respect to R_o the pore volume is $\pi\phi H R_o^2$. Substituting equations (1) and (2) and using these dimensionless variables yields, for oxidant,

$$\frac{\partial C_{ox}}{\partial t_D} + \frac{k}{\bar{k}}\frac{1}{2r_D}\frac{\partial C_{oxD}}{\partial r_D} - \frac{k}{\bar{k}}\frac{1}{2r_D}\frac{\alpha_\ell}{R_o}\frac{\partial^2 C_{oxD}}{\partial r_D^2} - \frac{1}{2r_D}\frac{R_o \alpha_t}{H^2}\frac{\partial}{\partial z_D}\left(\frac{k}{\bar{k}}\frac{\partial C_{oxD}}{\partial z_D}\right) = -\sum_{i=1}^{N} a_i \frac{k_i \Delta W_{iI}}{\frac{Q}{(\pi H R_o^2 \phi)}} \Delta W_{iD} C_{oxD} \rho_o \frac{(1-\phi)}{\phi} \qquad (8)$$

and for minerals i=1...N,

$$\frac{\partial C_i}{\partial t_D} + \frac{k}{\bar{k}}\frac{1}{2r_D}\frac{\partial C_i}{\partial r_D} - \frac{k}{\bar{k}}\frac{1}{2r_D}\frac{\alpha_\ell}{R_o}\frac{\partial^2 C_i}{\partial r_D^2} - \frac{1}{2r_D}\frac{R_o\alpha_t}{H^2}\frac{\partial}{\partial z_D}\left(\frac{k}{\bar{k}}\frac{\partial C_i}{\partial z_D}\right) = \frac{k_i C_{ox}^{Inj}}{\frac{Q}{(\pi H R_o^2 \phi)}} \cdot \Delta W_{iI} \rho_o \frac{(1-\phi)}{\phi} \Delta W_{iD} C_{oxD} \qquad (9)$$

Here we introduce

$$C_{ipk} = \Delta W_{iI}\rho_o \frac{(1-\phi)}{\phi} \quad (10)$$

and a dimensionless fluid phase mineral concentration $C_{iD} = C_i/C_{ipk}$. The quantity C_{ipk} is the peak solution concentration the mineral i would attain if it were entirely dissolved by one pore volume of lixiviant. Another way of looking at C_{ipk} is that it is the leachable mineral i concentration expressed in solution concentration units. The u's in equations (1) and (2) were eliminated by[14]

$$u = \frac{Q}{2\pi rH}\left(\frac{k}{\bar{k}}\right) \quad (11)$$

The ratio $(k/\bar{k})$ is a normalized permeability function (recall k can vary with z). Equations (8) and (9) can further be simplified using the above definitions to yield, for oxidant,

$$\frac{\partial C_{oxD}}{\partial t_D} + \frac{k}{\bar{k}}\frac{1}{2r_D}\frac{\partial C_{oxD}}{\partial r_D} - \frac{k}{\bar{k}}\frac{1}{2r_D}\frac{\alpha_\ell}{R_o}\frac{\partial^2 C_{oxD}}{\partial r_D^2} - \frac{1}{2r_D}\frac{R_o\alpha_t}{H^2}\frac{\partial}{\partial z_D}\left(\frac{k}{\bar{k}}\frac{\partial C_{oxD}}{\partial z_D}\right) = -\sum_{i=1}^{N} D_i V_i \Delta W_{iD} C_{oxD} \quad (12)$$

for minerals i=1...N,

$$\frac{\partial C_{iD}}{\partial t_D} + \frac{k}{\bar{k}}\frac{1}{2r_D}\frac{\partial C_{iD}}{\partial r_D} - \frac{k}{\bar{k}}\frac{1}{2r_D}\frac{\alpha_\ell}{R_o}\frac{\partial^2 C_{iD}}{\partial r_D^2} - \frac{1}{2r_D}\frac{R_o\alpha_t}{H^2}\frac{\partial}{\partial z_D}\left(\frac{k}{\bar{k}}\frac{\partial C_{oxD}}{\partial z_D}\right) = D_i \Delta W_{iD} C_{oxD} \quad (13)$$

In equations (12) and (13) there appears a dimensionless function $k/\bar{k}$, two dimensionless groups involving the dispersivities, α_ℓ/R_o and $R_o\alpha_t/H^2$, and the two dimensionless groups, D_i and V_i. The D_i are Damkohler numbers based on the injected oxidant concentration defined as $D_i = (k_i C_{ox}^{inj})/(Q/\pi HR_o^2\phi)$. The Damkohler numbers are the ratio of reaction rate of mineral i to the rate of convective transport. V_i, the oxygen requirement for mineral i, is $V_i = a_i C_{ipk}/C_{ox}^{inj}$. It is the ratio of the reactant needed to dissolve all of leachable mineral per unit volume of porous media to the fluid reactant available in the pore space.

The boundary conditions become in dimensionless form,

$$C_{oxD}(r_D,0) = 0\ ,\ \left(\frac{\partial C_{oxD}}{\partial r_D}\right)_{r_D=R/R_o} = 0\ ,$$

$$(C_{oxD})_{r_D=r_w/R_o} = 0\ ,$$

$$(C_{oxD})_{r_D=r_w/R_o} = 1 \quad (14a)$$

for the oxidant, and for the minerals

$$C_{iD}(r_D,0) = 0\ ,\ \left(\frac{\partial C_{iD}}{\partial r_D}\right)_{r_D=R/R_o} = 0\ ,$$

$$(C_{iD})_{r_D=r_w/R_o} = 0\ ,$$

$$W_{iD}(r_D,0) = 1 \quad (14b)$$

The parameters R/R_o and r_w/R_o introduced by this procedure do not affect the results of the calculations as long as they are large and small, respectively.

Pull Cycle

For this cycle Q = Q' is negative and the non-dimensional procedure is exactly as for the push cycle. Owing to the absolute value signs in equations (3) and (4), however, the signs of the dispersion terms are reversed. We maintain the Damkohler number based on the push cycle in what follows and treat the ratio Q/Q' as an additional parameter. The boundary conditions for the pull cycle are the same as in the push cycle except the conditions at the well become

$$\left(\frac{\partial C_{oxD}}{\partial r_D}\right)_{r_D=r_w/R_o} = \left(\frac{\partial C_{iD}}{\partial r_D}\right)_{r_D=r_w/R_o} = 0 \quad (15)$$

Equation (15) expresses the continuity of flux across the well bore boundary.

Method of Solution

Equations (9) and (10) are solved by a finite difference method. A totally implicit formulation has been used with up-wind differencing scheme so that the model is unconditionally stable.

Results

Restoration

In a previous work we considered a one-well test to measure those factors which determine the ease of restoration after in situ leaching[9]. The one-well test did not include observation wells as does the minitest. We determined that it is not possible to measure cation exchange capacity and selectivity simultaneously, though if one is known the other follows readily. Before proceeding to the consideration of leaching parameters we summarize how we will determine restoration parameters in the minitest. The conclusions presented here are based on other work[14,15].

Anion is a non-reacting species and is considered to be a tracer. The tracer's effluent history at the observation well is affected by the following parameters:

1) The layer permeability function determined by $k/\bar{k}$.

2) The longitudinal dispersivity α_ℓ which determines K_ℓ.

3) The transverse dispersivity α_t affecting K_ℓ.

The combined affect on the tracer of the above three quantities has been described by Lake and Hirasaki[14]. Under certain conditions, these quantities may be determined from the tracer's effluent at the observers[15]. The test well effluent is relatively independent of layering and α_t; thus, it is possible to determine α_ℓ from the test well (pull cycle) effluent and the other two from the observer well effluent. The pore volume can also be determined from tracer effluent history at the observers and compared with that calculated from thickness, porosity, and observation well distance. These parameters must be known prior to interpreting all exchanging or reacting species data.

The cation exchange parameters can be evaluated from the solution concentrations at the observation well using adsorbing species material balances as described by Hill and Lake[8].

Leaching Phase

The calculations reported here assume that, except for the oxidant, the composition of the leach solution remains essentially unchanged during the minitest. This assumption is valid provided a preflush designed to displace divalent cations from the vicinity of the test well bore is used. This step has a number of advantages including reducing precipitation of the leach solution and, as noted, provides important data which will help to characterize the reservoir and to define the restoration parameters. Preflushing is a recommended step in carrying out a minitest. We assume, further, that uranium is the only reacting species present. Other data that are used in all the simulator runs are $\alpha_\ell/R_o = 0.045$, $R_o\alpha_t/H^2 = 0.0$, and $Q_{production}/Q_{injection} = 4.0$.

For constant injection the normalized uranium solution concentration will depend primarily on V_u and D_u provided other minerals which compete for oxidant are absent. There are other variables such as dispersion which play a secondary role in fixing the uranium concentration and these will be ignored in the discussion to follow.

Figure 2 shows the normalized uranium solution concentrations measured at the observation well. Increasing the Damkohler number will, as seen, result in an increased amount of uranium dissolution between the injection and observation wells. This trend is evidenced by the greater solution concentrations observed at early times; however, it should be noted that the uranium concentration is normalized so that the area under each one of the concentration histories measured at the observation well must be approximately unity provided enough leach solution is injected to dissolve all of the uranium between the test and observation wells. Thus the uranium concentrations at the observation wells must intersect ultimately when plotted as a function of t_D. The area of the normalized uranium concentration curve can easily be shown to be given by

$$\int_0^\infty C_{uD}(t_D,1)dt_D = 1 - \frac{2\pi K_\ell H\phi}{Q}\int_0^\infty \left.\frac{\partial C_{uD}}{\partial r_D}\right|_{r_D=1} dt_D \qquad (16)$$

The second term on the right side of equation (16) represents the contribution to the flux owing to dispersion and this contribution is normally small compared to unity. Therefore the areas under the uranium histories should all be equal to one for t_D large. This condition can be used to verify the correctness of the C_{upk}, which defines the recoverable (using the leach solution tested) uranium, if an estimate of this quantity is available. If an estimate is not available, equation (16) can be used, but will in general require extrapolation of the observation well concentrations to long times. This may be difficult to do in practice with any precision. As will be seen better methods are available for determining C_{upk}.

Increasing the oxidant requirement number, V_u, tends to reduce the rate at which uranium is dissolved from the zone between the injection and observation wells. The oxidant requirement determines the rate at which the reaction front, which is arbitrarily defined as the position at which the concentration of uranium (solid phase) or oxidant changes most rapidly, progresses away from the injection well. For a homogeneous system the analysis presented by Hekim and Fogler[12] and by Bommer and Schechter[13] can be extended to a radial system. It can be shown without assumption that the dimensionless distance to which the reaction front has penetrated is given by

$$r_{fD}^2 = \left(\frac{1}{(1 + V_u)}\right)t_D \qquad (17)$$

Since V_u does not depend on the reaction rate, k_u, the position of the reaction front is independent of the Damkohler number; however, the width of the reaction zone is a sensitive function of the reaction rate. Thus for small Damkohler numbers, the width of the reaction zone may exceed the spacing between the observation and test wells. This situation is not desirable since oxidant penetrates beyond the observation wells almost from the initiation of the process. This can make it extremely difficult to extract the desired kinetic and stoichiometric quantities.

Assuming an adequate Damkohler number, the influence of the oxidant requirement number is best seen during the pull cycle since equation (17) indicates that as V_u increases, the time required to produce the maximum uranium concentration during the pull cycle will decrease.

This is true because as V_u decreases, the distance the reaction front penetrates increases. This anticipated relationship between the peak concentration arrival time and the oxidant requirement number is demonstrated in Figure 3.

The correlation between the peak arrival time and the oxidant requirement number should be useful in helping to establish the leachable uranium since all of the factors appearing in V_u are known from laboratory tests with the exception of ΔW_{uI}. The value of this correlation is greatly enhanced since it does not depend on the Damkohler number as shown by the results presented in Figure 4. This independence of the peak arrival time from the Damkohler number is not surprising since r_{fD} as defined by equation (17) is a function of V_u but not D_u. This is an important result since one procedure for evaluating the two important parameters could involve determination of V_u from knowledge of the peak arrival time and D_u by history matching the production effluent history. This method is discussed in a subsequent section.

To determine favorable conditions for amplifying the differences in uranium peak arrival times at the test well during the pull cycle, a series of computer runs have been performed. The peak arrival time is shown in Figure 5 as a function of V_u and t_D^{Inj}, the total number of leach solution pore volumes injected. This figure shows that for small values of V_u (less than 2), about 3 pore volumes of leach solution may be adequate to yield precise arrival times. However, for larger V_u, more leach solution is required to permit an adequate evaluation of V_u based on the measured peak arrival time.

There is a second requirement which has not been discussed which must be understood. If the Damkohler number is small, then, as seen in Figure 4, the maximum concentration is not sharply defined and this will tend to obscure the peak arrival time. If at all possible, the minitest should be conducted with the controllable system parameters fixed so that $D_u > 0.2$. This may mean increasing the injected oxidant concentration C_{ox}^{Inj} or decreasing the injection rate Q. Any of the procedures recommended in this paper for determining V_u and D_u are subject to reduced accuracy if the Damkohler number is allowed to become too small.

Figure 6 shows the oxidant production history as measured in the produced fluids at the test well during the pull cycle. This history has particular importance in cases for which there are other minerals which are competing for oxidant; however, an evaluation of the oxidant production history is of value even in the absence of competitors. By material balance we can write

$$C_{ox}^{Inj}\left[t_D^{Inj} - \int_{t_D^{Inj}}^{A} C_{oxD}\,dt_D\right] = a_u C_{upk} \int_{t_D^{Inj}}^{A} C_{uD}\,dt_D \qquad (18)$$

This equation can be used in two ways. If the stoichiometric coefficient is known, then by integrating the oxidant concentration as a function of t_D, a check of the uranium produced is available. Often the precise mineral form of the uranium ore is not known. Then by application of equation (18), the stoichiometric coefficient may be determined.

Non-uniform Permeabilities

All the results shown so far have been for a homogeneous reservoir $k = \bar{k}$. Since this condition is never encountered in a actual ore body, we must be able to interpret the minitest results in a heterogeneous ore body. Permeability heterogeneity can occur in any coordinate direction. Permeability variations in the vertical direction--ore body layering--have the most effect on fluid flow and is the only type considered here.

Figure 7 shows dimensionless uranium concentration at the observer (7a) and the test well (7b) for a two-layer, and a homogeneous ore body. Even though the permeability contrast is only two-to-one, the layering has completely obliterated the peak in the uranium effluent. Thus, it is evident that any analysis based on the peak arrival time at the observation wells will be subject to considerable error. The analysis procedure shown in the following section utilizes test well data.

Figure 7b shows that the effects of layering are, themselves, largely damped out in the test well effluent. This is because the high permeability layers flow equally well in both the push and pull cycles. The absence of permeability effects in the test well effluent make this data an important factor in analyzing layered ore bodies.

Type Curve Analysis

Another way to analyze the effluent from the observation wells is to use "type" curves, a procedure introduced in the petroleum literature to analyze pressure transient behavior in wells[6]. For the special case of a single reacting species in an observation well effluent the technique is readily applicable here also.

Consider the dimensionless effluent curves shown in Figure 8. These type curves were generated by the simulation model for three values of the Damkohler number and two values of the oxygen requirement factor. The curves are also for fixed values of the dispersion parameter and the layering function $k/\bar{k}$, quantities which were determined from the anion data (see above). Except for the log-log scale, which is a crucial feature in type curve analysis, the curves are the same as those in Figure 2. Note that six combinations of D_u's and V_u's result in curves with different shapes, also a crucial feature.

The type curve analysis proceeds as follows:

1) Plot the effluent uranium concentration (in kg/m^2) vs. time (in hours) on log-log axes having the same scale as the type curve, Figure 7. To facilitate step 2 this plot should be transparent.

2) Place the step 1 data plot on top of the type curve plot. Keeping the respective axes of both plots parallel slide the data plot until

the data points coincide with one of the type curves. This translation is permissible since the normalization of log-log curves results in curves having the same shape, but translated by linear distances equal to the logarithm of the normalization factors. Since the shape of each type curve is uniquely determined by the particular D_u and V_u, these quantities are now determined. Knowing D_u and V_u it is possible to calculate the rate constant k_u and the $a_u C_{upk}$ product from their respective definitions. The V_u value determined here may be compared with that determined from the arrival time analysis, or the type curves themselves may be generated for a fixed V_u.

3) With the plots aligned as in step 2, select a "match" point, a point which coincides on both plots. The match point is to determine the normalization factors of the type curve and can be anywhere on the plot. The y-axis normalization factor is then

$$C_{upk} = (C_u)_{MP}/(C_{uD})_{MP} \quad (19)$$

where the $(C_u)_{MP}$ and $(C_{uD})_{MP}$ are the y-coordinates of the match point on the data and type curve plots, respectively. Similarly, the pore volume may be found as

$$PV = \frac{(Qt)_{MP}}{(t_D)_{MP}} \quad (20)$$

where $(Qt)_{MP}$ and $(t_D)_{MP}$ are the x-coordinates of the match point on the data and type curve plot.

The PV value from equation (20) will be compared to that obtained from the tracer effluent. Since C_{upk} is now known, a_u follows from the information in step 2. Finally, the leachable uranium concentration ΔW_{uI} may be calculated from equation (10).

The type curve procedure thus allows the determination of k_u, a_u, ΔW_{iI}, and PV. The accuracy of this interpretation procedure depends heavily on there being a unique curve for each (V_u, D_u) pair in the type curves and the absence of competing minerals. If the competing minerals are present, the analysis is considerably complicated. More data (the competitor's effluent history) is available, however, so that the extension to this case is not impossible. Future publications will address this question.

Conclusions

For a single reacting species the effluent histories at the observation and test wells are determined by fluid flow properties, a Damkohler number, and an oxidant requirement factor. These factors contain the stoichiometric constant, the kinetic constant, and the recoverable uranium concentration. Procedures which unambiguously determine D_u and V_u, therefore, will determine quantities directly related to the leaching efficiency of a particular ore body.

Two interpretation procedures were presented: based on peak arrival time at the observation well, and type curve analysis. Both procedures rely on large Damkohler numbers so that the leaching phase of the minitest must be done at low rates or large oxidant concentrations. If permeability layering exists, the type curve analysis procedure is preferred for the observation well effluents. If the ore body is homogeneous both procedures are applicable and will be used to provide a redundancy of values.

Acknowledgments

The authors want to thank the Bureau of Mines for its support of this project under the terms of the United States Department of the Interior, Contract H0282016. Professor R. S. Schechter holds the Dula and Ernest Cockrell Chair in Engineering.

Nomenclature

a_i	=	stoichiometric coefficient of mineral i, mass of oxidant/mass of mineral i, M/M
C_i	=	mobile phase concentration of mineral i, M/L^3
C_{ox}	=	oxidant concentration in mobile phase, M/L^3
D_i	=	Damkohler number of mineral i, dimensionless
D_o	=	molecular diffusion coefficient, L^2/T
F	=	formation resistivity factor
H	=	total reservoirs thickness, L
k	=	permeability, L^2
$\bar{k}$	=	thickness weighted average permeability
K_ℓ	=	longitudinal dispersion coefficient, L^2/t
K_t	=	transverse dispersion coefficient, L^2/t
N	=	number of minerals
PV	=	pore volume, L^3
Q	=	volumetric injection or production rate, L^3/T
r_{fD}	=	dimensionless distance penetrated by the reaction front
r	=	radial distance from the test well, L
R	=	hypothetical boundary distance, L
R_o	=	observer distance from the test well, L
t_D^{inj}	=	dimensionless injection time
t_{iD}	=	dimensionless peak concentration arrival time of mineral i at the test well during pull cycle
u	=	flux at position r, L/t
V_i	=	oxidant requirement factor of mineral i, dimensionless
W_i	=	concentration of mineral i, mass/mass of rock
W_{iR}	=	unleachable concentration of mineral i, mass/mass of rock
ΔW_{iI}	=	leachable mineral i initially present, mass/mass of rock

Greek Symbols

ϕ	=	porosity
α_ℓ	=	longitudinal dispersivity, L
α_t	=	transverse dispersitivity, L
ρ_o	=	ore density, M/L^3

Subscripts

i=1...N	=	number of minerals
I	=	initial
D	=	dimensionless
ox	=	oxidant
u	=	uranium

References

1. Taylor, William R., Regulation of In Situ Uranium Mining for Purposes of Protecting Ground Water Quality, SPE-7536, presented at 53rd Annual SPE Meeting, Oct. 1-3, 1978.
2. Hill, A. D., Silberberg, I. H., Walsh, M. P., Breland, W. M., Humenick, M. J., and Schechter, R. S., Restoration of Uranium In-Situ Leaching Sites, Soc. Pet. Eng. J., Vol. 4, No. 20, August 1980, pp. 221-227.
3. Tweeton, Daryl R., Guilinger, Terry R., Breland, W. Michael, and Schechter, Robert S., The Advantages of Conditioning an Ore Body with a Chloride Solution before In Situ Uranium Leaching with a Carbonate Solution, SPE-9490, presented at 55th Annual SPE Meeting, Sept., 21-24, 1980.
4. Bommer, P. M. and Schechter, R. S., Mathematical Modeling of In Situ Uranium Leaching, Soc. Pet. Eng. J., Vol. 19, No. 6, December 1979, pp. 393-400.
5. Tatom, Aline, Schechter, R. S., and Lake, Larry W., Factors Influencing the In Situ Acid Leaching Rates of Uranium Ores, this volume, pp. 133-149.
6. Earlougher, Robert C., Jr., Advances in Well Test Analysis, SPE Monograph, Vol. 5, Dallas, 1977.
7. Jones, S. C. and McAtee, R. W., A Novel Single Well Field Test of a Micellar Solution Slug, Jour. Pet. Tech., November 1972, pp. 1371-1376.
8. Hill, H. J. and Lake, L. W., Cation Exchange in Chemical Flooding: Part 3 - Experimental, Soc. Pet. Eng. J., Vol. 18, No. 6, December 1978, pp. 445-556.
9. Kabir, M. I., Lake, L. W., and Schechter, R. S., Evaluation of the One Well Uranium Leaching Test, SPE-9486, presented at 55th Annual SPE Meeting, Sept. 21-24, 1980.
10. Lake, L. W., Pope, G. A., Carey, G. F., and Sepehrnoori, K., Isothermal, Multiphase, Multicomponent Fluid-Flow in Permeable Media. Part I: Description and Mathematical Formulation, Center for Enhanced Oil and Gas Recovery Research, The University of Texas at Austin.
11. Perkins, T. K. and Johnson, O. C., A Review of Diffusion and Dispersion in Porous Media, Soc. of Pet. Eng. J., Vol. 3, No. 1, March 1963.
12. Hekim, Y. and Fogler, H. S., On the Movement of Multiple Reaction Zones in Porous Media, AIChE J., Vol. 26, 1980, pp. 403.
13. Schechter, R. S. and Bommer, P. M., Optimization of Uranium Leach Mining, SPE-9488, presented at 55th Annual Fall SPE Meeting, Sept. 21-24, 1980.
14. Lake, L. W. and Hirasaki, G. J., Taylor's Dispersion in Stratified Porous Media, Soc. Pet. Eng. J., Vol. 21, August 1981.
15. Ravnaas, R. D., Field Scale Measurements of Dispersion Coefficients, M.S. Thesis, The University of Texas at Austin, 1981.

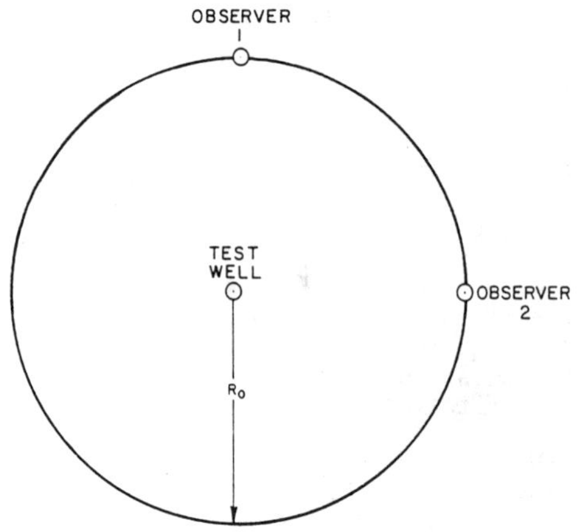

Figure 1 - Schematic illustration of a mini field test.

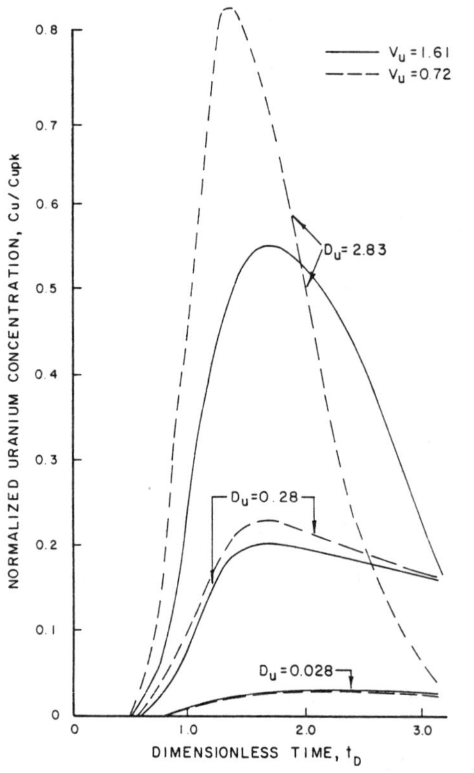

Figure 2 - Dimensionless uranium effluent histories at the observer during push cycle.

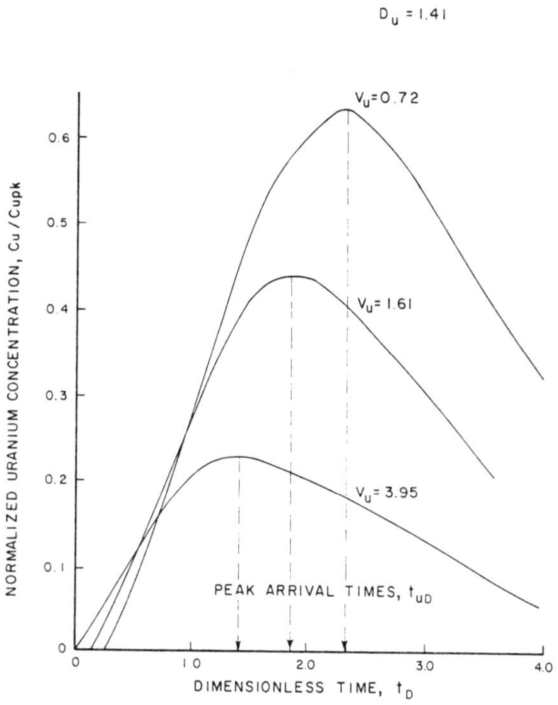

Figure 3 - Dimensionless uranium effluent histories at the test well during pull cycle for various oxidant requirements.

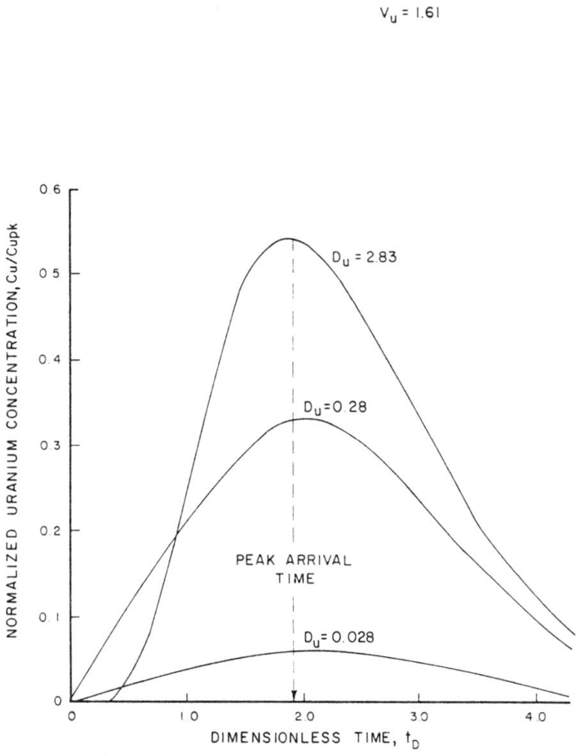

Figure 4 - Dimensionless uranium effluent histories at the test well during pull cycle for various Damkohler numbers.

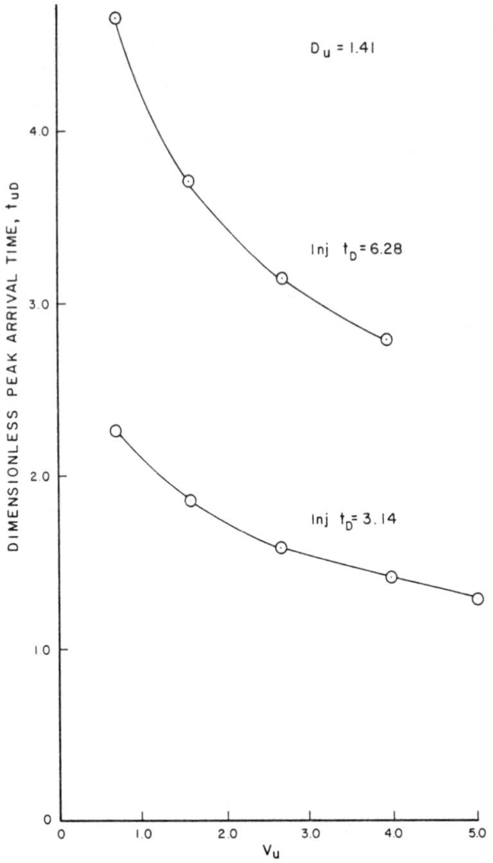

Figure 5 - Peak arrival times at the test well during pull cycle as a function of oxidant requirement for various dimensionless injection times.

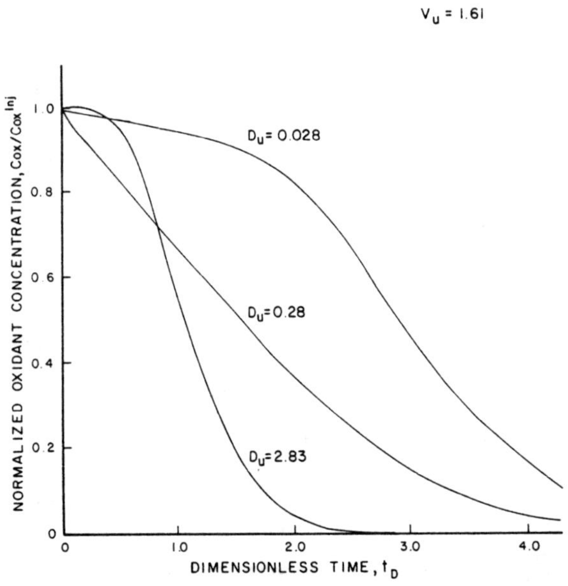

Figure 6 - Oxidant effluent histories at the test well during pull cycle.

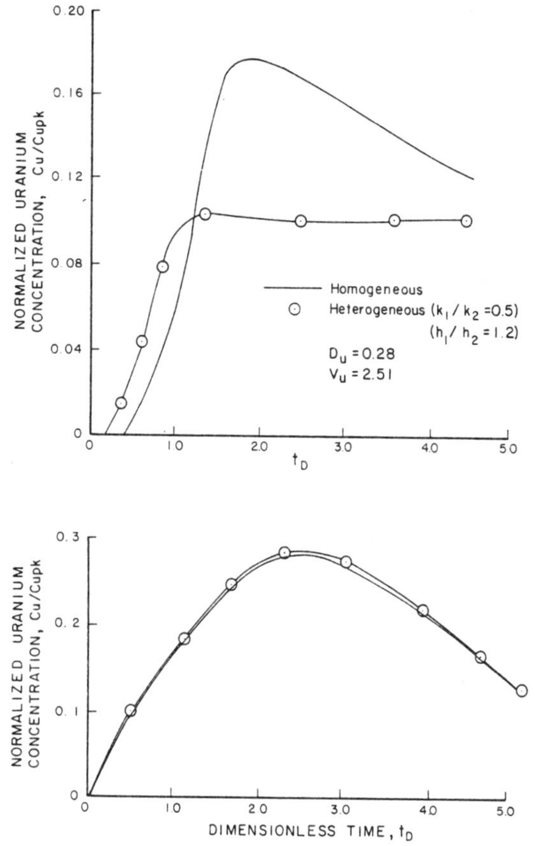

Figure 7 - Comparison of uranium effluent histories for a homogeneous and heterogeneous case at (top: 7a) observer and (bottom: 7b) test well.

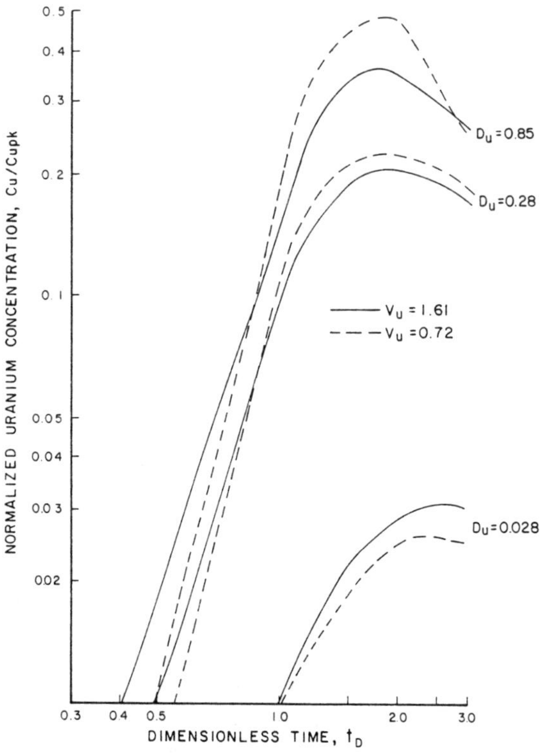

Figure 8 - Type curve for uranium effluent histories at the observer.

Chapter 9

A STATISTICAL METHODOLOGY FOR ASSAY OF GROUND–WATER QUALITY IN URANIUM SOLUTION MINES

Ray A. Rothrock
Exxon Minerals Company
Houston, Texas 77001

ABSTRACT

Ground-water quality is stochastic in nature. Therefore, the assessment of ground-water quality lends itself to statistical methods. The data base employed in this analysis is actual data from a proposed uranium solution mine. Assuming normal (Gaussian) distributions, standard statistical procedures are used to compare the means of data distributions for defining the baseline, for demonstrating injected chemical excursion, and for demonstrating restoration. Variance analysis is used to illustrate spatial and temporal changes in the ground-water quality. Although this study is based on a limited number of samples, the advantages of using this type of methodology are clear. The use of log-normal distributions might improve this methodology.

INTRODUCTION

The *in situ* mining of uranium is the process whereby chemical lixiviant is injected by means of wells into, and flowed through, an ore body. The lixiviant reacts with uranium which has been previously oxidized making the uranium soluble. The solution is then brought to the surface by means of producing wells.

In situ mining disturbs the ground-water quality in the immediate area. Since ground water is a natural resource used by man, care must be taken not to allow the lixiviant to escape the local mining area. Also, when the ore body is mined out, the area should be restored to as near its original state as is reasonably achievable. The return of ground-water quality to its original state can occur naturally over a long period of time or be enhanced using appropriate technological means.

To monitor and restore an *in situ* mined area, the original or pre-mining condition of the ground water must first be known. This involves the collection of water samples from the area to be mined and its surroundings for some time prior, during, and after the mining operation. Since ground-water quality is stochastic in nature, the analysis of the ground-water samples lends itself to statistical procedures. This paper illustrates some of the statistical procedures (using real data) which can be used for ground-water analysis.

Statistical methods are used to compare the means of sample distributions for both excursion detection and demonstration of restoration. Variance analysis is used to indicate temporal and spatial differences in the chemical constituents of ground water. Also, the paper contains a discussion on appropriate sample size.

The data for the example calculations are actual ground-water samples of a potential uranium solution mine taken prior to any mining or testing. While this data is quite limited in quantity, it does provide a demonstration of how statistical concepts can be applied to data analysis.

ASSUMPTIONS AND BASES

Most important to any analysis, statistical or otherwise, are the assumptions made. It is necessary to make certain assumptions in order to use statistical methods of analysis. To say water quality is stochastic in nature is to say that variations of water quality are governed in some manner, totally or partially, by chance. Ground-water quality is greatly affected by the surrounding materials, i.e. soil and rock. The composition can change with location and with time if the water is moving or if chemical reactions are occurring.

Frequently the assumptions made for a statistical analysis inherently put restraints on the data in order for the assumptions to remain valid. Furthermore, the data gathering method or experiment should be developed with the use of statistics in mind. This can help eliminate bias in the data which might invalidate an assumption. Unfortunately, the data used in this work does not comply with all the constraints required by the assumptions yet to be addressed. Furthermore, this analysis was performed subsequent to the taking of samples.

The three major assumptions are: the data are random, the data are normally distributed, and the sample size is adequate to perform the tests. These will each be discussed in some detail.

The first assumption that the data are random and independent when measured is required if one is to use parametric statistics for analysis. Even though chance probably plays a major role in determining ground-water quality at a specific place and time, the fact that it is not totally by chance partially invalidates this assumption. Sanders (1980) gives an excellent discussion of the variability and randomness of water quality for rivers or streams. The precepts developed by him are applied and assumed to hold for ground water in this paper.

The second assumption of "normally" distributed data implies the first assumption of randomness. There are several types of distributions which could be assumed to represent the data besides the normal distribution. An example of a normal distribution is illustrated in Figure 1, "Normal or Gaussian Distribution Function." The normal distribution was primarily assumed for simplicity of calculation. Some regulatory agencies are recommending other distributions or nonparametric tests for use in ground-water quality assessment instead of a normal distribution. One such test using nonparametric statistics is called the Wilcoxon-Mann-Whitney test. This test is quite powerful for comparing two small data sets. Another distribution that would be more appropriate is the log-normal distribution. Since no negative values of ground-water quality can occur, log-normal would be a more appropriate method if parametric statistics are used.

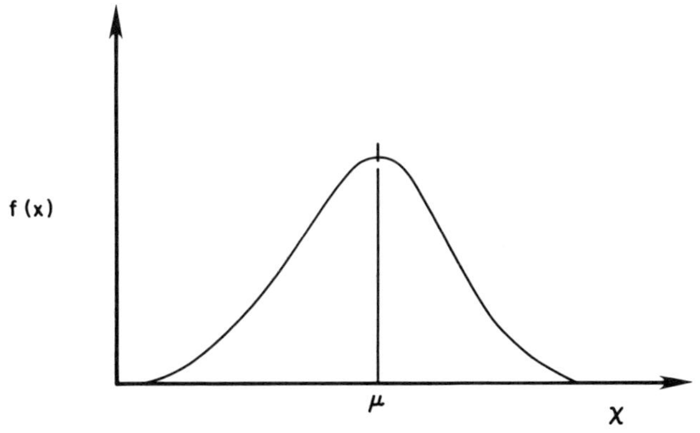

f(x) = Probability distribution function of random observation

X = Random observation

μ = Mean of the distribution

FIGURE 1 - NORMAL OR GAUSSIAN DISTRIBUTION FUNCTION

TABLE 1

GROUND WATER CHEMICAL CONCENTRATION DATA

Chemical Concentrations (mg/l or 10^{-3} kg/m^3)

Standard Chemicals	Hole P-7				Hole P-9			
	7/78	8/78	11/78	12/78	7/78	8/78	11/78	12/78
Aluminum	8.1	3.7	0.7	8.9	16.0	4.4	0.7	8.3
Arsenic*	0.01	0.01	0.01	0.01	0.01	0.02	0.01	0.01
Barium*	0.2	0.1	0.1	0.1	0.2	0.1	0.01	0.1
Boron	0.8	0.9	0.7	0.7	0.8	0.7	0.8	0.8
Cadmium*	0.005	0.0001	0.003	0.001	0.001	0.001	0.001	0.001
Chloride	21.0	33.0	34.0	33.0	40.0	36.0	39.0	37.0
Chromium	0.026	0.010	0.019	0.009	0.009	0.008	0.008	0.004
Cobalt*	0.01	0.01	0.01	0.01	0.01	0.01	0.01	0.01
Copper	0.055	0.003	0.010	0.009	0.007	0.006	0.005	0.010
Cyanide*	0.1	0.1	0.1	0.1	0.1	0.1	0.1	0.1
Fluoride	0.82	0.86	0.90	0.94	0.84	0.82	0.86	0.88
Iron	16.0	2.8	2.3	3.1	28.0	4.0	1.5	2.8
Lead	0.014	0.017	0.013	0.011	0.022	0.12	0.019	0.016
Manganese	0.31	0.097	0.096	0.14	0.48	0.27	0.090	0.099
Mercury*	0.0004	0.0004	0.0004	0.0004	0.0004	0.0004	0.0004	0.0004
Molybdenum	0.004	0.002	0.004	0.004	0.085	0.025	0.010	0.009
Nitrate	0.6	0.1	2.5	0.3	0.6	0.2	3.6	0.7
Phenols	0.006	0.030	0.003	0.005	0.002	0.020	0.005	0.010
Selenium	0.01	0.01	0.01	0.01	0.01	0.01	0.01	0.01
Silver*	0.01	0.01	0.01	0.01	0.01	0.01	0.01	0.01
Sulfate	702	646	749	682	692	730	737	690
TDS	1,470	1,450	1,540	1,490	1,480	1,480	1,791	1,500
pH	8.43	8.32	8.31	--	8.51	8.14	8.45	--
Zinc	0.05	0.07	0.11	0.05	0.04	0.71	0.29	0.20
Nickel	--	--	--	0.01	--	--	--	0.01
Uranium	0.051	0.023	0.033	0.037	0.384	0.287	0.301	0.169

	Hole P-19				Hole Ø-10			
	7/78	8/78	11/78	12/78	7/78	8/78	11/78	12/78
Aluminum	4.6	3.1	0.8	6.5	1.8	3.2	0.5	4.0
Arsenic*	0.01	0.01	0.01	0.01	0.01	0.01	0.01	0.01
Barium*	0.1	0.1	0.01	0.1	0.2	0.1	0.01	0.1
Boron	0.7	0.7	0.6	0.7	0.9	0.8	0.9	0.6
Cadmium*	0.002	0.001	0.001	0.001	0.009	0.002	0.001	0.001
Chloride	45.0	44.0	44.0	42.0	36.0	33.0	39.0	38.0
Chromium	0.008	0.006	0.012	0.005	0.010	0.010	0.008	0.006
Cobalt*	0.01	0.01	0.01	0.01	0.01	0.01	0.01	0.01
Copper	0.010	0.006	0.006	0.013	0.006	0.004	0.006	0.012
Cyanide*	0.1	0.1	0.1	0.1	0.1	0.01	0.1	0.1
Fluoride	0.78	0.86	0.90	0.90	0.84	0.82	0.86	0.88
Iron	0.25	1.6	2.8	3.1	3.7	2.3	1.7	5.2
Lead	0.049	0.090	0.035	0.011	0.097	0.075	0.024	0.023
Manganese	0.57	0.12	0.082	0.17	0.21	0.11	0.082	0.15
Mercury*	0.0004	0.0005	0.0006	0.0004	0.0005	0.0004	0.0004	0.0014
Molybdenum	0.007	0.003	0.007	0.004	0.007	0.001	0.007	0.003
Nitrate	0.3	0.1	1.1	0.9	0.5	0.1	3.1	0.5
Phenols	0.001	0.016	0.003	0.004	0.003	0.008	0.015	0.005
Selenium	0.01	0.01	0.01	0.01	0.01	0.01	0.01	0.01
Silver*	0.01	0.01	0.01	0.01	0.01	0.01	0.01	0.02
Sulfate	708	725	731	684	702	704	711	691
TDS	1,530	1,460	1,410	1,420	1,590	1,490	1,740	1,420
pH	8.44	8.14	8.37	--	8.64	8.33	8.39	--
Zinc	0.08	0.28	0.16	0.13	0.30	0.35	0.16	0.11
Nickel	--	--	--	--	--	--	--	0.01
Uranium	0.235	0.030	0.039	0.027	0.041	0.028	0.054	0.021

* Some data were listed as "<0.01." These data are assumed to be "0.01."

The final assumption is most critical for any statistical test. This assumption is adequate sample size. Regardless of parametric or nonparametric statistics, one must have a reasonable data base from which to start. Unfortunately the data in this work does not conform to minimum sample sizes. For calculational purposes, the data set will be assumed large enough. A discussion of data size requirements dealing with this problem for normal statistics is given in a subsequent section.

The analysis which follows deals specifically with the chemical constituents of ground-water samples taken from the area of a potential uranium solution mine. The analysis does not include all the constituents of a water sample. The data included in this paper contain most of the chemicals about which a regulatory agency may be concerned.

Table 1, "Ground-Water Chemical Concentration Data", contains the results of a water sampling program for four wells taken over a six month period. These data are given in units of mg/l (10^{-3} kg/m^3), as are all units of chemical concentration throughout this work. This sampling program was not developed with statistical analysis in mind.

GROUND-WATER QUALITY ANALYSIS

Base Line

Base line chemical concentrations are used throughout the mine life as the standard for comparison for excursion and restoration. It is important to have a good data base for the base line since these comparisons will be made against it. Simple averaging and standard deviation calculations are employed to characterize the data with respect to a normal distribution.

The average and standard deviation were calculated using equations (1) and (2), respectively for each of the chemicals listed in Table 1. The standard statistical equations are:

$$\bar{X} = \frac{1}{N}\sum_{i=1}^{N} X_i \tag{1}$$

$$S^2 = \frac{1}{N-1}\sum_{i=1}^{N} (X_i - \bar{X})^2 \tag{2}$$

where $\bar{x}$ is the average, x_i is an observation i, N is the number of observations, and S is the standard deviation.

The data were lumped into one group for this calculation. This assumes no temporal or spatial variance in the data which subsequently will be proven false given the stated assumptions.

The results of calculating the average and standard deviation for each chemical are given in Table 2, "Average and Standard Deviation of the Data". This table also shows the averages for each well over all the dates sampled. A similar table could be produced grouping or classifying the data by date.

TABLE 2

AVERAGE AND STANDARD DEVIATION OF THE DATA

Chemical	Well Average A	B	C	D	Overall Average	Overall S
Aluminum	5.4	7.4	3.8	2.4	4.7	4.1
Arsenic*	0.01	0.01	0.01	0.01	0.01	0.0025
Barium*	0.10	0.10	0.08	0.10	0.10	0.06
Boron	0.8	0.8	0.7	0.8	0.8	0.1
Cadmium*	0.003	0.001	0.001	0.003	0.002	0.002
Chloride	30.0	38.0	44.0	37.0	37.0	6.0
Chromium	0.016	0.007	0.008	0.009	0.010	0.005
Cobalt*	0.01	0.01	0.01	0.01	0.01	0.0
Copper	0.019	.007	0.009	0.007	0.011	0.012
Cyanide*	0.1	0.1	0.1	0.1	0.1	0.0
Fluoride	0.88	0.85	0.86	0.85	0.86	0.04
Iron	5.8	9.1	1.9	3.2	5.0	7.1
Lead	0.014	0.044	0.046	0.055	0.040	0.036
Manganese	0.161	0.235	0.236	0.138	0.192	0.147
Mercury*	0.0004	0.0004	0.0006	0.0007	0.0005	0.0003
Molybdenum	0.004	0.032	0.005	0.005	0.011	0.020
Nitrate	0.9	1.3	0.6	1.1	1.0	1.1
Phenols	0.011	0.009	0.006	0.008	0.009	0.008
Selenium	0.01	0.01	0.01	0.01	0.01	0.0025
Silver*	0.01	0.01	0.01	0.01	0.01	0.0
Sulfate	695	712	712	702	705	26
TDS	1,488	1,563	1,455	1,560	1,516	108
pH	8.35	8.37	8.32	8.45	8.37	0.14
Zinc	0.07	0.31	0.16	0.23	0.19	0.17
Nickel**	0.01	0.01	0.01	0.01	--	--
Total Uranium	0.036	0.285	0.036	0.036	0.123	0.123

* Data contain items such as "<0.01." These are interpreted for averaging purposes to be "0.01."

** Data taken only on last date for each hole. Average and Standard Deviation are meaningless.

Since ground-water quality is stochastic, another sample could be made which would probably have a different average and standard deviation even though the samples are from the same base population. This could result purely from chance and demonstrates the need to compare the means of distributions for test. If the data sets were large enough, one could demonstrate statistical indifference thereby allowing one to pool all the data into the base line. The comparison of data sets is demonstrated shortly.

The mean of the base line cannot be exactly determined but the average can be considered equal to the mean if a large enough data set is taken. However, for a stated level of confidence, an interval for the mean can be determined. This interval represents the mean of the population from which the samples were drawn. This more truly represents the "average" concept for which many regulators strive when setting regulations regarding the definition of base line.

The t-test methodology can be used to determine the interval for the mean. The interval on which the mean is defined depends upon the average, standard deviation, sample size, and confidence level. The confidence level is a measure of the chance of being wrong in stating that the mean lies within the interval calculated. Literally, the probability of being wrong is one minus the confidence level. A confidence level of 95% indicates a 1 in 20 chance of being wrong.

The lower limit (LL) of the interval is defined by

$$LL = \bar{x} - tS/\sqrt{N} \quad (3)$$

and the upper limit (UL) is defined by

$$UL = \bar{x} + tS/\sqrt{N} \quad (4)$$

where t is the appropriate value from the t-test table for normal distributions and a given level of confidence. A statistics text book (Ostle and Mensing, 1975) contains a discussion of t values and test methodology.

The mean can be bracketed by a UL and LL, as in this case, or by being no higher than UL or lower than LL. The former case is known as a two-sided interval (TSI) and the latter two are known as one-sided intervals (OSI). Each case has a specified value for t. The appropriate application of TSI or OSI depends on the problem. The TSI is used here for base line.

t for TSI is written as

$$t_{1-\alpha/2,m} \quad (5)$$

and for OSI as

$$t_{1-\alpha,m} \quad (6)$$

M is the degrees of freedom which is equal to N-1. α is the probability of being wrong (1 - α is the confidence level associated with it).

Values of t are given in Table 3, "Select Values of the t Statistic," for various levels of confidence for both TSI and OSI cases. From Table 3, the value of t is seen to increase with confidence. From equations (3) and (4), the LL and UL for larger values of t become lower and higher, respectively. This follows from the logic; if for a higher confidence the mean, in fact, lies between LL and UL, then the difference between LL and UL must become larger to accommodate the increased confidence (or lower probability of being wrong.)

The results of the calculations for each of the chemicals given in Table 2 using the t-test for bracketing the mean are given in Table 4, "Summary of t-Test Results for Ground-water Quality Mean Calculations." Note that some of the LL results are negative. This illustrates either the error in assuming a normal distribution of concentrations for some of the chemicals, assuming no time or space dependence, inadequate sample size, or any combination of these. Discussion of temporal and spatial variation as well as sample size is given subsequently.

The interpretation of these results is straight forward. Given the assumptions, if another sample is made of the ground-water quality, then the mean of that sample has a 1 - α chance of being within the specified limits if nothing has changed. In light of the insufficient data base, a change will most likely occur. If the mean is different than before, then there is a chance of the difference being a random variation or induced by other means (man-made).

TABLE 3

SELECTED VALUES OF THE t STATISTIC

Confidence Level	α	t Statistic TSI*	OSI**
95%	.05	2.131	1.753
99%	.01	2.947	2.602
99.9%	.001	4.073	--

* TSI is written as $t_{.975,15}$ for 95% confidence.

**OSI is written as $t_{.95,15}$ (upper limit only) for 95% confidence.

TABLE 4

SUMMARY OF T-TEST RESULTS FOR GROUND-WATER QUALITY MEAN CALCULATIONS

	α = 0.05	m = 15		α = 0.01	m = 15		α =0.001	m = 15
	TSI		OSI	TSI		OSI	TSI	
Chemical	LL	UL	UL	LL	UL	UL	LL	UL
Aluminum	2.5	6.9	6.5	1.7	7.7	7.4	0.5	8.9
Arsenic*	0.01	0.01	0.01	0.01	0.01	0.01	0.01	0.01
Barium*	0.07	0.13	0.13	0.06	0.14	0.14	0.04	0.16
Boron	0.75	0.85	0.84	0.75	0.87	0.87	0.70	0.90
Cadmium*	0.001	0.003	0.013	0.001	0.003	0.003	-.00004#	0.004
Chloride	33.8	40.2	39.6	32.6	4.14	40.9	30.9	43.1
Chromium	0.007	0.013	0.012	0.006	0.014	0.013	0.005	0.015
Cobalt*	0.01	0.01	0.01	0.01	0.01	0.01	0.01	0.01
Copper	0.005	0.017	0.016	0.002	0.020	0.019	0.001#	0.023
Cyanide*	0.1	0.1	0.1	0.1	0.1	0.1	0.1	0.1
Fluoride	0.84	0.88	0.88	0.83	0.89	0.89	0.82	0.90
Iron	1.2	8.8	8.1	-0.2#	10.2	9.6	-2.2#	12.2
Lead	0.021	0.059	0.056	0.013	0.067	0.063	0.003	0.077
Manganese	0.114	0.270	0.256	0.084	0.300	0.288	0.042	0.342
Mercury*	0.0003	0.0007	0.0006	0.0003	0.0007	0.0007	0.0002	0.0008
Molybdenum	0.0003	0.022	0.020	-0.0037#	0.026	0.024	-.0094#	0.031
Nitrate	0.4	1.6	1.5	0.2	1.8	1.7	-0.1#	2.1
Phenols	0.005	0.013	0.013	0.003	0.015	0.014	0.001	0.017
Selenium	0.01	0.01	0.01	0.01	0.01	0.01	0.01	0.01
Silver*	0.01	0.01	0.01	0.01	0.01	0.01	0.01	0.01
Sulfate	691	719	716	686	724	722	679	931
TDS	1,458	1,574	1,563	1,436	1,596	1,586	1,406	1,626
pH**	8.28	8.46	9.44	8.24	8.50	8.48	8.19	8.55
Zinc	0.10	0.28	0.26	0.006	0.32	0.30	0.02	0.36
Nickel***	--	--	--	--	--	--	--	--
Uranium	0.044	0.176	0.164	0.019	0.201	0.190	-.015#	0.235

* Data contain items such as "<0.01". These are interpreted as "0.01".

** Three data points per well. $t_{.975,11} = 2.201$; $t_{.995,11} = 3.106$; $t_{.9995,11} = 4.437$; $t_{95,11} = 1.796$; $t_{.99,11} = 2.719$

*** Inadequate data for statistical methods.

\# Lower limit result is negative. This is, in theory, impossible for a normal distribution of positive data.

Demonstrating Restoration

Restoration criteria are usually defined as those chemical concentrations to which the operator must return the ground water before plugging wells and reclaiming the mine site. These criteria are usually set prior to *in situ* mining or may be dictated by regulation. These are based on the base line data and are usually specified as a return of the quality to the average value or better. As previously discussed, such a requirement is not well founded when applied to stochastic processes.

When proving statistically that ground water has been returned to its original state, one proves that upon comparing the base line data and the restored ground-water data, no difference can be seen between the two distributions, and, in fact, can be considered as one. This is accomplished by comparing the means and variances. With the assumptions previously stated, this would require large amounts of data.

Another definition for restoration might be that if it can be shown that a sample mean has a high chance of lying below a certain value, then the field may be considered restored. This value, of course, would be determined prior to disturbing the ground water or be set by regulation. This latter definition of restoration may be more applicable than the former definition to *in situ* mining since the circulation of ground water between two wells would disturb the original chemical equilibrium in the ground water and probably make restoration to the original state impossible.

Consider first the comparison of two samples (base line and restored) to test for statistical difference. This procedure involves the variance of the combined sample and the difference between the means.

The test is

$$\frac{|\bar{x}-\bar{y}|}{S_{xy}\left(\frac{1}{N_x}+\frac{1}{N_y}\right)^{1/2}} > t_{1-\alpha, N_x+N_y-2} \qquad (7)$$

where $\bar{x}$ is the base line sample average, $\bar{y}$ is the restored sample average, N_x is the number of base line observations, N_y is the number of restored observations, and S_{xy} is the combined sample standard deviation.

Sxy is calculated by

$$S_{xy}^2 = \frac{S_x^2(N_x-1)+S_y^2(N_y-1)}{N_x+N_y-2} \qquad (8)$$

S_x is the standard deviation of the base line sample. S_y is the standard deviation of the restored sample.

If equation (7) is true, then the two samples are statistically different. If equation (7) is false, then there is no statistical difference between the samples.

As a sample calculation consider the aluminum data from the ground-water base line. From Tables 1 and 2, $\bar{x}$ = 4.7, S_x = 4.1, and N_x = 16.

Since no real restoration data exist at this time, for illustration assume $\bar{y}$ = 7.5, S_y = 5.0, N_y = 14.

S_{xy} is calculated as

$$S_{xy} = \left[\frac{(4.1)^2(15)+(5.0)^2(13)}{16+14-2}\right]^{1/2} = 4.54 \qquad (9)$$

Assume a 95% confidence is desired. The t statistic is $t_{.95,28}$ = 1.73 for OSI. OSI is appropriate since this is a test for restoration.

Substituting the appropriate values into equation (7) yields

$$\frac{|4.7-7.5|}{4.54(1/16+1/14)^{1/2}} = 1.685 \overset{?}{>} 1.73 \qquad (10)$$

Equation (10) is false indicating that no significant statistical difference exists at the 95% confidence level. If all the assumptions were valid, then one could conclude that restoration had been achieved.

The actual normal distributions for this example can be calculated using the average and standard deviation values. Figure 2, "Comparison of Base Line and Restored Distributions," is a graphic representa-

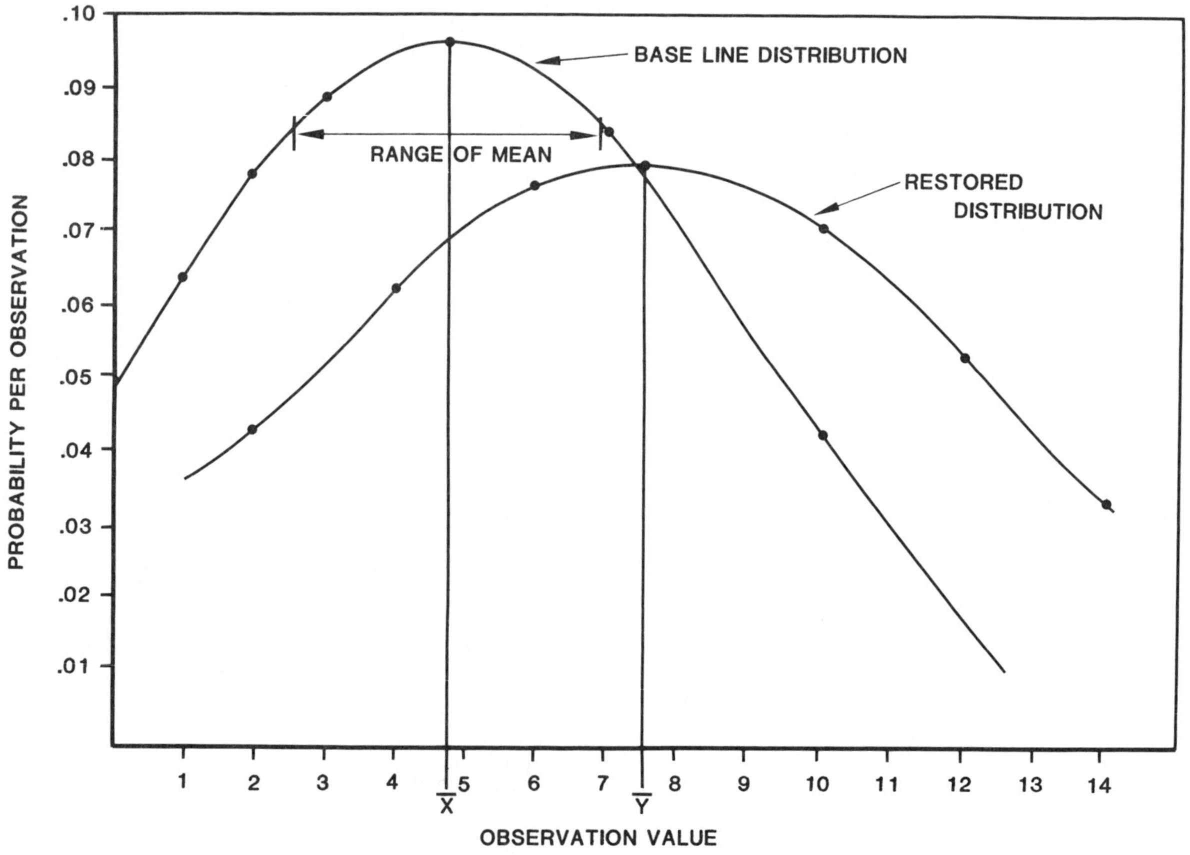

FIGURE 2 - COMPARISON OF BASE LINE AND RESTORED DISTRIBUTIONS tion of the actual distributions. From Figure 2 note that in the original case, the restored average is not below or equal to the base line average, yet they are not statistically different. From equation (7), it is important to note that if S_{xy} had been smaller (i.e. less variability in the data), then the test might have been true indicating statistical difference. Also, if $\overline{y}$ were 7.6, then the result would have been 1.75 and the test true, indicating that the samples are statistically different.

Demonstrating Excursion

The second method previously discussed of comparing a mean to a value is useful in determining excursions in the mine area. Excursion criteria are usually specified as a value. The t-test is used again for this calculation. The test is

$$\bar{x} < M_o - t_{1-\alpha, m} S/\sqrt{N} \qquad (11)$$

where M_o is the excursion limit or target value.

The test statement associated with equation (11) is the population mean is not smaller than the target value. This test statement is rejected if equation (11) is true. The statement is accepted if equation (11) is false. This form of test and the one used for demonstrating restoration is referred to in the literature as testing hypotheses by means of samples.

Using the previously assumed restored data as excursion data, assuming the target value $M_o = 6$, and $t_{.95,13} = 1.77$, one can substitute the data into equation (11). This yields

$$7.5 < 6 - 1.77(5)/\sqrt{14} = 3.6 \qquad (12)$$

Equation (12) yields a false condition. Therefore, the excursion mean is not smaller than the target value at the 95% confidence level. Again, if all the assumptions were valid, then one could state that an excursion has been demonstrated.

The excursion distribution is graphically illustrated in Figure 3, "Comparison of an Excursion Mean to a Target Value." From equation (11), it can be seen that the average must always be at least the target value. In Figure 3, note that the average is greater than the target value.

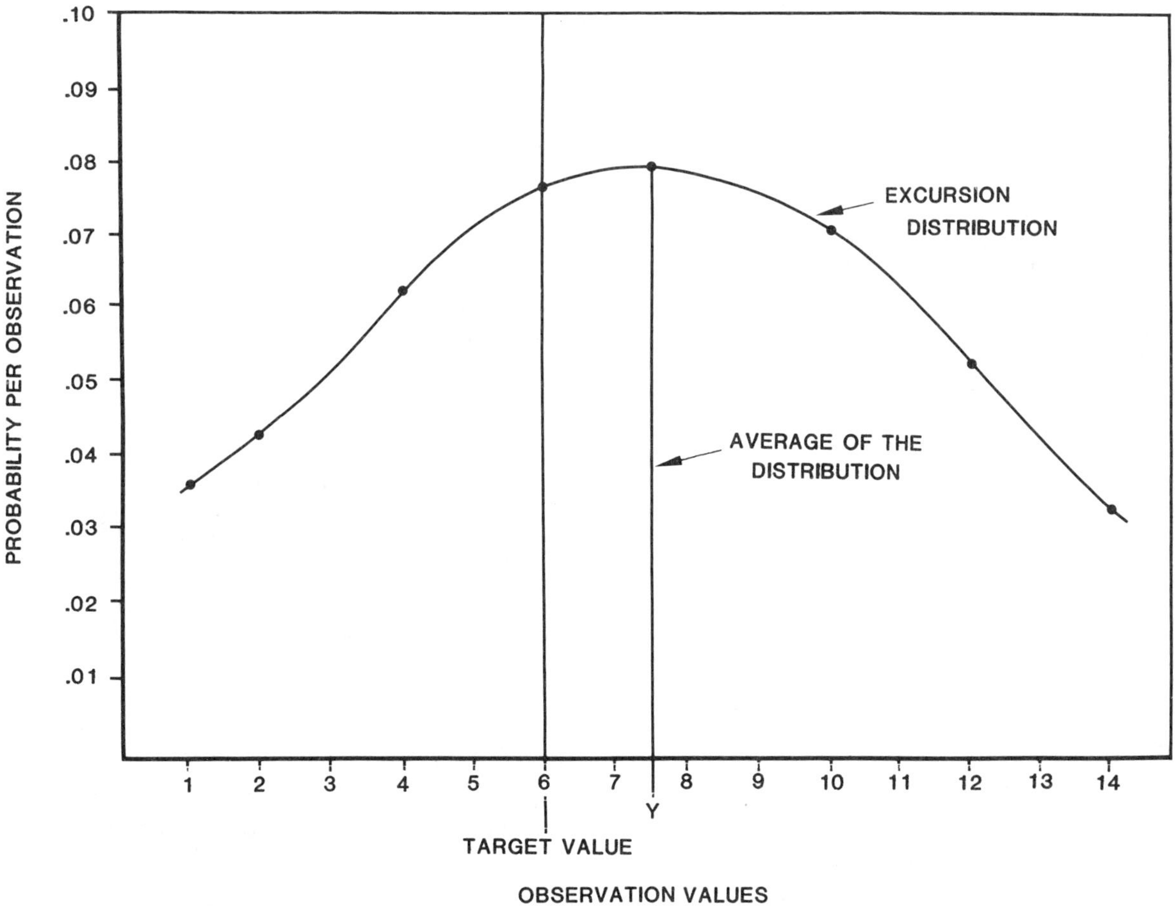

FIGURE 3 - COMPARISON OF AN EXCURSION MEAN TO A TARGET VALUE

SPATIAL AND TEMPORAL VARIANCE

In the previous analysis, the assumptions were made that no spatial or temporal variation exists in the data. These assumptions were necessary to simplify the statistical test and maintain a large sample size. In nature there is no ideal non-varying system, so one can expect the ground-water quality to vary spatially and temporally. The questions which arise are how much does it vary and what components of the water quality vary. It may turn out that the variation is insignificant. To answer these questions, variance analysis can be used.

Variance analysis can be used when more than two samples from the same population are tested for significant difference. The variance among the means of the classification is compared with the variance within the classification. In this analysis, classifications have been made by well and by date. That is to say the data can be classed either by all measurements on a well by well basis for all dates, or all measurements on a date by date basis for all wells. If the variance among the means is significantly greater than the variance by classification, the classifications represent different populations or the water quality between wells (or dates) are not from the same statistical population.

Variance analysis is made simple by use of the F-table. The F-table provides critical values which are compared to the ratio of the variances previously discussed. In using the F-table, care must be used in determining the degree of freedom. Table 5, "Average Data for Aluminum Grouped by Well and by Date," contains the results of averaging the aluminum data classified by well and by date.

TABLE 5
AVERAGE DATA FOR ALUMINUM
GROUPED BY WELL AND BY DATE

Well	Date 7/78	8/78	11/78	12/78	Average by Well
P-7	8.1	3.7	0.7	8.9	5.4
P-9	16.0	4.4	0.7	8.3	7.4
P-19	4.6	3.1	0.8	6.5	3.8
Ø-10	1.8	3.2	0.5	4.0	2.4
Average By Date	7.6	3.6	0.7	6.9	4.7

First calculate the variance analysis by date. Calculation of the variance within the means is simply the sum of the square of the differences between the data and the average for each date respectively, divided by the appropriate degree of freedom. In this work, the average was substituted as an approximation of the mean. Results of the calculation are:

Date	Average	$(\bar{x}-x)^2$
7/78	7.6	113.45
8/78	3.6	1.06
11/78	0.7	0.0475
12/78	6.9	14.53
		129.09

Since the four averages, one for each date, were used to calculate $(\bar{x}-x)^2$, the degree of freedom, m_1, is determined by

$$m_1 = N-4 \tag{13}$$

The variance within the means is calculated as

$$\text{VARIANCE WITHIN THE MEANS} = \frac{129.09}{12} = 10.76 \tag{14}$$

The variance among the means is calculated by the sum of the square of the difference of the average from the other averages (by date) for each of the dates taken one at a time, divided by the appropriate degree of freedom.

The calculation results are:

Date	Average	$(\bar{x} - x)^2$
7/78	7.6	64.10
8/78	3.6	35.30
11/78	0.7	94.46
12/78	6.9	49.82
		243.68

Since the average of the test by date is used to estimate the true mean of test, one degree of freedom is lost from the four original classifications. The degree of freedom, is determined by

$$m_2 = 4-1=3 \tag{15}$$

Thus, the variance among the means is calculated as

$$\text{VARIANCE AMONG THE MEANS} = \frac{243.68}{3} = 81.23 \tag{16}$$

F is calculated as the ratio of these two variances from equations (16) and (14) as

$$F = \frac{\text{VARIANCE AMONG THE MEANS}}{\text{VARIANCE WITHIN THE MEANS}} = \frac{81.23}{10.76} = 7.55 \quad (17)$$

This F value is then compared to tabulated values of F for the appropriate level of confidence and degrees of freedom. F is written $F_{1-\alpha}(M_1, M_2)$.

Some F values for normal distributions for this problem are $F_{.95(3,12)} = 3.49$, $F_{.99(3,12)} = 5.95$, and $F_{.999(3,12)} = 10.8$.

Given the stated assumptions, comparing the calculated F value to the tabulated F value yields the following conclusion: because $F_{CALC} > F_{.95(3,12)}$, there is significant statistical difference at the 95% confidence level. In fact, $F_{CALC} > F_{.99(3,12)}$ so there is significant statistical difference at the 99% level. However, here again the lack of adequate sample size greatly increases the chances for statistical difference. Therefore one might expect difference at even a level such as 95% on such a small sample. This means that there is a variation by date for aluminum at the 99% confidence level. But at the 99.9% level there is no statistical difference between the classifications. This level of confidence is extremely high. For a sample size as small as this one and with the mean and variance approximated by the average and standard deviation, such a confidence level is of no real value.

Considering the aluminum data classified by well, F is calculated

$$F = \frac{110.96/3}{197.74/12} = 2.24 \quad (18)$$

Comparing this to the tabulated F values indicates no statistical difference at the 95% confidence level. In other words, there is no significant statistical difference between wells in the aluminum data at the 95% confidence level.

The variance analysis was performed on all chemicals in Table 1 except the ones that had data in the "less than" category or whose data was constant such as selenium.

Graphic illustration of the results is shown in Figure 4, "Summary of Variance Analysis." From Figure 4, significant statistical difference can be quickly observed. To interpret the graph, if a bar exceeds a level of confidence, then for that chemical in the classification, significant statistical difference exists between the groups within that classification for the level of confidence which has been exceeded.

In statistics, a 95% confidence level is reasonable. Figure 4 shows temporal differences for 10 out of 16 chemicals and spatial differences for only 4 out of 16 at the 95% confidence level. Also, note that if a chemical exhibited statistical difference

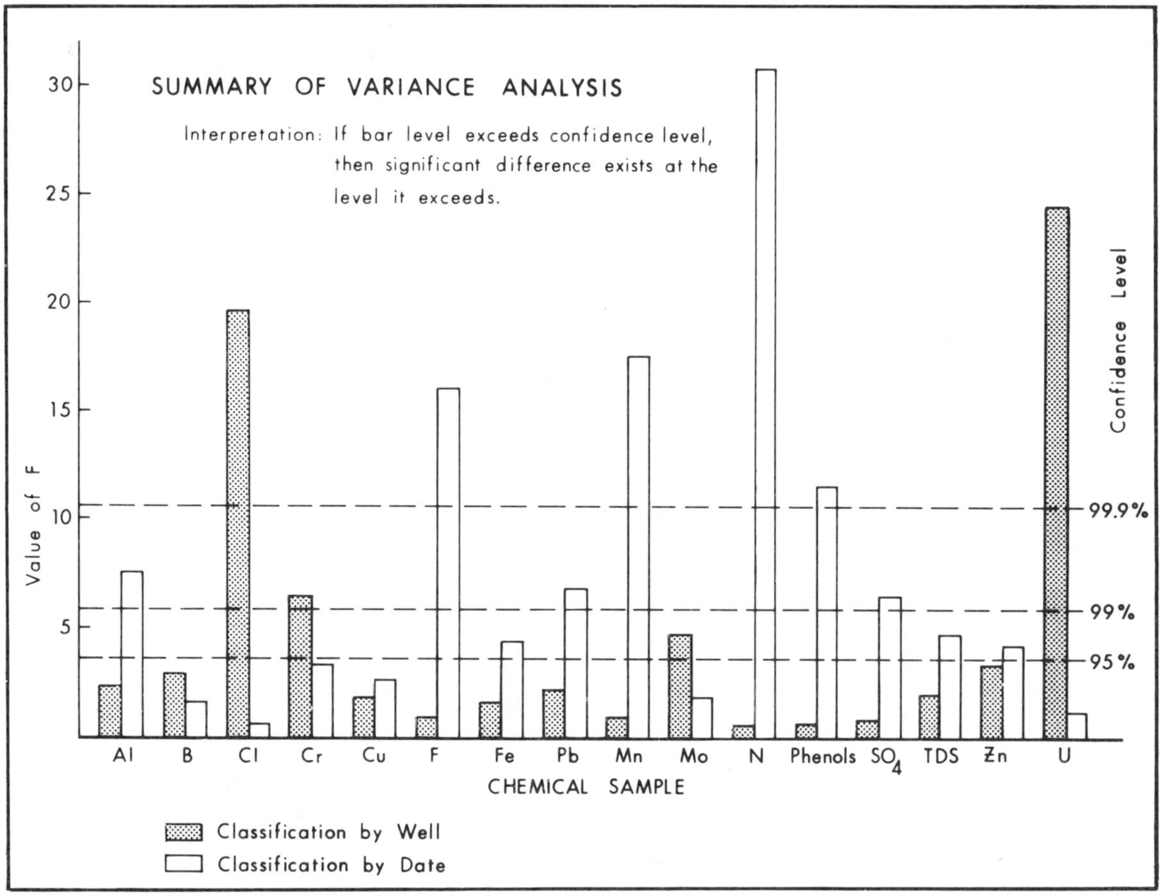

FIGURE 4 - SUMMARY OF VARIANCE ANALYSIS

in one class, the other class did not. Only two chemicals, boron and copper showed no difference in either class at the 95% confidence level. Recalling Table 5, for aluminum, the temporal effect can be noted by observation, but it must be analytically determined to be sure.

SAMPLE SIZE

Sample size is one of the most important questions an analyst addresses when preparing to use statistics. This question is as fundamental as assuming a normal distribution. Before beginning an analysis some relevant questions must be considered.

- What is the experimental design?
- What is the test statement or hypothesis?
- What is being estimated?
- What level of confidence is required?
- How large a difference or error is detectable and with what probability?
- What is the variability of the data?

Many of these questions may not have precise answers. However, any answer provides insight into the problem.

Each of the chemicals analyzed for ground-water quality could have different variability affecting the sample size required for each chemical. Careful investigation in this area could yield sizeable economic benefit because of the high costs of water analysis.

Sample size is also affected by the degree of confidence required. This, in conjunction with data variability, provides some measure of sample size requirements. In previous example calculations the problem was dealt with by assuming enough observations were made. If the problem is worked backwards, that is, by assuming a difference between the base line and the restored averages in conjunction with the standard deviation of the data, then one can formulate the following:

$$D = \frac{\Delta}{S} \tag{19}$$

Δ is the difference between the base line average and the restored average and S is the standard deviation. Working in reverse with the t-test statistic yields the results shown in Table 6, "Sample Size Requirements for Various Levels of Confidence when Comparing Means of Distributions."

Table 6 gives minimum sample sizes for various combinations of confidence as a function of D for normal distributions.

Table 6 contains two confidence factors, α and β. As previously stated, α is the probability of rejecting a test statement or hypothesis that is actually true. β, however, has the opposite meaning. β is the probability of accepting a test statement that is actually false. Both play an important role in setting criteria and sample size requirements. These values of α and β are referred to in the literature as producer's and buyer's risk, respectively.

The trends in Table 6 are quite logical. For a certain standard deviation in the data, if a small difference between the sample averages is required, then the number of observations becomes large. Likewise for a stated difference between sample averages, if the standard deviation becomes small, then the required number of observations becomes smaller.

In general, less data variability means fewer data points for a set level of confidence. Also, the level of confidence (α and β) greatly affects the number of observations. For example, to demand a 1 in 100 chance of error [β = .01] as opposed to 1 in 20 chance [β = .05] increases the required number of observations.

If the data conformed to the assumption of normality and randomness, then Table 6 would indicate that large data sets are required to make an analysis. Care must be taken when setting up a sampling program to insure an adequate sample size. If parametric normal statistics are used, the data size requirement may be economically prohibitive.

TABLE 6

SAMPLE SIZE REQUIREMENTS FOR VARIOUS LEVELS OF CONFIDENCE WHEN COMPARING MEANS OF DISTRIBUTIONS

	NUMBER OF OBSERVATIONS, N			
	RELATIVE CONFIDENCE AND RISK			
RATIO OF THE AVERAGE DIFFERENCE TO STANDARD DEVIATION	BEST α=0.01 β=0.01	BETTER α=0.01 β=0.05	GOOD α=0.15 β=0.10	MEDIUM α=0.10 β=0.20
Minimum Acceptable D and N	0.5/100	0.4/115	0.3/119	0.25/10
D = 0.5	100	75	44	27
D = 0.6	71	53	32	19
D = 0.75	47	36	21	13
D = 0.80	41	32	19	12
D = 1.00	28	22	13	8
D = 2.00	10	8	5	-

Note: $D = \frac{\Delta}{S} = \frac{|O - A|}{S}$ O = Observed Average; A = Base Line Average; S = Standard Deviation

α = Error probability of rejecting a true hypothesis

β = Error probability of accepting a false hypothesis

Data for normally distributed samples only.

SUMMARY AND CONCLUSIONS

The assessment of ground-water quality for in situ mining operations can be performed using statistical methods. It must be recognized that determining ground-water quality is not exact and that ground-water quality is stochastic in nature, most likely governed only partially by chance. By assuming normal distributions and considering sizable data bases, ground-water quality data can be statistically tested. Demonstrating the restoration of ground-water quality to its original state when mining operations are complete can be a straight forward procedure by using techniques for comparing data distributions. Restoration of ground water to an average value based on a limited sample may not be well founded given the stochastic nature of ground-water quality. Illustrated in the examples in this paper is the need for large data bases if normal parametric statistics is used. Other distributions or nonparametric statistics could be used that would enhance the results of the methodology illustrated in this paper. Also, most certainly nonparametric statistics would reduce the sample size requirement compared to normal parametric statistics. Sound procedures and good data bases are a must for the appropriate and proper use of statistics for ground-water quality assessment.

BIBLIOGRAPHY

Colchin, M. P., "Sampling of Ground Water Base Line and Monitoring Data for In Situ Processes," CRWR-157, EHE 78-01, June 1978, Center for Research in Water Resources, University of Texas, Austin, Texas.

Fryer, H. C., Concepts and Methods of Experimental Statistics, Allyn and Bacon, Inc. Boston, 1966.

ISO Standards Handbook 3, International Organization for Standardization, 1st ed., ISBN 92 67 10026 2, 1979.

Ostle, B., and Mensing, R. W., Statistics in Research, 3rd, ed., Iowa State University Press, 1975.

Sanders, T. G., "Water Quality As A Stochastic Process," Preprint 80-089, American Society of Civil Engineers, April, 1980.

Sanders, T. G., and Ward, R. C., "Relating Stream Standards To Regulatory Water Quality Monitoring Practices," Reprint, Establishment of Water Quality Monitoring Programs, American Water Resources Association, Everett, L. G., and Schmidt, K. D., ed., June, 1978.

Ward, R. C., "Regulatory Water Quality Monitoring: A System Prospective," Reprint, Water Resources Bulletin, American Water Resources Association, Vol. 15, No.2, April, 1979.

Chapter 10

DESIGN CRITERIA FOR IN SITU MINING OF HARD ROCK ORE DEPOSITS

C. R. McKee
R. H. Jacobson
S. C. Way
In Situ, Inc.
Laramie, Wyoming 82070

M. E. Hanson
Lawrence Livermore Laboratory
Livermore, California 94550

K. Chong
University of Wyoming
Laramie, Wyoming 82071

ABSTRACT

To create conditions for economic rates of mineral recovery in a deep, hard rock mass, enhancement of naturally occuring permeability is necessary, and may be achieved through explosives detonated in boreholes. Previous theories of explosively created permeability are reviewed and found to relate permeability solely to the degree of fracturing, neglecting porosity. A theory is presented that includes both fracture density and porosity, and its results are applied, together with a knowledge of explosive behavior, to derive design parameters for in situ mining.

The theory is verified through postshot permeability measurements around cavities, and is extended from one explosive to include multiple explosive detonations. The postshot cavity radius is the single most important design parameter, controlling the distance over which permeability is enhanced. Explosive spacing is limited to a scaled distance of 10–15 times the postshot cavity radius. Explosive properties are reviewed; those explosives that perform best are those that produce the largest volumetric expansion in the final cavity pressure range.

Well flow rates are calculated for single and multiple-explosive detonations, and in situ solution mining of a model ore body is used to illustrate that matrix dissolution can chemically enhance the explosively-induced permeability.

1.0 INTRODUCTION

In-situ mining of hard rock deposits will only become successful as we understand the engineering parameters involved and the design limitations they impose. As recognized by Wadsworth[1] the most critical area is the enhancement of permeability in a hard rock ore deposit. Along with creating permeability, the appropriate flow and fragment size conditions must be generated to insure economic rates of mineral recovery. In this article our interest is confined to true in-situ processes in which boreholes are drilled from the ground surface or a mine drift to deeper strata. We assume that fractures and permeability enhancement will be due to explosives detonated in boreholes. Other techniques such as hydraulic fracturing do not produce the necessary size distribution to promote economic extraction rates unless the deposit is already highly fractured and interconnected.[2] Notable exceptions where hydraulic fracturing is applicable are many oil, gas, and salt deposits. For many in-situ processes, an array of explosive charges offers the opportunity to create a distribution of many fractures not achievable by other techniques. Such a situation would be suitable, for example, to in-situ mine hard rock gold, uranium, copper, or oil shale deposits and may enhance oil and gas recovery under certain conditions.

Our experience with in-situ uranium mining of permeable sand deposits has shown that well spacing ranges from 40 to 70 ft. Permeability varies from 150 millidarcies (md) to 10 darcies, with more typical values in the range of a few hundred millidarcies. Injectivities and productivities of wells are as low as a few gallons per minute (gpm) up to as much as 60 gpm. The depth of these deposits is generally less than 1000 ft. The water table is usually shallow. The latter point is often not greatly appreciated but plays an important role in determining well injectivity and productivity, and the ratio of injection to production wells.

Figure 13 (to be discussed later) shows permeability vs. depth for hard rock ore deposits. At depths to a few hundred feet some hard rock deposits may satisfy the permeability requirement. However, at 1000 ft the permeability is generally less than 10 md. For these deposits it is clear that some form of stimulation is required to make the rock more permeable such that it conforms with our previous experience for economic extraction. Since the bulk of the world's metals and other energy resources resides in hard rock, it is necessary to examine alternative mining methods.

The purpose of our article is, then, to review and extend the explosive permeability theory of McKee and Hanson[3,4,5] and to apply its results together with explosive behavior and obtain design parameters suitable for in-situ mining.

2.0 EXPLOSIVE PHENOMENOLOGY

Direct observation of explosion phenomenology in earth media is limited to a few dynamic stress and velocity histories at selected points from the explosive source and postshot excavation of the region around the blast-generated cavity. Knowledge of the processes involved in the shock propagation and fracturing of the surrounding rock are deduced from numerical computer calculations. These models contain the basic physics and use medium property model parameters developed from laboratory measurements on small rock samples. The computer models have been verified by comparison with field data and the observations noted above. Agreement is good for final cavity size and peak stress and velocity vs. distance (i.e., attenuation of the shock wave strength). Improvements in predictive capability are still needed for stress wave form and far field displacement.

To verify the theory developed later, we rely on postshot permeability measurements taken around cavities formed by detonating chemical and nuclear explosives. Accordingly, we discuss each explosive type separately.

Detonation of a tamped chemical high explosive in a borehole deep below the ground surface results in a peak stress into the rock of slightly greater than 100 kbars which may vaporize the water in the rock for a few centimeters beyond the borehole wall.

Detonation of a nuclear explosive deep below the ground surface produces an extremely high temperature/high pressure spherical ball of plasma. A high intensity shock wave is generated, propagating into the surrounding rock. For a nuclear device with the equivalent energy of one kiloton TNT, a sphere of rock 2 m in radius is vaporized, 4 m in radius is melted, and all water in the rock is vaporized out to a radius of approximately 6 m. This interval separates the solid rock from the cavity gases and fluids and defines the initial cavity boundary. The peak acceptance stress of the shock wave at this range is approximately 100 kbars (10 GPa).

With either chemical or nuclear explosives the shock wave is attenuated by energy deposition into the rock and geometrical attenuation. The geometrical attenuation factor accounts for the divergence of the surface area of the wave front as it propagates. Thus the stress (α) and the peak particle velocity (v) being proportional to the square root of the energy per surface area attenuates at $r^{-\alpha}$, where r is the distance from the source and $\alpha = 1$ for spherical geometry, 1/2 for cylindrical geometry, and 0 for plane geometry.

Energy deposition into the rock varies as a function of distance from the source. Because the final cavity size is proportional to the square root of the explosive energy in a borehole, it is convenient to use the cavity size as a scale parameter for distance measurements. The shock wave deposits energy into the rock by fracturing, compaction, thermal and kinetic energy. The kinetic energy is then degenerated into the first three by the shear stress in the rock.

Near the cavity wall the rock is pulverized by the shock wave such that even the quartz grains within the rock matrix are crushed. The shock wave fractures the rock in shear deformation out to a distance of approximately 3 (spherical) to 12 (cylindrical) cavity radii. The degree of damage (fracturing) decreases exponentially with distance from the source. Beyond the shear fracture region the shock wave displaces the rock such that tensile fractures develop due to the divergence of the displacement, creating a tensile hoop stress. The tensile fracture region extends out to about 5 to 7 cavity radii for spherical events, and to 10 to 24 cavity radii for cylindrical boreholes. These distances will vary due to the depth of burial and the residual strength of the rock. The tensile fractures are usually short isolated fractures that do not increase the permeability.

The initial stages of cavity growth are the result of momentum deposited in the rock by the shock wave. Only in the late stages of cavity growth when the cavity has achieved full size does the cavity gas pressure begin to dominate the expansion process. Depending on the shear strength, the cavity may either expand past the equilibrium point or slowly to the equilibrium point, equilibrium being determined by the cavity pressure in balance with the external stresses. In spherical geometry, shortly after maximum cavity size has been achieved, the entire region within 2 cavity radii rebounds simultaneously toward the detonation center. This tends to lock in a compressive hoop stress in the media around the cavity and preserve its dimensions.

Chemical and nuclear explosives differ in the magnitude of energy release. Cavities sprung by nuclear explosives are larger than those created by chemical explosives. However, the stress accepted by the medium is similar, since the bulk of the energy in a nuclear explosive goes into vaporizing and melting rock. The theory described in the next section is justified using both chemical and nuclear explosives and, we believe, works because of the common features exhibited by the stress wave when propagating into the rock.

3.0 EXPLOSIVELY CREATED PERMEABILITY

3.1 Previous Theories

Attempts to predict explosive-generated permeability have previously been based on correlating the degree of fracturing or damage caused by the stress wave to the enhanced permeability. The main proponents of these theories were Laspe,[6] Terhune and Shaw,[7] and Butkovich.[8]

Laspe's theory attempts to correlate surface area created from the blast with energy absorption. Laspe assumed that particles in sedimentary rocks are cemented together with a statistical distribution of bonding strengths. A distribution of bond strengths was postulated to give rise to an exponential absorption of energy with distance from the shock analogous to absorption of radiation in a solid.

Laspe suggested that if a material's specific surface energy for fracturing is given, a mean particle diameter as a function of distance can be calculated. Based on mean particle diameter, explosive fracture effects are then divided into a pulverized zone, a fractured zone, and a competent zone. Laspe then simplified the treatment of the pulverized zone by assuming it to be similar to a 300 micron (50 mesh) fracture proppant sand. He resorted to a semi-empirical approach for treating the secondary fractured zone. Distribution of permeability is obtained by assuming its decay rate is the same as that for energy, namely exponential. Empirical constants are adjusted to obtain agreement with observation.

A second theory due to Terhune and Shaw is considerably more elaborate than Laspe's. Laspe relied mainly on empirical predictions, which did not use detailed material property data, depth of burial, or explosive source parameters. In contrast, the model of Terhune and Shaw[7] was a detailed mathematical calculation of the nonlinear elastic continuum equations describing the response of rock. They also included a fairly complete simulation of the explosive detonation process. Moreover, extensive experimental rock property measurements were used as input to the model. An attempt was made to correlate permeability with failure characteristics of the rock. In particular, they defined a quantity termed "crack number," which referred to the number of times a zone in the calculations exceeded the experimentally measured failure criterion. They found that the "crack number" was a function of zone size. However, they were able to locate a range for which the "crack number" was reasonably insensitive to this quantity.

Terhune and Shaw observed that fluid loss during postshot drillback occurred whenever the "crack number" exceeded 5. This corresponded to the limit of shear failure on compressive loading and indicated a correlation with enhancement of permeability. They also reported some agreement between inferred permeability distributions from pressure testing and "crack number." However, they concluded that the data were insufficient at that time to attempt to develop a consistent criterion relating crack number to a permeability magnitude.

Another version of the computer model of Terhune and Shaw was used by Butkovich[8] to correlate fracturing effects as well as permeability around detonations. This version, called SOC73, was modified by Schatz.[9] Schatz, instead of computing "crack number" as a measure of fracture damage, used a quantity termed "failure shear strain." Failure shear strain is based on separating the total strain into elastic and "plastic" parts. As the material deforms, the elastic part of the strain is computed from relations based on Hooke's law. If the material fails, the portion of the strain not associated with elastic strain is the failure-associated strain, or "failure shear strain." Failure shear strain is accumulated in a given zone with time and serves as a damage index which can be interpreted in terms of fracture intensity. Butkovich found that the limit of observable fracturing for a cylindrical borehole shot in coal correlated approximately with a value 0.01, for failure shear strain. This corresponded to a distance of 8.5 postshot cavity radii. Butkovich indicated that if the failure shear strain was cubed it displayed the same decay characteristics as the measured permeability distribution.

The above models all have a common approach: They assume that permeability is solely related to the degree of fracturing. High fracture intensity, according to these theories, should result in high permeability. This is not correct and, in fact, can result in lowering permeability. Permeability depends on both the degree of fracturing and porosity. Attempts to correlate permeability to experimental data for use in predicting permeability without considering porosity must therefore be regarded as theoretically unsound. This is apparent, since none of these approaches can predict the behavior of permeability around an explosive without matching arbitrary parameters to experimental measurements.

A theory that includes both the effects of fractures and porosity was devised by McKee and Hanson.[3,4] Their model appears to be valid for the case in which all fractures are explosively created. This theory was recently extended to include the initial fracture density, nonlinear dissipation, and multiple explosives,[5] and is summarized in the following section.

3.2 Theory of Explosively Created Permeability

To obtain an expression for permeability in terms of known parameters in rock mechanics, a constitutive equation is required. The approach of McKee and Hanson[3,5] was to begin with the widely accepted Blake-Kozeny-Carmen theory.[10] For more detail the reader is referred to their papers.

In this theory a wide range of pore geometry[4] results in the expression

$$k = \frac{\phi^3}{CT(1-\phi)^2 S^2}, \qquad (3.1)$$

where C = 3 for fractures, T is the tortuosity, ϕ the porosity, and S the specific surface area.

In the case of explosively created fractures the porosity is much less than one and can be neglected in the denominator. The fracture density (n) can be written in terms of the specific surface area as

$$S = \frac{2nh}{1-\phi}. \qquad (3.2)$$

Substituting equation (3.2) into (3.1) and assuming the fractures are pervasive (h = 1), we obtain

$$k = \frac{\phi^3}{4CTn^2}. \qquad (3.3)$$

This clearly demonstrates the dependence on both the porosity and the fracture density. Equation (3.3) illustrates the problem with previous theories and their exclusive reliance on attempting to correlate permeability with fracture density. In fact, the above equation indicates that for a given porosity an increase in the number of fractures per unit area will actually decrease permeability. This is due to increased surface area and greater drag on the fluid traveling through the medium.

Hence it is not sufficient simply to create cracks: Porosity must be placed in these fractures. Distribution of a given amount of porosity over a greater number of cracks will reduce the permeability.

To link explosive effects to either of the permeability equations above, we view the explosion as occurring in two stages. The first is dominated by a large-amplitude stress wave, and the second involves an expansion of the cavity by high-pressure gases from the detonation. The effects of the first stage on the medium are of a dynamic nature, while those of the second stage extend over a much longer time interval and can be regarded as a quasi-static process. To obtain a description of permeablity, these processes must be related to the fundamental equations for permeability.

The dynamic stress wave will cause all flaws whose strengths are less than the magnitude of the locally applied stress to extend. A relation between the growth of flaws or the increase in specific surface and energy can be obtained from comminution theory. Several comminution relations have been proposed. The one most applicable to our situation is Rittinger's Law, which states that an increase in specific surface area is directly proportional to the energy input:

$$S \propto E . \qquad (3.4)$$

Rittinger's Law has been substantiated by the general scaling laws of Langefors and Kihlstrom,[11] which have been verified for burden dimensions varying between 0.01 and 10 m with a 10^7 variation in explosive charge. Rittinger's Law has been further substantiated by Felts and others[12] in the laboratory, and theoretically by Rose.[13]

Creating fractures does not in itself generate permeability. This is because the stress wave propagates at the compressional-wave velocity (c_p), while fractures can grow in a rectilinear path at a maximum velocity of 1/3 c_p. Hence, the stress wave will inevitably outrun the fractures it generates. New fractures will then be initiated on other flaw sites in the material. At this moment in the process, the medium consists of a noninterconnected system of fractures with essentially no new porosity.

The second stage of the essentially continuous explosion process is dominated by the quasi-static expansion of the gas in the cavity. The cavity void space is produced by irreversible pressure-volume work of the explosive gases. Void space is created by irreversible radial compression and by the tangential tension of the surrounding rock. The fracture porosity will be proportional to the first invariant of the strain tensor,[14]

$$\phi \propto \Delta , \qquad (3.5)$$

where Δ is the first invariant and contributes to porosity only when it assumes positive values (dilatation).

Figure 1 shows the postshot configuration. The cavity's initial radius is r_o. The explosion springs the cavity to a larger radius r_c. The material initially between r_o and r_c has been compressed into a narrow region beyond r_c. This material, in trying to recover a fraction of its original size, is locked in. The tangential stress in the locked-in region gives rise to a compressive radial stress or effective pressure.

The power for plane harmonic waves[14] and for cylindrical or spherical waves with dissipation[15] can be expressed in terms of particle velocity or stress squared. We may, therefore, write

$$E \propto \bar{\dot{u}}^2 \propto \bar{T}^2. \qquad (3.6)$$

Even for high stress levels, equation (3.6) is still a good approximation. Other criteria will be discussed below. For the case of pervasive fracturing (h = 1) and small porosity, surface area is directly proportional to fracture density, or

$$S \simeq 2n. \qquad (3.7)$$

Denoting the explosion-induced fractures by n_e and the naturally occurring ones by n_o, the total specific surface or fracture density becomes

$$n_t = n_o + n_e , \qquad (3.8)$$

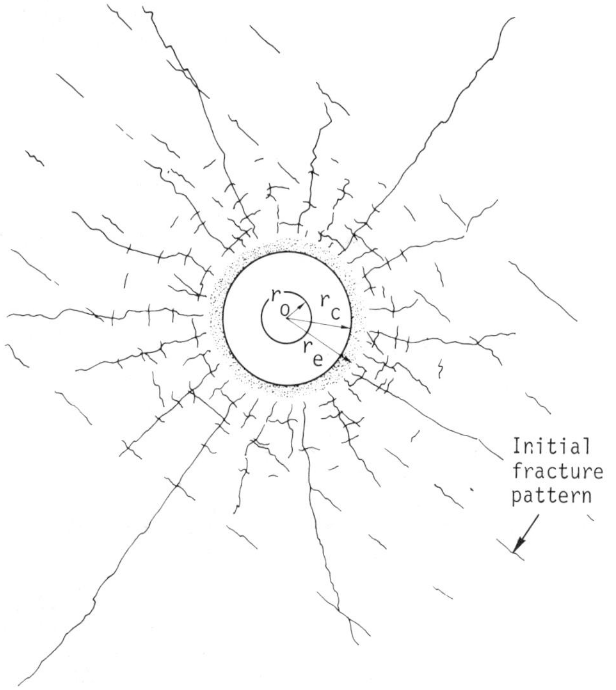

FIGURE 1. An illustration of the postshot configuration for an explosive fired in a borehole. The sketch includes an initial fracture pattern with the explosive patterns. The preshot radius is r_o, the postshot radius is r_c, and the limit of significant compaction is r_e.

where

$$n_e = bE \tag{3.9}$$

and b is a constant of proportionality.

The second or quasi-static stage can be satisfactorily treated by using the bilinear elastic static solutions for dilatation around a pressurized cavity,[3] which can be substituted into equation (3.5) to yield

$$\phi \propto \frac{f - \alpha}{r^{1 + \alpha}}, \tag{3.10}$$

where f = 1 for cylindrical and 2 for spherical symmetries, and

$$\alpha = f - \varepsilon, \tag{3.11}$$

where ε is a small positive number related to the ratio or elastic constants in compression and tension.[3]

If no bilinearity in the elastic constants is present, then from equations (3.10) and (3.11), dilatation, and hence porosity, would be identically zero. Bilinearity implies the existence of cracks and flaws that are necessary to create permeability. The use of equations (3.10) and (3.11) is supported by displacement measurements reported by Borg.[16]

Combining equations (3.5), (3.8), and (3.9), we obtain the expression

$$k_e \propto \frac{\Delta^3}{(n_o + bE)^2} \tag{3.12}$$

for the explosively created permeability. The effect of preshot permeability is included by adding to it the explosively generated permeability to obtain

$$k_t = k_e + k_o , \tag{3.13}$$

where k_o is the preshot permeability.

Some limiting cases to equations (3.12) and (3.13) are evident. From equation (3.13), if initial permeability is high, then the explosive's contribution to the total permeability can be negligible. Secondly, if initial fracture density is large compared to that generated by the explosion, then substituting equations (3.10) and (3.11) into (3.12), we obtain

$$k_e \propto 1/r^6 \text{ (cylindrical geometry)} \tag{3.14}$$

$$k_e \propto 1/r^9 \text{ (spherical geometry)} , \tag{3.15}$$

where ε is small compared to f.

Hence, the effect of the explosive decays strongly with distance. On the other hand, if the material is relatively competent or if explosion-induced fractures dominate the initial fracture density, we obtain

$$k_e \propto \frac{\Delta^3}{E^2} . \tag{3.16}$$

From Selberg,[17] energy decays asymptotically in the elastic limit as

$$E \sim 1/r^2 \text{ (dynamic, spherical geometry)} \tag{3.17}$$

$$E \sim 1/r \text{ (dynamic, cylindrical geometry)}. \tag{3.18}$$

Substituting for dilatation from equation (3.10), we find (for $n_e >> n_o$ and in the elastic limit)

$$k_e \propto 1/r^4 \text{ (cylindrical geometry)} \tag{3.19}$$

$$k_e \propto 1/r^5 \text{ (spherical geometry)} . \tag{3.20}$$

In the general case, energy will obey a power law decay

$$n_e = bE \sim 1/r^{f+m}, \tag{3.21}$$

where f is defined as before (f = 1, 2 for cylindrical and spherical geometry, respectively). In the perfect elastic case, m = 0 (asymptotic limit). However, in general, m is greater than zero. For a competent formation ($n_e >> n_o$), permeability will have the form

$$k \propto \frac{r^{2m}}{r^{3 + f}} . \tag{3.22}$$

If m = 0, then we recover equations (3.19) and (3.20). Real materials will exhibit dissipation and deviate from elastic behavior (i.e., m will be greater than zero). From equation (3.22) we see that explosion-created permeability can decay much more slowly than for the perfectly elastic case or for the case where the initial fracture pattern dominates. A physical interpretation of this is readily apparent. For the perfectly elastic case, we note that fractures must exist as continuous rays emanating from the cavity. On the other hand, a real material will exhibit a fracture pattern having a distribution of fracture lengths. For a dissipative material, fracture density decreases more rapidly with distance from the cavity than in the elastic limit. There is always competition between fracture density and porosity. When fracture density decreases with distance from the borehole, as in a real material, surface area declines more rapidly, tending to compensate for the decrease in porosity. Hence, the result is a slower decrease in permeability away from the explosive cavity. This results because an increase in surface area increases drag, and hence resists flow. On the other hand, making porosity larger increases the cross-sectional area to flow, thus decreasing resistance.

4.0 EXPERIMENTAL RESULTS FROM SINGLE EXPLOSIVE DETONATIONS

4.1 Hardhat (Spherical Geometry)

The Hardhat event was a 5-kt nuclear explosion in granite. Boardman and Skrove[18] performed extensive air permeability measurements around the Hardhat nuclear chimney. Their raw data show considerable scatter, some of it attributable to their measurement technique. We have applied a selection criterion[3] to their data to obtain a consistent set. As discussed in reference (3), the stress wave decayed approximately as the elastic case. Initial fracture density was taken to be small compared to explosively created fractures. Borg[16] reports that preshot material was broken by fractures, faults, and shear zones, most of which were cemented by minerals of a later origin. She further states that the consensus among geologists is that joints and fractures are short-lived, i.e., terminated within a few yards. The typical flaw size is, therefore, small compared to the cavity radius (20 m) and justifies ignoring the effects of the existing fractures. Figure 2 shows the comparison between permeability measurements and equation (3.20), the theoretical prediction for an elastic material. Preshot permeability measurements were not performed. However, other initial permeability was estimated to be on the order of tenths to several millidarcies. Agreement between the theoretically predicted slope and the experimental data is quite good. Extrapolating the permeability trend line from the theory yields 65 darcies for k_c, the permeability at $r = r_c$ (the cavity wall).

4.2 Hoggar (Spherical Geometry)

Delort and Supiot[19] report results of permeability measurements at the French Hoggar test site in the Sahara. Their measurements give a detailed picture of the permeability distribution around the postshot cavities and resulting chimneys.

In view of the fact that explosion-induced fractures are much more numerous than the natural fracture density, we ignore n_o in equation (3.8). Beyond 4 to 5 cavity radii, the explosion-induced fractures will, with increasing distance, blend into the natural fracture system by using equation (3.13). The undisturbed permeability appears to be on the order of 5 X 10^{-3} μm^2 (5 md).[19]

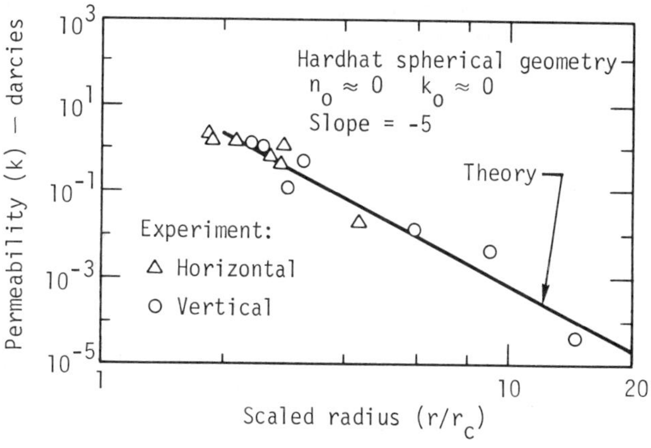

FIGURE 2. Comparison of theoretical and measured permeability, k, as a function of distance scaled in terms of the cavity radius r_c. Permeability was measured in both horizontal and vertical holes and is independent of direction. The experiment was Hardhat, a 5-kt nuclear explosion in granite at the Nevada Test Site.

Explosion-created fracture density can be predicted by combining equations (3.6) and (3.9), to obtain

$$n_e \propto \dot{u}^2 . \tag{4.1}$$

Terhune's analysis[20] indicated that particle velocity attenuates with distance as

$$\dot{u} \sim 1/r^{1.6} , \tag{4.2}$$

in agreement with Hoggar measurements.

Equation (4.2) is an asymptotic result, valid for distance greater than 1.5 cavity radii. Nearer the explosive, attenuation is more rapid.

Substituting equation (4.2) into equation (4.1), we find that the fracture density should decay asymptotically as

$$n_e \propto 1/r^{3.2} , \tag{4.3}$$

which corresponds to a value for m of 1.2 from equation (3.13). We recall that $m > 0$ implies greater energy dissipation due to inelastic effects.

Substituting equation (4.3) into equation (3.22), with $f = 2$, $m = 1.2$, yields

$$k_e \propto 1/r^{2.6} . \tag{4.4}$$

Substituting into equation (3.13) with $k_o = 5$ md and scaling r with respect to the cavity radius, we find

$$k_t = k_c \left(\frac{r_c}{r}\right)^{2.6} + 5 \quad \text{(in md)}, \tag{4.5}$$

where k_c is the intercept at the cavity wall. An approximate fit to the Hoggar data given by Delort and Supiot[19] gives a value of

$$k_c = 1700 \text{ md.} \tag{4.6}$$

The predictions of this theory are in good agreement with the few available observations, as shown in Figure 3. Data points were taken from reference (19).

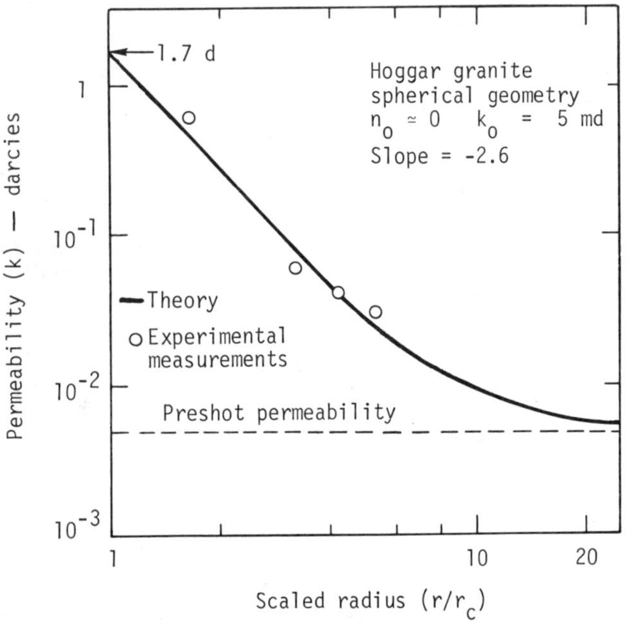

FIGURE 3. Comparison of theoretical and measured permeability values. The log of permeability, k, is given as a function of distance scaled in terms of cavity radius r_c. The data are taken from a series of nuclear detonations in the Sahara by the French.

4.3 Kemmerer Coal (Cylindrical Geometry)

As part of a coal gasification program, Lawrence Livermore National Laboratory conducted an explosive test in a coal seam near Kemmerer, Wyoming.[21] A cylindrical, high-explosive charge, 5.5 m long, 0.1 m in diameter, and weighing 57 kg, was detonated in a vertical borehole. The center of the explosive charge was located 15 m below the ground surface in a 26-m thick sub-bituminous coal seam. The postshot cavity radius was estimated from calculations to be 0.17 m. The coal was highly jointed with two joint sets normal to each other. The coal seam was bedded with the bedding dipping at 18°. Mechanical properties as indicated from the logs did not vary appreciably from hole to hole. Permeability tests were conducted in situ from additional holes drilled around the shot hole. Packers were used to insulate sections of the measurement holes and pressure-testing was performed by using water injection until a quasi-steady-state condition was attained. Permeability was then measured at steady state.

Figure 4 shows the results of these measurements with the theoretical prediction equation (3.19) plotted as a straight line. The horizontal lines in the boxes are values of reported measurements in the same hole. These represent the uncertainty in the permeability. The width of the boxes along the radial direction is a measure of the uncertainty of the position of the hole at the permeability measurement location. A least-squares fit to the data[5] indicates that m = 0.5 and a permeability decay which is proportional to $1/r^3$. For this case, k_c = 427 darcies. Further details of the experiment are reported elsewhere.[8,21]

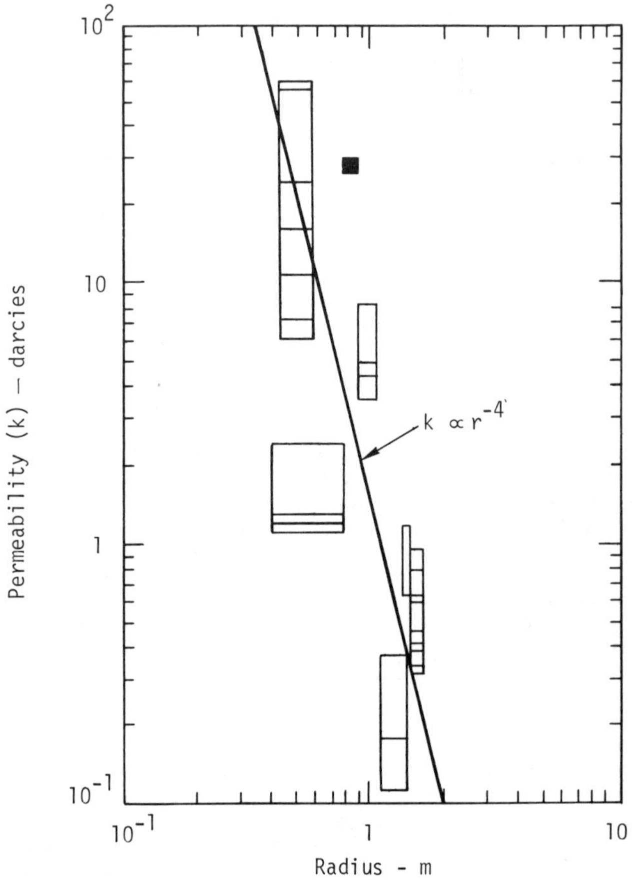

FIGURE 4. Postshot permeability vs radius for a chemical explosive detonated in a coal seam. The emplacement geometry possessed cylindrical symmetry. Permeability is in darcies, while the radial distance in meters is measured from the axis of symmetry. The small box on the upper right represents two measurements in the same borehole.

Despite a large variation in explosive yield and rock type between the Hardhat nuclear event and the Kemmerer coal experiment, the predictions of theory in both the spherical and the cylindrical symmetries are in agreement with the field measurements. When the rock behaves in a more inelastic manner, an analysis of the decay characteristics of the dynamic wave is required to provide a prediction of permeability decay away from the explosive cavity. In general, the inelastic characteristic will cause the amplitude of the dynamic wave to decay more rapidly than 1/r in the cylindrical case or $1/r^2$ for the spherical geometry. If the material is fairly competent and there is not too much porosity in the existing fracture pattern, then the inelastic behavior will result in the permeability decaying less rapidly as compared to the elastic case.

The data for this section are summarized in Table 1. The cavity wall permeability varies from approximately 2 to 400 darcies and is increased several hundredfold over the initial permeability. We note that

TABLE 1. PERMEABILITY ENHANCEMENT FACTOR

Site	Rock type	Explosive	k_o	k_c	k_c/k_o	m
Hardhat	Granodiorite	Nuclear	.001 d	65 d	65000	0
Hoggar	Granite	Nuclear	.005 d	1.7 d	340	1.2
Kemmerer	Coal	Chemical	1 d	427 d	427	0.5

a chemical explosive detonated in coal produces the highest cavity wall permeability. A nuclear explosive detonated in Hoggar granite--one of the hardest known granites having very few preshot fractures--resulted in the lowest cavity wall permeability, although the decay rate was much slower. Admittedly, the explosive results are limited. However, for design purposes we assume that k_c = 60 darcies and m = 0.5 are appropriate values for hard rock ore bodies. For a given ore body, it is obviously desirable to test these assumptions against field trials.

5.0 COMPARISON OF SINGLE AND MULTIPLE DETONATIONS

The effect of multiple detonations is a straightforward extension of the theory for single charges if certain assumptions are made. These are: (1) that energy fluxes from a number of detonations can be superimposed, (2) that the fracture surface area is directly proportional to cumulative energy input, and (3) that the elastic potential can be superimposed from many cavities to result in a total dilatation which, when positive, is proportional to the porosity increase. The third assumption is not strictly correct unless the bilinear effect is small. The three explosive events discussed above (Hoggar, Hardhat, and Kemmerer coal) tend to substantiate assumption three. We can conclude, therefore, that dilatation, to a close approximation, behaves linearly and that superposition is valid.

The total permeability from a single charge is

$$k = k_c \frac{(\phi/\phi_c)^3}{(n/n_c)^2} + k_o , \tag{5.1}$$

where ϕ is the explosion-created porosity (we are omitting the subscript t on k, the total permeability). Equation (5.1) is derived by substituting equation (3.3) into equation (3.13) and introducing k_c, ϕ_c, and n_o as proportionally and normalizing constants. Rewriting the fracture density to normalize it to n_c, we obtain

$$\frac{n - n_o}{n_c} = r_d^{-f-m}. \tag{5.2}$$

Similarly, from equations (3.10) and (3.11) we obtain

$$\frac{\phi}{\phi_c} = r_d^{-(f+1)} , \tag{5.3}$$

where α has been obtained from equation (3.11) with $\varepsilon << 1$.

Equation (5.1) can also be written in dimensionless form as

$$\frac{k - k_o}{k_c} = \frac{(\phi/\phi_c)^3}{(n/n_c)^2} = \frac{r_d^{-3f-3}}{(r_d^{-f-m} + n_o/n_c)^2} , \tag{5.4}$$

where $r_d = r/r_c$, and r_c is the cavity radius. Permeability scales with the postshot cavity radius r_c. The larger r_c, the greater the distance over which the permeability will be enhanced.

Figure 5 shows the compared effects of fracture density from equation (5.2), porosity from equation

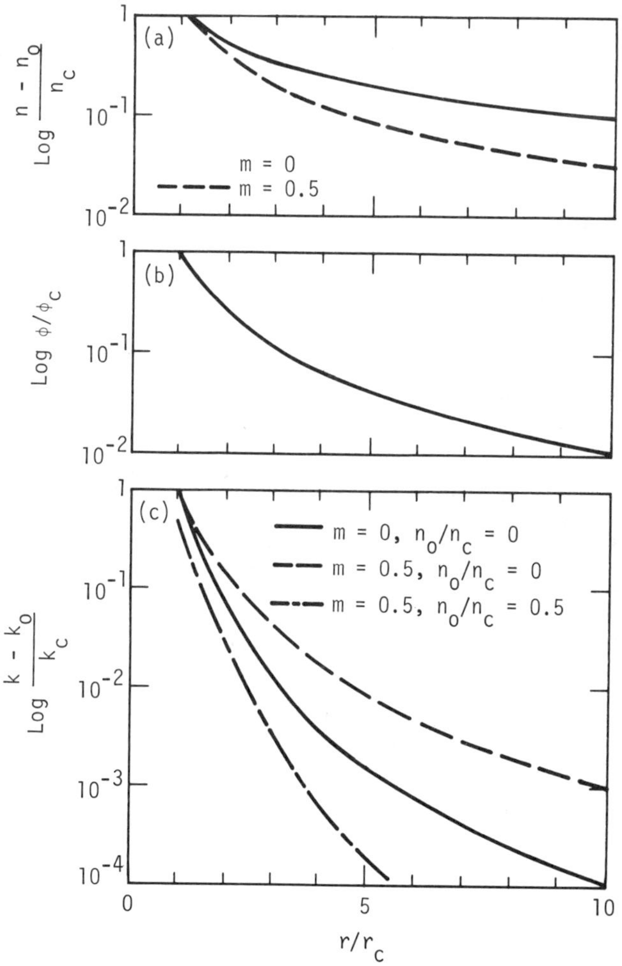

FIGURE 5. (a) Explosion-induced fracture-density enhancement, normalized to the fracture density at the cavity wall as a function of scaled distance. The solid curve represents a semielastic medium (m=0) and the dashed line a dissipative medium (m=0.5). (b) Explosion-induced porosity, normalized to porosity at the cavity wall as a function of scaled distance. (c) Explosion-induced permeability enhancement, normalized to the permeability at the cavity wall. Three cases are shown: a semielastic medium with zero initial fracture density, a dissipative medium (m=0.5) with zero initial fracture, and a dissipative medium with an initial fracture density of half the explosion-induced fracture density at the cavity wall (m=0.5).

(5.3), and scaled permeability from equation (5.6) for a single explosive charge in different media. Media are not quantified by rock type in a geologic sense, but rather in terms of mathematical parameters such as energy dissipation and initial fracture density required for the above equations. For hard rock deposits the term n_o/n_c is negligible, the exception being those which are relatively shallow and hence more highly weathered and fractured.

Permeability is the result of substituting the expressions for fracture density and porosity into equation (5.1) for permeability, yielding the formula

$$k = k_c \frac{\left(r_{1f}^{-2} + r_{2f}^{-2}\right)^3}{\left(r_{1f}^{-1-m} + r_{2f}^{-1-m} + n_o/n_c\right)^2} + k_o, \qquad (5.5)$$

where the variables are defined in Figure 6, k_c is the permeability due to a single explosion, and the distances are scaled in terms of r_c. The expression is valid when the two explosives have the same strength.

The curves in Figure 7 are obtained by evaluating equation (5.5) along the line of center (x-axis) as done for Figures 7a and 7b.

Permeability again declines rapidly due to the fact that porosity enters in the equations for permeability to the third power. We again see the danger of trying to correlate permeability with fracture density alone. Such a correlation would lead one to expect a more gradual decline in permeability with distance, which is incorrect.

All three curves show a slight decrease in permeability as the cavity wall is approached. In Figure 7c, it is only noticeable in the bottom curve. The decrease for the top two curves in Figure 7c is due to the fact that porosity decays more rapidly than fracture density, resulting in a higher contribution to surface area as each hole is approached. The decrease in permeability for the bottom curve is due to increased surface area from the presence of an initial fracture density. The values at $r/r_c = 5$ are doubled over that for a single charge for the top two curves. However, permeability for the bottom curve is enhanced by a factor of 5. This is due primarily to the increased porosity from the second explosive. Enough porosity is added to compensate for initial fracture density.

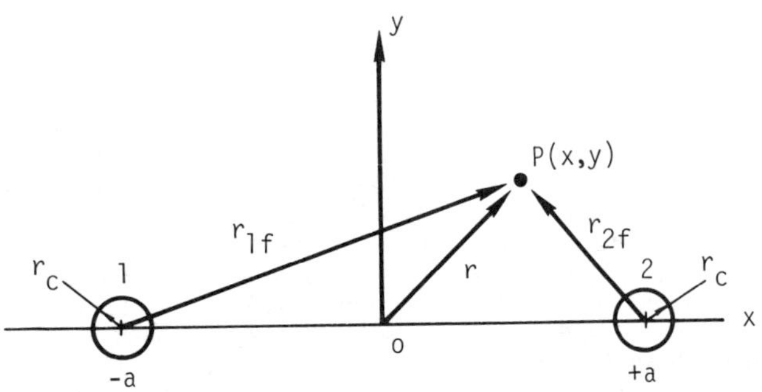

FIGURE 6. Schematic illustrating the coordinate system for superposition of single charge effects to simulate multiple detonations.

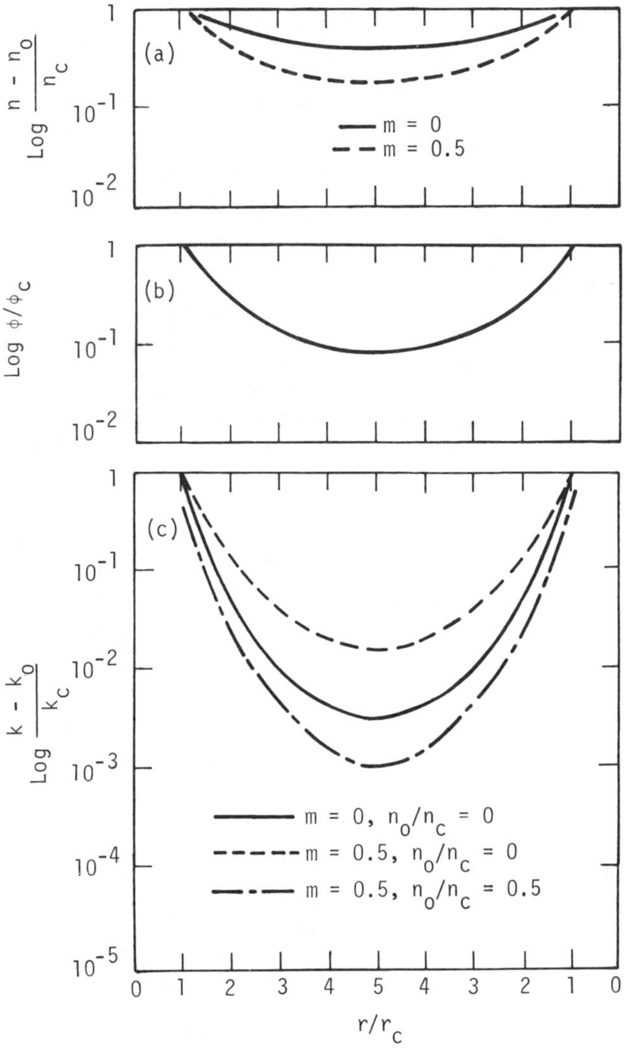

FIGURE 7. Parameter study for multiple detonations showing variations along the x-axis in terms of scaled distance. r_c is the cavity radius of a single charge. Explosive centers are separated by a distance 10 r_c. (a) Explosion-induced fracture-density enhancement between two cylindrical charges, normalized to the fracture density at the cavity wall of a single charge. The upper curve is for a semielastic medium (m=0) and the lower for a dissipative medium (m=0.5). (b) Explosion-induced porosity normalized to the induced porosity at the cavity wall between two cylindrical charges at 10 r_c spacing. (c) Explosion-induced permeability enhancement between two cylindrical charges.

Of the three cases shown, only the case m = 0 and m = 0.5 is relevant to in-situ mining. The third case is applicable only when the initial fracture density is comparable to that created by the explosion at the cavity wall. The only materials possessing this characteristic would be semiconsolidated sands, which are not good candidates for explosive stimulation. In a real rock the dissipation factor m is estimated to be approximately 0.5.

The explosive permeability distribution has one important property owing to its nonuniformity. From Figure 7 we note that the permeability is high near the wellbore and decreases to its lowest value midway between the two holes. This is ideal from the standpoint of well injectivity and productivity because most of the pressure drop in an ore body having uniform permeability occurs near the wellbore. The higher permeability near the wellbore in the case of explosives lowers the pressure drop and assists in enhancing circulation through the medium. For this reason the permeability at the midpoint can be lower than the values which we are accustomed to for permeable sands.

Table 2 shows the midpoint permeablity for two and four explosives. In the case of four explosives a square grid is assumed and the spacing is the diagonal spacing with the midpoints being the center of the square pattern. The permeability is doubled at the center of the square pattern. The square pattern corresponds to a direct line drive, with $d/a = 1$. From

TABLE 2. PERMEABILITY MIDPOINT BETWEEN 2 EXPLOSIVES AND AT THE CENTER OF A 4-EXPLOSIVE RECTANGULAR PATTERN WITH k_c = 60 DARCIES AND k_o = 3 MD.

Number of explosives	Spacing	m	Permeability (md)
2	10 r_c	0	183
2	10 r_c	0.5	963
2	15 r_c	0	5.4
2	15 r_c	0.5	40
4	10 r_c	0	363
4	10 r_c	0.5	1920
4	15 r_c	0	7.8
4	15 r_c	0.5	75

the table, permeability values range from a high of approximately 10 darcies to a low of 5 md. Permeability is sensitive to m, the stress wave dissipation exponent. The larger m, the greater the distance for permeability enhancement from the shot hole. However, this is achieved because of a lower fracture density. While permeability for two explosives spaced 15 r_c apart is 40 md when $m = 0.5$, this is still sufficient to obtain adequate flow rates, as will be shown in the hydrology section.

In summary, the results of this section indicate that explosive spacing is limited to a scaled distance of 10-15 times the postshot cavity or sprung hole radius. The dissipation factor m will be greater than zero and probably nearer 0.5 for hard rock ore bodies. The cavity wall permeability is estimated to be 60 darcies. We note that of the design parameters discussed above the postshot cavity radius is the most critical. This is because it controls drillhole and explosive spacings which are expected to have the most effect on the overall process economics. Accordingly, we address explosives and their effect on r_c and explosive spacing.

6.0 EXPLOSIVE SELECTION TO MAXIMIZE PERMEABILITY ENHANCEMENT

As can be seen from the previous sections the single most important scaling parameter is the postshot cavity radius or sprung hole diameter. From Figures 5 and 7 it can be seen that permeability enhancement is considerably diminished at a distance of 6-10 r_c from the center of the hole. Accordingly, the larger the postshot cavity radius r_c, the greater will be the distance over which the permeability is enhanced. This leads us naturally into the question of which explosive to use to produce the largest postshot cavity radius r_c, and therefore the maximum extent of permeability enhancement.

Rock blasting and explosives are treated exhaustively in a number of articles and books. Virtually all the applications are related to the proximity of a free surface. Reflection of the compressional stress from the free surface creates tensional stresses which spall the rock near the free surface and provide relief to enhance breakage. In-situ applications in which the explosive is deeply buried and hence far removed from a free surface have received little attention by comparison.[22]

In formulating a model for explosively expanding the hole from its initial dimensions to a larger diameter we must first understand explosive properties in addition to the rock properties. Explosives are classified into two categories: ideal and nonideal. Ideal explosives are characterized by a rapid release of energy. The chemical reactions occur on a time scale which is small compared to the time required for gases to expand and the stress wave to propagate into the surrounding rock. Ideal explosives are generally a homogeneous material consisting of one component. TNT, nitromethane, and PETN fall into this category.

Nonideal explosives, on the other hand, require a much longer time to release their energy. Such explosives are multicomponent systems, with fuel and oxidizer usually occurring as fine particulates in physically separated phases. Ammonium nitrate and fuel oil (ANFO) is one example.

The detonation velocity for both ideal and nonideal explosives generally lies in the range of 0.5 to 1.0 cm/microsecond. However, the time for total chemical reaction may be on the order of a microsecond for an ideal explosive such as PETN. A nonideal explosive such as ANFO reacts totally in approximately 50-100 μs;[23] the cavity is fully formed in about 3 milliseconds.[22] Their energy content, E_o, is released in a different manner, as illustrated in Figure 8.[24] In Figure 8a (ideal explosive), all the energy is released in the time required for the detonation wave to transverse the explosive. The nonideal explosive releases an initial amount of energy as the detonation wave passes which is significantly less than the total content, as shown in Figure 8b. This has the effect of slowing the decline of pressure in the cavity so that it can perform more work against the rock and expand the hole to a larger radius. The slower rate of energy release also provides a better impedence match to the rock.[25] It can also result in gas fracturing to appreciable distances from the cavity.[26]

The rapid release of energy from an ideal explosive tends to be manifested in higher frequency components in the stress wave which attenuates rapidly by dissipating its energy in the production of fines. A slower release of energy results in a longer wavelength, greater energy acceptance by the rock, and production of more desirable block sizes.

In enlarging the hole or cavity from its initial dimensions to its final sprung radius, the explosive gases continue to expand until they reach equilibrium with their surroundings. This problem was first studied by Butkovich.[27] He examined the output of extensive computer calculations and compared their productions with observed events from underground nuclear

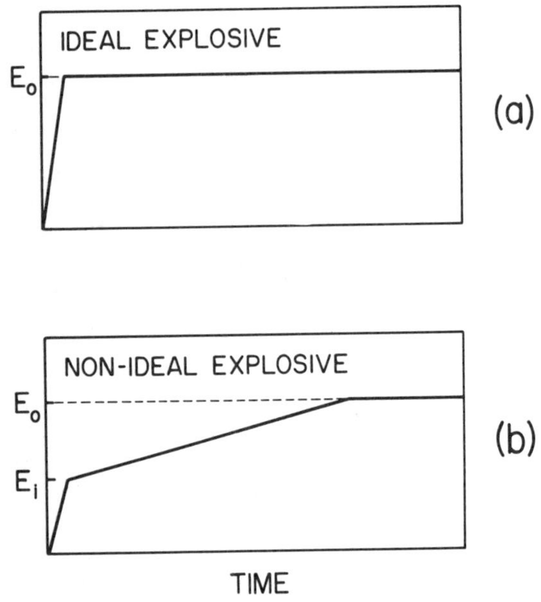

FIGURE 8. Energy release rate for ideal (a) and nonideal (b) explosives.

explosions. He found that the gases generated by the explosive expand to an equilibrium state which is determined by the overburden pressure (ρgh) and the strength properties of the medium.[22,27,28] The failure radius, extent of fracturing, and permeability enhancement were found to depend on the final cavity size, in agreement with our results. Several high explosives were compared by Butkovich[22] with different detonation velocities and C-J pressures. He concluded that there was no significant difference in the failure radius due to these quantities: "The main characteristic of the explosive in terms of high explosive fracturing is the value of v/v_o at the final cavity pressure." v/v_o is the volume expansion factor of the explosive, and hence the cavity, over its original specific volume v_o.

The equation describing the adiabatic expansion of an explosive is the JWL expression[29]

$$P_s = Ae^{-R_1V} + Be^{-R_2V} + CV^{-(\omega+1)} , \qquad (6.1)$$

where P_s is the pressure under constant entropy or adiabatic conditions, V is the relative value v/v_o, and the remaining constants are parameters appropriate to individual explosives. These parameters are listed in Table 3.[25,30] The qualitative composition of some of the explosives is given in Table 4. Expansion adiabats for the explosives listed in Table 4 are graphed in Figures 9 and 10. The explosives in these figures which perform best are those that produce the largest volumetric expansion in the final cavity pressure range, which is generally below 1 kb (10^8 PA).[5,6]

Examining the effectiveness of the explosives given in Figures 9 and 10 demonstrates that those with an ammonium nitrate base, added oxidizer, and aluminum performed best. These explosives have the highest density and energy content. While military-grade explosives such as TNT and PETN have higher initial pressures at $v/v_o = 1$, they lose their strength to expand the cavity in the critical range in comparison to DuPont's EL-836 and IRECO's DBA-65-T2. Figure 11 depicts the expansion curves in final cavity pressure range. The superiority of these two explosives is even more apparent.

For explosives in boreholes the initial (r_o) and postshot hole radius (r_c) are related by

$$r_c = r_o \left(\frac{v}{v}\right)^{1/2} , \qquad (6.2)$$

From the final cavity pressure p_f, the value v/v_o is determined using Figure 12 and substituting into equation (6.2). As previously discussed, the final explosive gas pressure at which the cavity stops growing is the sum of overburden pressure and some strength effect of the rock. Butkovich did not quantify the strength effect term to enable prediction of p_f from fundamental rock properties. This can be done by constructing a simplified model of the process, which we discuss below.

To obtain the final cavity pressure we assume the cavity growth ceases when the pressure exerted on the cavity wall falls just below the yield strength of the rock. An equilibrium solution for the stresses around a pressurized cavity gives for the stress[14]

$$\sigma_r = p_{ob}\left(1 - \frac{r_c^2}{r^2}\right) + p_c \frac{r_c^2}{r^2} \qquad (6.3)$$

$$\sigma_\theta = p_{ob}\left(1 + \frac{r_c^2}{r^2}\right) - p_c \frac{r_c^2}{r^2} \qquad (6.4)$$

where σ_r = radial stress
σ_θ = tangential stress
p_{ob} = overburden pressure
r_c = cavity radius
r = distance from the center of the cavity
p_c = final cavity pressure.

There are at least seven failure criteria which can be considered.[32] The one most frequently used is that of maximum distortional strain energy, also known as the maximum octahedral shear stress criterion, which is attributed to Von Mises.[14,32] For our use it has the form

$$(\sigma_r - \sigma_\theta)^2 + (\sigma_\theta - \sigma_z)^2 + (\sigma_z - \sigma_r)^2 = 2\sigma_o^2 . \qquad (6.5)$$

where σ_o is the yield stress in uniaxial compression.[32] This formula has been used by Scott and others to explain rupture around boreholes.[33] Evaluating the stresses at $r = r_c$ yields

$$\sigma_r = p_c$$

$$\sigma_\Theta = 2p_{ob} - p_c . \qquad (6.6)$$

Setting σ_z equal to the overburden pressure p_{ob} and substituting into equation (6.5) yields

$$p_c = p_{ob} + \frac{\sigma_o}{\sqrt{3}} , \qquad (6.7)$$

TABLE 3. EQUATION OF STATE PARAMETERS FOR EXPLOSIVES CITED

Name	C-J Parameters					JWL-EOS Coefficients					
	P Mbar	D M/sec	E_0 Mb•cc/cc	ρ g/cc	Γ	A	B	C	R_1	R_2	ω
ANFO	.060	4650	.0325	0.850	2.063	0.4760	.00524	.00720	3.5	.9	.31
POURVEX & POURVEX EXTRA	.130	6100	.0450	1.360	2.893	3.2207	.07769	.00324	4.7	1.4	.16
POURVEX III (EL-836)	.135	5790	.0920	1.520	2.775	2.8123	.02507	.01445	4.5	1.1	.20
DBA-65-T2	.120	5400	.0800	1.520	2.694	2.1467	.02157	.01295	4.3	1.4	.20
UNIGEL	.120	5760	.0510	1.262	2.490	1.9070	.07580	.00627	4.4	1.4	.23
AQUANAL	.055	3700	.0550	1.430	2.559	0.9123	.00407	.00746	4.4	1.0	.16
NITROMETHANE			.051	1.128		2.092	.05689	.0077	4.4	1.2	.30
TOVEX 600	.91	5808		1.572							
TNT	0.210	6930	0.0600	1.630	2.727	3.738	0.03747	0.00734	4.15	0.90	0.35
PETN	0.335	8300	0.1010	1.770	2.640	6.170	0.16926	0.00699	4.40	1.20	0.25

TABLE 4. QUALITATIVE COMPOSITION OF SELECTED COMMERCIAL BLASTING AGENTS[4]

Name	Mfgr	Density gm/cc	Ammonium nitrate	2nd oxidizer	Sensi-tizer	Aluminum	H_2O	Carbonaceous materials
POURVEX III (EL-836)	DuPont	1.52	X	X	X	X	X	X
DBA-65-T2	IRECO	1.52	X	X	X	X	X	X
AQUANAL	Atlas	1.43	X		X	X	X	X
POURVEX	DuPont	1.36	X	X	X		X	X
ANFO	Gulf	.85	X					X
UNIGEL	Hercules	1.28	X	X	X			X

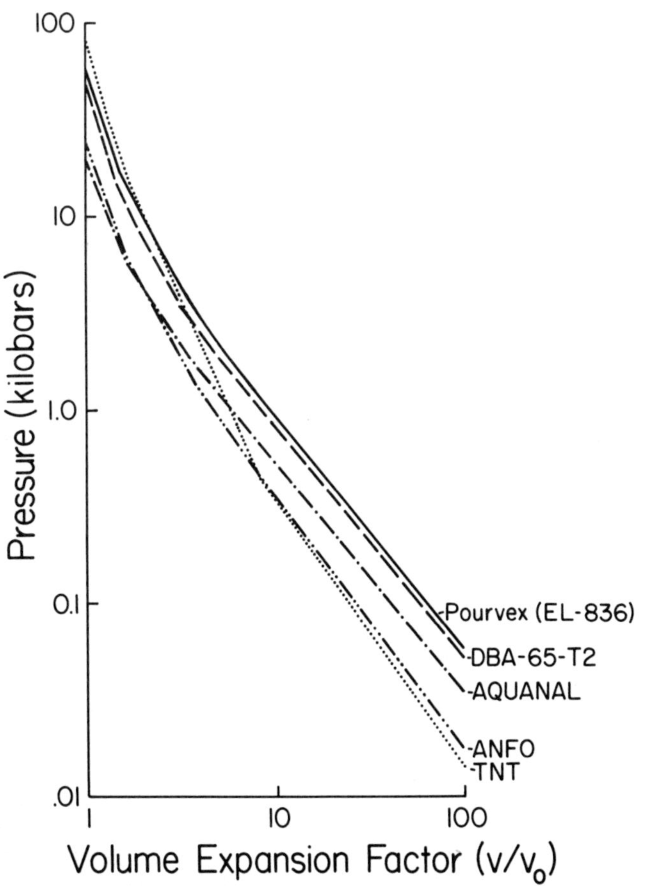

FIGURE 9. Explosive expansion adiabats.

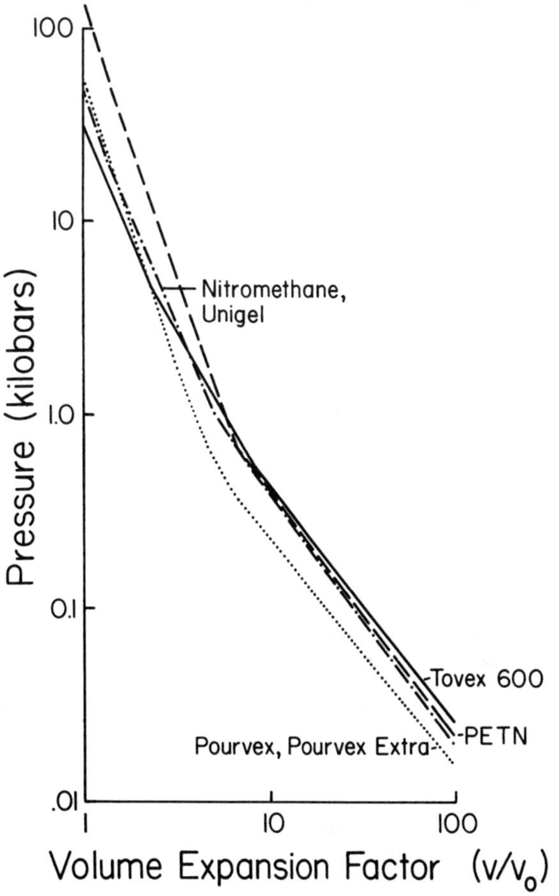

FIGURE 10. Explosive expansion adiabats.

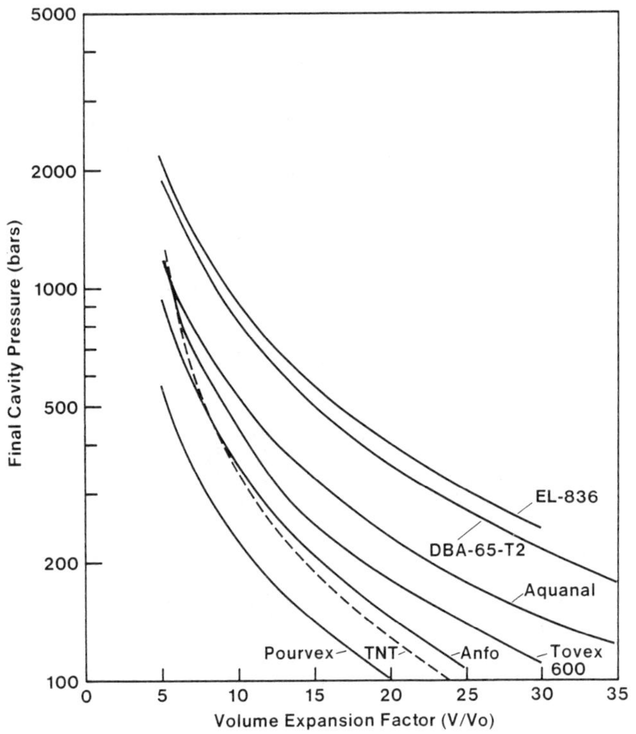

FIGURE 11. Explosive expansion curves, final cavity pressure range.

where the second term on the right is the shear strength of the rock.[32] The final cavity pressure at which expansion stops is therefore equal to the sum of the overburden pressure plus the shear strength of the rock. This is in general agreement with Butkovich's findings discussed earlier. If the explosive's expansion adiabat is represented by

$$p\left(v/v_o\right)^{\gamma} = c \ , \qquad (6.8)$$

where γ and c are determined for each explosive from Figure 11, then the cavity radius is given by the expression

$$r_c = \left(\frac{c}{p_{ob} + \frac{\sigma_o}{\sqrt{3}}} \right)^{\frac{1}{2\gamma}} r_o \ . \qquad (6.9)$$

As an example, let us consider a copper/uranium ore body with an average compressive yield strength of 0.9 kb at a depth of 675 ft. If the overburden density is 1.1 gm/cc, then the overburden pressure is 0.05 kb. Substituting into equation (6.7) results in a final cavity pressure of 0.57 kb. Returning to Figure 11, we can find the volume expansion factor appropriate to each explosive (Table 5).

From Figures 5 and 7 the power law range for permeability decay will limit the spacing of explosives. Depending on k_c (explosively created permeability at the cavity wall), and k_o (initial permeability), the spacing of explosives can probably be no greater than ten to fifteen times the postshot cavity radius apart.

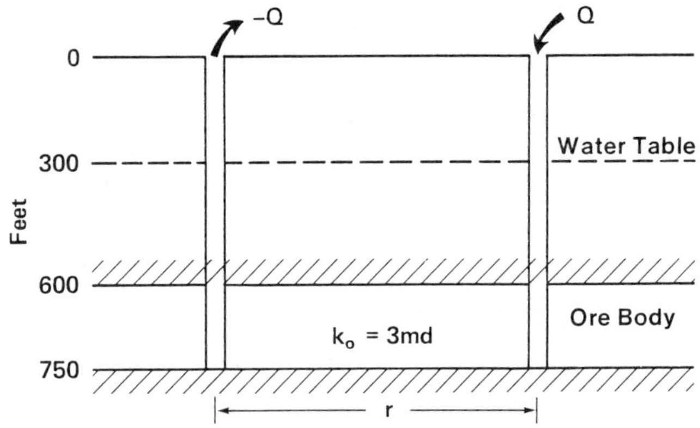

FIGURE 12. A model ore body used to illustrate the hydrologic concepts involved in explosive permeability enhancement. The effects of stimulating one well and then two wells are examined as to their injectivity and productivity.

TABLE 5. VOLUMETRIC EXPANSION FACTOR FOR COPPER/URANIUM ORE BODY HAVING POSTSHOT CAVITY PRESSURE OF 0.57 KB.

Explosive	v/v_o	r_c/r_o
EL-836	15.0	3.9
DBA-65-T2	13.5	3.7
AQUANAL	9.2	3.0
TOVEX 600	7.8	2.8
ANFO	7.0	2.6
TNT	7.2	2.7
POURVEX	5.0	2.2

Presented with this fact, two alternatives are available. The first is to drill a large-diameter hole, in the range of 24-30 in., with a drill such as the Ingersoll Superdrill. Or, one might drill a smaller hole approximately 10 in. in diameter and spring the hole to 3.25 ft (39 in.) in diameter using a high-energy explosive such as EL-836, or an equivalent product. This could be followed by reloading with a less expensive ammonium nitrate water-gel type of explosive or ANFO if conditions permit. The follow-up step with ANFO would spring the cavity to 8.5 ft in diameter (r_c = 4.25 ft). In view of the cost of EL-836 ($1.10/lb)[26] vs. ANFO ($0.10/lb)[26] the second approach may be more cost-effective. These two steps are correctly summarized in Tables 6 and 7, which give hole spacing for each type of explosive. This is based on a final cavity pressure of 0.57 kb from our previous example, and spacings of 15 r_c and 10 r_c as a function of initial hole diameter. Table 6 shows spacings of up to 95 ft using the highest energy explosive, with typical spacings in the range of 40-60 ft. Table 7, based on 10 r_c, is more conservative, with spacings of approximately 40 ft for the higher energy explosives. In each table the explosive boreholes must be loaded over a length which is approximately twice the well spacing, otherwise end effects are important and will reduce the spacing, the amount of reduction to be determined from more detailed computer calculations.

These calculations for the example problem demonstrate (a) that explosives must be loaded in large-diameter boreholes created either by direct drilling or by initially drilling smaller holes which are sprung and then reloaded; (b) that unless holes are sprung to larger dimensions and reloaded the practical

TABLE 6. HOLE SPACING FOR SELECTED EXPLOSIVES VS. HOLE DIAMETER. SPACING BASED ON 15 TIMES THE POSTSHOT SPRUNG HOLE RADIUS (r_c) AND CAVITY PRESSURE OF 0.57 KB.

	Diameter			
Explosive	10 in.	24 in.	30 in.	39 in.
EL-836	24 ft.	59 ft.	73 ft.	95 ft.
DBA-65-T2	23 ft.	56 ft.	69 ft.	90 ft.
AQUANAL	19 ft.	45 ft.	56 ft.	73 ft.
TOVEX 600	18 ft.	42 ft.	53 ft.	68 ft.
ANFO	16 ft.	39 ft.	49 ft.	63 ft.
TNT	17 ft.	41 ft.	51 ft.	66 ft.
POURVEX	14 ft.	33 ft.	41 ft.	54 ft.

TABLE 7. HOLE SPACING FOR SELECTED EXPLOSIVES VS. HOLE DIAMETER. SPACING BASED ON 10 TIMES THE POSTSHOT SPRUNG HOLE RADIUS (r_c) AND CAVITY PRESSURE OF 0.57 KB.

	Diameter			
Explosive	10 in.	24 in.	30 in.	39 in.
EL-836	16 ft.	39 ft.	49 ft.	63 ft.
DBA-65-T2	15 ft.	37 ft.	46 ft.	60 ft.
AQUANAL	13 ft.	30 ft.	38 ft.	49 ft.
TOVEX 600	12 ft.	28 ft.	35 ft.	46 ft.
ANFO	11 ft.	26 ft.	33 ft.	42 ft.
TNT	11 ft.	27 ft.	34 ft.	44 ft.
POURVEX	19 ft.	22 ft.	28 ft.	36 ft.

well spacing is limited to between 40 and 60 ft. If the initial permeability is in excess of 10 md and the initial shot hole diameter is large, then spacings as great as 90 ft may be achieved. Grade and thickness are then the primary considerations. While detailed economics are not given it is apparent that in-situ mining using chemical explosives is not necessarily for low-grade deposits. It is in fact like any other form of mining in which high grade and sufficient thickness lead to an economically viable mine.

7.0 HYDROLOGIC DESIGN CONSIDERATIONS

The equation to calculate the well flow rate, after the formation permeability is enhanced by a single-hole detonation, is[33]

$$\Delta h = \frac{Q}{2\pi T_o}\left[\ln\left(\frac{r_e}{r}\right) + \frac{1}{\alpha}\ln\frac{k_c\left(\frac{r_e}{r_c}\right)^{-\alpha} + k_o}{k_c\left(\frac{r}{r_c}\right)^{-\alpha} + k_o}\right] . \quad (7.1)$$

Setting $r = r_o$ (the preshot hole radius), in agreement with standard practice, gives the formula for skin enhancement due to the explosive

$$\Delta h = \frac{Q}{2\pi T_o}\left[\ln\left(\frac{r_e}{r_o}\right) + \text{skin}\right] , \quad (7.2)$$

where r_e is the effective radius, at which Δh tends to zero, and is given by

$$r_e^{\,2} = \frac{2.25\,T_o t}{S_o} , \quad (7.3)$$

where T_o and S_o are the undisturbed transmissivity and storage coefficient.

The skin factor indicates the degree of improvement in the well flow rate due to the explosive. For virtually all practical cases the skin factor will be negative unless in-situ stresses are large enough to make r_e less than r_c. The negative skin has the effect of reducing the pressure drop for a given flow rate as compared to preshot conditions.

Let us now consider the case shown in Figure 12 and use it as a model ore body to illustrate the hydrologic concepts involved in explosive permeability enhancement. Its permeability is taken as 3 md based on the correlation given in Figure 13.[34] Other ore body characteristics are

Depth to top of ore body	600 ft
Ore body thickness	150 ft
Depth to water table	300 ft
Ore body storage coefficient	10^{-5}
Preshot borehole radius	1 ft
Postshot cavity radius (r_c)	3 ft
Well spacing	10 r_c and 15_c

For each postshot hole, permeability has the form

$$k = k_c\left(\frac{r}{r_c}\right)^{-\alpha} + k , \quad (7.4)$$

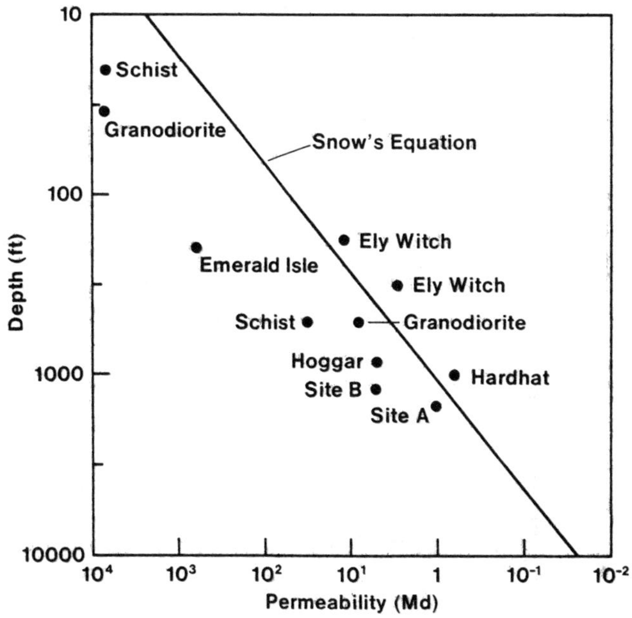

FIGURE 13. Permeability vs. depth for hard rock ore deposits. Most formations and ore bodies studied follow Snow's trend line but with generally higher permeability for a given depth. At depths of 1000 ft permeabilities in the range of 1-10 md may be expected.

with $\alpha = 4$, for elastic behavior, and $\alpha = 3$ to include the dissipation characteristics of real explosives.

In the case of two-well explosive detonations, there is one injection well and one production well separated by a distance of r. The fluid is circulated between the two wells. Applying the methods of superposition as an approximation, we can use previous equations to estimate the flow rate

$$\Delta h = \frac{Q}{2\pi T_o}\left\{\ln\left(\frac{r_e}{r}\right) + \text{skin} - \frac{1}{\alpha}\left[\frac{k_c\left(\frac{r_e}{r_c}\right)^{-\alpha} + k_o}{k_c\left(\frac{r}{r_c}\right)^{-\alpha} + k_o}\right]\right\}. \quad (7.5)$$

Using the same assumptions, we have estimated the flow rates for the cases of $r = 10\ r_c$ and $15\ r_c$.

The results of the calculations for single and two-well detonations are given in Figures 14 and 15. The injectivity in gpm is plotted as a function of the permeability enhancement factor k_c/k_o. Pumping rates for the single explosively stimulated well are in the range of several gpm as compared to a preshot rate of 1.5 gpm. With two wells working as an injection-production pair, the flow rate is considerably more than doubled. This is because the second well's enhanced permeability overlaps that of the first well and provides a pressure relief. The effect of confinement is contained in the third term of equation (7.5). This illustrates the importance of proper interpretation if the test uses only one explosive and the desirability of an additional shothole to obtain a better idea of the potential flow rates. To obtain an idea of the flow rates we refer to Table 1. While the data in the table rely mainly on nuclear explosives, we recall from section 2 that the acceptance stress into the rock is comparable with chemical explosives. Permeability enhancement effects should therefore bear some similarities.

If these assumptions are acceptable, while recognizing the limitations of the data, one could expect ratios of k_c/k_o in the range of 300 to several thousand. This implies that for the model ore body flow rates should be enhanced from ten gpm up to as much as 100 gpm. These rates are similar to those occurring in conventional permeable ore bodies in Texas and Wyoming and lend encouragement to the feasibility of in-situ mining of hard rock ore bodies.

8.0 IN-SITU MINING APPLICATIONS

A complex, deep-seated copper/uranium ore deposit will illustrate one in-situ scheme for metal recovery after blasting. Core samples from this deposit revealed the following characteristics:

1. Depth to ore	300 m
2. Ultimate compressive strength plus overburden stress	1.0 kb
3. Copper content	1-3% Cu
4. Uranium content	0.02-0.06% U_3O_8
5. Major minerals	Quartz, Magnatite, Hematite, Pyrite, Chalcopyrite, Feldspar, Mica
6. Minor minerals	Sphalerite, Bannerite, Gold, rare earths
7. Pyrite/Chalcopyrite ratio	Approx. 1.0
8. Fe/Cu ratio from sulfide minerals	Approx. 1.0

For this depth the cavity size would be only slightly reduced over that shown in the previous section.

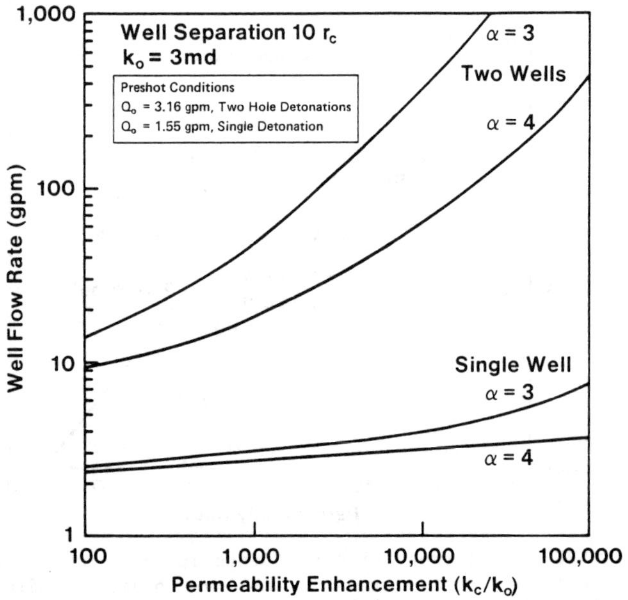

FIGURE 14. Injectivity-productivity for model ore body vs. explosive permeability enhancement, well separation 10 r_c.

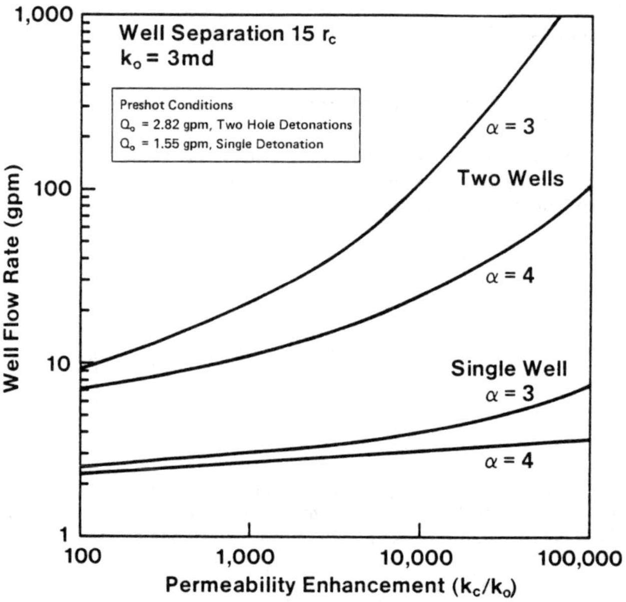

FIGURE 15. Injectivity-productivity for model ore body vs. explosive permeability enhancement, well separation 15 r_c.

8.1 Hydrometallurgy

Samples of a hard rock copper/uranium ore body were ground to -200 mesh for preliminary autoclave leach tests. The lixiviant[36,37] was water, adjusted to an initial pH of 2.0 with sulfuric acid, and a high concentration of dissolved oxygen. Final pH after leaching tests varied from 3 to 4. Final condition of ground samples showed almost complete conversion to fine particles of hematite and quartz. Results of four sample leach tests are shown in Table 8.

TABLE 8. LEACH TEST RESULTS

Test no.	Temp (°C)	Agitation rate (rpm)	Oxygen press. (psi)	Leach time (days)	Copper recovery (% Cu)	Uranium recovery (% U_3O_8)
1	70	1000	800	5	56	99
2	90	400	500	19	69	70
3	90	400	1000	7	91	88
4	90	None	1000	2	--	--

Test #4 was a leach test primarily on the host rock material to determine the rate of disintegration. Five sample cubes of 1 cm on a side were used. After one day almost total disintegration of the cubes was observed, and the product was primarily fine and coarse hematite and quartz particles.

The results of these and other tests indicate that in-situ solution mining of this particular ore using oxygen at high pressure can potentially leach at an economical rate, and that the host rock would disintegrate at a rate fast enough to keep pace with sulfide leaching. This would provide chemical enhancement of permeability above that induced by the explosives, provided excessive fines were not generated in the process.

8.2 Well Completion

As mentioned previously, we assumed that fairly large diameter blastholes could be drilled to moderate depths, economically, using air percussion techniques as exemplified by the Ingersoll Rand Superdrill.

One possible technique is presented to complete injection and production wells that should be relatively inexpensive. After detonation the blastholes above the mineralized zone are shown in Figure 16 to be cased with either PVC or fiberglass pipe (6-12 in. diameter). This outer casing would be secured in place with acid-resistant concrete.

Inside the injection well casing are shown two pipes. The inner pipe is a retrievable perforated stainless steel downhole injector pipe. Around it is shown a fiberglass pipe for protection of the s.s. injector. The inner two pipes are set in place as deep as possible below the overburden using high-pressure water jetting, followed by insertion of the injector pipe. In actual practice, injection and production wells should be interchangeable to reverse flow of lixiviant for better utilization of oxygen, thereby increasing overall metal recovery.

8.3 Wellfield Pumping Facility

Figure 17 shows a high-pressure circulation pump moving the lixiviant through the induced-fractured ore zone. This high-pressure circulation loop is developed in line with a filtration unit; such a filtration system is important to any in-situ project since considerable solid waste (i.e., sand and hematite) will be recovered during in-situ leaching and should be disposed of, continuously, with a minimum loss of dissolved oxygen.

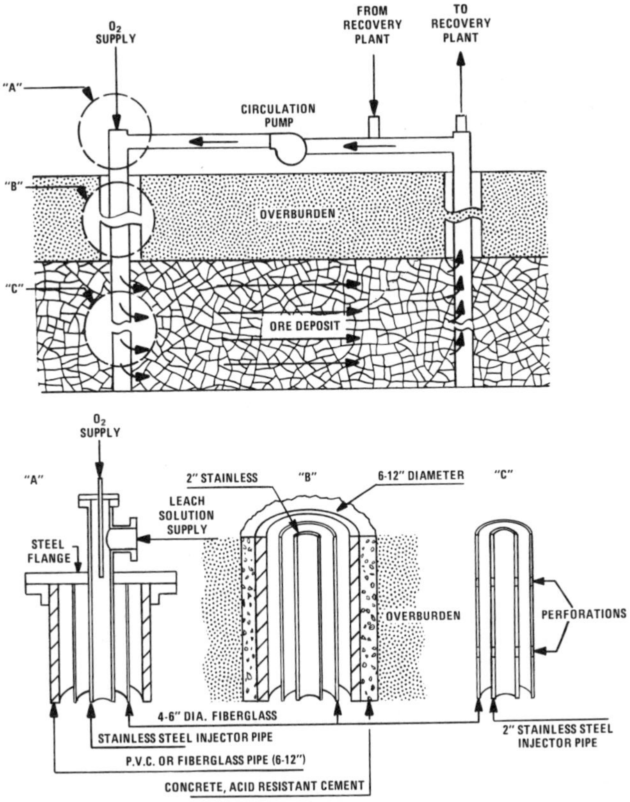

FIGURE 16. Well completion.

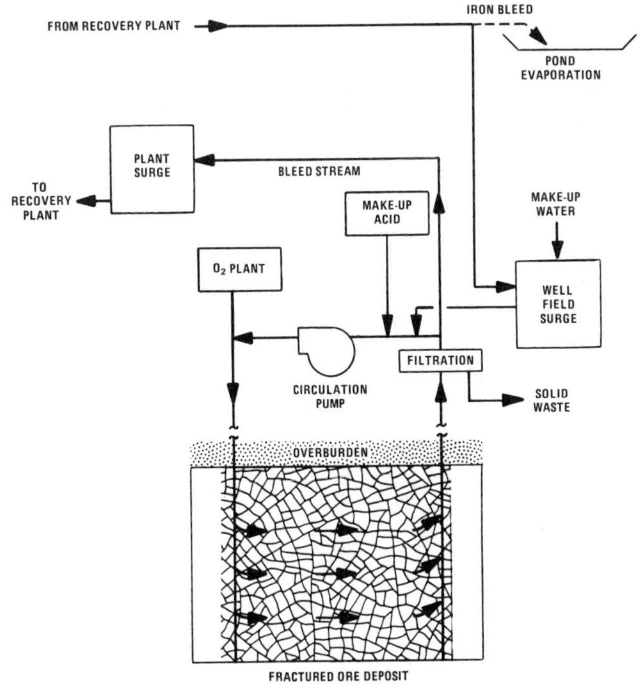

FIGURE 17. Wellfield pumping plant.

8.4 Metallurgical Facility

Figure 18 shows a bleed stream being diverted to a metallurgical plant when metal values reach a desired level. This bleed stream is simultaneously replaced with barren lixiviant from wellfield surge.

Figure 18 is a simplified flow diagram for the recovery of both uranium and copper. Uranium is recovered by ion exchange using an acid elution scheme followed by peroxide precipitation. The uranium peroxide slurry is pumped to a vacuum dryer. The copper is recovered by the more commonly used solvent extraction method. Any organic reagent loss returned to the wellfield system will be oxidized in situ.

If gold and/or silver are present in this type of ore, their recovery is now possible using any one of several lixiviants developed by In-Situ, Inc., that are environmentally acceptable. The low-pH lixiviant used for leaching copper and uranium is first transferred to a new pattern, and water with the new gold or silver leach reagent is added. No oxygen is needed in this new leach circuit. Gold or silver are recovered using a new ion exchange process. Resin elution is fast and complete in less than 10 bed volumes, and the pregnant eluate is ideally suited for direct electrolytic recovery.

In summary, the practical aspects of in-situ mining are presented whereby blasting induces the initial permeability necessary for startup of a leach circuit. A powerful, inexpensive lixiviant is shown to be capable of leaching the metal sulfides, thereby increasing the fractured permeability of the host rock matrix. Feasible methods for drilling, well completion, leaching, and metal recovery are discussed to demonstrate one approach to develop the in-situ mine.

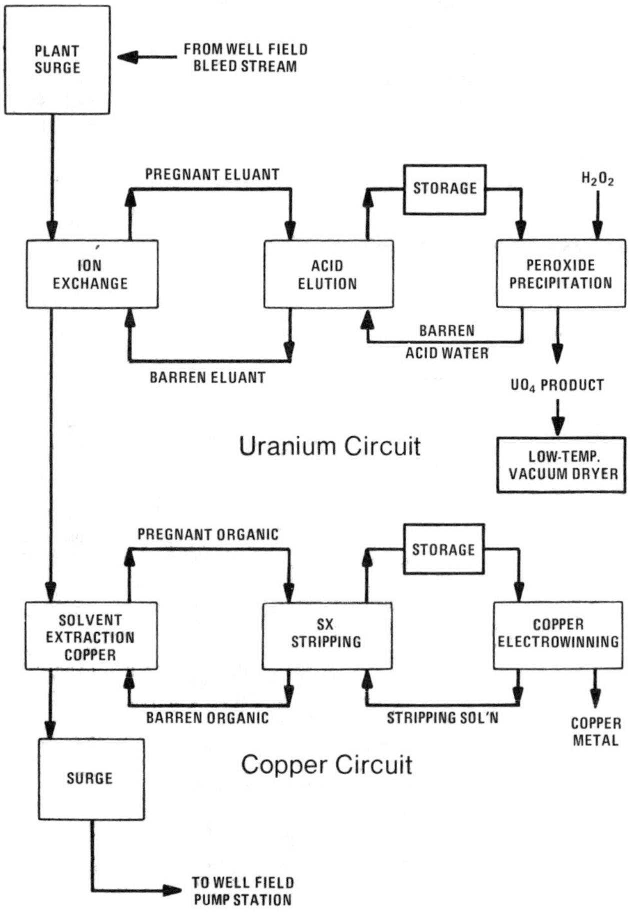

FIGURE 18. Recovery plant flow diagram.

ACKNOWLEDGMENTS

The authors express their appreciation to Ted Butkovich for his many helpful comments and useful discussions, and to Jane Reverand for preparing the manuscript.

REFERENCES

1. Wadsworth, M. E., "Interfacing Technologies in Solution Mining," Mining Engineering, Vol. 29, No. 12, Dec. 1977, pp. 30-33.

2. Hanson, M. E., Terhune, R. W., and McKee, C. R., "Explosion Phenomenology and Permeability Enhancement in Earth Media," Proceedings, 1981 Explosives Conference of the International Association of Drilling Contractors, June 9-11, International Association of Drilling Contractors, Houston, TX, 1981.

3. McKee, C. R., and Hanson, M. E., "Explosively Created Permeability from Single Charges," Society of Petroleum Engineers Journal, Vol. 15, No. 6, Dec. 1975, pp. 495-501.

4. McKee, C. R., Hanson, M. E., and Terhune, R. W., "Permeability from Single and Multiple Detonations of Explosive Charges," World Mining and Methods Technology, Weiss, A., ed., Vol. 1, AIME, New York, 1972, p. 365.

5. McKee, C. R., Hanson, M. E., and Terhune, R. W., "Permeability from Single and Multiple Detonations in Boreholes," In-Situ, Vol. 1, No. 1, 1977, pp. 37-73.

6. Laspe, C., "How to Predict Explosive Stimulation Results," Petroleum Engineering, May 1971, p. 68.

7. Terhune, R., and Shaw, J., "Calculation of Rock Fracturing from Multiple Nuclear Explosive Sources," Rept. UCRL-74017, 1972, Lawrence Livermore Laboratory, Livermore, CA.

8. Butkovich, T., "Correlations between Measurements and Calculations of High-Explosive-Induced Fractures in a Coal Outcrop," International Journal of Rock Mechanics and Mining Science, Vol. 13, 1976, pp. 45-51.

9. Schatz, J., "SOC73, A One-Dimensional Wave Propagation Code for Rock Media," Rept. UCRL-51689, 1974, Lawrence Livermore Laboratory, Livermore, CA.

10. Carmen, P., "Fluid Flow through Granular Beds," Transactions, Inst. Chem. Engrs., London, Vol. 15, 1937, p. 150.

11. Langefors, U., and Kihlstrom, B., The Modern Technique of Rock Blasting, John Wiley & Sons, New York, 1967.

12. Felts, L., Clark, G., and Yanick, J., "A Laboratory Method of Determining the Thermodynamic Efficiency of High Explosives," Trans. AIME, March 1956, p. 318.

13. Rose, H. E., "A Comprehensive Theory of the Comminution Process," Proceedings, 2nd European Symposium on Size Reduction (1966), Decima-Monograph No. 993-1026, Verlag Chemie, Amsterdam, 1967.

14. Jaeger, C., and Cook, N. G. W., Fundamentals of Rock Mechanics, 3d ed., Chapman & Hall, London, 1979.

15. Hanson, M., and McKee, C., "SAFE (Synergetic Acoustic Fracturing Effect)," Rept. UCRL-51822, 1975, Lawrence Livermore Laboratory, Livermore, CA.

16. Borg, I. Y., "Some Shock Effects in Granodiorite to 270 Kilobars at the Piledriver Site," Geophysical Monograph Series Vol. 16, American Geophysical Union, Washington, D.C., 1972.

17. Selberg, H. L., "Transient Compression Waves from Spherical and Cylindrical Cavities," Arkiv for Fysik, Vol. 5, 1952, p. 97.

18. Boardman, C. R., and Skrove, J., "Distribution in Fracture Permeability of a Granite Rock Mass Following a Contained Nuclear Explosion," Trans. AIME, Vol. 237, 1966, p. 619.

19. Delort, F., and Supiot, F., "Nuclear Stimulation of Oil Reservoirs," Proceedings, Symposium on Engineering with Nuclear Explosives, Conf. 700101, American Nuclear Society, Hinsdale, IL, Vol. 1, p. 649, 1970.

20. Terhune, R. W., "Comparison of Computer Model Predictions of Nuclear Explosion Effects in Hoggar Granite with Experimental Measurements," Rept. UCRL-74306, 1972, Lawrence Livermore Laboratory, Livermore, CA.

21. Hearst, J. R., "Fractures Induced by a Contained Explsion in Kemmerer Coal," Rept. UCRL-57790, 1975, Lawrence Livermore Laboratory, Livermore, CA.

22. Butkovich, T. R., "Some Studies of Deeply Buried High Explosive Rock Fracturing," Rept. UCRL-16526, 1974, Lawrence Livermore Laboratory, Livermore, CA.

23. Cook, M. A., The Science of Industrial Explosives. IRECO Chemicals, Salt Lake City, UT, 1974.

24. Larson, D. B., et al., "In-Situ Coal Gasification, Progress on the Experimental and Calculational Coal Fracture Program, July-November 1975," Rept. UCID-16651, 1975, Lawrence Livermore Laboratory, Livermore, CA.

25. Finger, M., et al., "Characterization of Commercial, Composite Explosives," Proceedings, 6th Detonation Symposium, Office of Naval Research Report No. ACR-221, 1976, pp. 729-739.

26. Coursen, D. L., DuPont Research, Potomac River Laboratory, Martensburg, WV, private communication, November 1981.

27. Butkovich, T. R., "Use of High Explosives for Springing Underground Cavities," Rept. UCID-15145, May 1967, Lawrence Livermore Laboratory, Livermore, CA.

28. Butkovich, T. R., "The Gas Equation of State for Natural Materials," Rept. UCRL-14729, 1967, Lawrence Livermore Laboratory, Livermore, CA.

29. Lee, E. L., and Hornig, H. C., "Adiabatic Expansion of High Explosive Detonation Products," Rept. UCRL-50422, 1968, Lawrence Livermore Laboratory, Livermore, CA.

30. Finger, M., et al., "The Effect of Elemental Composition on the Detonation Behavior of Explosives," Proceedings, 6th Detonation Symposium, Office of Naval Research Report No. ACR-221, 1976, pp. 710-722.

31. Michaud, L., "Explosions nuclaires souterraines des rayons de carite," Rept. CEA-R-3594, 1968, Commissariat a l'Energie Atomic.

32. Jaeger, J. C., Elasticity, Fracture, and Flow, John Wiley & Sons, New York, 1956.

33. Scott, P. P., Bearden, W. G., and Howard, G. C., "Roof Rupture as Affected by Fluid Properties," Trans. AIME, Vol. 198, 1953, pp. 111-124.

34. McKee, C. R., and Way, S. C., "Pressure Response Equations for Explosively Stimulated Wells," to be published 1982.

35. McKee, C. R., "Permeability of Hard Rock Ore Deposits," to be published 1982.

36. Pitt, C. H., and Wadsworth, M. E., "An Assessment of Energy Requirements in Proven and New Copper Processes," Rept. DOE/CS/40132, 1980, NTIS, Springfield, VA.

37. U.S. Energy Research & Development Administration, Kennecott Copper Corp., and Lawrence Livermore Laboratory, "In Situ Leaching of a Nuclear Rublized Copper Ore Body," 2 vols., Rept. NVO-155, June 1975, U.S. Energy Research & Development Administration.

Chapter 11

FEASIBILITY OF IN SITU LEACHING OF METALLIC ORES OTHER THAN COPPER AND URANIUM

G. M. Potter
Tucson, Arizona 85731

C. K. Chase
Mountain States Engineers
Tucson, Arizona 85731

P. G. Chamberlain
U. S. Department of the Interior
Bureau of Mines
Minneapolis, Minnesota 55417

ABSTRACT

Under contractual arrangements with the Bureau of Mines, Mountain States Research and Development, Inc. (MSRD) completed a technological investigation to determine the feasibility of in situ mining of metal commodities other than copper and uranium. The study was based upon available literature and experience. The selection of potential candidates for in situ mining was influenced by economic importance, probable amenability to in situ methods, and environmental considerations. Four metals (or metal pairs) were chosen for feasibility studies: aluminum (with a copper byproduct), gold-silver, and manganese were recommended for further study and action; cobalt-nickel was also recommended contingent upon preliminary laboratory leaching and recovery investigations and field geohydrological work. Mining and recovery plans and an economic evaluation are included. Other metals with potential for in situ leaching are lead/zinc, molybdenum, and vanadium.

Introduction

In the last decade, solution mining has become a significant mining technique for copper, uranium, gold and silver. Although dump and heap leaching operations predominate, in situ leaching operations are increasing, particularly for uranium, as knowledge and experience with this mining technique overcome operational problems. Other metals have been leached as secondary minerals during copper leaching operations. No in situ leaching operations are currently targeted, however, for commodities other than copper and uranium. Since in situ leaching has proven successful for uranium and copper and since in situ leaching offers the advantages of potentially low environmental damage, safe working environment, less capital requirements, and shorter startup times, a detailed investigation of in situ leaching of "other" metals was initiated by the Bureau of Mines under its Mineral Resources Technology Program.

As a first step in this investigation, the Bureau awarded a contract to Mountain States Research and Development, Tucson, Arizona, for determining the feasibility of in situ leaching metals other than copper and uranium. Since laboratory tests on the leachability of various metallic minerals abound, the contract was basically a literature survey to determine which metals could be most readily leached in an in situ environment, the leaching solutions required, the reactions involved, effects of host rock, and environmental factors to be considered. The Contractor was also expected to project commodity supply/demand information to establish priority minerals for domestic exploration by in situ leaching.

Based on the information gathered, four commodities were selected as the most desirable candidates for in situ leaching development. After detailed data were gathered on geologic settings of typical deposits of each commodity, evaluations were made of topographical, hydrological, and climatic conditions that might affect an in situ leaching operation. The Contractor then developed a conceptual design for a leaching operation for each commodity based on the so-called "typical" deposit. Each conceptual design was subjected to an economic analysis. Finally, the Contractor identified problems that still must be resolved before commercial operations could be developed and provided recommendations on suitable research for solving the problems. This paper summarizes the Contractor's findings (Potter et al, 1981) regarding feasibility of in situ leaching of metals other than copper and uranium.

Description of In Situ Mining Technology

The technology of in situ mining has been presented in an article by Dr. Roshan B. Bhappu, (1978), in a paper covering leaching techniques, controlling parameters in leaching and metal recovery phases, and research needs.

In situ mining, also called chemical or solution mining, is defined as the in-place extraction of metals from ores located within the confines of a mine (unfractured or fractured ore, stope filling, caved material, and ores in permeable zones). In situ mining technology comprises the preparation of ore for subsequent in-place leaching, the flow of leaching solutions through rock masses and within rock pores, the dissolution of minerals with inexpensive, regenerable reagents under in-place conditions, the preparation and regeneration of such solutions, and the recovery of marketable metals or compounds from the solutions. In situ mining requires the interdisciplinary application of geology, geohydrology, mining, metallurgy, and environmental science and technology in achieving economical metals production.

In situ leaching situations may be conveniently classified in three types, namely (1) deposits above water table, (2) deposits under water table but near surface, and (3) deep deposits, normally under the water table.

Notable advantages of in situ methods include lower operating costs, minimal surface disturbance, greatly reduced subsidence, and improved safety. A main disadvantage is the need for strict solution control.

Selection of Mineral Commodities

As an aid in the selection of the four prime commodities (or pairs) requested by the Bureau of Mines for the preparation of in situ mining and recovery plans, MSRD tabulated key factors relating to potential candidates. Key factors listed were (1) economic and strategic importance, (2) availability of U.S. deposits, (3) extractive metallurgy (leaching and recovery), (4) environmental considerations, and (5) suitability for in situ processing. Commodities used mainly as nonmetallic or industrial minerals or commonly mined as soluble salts were not included. Certain rare earths and radioactive materials were excluded because they are used in small quantities. Iron, still heavily mined from U.S. open pits, was omitted.

The Bureau of Mines data on present U.S. commodity demand and projected needs were helpful in assessing their economic importance. Bureau of Mines information was also useful in determining the strategic importance of various commodities to the U.S. economy. The probable availability of U.S. deposits of various metals was determined by survey of government and general literature. Similarly, all readily available background on geological considerations was reviewed. The dissemination of the ore mineral, accessibility of leaching solutions through interconnected pores and fractures, physical dimensions of leachable zones, and competence of enclosing rocks were reviewed. Of particular importance were geohydrological considerations affecting the introduction and recovery of solutions, without losses in surrounding rocks.

The economic extraction of metallic commodities is dependent upon the solubility of the ore minerals in inexpensive solutions and the accessibility of the ore minerals to the leaching solution. The reactivity of gangue minerals to the selected solution directly affects reagent consumption. A marketable product must be economically recoverable from leach liquors. Solution control is essential for environmental reasons (to prevent escape of polluting substances to groundwaters) and for economic considerations. The general suitability of candidate commodities for in situ mining greatly depends upon the economics of the foregoing.

By applying the criteria already described, a list of eight prime possibilities was prepared, namely:

Aluminum	Molybdenum
Gold-Silver	Cobalt-Nickel
Lead-Zinc	Vanadium
Manganese	Tungsten

From this list the final commodities for developing a leach mining plan were selected. They are aluminum, gold-silver, manganese, and cobalt-nickel. The lead-zinc pair remains a strong possibility for future consideration, and additional deposit review is recommendable. Molybdenum, although not chosen for one of the four immediate mining studies, remains a good future possibility. Vanadium may also be a good potential candidate, although many of the deposits are carboniferous and are preferably roasted before leaching. Tungsten was deferred because many of the host rocks have low permeability and the probable choice of caustic leaching solutions might tend to attack silicates, with attendant solution jelling problems.

Once the list of candidate commodities was narrowed to four, additional evaluation of domestic deposits led to the selection of a typical deposit for each that would serve as the basis for a conceptual design. The four deposits selected and the in situ leaching system designed for each are discussed in the following section.

Conceptual Designs for Leaching

Aluminum

The typical aluminum site chosen was the Carlota Mine area near Globe, Arizona, where a weathered porphyritic copper orebody capping offers potential aluminum production from in situ mining plus welcome copper byproduct revenue.

In the Carlota area, Gila conglomerate contains copper deposits related to volcanic activity. The conglomerate is made up of sands, tuffs, basalts, limestones, dacite, and diabase. The cap rock has been altered sufficiently to provide acid soluble alunite and other aluminum minerals, but is still competent enough to withstand leaching without greatly decreasing permeability. A 42-million ton block of leachable ore is assumed.

The deposit would be prepared for leaching by driving drifts and block caving the ore. Since caving merely has to create enough fracturing for solution flow, only 10 to 15 percent of the ore would be removed to promote caving. Aluminum and copper would be leached by percolating dilute sulfuric acid through the deposit by injection holes drilled from the surface. The pregnant solution would drain into collection drifts beneath the orebody for pumping to the surface. Since the water table is within 200 ft (60 m) of the surface, leaching solutions will be readily controlled by a net groundwater drainage into the rubblized ore. The aluminum will be recovered from the effluent solution by means of a solvent extraction process such as developed on pilot scale by the Bureau of Mines (George, et al, 1968). Copper is also recovered from the solutions by solvent extraction prior to the aluminum recovery step.

The mining and processing plan as illustrated in figure 1, provides for the production of 15,860 metric tons of alumina and 8,400 metric tons of copper per year for about 23 years. In the economic evaluation of the plan, assumed alumina prices of $200 to $280 per metric ton and copper prices of $1 per pound ($2.20/kg) was used in the computer program. Preproduction costs of $13 million and $36 million depreciable capital are estimated. Discounted cash flow (DCF) returns on investment varies from 7.5 percent (10 years payout) to 9.5 percent (9.4 years payout). Although return on the investment is somewhat low because of relatively low prices for alumina, the alumina candidate is recommended for further study and action as a strategic material and probable future production necessity.

Research needs include further areal geohydrological investigation for solution control plus bench and pilot scale confirmation of the Bureau solvent extraction flowsheet for the recovery of alumina (and copper) from sulfate liquors.

Gold-Silver

To be treated by in situ leaching, gold-silver bearing rock should be competent, porous, relatively cyanicide-free, and contain fine-sized gold and silver particles. Clay content must be low enough so that the ore remains permeable to percolating solutions during leaching. Since numerous deposits are being heap leached, technical data on the hydrometallurgical aspects of leaching gold/silver are readily available (for instance, Chamberlain and Pojar, 1981).

The Gold Quarry deposit near Carlin, Nevada, was chosen as a representative shallow gold-silver site. Like the ore at the Carlin mine, the Gold Quarry ore is fine grained and disseminated. The 3.5 million ton (3.2×10^9 kg) deposit is in the Tuscarora Mountains and contains 0.066 oz/ton gold (2×10^{-4} kg/ton) in iron stained, sheared and fractured quartzite and chert.

It is proposed that the orebody be broken in-place by controlled blasting to prepare it for in situ leaching (figure 2). The mining plan calls for 9-inch (0.23 m) vertical blasting holes to be drilled on an approximate 12-foot (3.7 m) grid pattern. The blast holes range from 100 to 200 feet deep (30 to 60 m). After the ore is fragmented, benches would be constructed to prevent runoff and sprays would be used to apply the leaching solution. A diversion ditch must be constructed so that run-off during rains would not interfere with leaching operations. Actual leaching of a 440,000 square foot (41,000 m^2) area will require 2,000 gpm (0.13 m^3/sec) of solution. To insure against solution losses, an underground collecting system would be provided, consisting of drifts under the broken zone with drill holes to the broken zone, all leading to a pump shaft. An alternative would be to simply pump the pregnant solution to the surface through vertical recovery wells. Solution strengths may range from 0.05 to 0.1 percent NaCN, depending on ore requirements, and the pH may be maintained at about 10.5 with lime (CaO) or caustic soda (NaOH). The rate of solution application by sprays may be 5 gallons per square foot (0.2 m^3/m^2) per 24 hours. Adequate storage should be provided for pregnant solution between the leaching step and the gold recovery step that follows.

A five- to eight-stage countercurrent column using minus 10-plus 20-mesh activated carbon for gold recovery from the leach liquor would be expected to produce a 0.0002-ounce per ton barren solution from 0.10 to 0.20 ounce per ton pregnant solution and to load the carbon to at least 400 ounces per ton, while flowing at the rate of 15 to 20 gallons per minute per square foot of column cross section. The loaded carbon would be stripped, and the gold and silver electrowon on steel wool for subsequent smelting to marketable doré.

The economic evaluation of the system was highly attractive. The production of 137,800 troy ounces of gold (silver ignored) was evaluated in a 3-year life period using prices of \$400, \$450, \$500, \$550, and \$600 per ounce. The resulting estimated DCF return on investment ranged from 64.6 percent to 98.1 percent, while the payback ranged from 0.93 to 0.60 years. A 60 percent recovery over 4 years was assumed in the economic analysis.

The potential impact on groundwater of in situ leaching with cyanide has caused hesitance among potential leaching operators. The public attitude towards cyanide would probably make it impossible to gain approval for leaching where potable aquifers could directly be contacted by the cyanide. For nonpotable water supplies, however, experience with sulfuric acid and caustic leaching of copper and uranium has shown that solutions can be controlled at or beneath the water table by proper pumping programs.

Manganese

The Artillery Peak area of west central Arizona was selected for the typical manganese in situ site. About 175 million tons of 3.5 to 4 percent manganese ore is indicated in the region. Early Pliocene alluvial fan and playa deposits form the principal manganese-bearing formations; basalt cappings are common.

Test work by the Bureau of Mines (Bender and Rampacek, 1957; Dean, et al, 1934; Rampacek, et al, 1959; and Wyman and Ravitz, 1947) and private individuals indicates that the Artillery Peak ore is generally permeable to leaching solutions and that the manganese minerals respond to leaching with sulfurous acid. The essential reaction is:

$$MnO_2 + SO_2 \rightarrow MnSO_4$$

Two plans were selected for breaking the initial orebody for in situ leaching (figure 3):

1) Blasting in place for portions of the orebody that have little or no overburden.

2) Breaking by block caving for that portion of the orebody covered by basalt.

Elemental sulfur will be burned and the SO_2 absorbed in water to make the sulfurous acid leaching solution. Ore broken in place by blasting or caving will be leached by percolating the solution down through the ore. The solution will be distributed to the ore by vertical injection wells from the surface. Solutions will be recovered by a series of drifts under the broken ore. The underlying materials probably would have a low enough permeability to permit solution recovery with little need for lining, except for various sumps and drains in the collection drifts. Effluent solutions are tentatively estimated to contain 10 to 12 percent $MnSO_4$.

The preferred manganese recovery method will be by solar evaporation to make the commercially salable product, manganese sulfate. Alternative recovery methods are precipitation of manganese carbonate or hydroxide or direct electrolysis to make metal and/or oxide, (with increased capital costs compared to those of solar evaporation).

In an economic evaluation of the system, a capital cost of \$13.3 million was assumed. Mining and operating costs over a 12-year life were estimated to be \$59 million. An average of 103,000 tons per year of manganese sulfate (75 to 78 percent $MnSO_4$) at a price range of \$100 per ton to \$300 per ton was assumed. The DCF return ranges from 4.6 percent (7.94 years payout) to 45.7 percent (1.76 years payout).

Although the sulfurous acid dissolution of Artillery Peak ores has been fairly well proved by the Bureau and private small scale tests, a continuous field test (40,000-ton block) is recommended to further prove attainable solution strength in $MnSO_4$ and to indicate overall extraction at a given rock size. Column leaching tests with sulfurous acid solutions and solar evaporation of the solutions to determine $MnSO_4$ product grade are also recommended. Further development of groundwater and solution control information is suggested. As already mentioned, a thorough market survey of the probable future needs and price trends for various manganese products is required. Depending upon the results of the market survey, manganese recovery flowsheets to make products other than manganese sulfate may have to be developed.

Cobalt-Nickel

The Diamond Flat area of Northern California was selected as a tentative site for in situ leaching of cobalt and nickel. This selection was made on the basis of hypothetical assumptions in the absence of a great deal of chemical and geohydrological information. Nevertheless, the cobalt-nickel pair is recommended for further study because of its strong strategic and economic attractiveness. Bench scale metallurgical studies and field geohydrological reconnaissance are recommended to prove or disprove the viability of the candidate site.

Cobalt-nickel lateritic deposits in the Diamond Flat area in the northwest corner of California are derived from the olivine contained in peridotite through weathering in a warm, humid climate. The bedrock is chiefly peridotite which has been weathered to serpentine in places. Magnesia and silica have been decomposed and leached out of the peridotites. Chromium, alumina, iron, nickel, and cobalt are concentrated. The deposits are generally shallow, but it is believed that there would be strong environmental objections to open pit mining of the material for transport to a processing plant. Thus, the in situ process may be favored because of a relatively low environmental effect. Careful study of the typical bedrock will be needed to determine the feasibility of solution control. A test block containing 0.68 percent nickel and 0.08 percent cobalt has been identified.

Prior art in the hydrometallurgical treatment of cobalt-nickel laterites does not favor in situ mining. Mined ores have been treated by reduction roasting and ammoniacal leaching or pressure hot acid leaching to attain high extractions. Such methods are not practically adaptable to in situ needs. After consideration of various possible leaching reagents, a relatively strong sulfuric acid solution has tentatively been selected for the hypothetical leaching system for Northern California laterites. Although Canterford (1975) has studied laterite leaching with acids, apparently the temperatures and concentrations are outside the conjectural limits proposed for the California in situ work. Therefore, a sulfuric acid leaching solution of about pH 1.0 is assumed.

A simple scheme for leaching the ore from the surface is proposed (figure 4). Since the ore is natually permeable, the only ore preparation required would be to rip the ground surface to prevent stagnate leach solution ponding. A leaching solution of 10 to 50 grams of sulfuric acid per liter would be applied to the block by spraying, ponding, or injection holes at the rate of 45 gpm. A shaft and solution recovery drift would be run in bedrock under the ore and 3-inch drill holes would be drilled up into the ore to provide solution collection. Alternatively, solution could be recovered by pumping from vertical recovery wells. The cobalt and nickel would be recovered by solvent extraction using LIX63-HDNNS (dinonylnapthalene sulfonic acid) mixture reported by Osseo-Asare (1978). Monitor holes would be drilled to check for any solution loss.

The economic outlook for the Northern California cobalt-nickel in situ site was evaluated by means of computer analysis assuming an annual production of 4,600,000 pounds (2,100,000 kg) of nickel and 545,000 pounds (250,000 kg) of cobalt with nickel prices at \$3, \$3.50, and \$4 per pound (\$6.60, \$7.70, and \$8.80 per kg) and cobalt prices at \$12, \$25, and \$37 per pound (\$26.50, \$55.10, and \$81.60 per kg). The lowest DCF return on investment was estimated to be 40.5 percent and the highest 74.1 percent. The payback time ranged from 1.66 years down to 0.91 years.

The cobalt-nickel in situ proposal is, at the time of preparing this report, highly conjectural. Nevertheless, further study of the cobalt-nickel possibility is recommended because of the high strategic and economic potential involved. Some major areas exist in which public reports are apparently lacking or in which often-conflicting individual views exist. Some research and development needs are listed herewith:

1) Laboratory leaching tests in which the response of various-sized lateritic ore particles to sulfuric acid solution leaching is determined, preferably at ambient temperatures.

2) Field studies of the underlying rock formations in the lateritic areas to delineate solution control problems.

3) Contingent upon results of points 1 and 2, an examination of the solvent extraction of the undoubtedly highly impure sulfate leaching effluents.

Conclusions

In preparing comparative evaluations of possible candidate metals for in situ leaching, the metal's economic and strategic significance and available leaching and recovery process technology were considered.

Listed in order of comparative readiness in the terms of in situ technology and economic and strategic importance, the manganese and the gold-silver projects may be classified as about equal in priority. Manganese as it occurs in the Artillery Peak, Arizona deposits has been indicated to be leachable with sulfurous acid and recoverable by solar evaporation. The product, manganese sulfate, has significant demand, manganese is strategic, and the economic outlook is favorable.

Fine grained, oxidized gold and silver deposits as exemplified in the Gold Quarry, Nevada site, should be readily leached and recovered as indicated by many successful heap leaching operations. Both are enjoying a currently unrestricted market and satisfactory price. Environmental concerns will dictate a cautious approach to site selection and strong measures to control leaching solutions.

The aluminum leaching concept based on the Carlota Mine, Arizona, is given third priority because further evaluation in the laboratory is believed desirable to confirm various leaching and recovery processes. Also, more detailed cost analysis is needed. Although the present economic outlook is not as favorable as manganese, gold-silver, or cobalt-nickel, the strategic importance of alumina is high.

The highly strategic and potentially economically-favorable cobalt-nickel project is nevertheless given fourth rank in priority because of the need for basic laboratory research on dissolution and recovery. Geohydrological reconnaissance is needed in more detail to confirm or disprove solution control potential.

References

1. Bender, F. N. and Rampacek, C., 1957, "Percolation Leaching of Manganese Ores With Sulfur Dioxide," BuMines RI 5323.
2. Bhappu, R. B., 1978, "In Situ Mining Technology," Underground Space, Vol. 2, Pergamon Press, pp. 211-221.
3. Canterford, J. H., 1975, "The Treatment of Nickeliferous Laterites," Min. Sci. Eng., Vol. 7, No. 1, pp. 3-17.
4. Chamberlain, P. G. and Pojar, M. G., 1981, "Gold and Silver Leaching Practices in the United States," BuMines IC, in press.
5. Dean, R. S., Leaver, E. S., and Joseph, T. L., 1934, "Manganese; Its Occurrence, Milling and Metallurgy. Part III," BuMines IC 6770.
6. George, D. R., Tame, K. E., Crane, S. R., and Higbie, K. B., 1968, "Recovery and Production of Alumina From Waste Solutions by Solvent Extraction," presented at AIME Annual Meeting, New York, NY, February 25-29, 1968.
7. Osseo-Asare, K., 1978, "Extraction of Nickel and Cobalt From Acidic Solutions Using LIX63-HDNNS Mixtures," AIME Annual Meeting, 1978, Denver, Colorado, SME Preprint 78-B-61.
8. Potter, G. M., Nordhausen, E. A., Fletcher, J. B., Lake, J. L., Rouse, J. V., and Wojtasiak, F., 1981, "Technological Investigation to Determine the Feasibility of In Situ Leaching Metallic Ores Other Than Copper and Uranium," Final Report on BuMines Contract J0295032, Final Report to be distributed by NTIS, Springfield, Virginia.
9. Rampacek, C., Fuller, H. C., and Clemmer, J. B., 1959, "Operation of a Dithionate-Process Plant for Leaching Manganese Ore From Maggie Canyon Deposit, Artillery Mountains Region, Mohave County, Arizona," BuMines RI 5508.
10. Wyman, W. F. and Ravitz, S. F., 1947, "Sulfur Dioxide Leaching Tests on Various Western Manganese Ores," BuMines RI 4077.

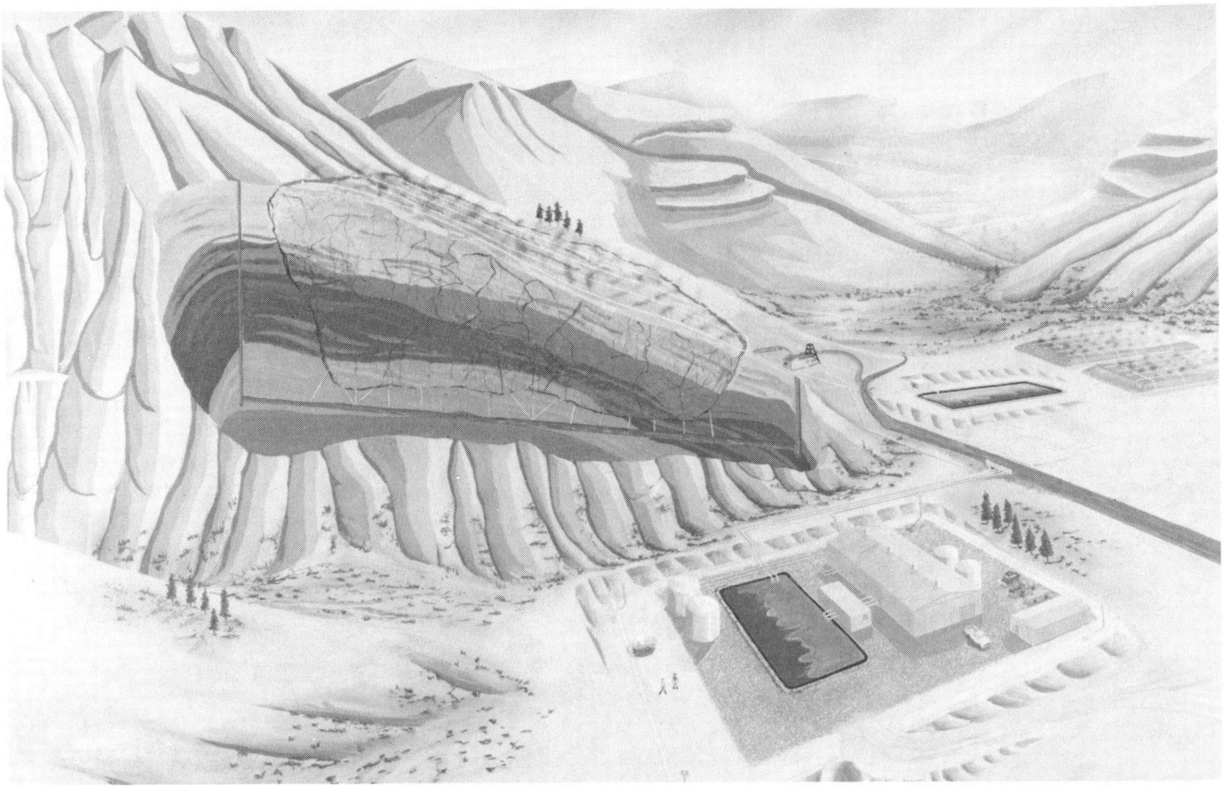

Figure 1. Conceptual design of system for in situ leaching of alumina.

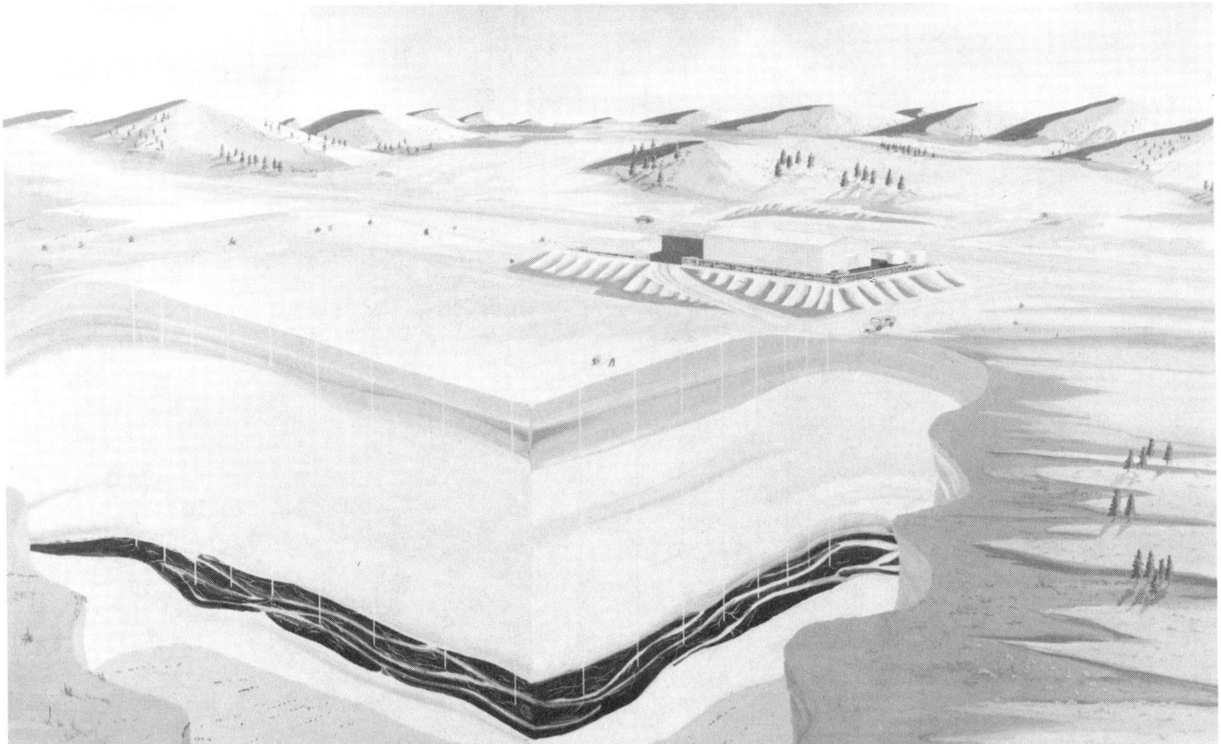

Figure 2. Conceptual design of system for in situ leaching of gold and silver.

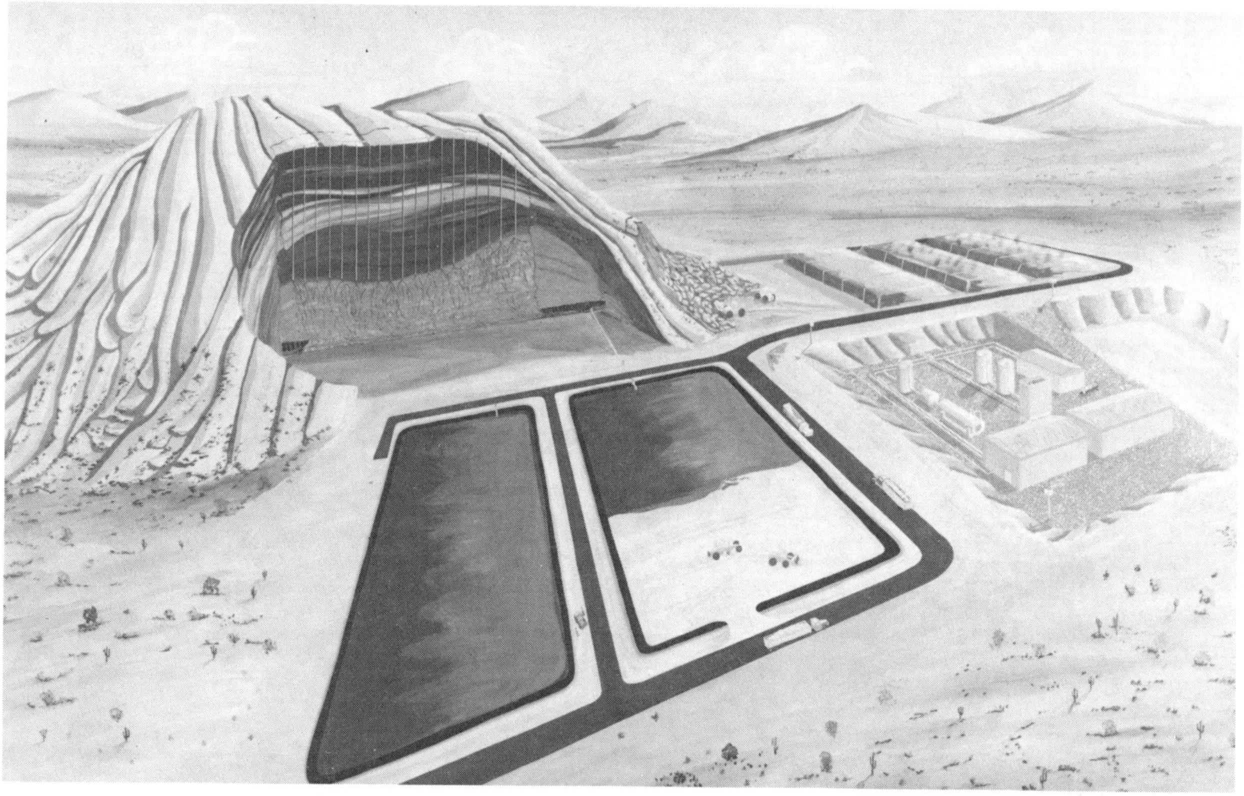

Figure 3. Conceptual design of system for in situ leaching of manganese.

Figure 4. Conceptual design of system for in situ leaching of cobalt and nickel.

Chapter 12

FACTORS INFLUENCING THE IN SITU ACID LEACHING RATES OF URANIUM ORES

Aline Tatom
Exxon Research and Engineering
Florham Park, New Jersey 07932

R. S. Schechter
Larry W. Lake
Departments of Chemical and Petroleum Engineering
University of Texas at Austin
Austin, Texas 78721

ABSTRACT

This paper presents numerical and experimental results dealing with the application of ferric ion oxidation of uranium in acidic media. Conclusions are derived from matching numerical results to experimental data taken on one-dimensional laboratory leaches using ore from a South Texas coastal plain deposit. Since the numerical results can be made quite general, one can extract and quantify the effects of oxidant concentration, flow velocity, mass transfer, chemical kinetics, and competitive oxidant consumers (pyrite and molybdenite) on the experimental results. Results presented here suggest that the leaching of uranium is mass transfer controlled while that of pyrite is kinetically controlled.

Introduction

Uranium deposits in the United States consist mainly of sandstone-type ores located on the Colorado Plateau, in the Wyoming Basins, and in the Gulf Coastal Plains.[1] They are believed to have been created by the downdip migration of oxidizing groundwaters carrying uranium[2] which was precipitated when the waters encountered reducing conditions, forming roll-front or tabular deposits. Grades of these deposits vary from less than 0.05 to 0.5 wt% and commonly average 0.2 to 0.3 wt%.

These uranium deposits have traditionally been exploited by conventional methods: open pit, strip, and underground mining. Since these methods consist of bringing the ore to the surface for uranium extraction, their profitability is limited by the grade of the ore and the depth of the deposit. A possibly more economical alternative for exploiting low grade or deep deposits is in situ leach mining (also called solution mining). This process consists of injecting an oxidizing solution (either acidic or alkaline) into the formation, allowing it to solubilize the uranium, and pumping the uranium-rich solution back to the surface, there it is then processed in a conventional uranium recovery system. Deposits amenable to solution mining techniques are usually located below the water table and must be confined by impermeable layers to minimize the escape of leach solution from the mineralized zone since loss of leach solution presents potential economic and/or environmental problems.

Purpose and Scope

Among the many variables which influence uranium production rates and recovery in in situ leaching are well placement, formation characteristics, the choice and concentration of oxidant, mineral concentration of uranium and of competitors, and the kinetics of the system[3] (here the word "kinetics" is used in a broad sense to mean the factors which affect the dissolution rates of the various minerals). The purpose of this work is to obtain quantitative information concerning the kinetics of uranium and of the other minerals present. To limit the scope of the work, a single system was chosen: acid leaching with ferric iron as the oxidant.

Our approach is to use a numerical simulation model to interpret laboratory flow experiments in linear flow. The numerical model can simulate -- and, hence, indicate the relative importance of -- the combined effects of chemical reaction rates, solid-liquid mass transfer, dispersion, cation exchange, and the presence of other minerals on the uranium dissolution rate. All of the processes of interest are time dependent and often competitive. Thus, it would appear difficult, if not impossible, to interpret the effluent histories of various metal ions and oxidant without benefit of a model.

Prior Work

Acid leaching has been used in the uranium milling industry since the mid-forties and has accounted for a majority of the production since that time. It is preferred over carbonate leaching because of its greater efficiency in recovering uranium. Consequently much research has been done on the factors which influence the kinetics of the oxidation of uranium, as well as those which influence the leaching process as a whole. This knowledge can be applied to the study of in situ leaching. The following is a survey of the uranium kinetics literature with particular regard to the oxidation of uranium by iron Fe(III) in acidic solutions, and of literature pertaining to acid leaching in general. The kinetics literature for pyrite and molybdenite, minerals commonly found in uranium-bearing ores, is also briefly reviewed.[4] A more comprehensive review is available elsewhere.

Uranium occurs in reduced ores in the tetravalent form in oxide minerals such as uraninite and pitchblende (mainly UO_2). In order to be dissolved by acid, the uranium must first be oxidized to the hexavalent form, which then dissolves as various uranyl sulfate species such as UO_2SO_4, $UO_2(SO_4)_2^{2-}$, $UO_2(SO_4)_3^{4-}$, and UO_2^{2+}. Ferric iron has been found to be an effective oxidant for this purpose. The stoichiometry of the reaction is universally agreed to be

$$UO_2 + 2Fe(III) \rightarrow UO_2^{2+} + 2Fe(II) \tag{1}$$

Rhees[5] found that other oxidants such as sodium chlorate and manganese dioxide were ineffective in dissolving UO_2 except in the presence of iron. Thus most of the research concerning the oxidation of tetravalent uranium in acid solutions has focused on the use of Fe(III) as the oxidant. In prior studies the acidic media have been acidic perchlorate and sulfate,[6,7] perchloric and sulfuric acid,[8-12] and sulfuric acid alone[13].

Of the above-cited works that of Nichol et al.[11] is particularly relevant here as these authors studied the effect of Fe(II) as well as Fe(III) on the uranium oxidation process. The uranium oxidation rate was found to have three regimes depending on the Fe(III):Fe(II) ratio and the absolute Fe(III) concentration. The regimes were explained in electrochemical terms as follows (see Figure 1):

Regime 1 - At very low ferrous Fe(II) concentrations the rate of oxidation of Fe(II) at a solid surface is small and the UO_2 oxidation rate is dependent on a fractional power of Fe(III) and independent of the ferrous concentration.
Regime 2 - At higher ferrous concentrations the Fe(II) oxidation becomes significant, and the UO_2 oxidation rate becomes dependent on the 1/2 power of the Fe(III):Fe(II) ratio.
Regime 3 - At very high Fe(II) concentrations the UO_2 oxidation rate becomes dominated by the Fe(II) oxidation and becomes a function of the Fe(III):Fe(II) ratio to the power of 1.1.

The three regimes are predicted by an electrochemical model. The reader is referred to the original paper[11] for a detailed explanation of the mathematical treatment.

In addition to the above, Fe(II) can affect the heterogeneous oxidation of UO_2 in two other ways:

1) absorbing on the solid surface, as do other cations, and
2) inhibiting the desorption of reaction products (which are present at the most reactive sites) as proposed by Laxen.[10]

Since Nicol et al. explained the behavior of a system containing Fe(II) using only the effect shown in Fig. 1 it would seem probable that this is the more important effect. It is the only effect considered here.

The apparent activation energy for reaction (1) ranges from 16.2 to 22.7 kcal/mole, with the larger figures generally believed to include a contribution from the heat of hydrolysis.[6,7,10-13]

The chemical leaching of pyrite (FeS_2) by an oxidant has been applied to the production of sulfuric acid (and simultaneous leaching of uranium ores) in a pressure reactor[14,15] and to the desulfurization of coal containing pyritic sulfur.[16,17] Since these are high pressure and/or high temperature applications which usually involve oxygen, they are not very relevant to the study of in situ leaching of pyritic uranium ores by acidic ferric sulfate solutions. Garrels and Thompson[18] conducted a study of the oxidation of pyrite in conditions closer to those existing for in situ leaching. The reaction equation is:

$$FeS_2 + 14Fe(III) + 8H_2O \rightarrow 15Fe(II) + 2SO_4^{2-} + 16H^+ \qquad (2)$$

Garrels and Thompson studied the rate of oxidation of pyrite in acidic ferric sulfate solutions and postulated a reaction mechanism whereby (2) is preceeded by an adsorption step in which both Fe(III) and Fe(II) are adsorbed on the pyrite surface. The reaction rate would then be proportional to the fraction of the surface occupied by ferric ions. The resulting reaction rate was found to agree with experimental data from three sources.

Molybdenite (MoS_2) probably reacts with ferric iron according to the following equation:

$$MoS_2 + 18Fe(III) + 12H_2O \rightarrow$$
$$18Fe(II) + HMoO_4^- + 7H^+ + 2H_2SO_4 \qquad (3)$$

As with pyrite, most of the oxidizing capacity of the solution goes to oxidize the sulfur to sulfate. The literature of molybdenite oxidation is sparse[19-21] and indicates an as yet unquantified tendency for Fe(II) to inhibit reaction (3).

All of the work cited above was on batch systems. In this paper we apply reaction expressions for reactions (2)-(3) to flow systems.

Mathematical Model

Mass Balance - For conciseness we adopt the following index conventions for the 5 chemical species considered here: 1 = Uranium, 2 = Fe(II), 3 = Fe(III), 4 = Pyrite, and 5 = Molybdenite. We use superscript bars to denote solid phase concentrations and titles to denote liquid concentrations at a liquid-solid interface.

The mass balance for single phase, one-dimension flow in permeable media is[22]

$$\phi\left(\frac{\partial C_i}{\partial t}\right) + u\frac{\partial C_i}{\partial x} - \phi K_\ell \frac{\partial^2 C_i}{\partial x^2} = \phi R_i, \quad i = 1\ldots5 \qquad (4)$$

Where R_i is the global reaction rate of species i, and C_i is the bulk concentration of species i. The remaining quantities are defined in the Nomenclature. Equation (4) may be made dimensionless by introducing $x_D = x/L$ and $t_D = ut/\phi L$ as a dimensionless length and time, respectively:

$$\frac{\partial C_i}{\partial t_D} + \frac{\partial C_i}{\partial x_D} - \frac{1}{N_{Pe}}\frac{\partial^2 C_i}{\partial x_D^2} = \frac{L\phi}{u} R_i, \quad i = 1\ldots5 \qquad (5)$$

where $N_{Pe} = uL/K_\ell\phi$ is the dimensionless Peclet number. As defined, t_D is the total fluid volume injected into the permeable medium in pore volumes. The analogous mass balance for the solid phase is

$$\frac{\phi\bar{C}_i}{\partial t_D} = -\frac{L\phi}{u} R_i, \quad i = 1\ldots5 \qquad (6)$$

Note that $\bar{C}_i$, the solid phase concentration, is in units of mass of i per unit pore volume. In the next few paragraphs we discuss the origin of the global reaction rate, adsorbed concentration, and dispersion terms.

Rate process - Phase boundaries, and therefore boundary layers, are inherent to heterogeneous systems. For this reason, the physical process of mass transfer through a boundary layer must be considered in modeling such systems. This is usually done by including the mass transfer phenomena in the rate expressions through a global, or overall, rate expression. Such expressions enable the same equations as are used for homogeneous media to be used for heterogeneous media.

The processes which must be included in the global rate equations are:

1) transport of reactants from the bulk fluid to the fluid-solid interface,
2) adsorption of reactants onto reaction sites on the solid,
3) chemical reaction of adsorbed reactants to adsorbed products,
4) desorption of adsorbed products, and
5) transport of products from the fluid-solid interface to the bulk fluid stream.[23]

Steps 2 through 4 are usually combined into the chemical reaction rate equation. Since these processes occur in series, the rates of the individual processes are equal at steady state, and the global rate is determined by the slowest step. Uranium leaching is not a steady state process because of changing bulk mineral concentrations. However, instantaneous changes in these concentrations are small enough to warrant the assumption of a quasi-steady state existing on a local scale.

The global rate can thus be expressed in terms of each of the above processes. For the heterogeneous reaction

$$A(solid) + B(aqueous) \rightarrow C(aqueous) + D(aqueous) \qquad (7)$$

the global rate of reaction of A is:

$$R_A = \gamma_{AB} h_A(C_B - \tilde{C}_B) =$$
$$-\gamma_{AC} h_A(C_C - \tilde{C}_C) = -\gamma_{AD} h_A(C_D - \tilde{C}_D) \qquad (8)$$

where R_A is the chemical reaction rate of A, and h_A is the mass transfer coefficient per unit pore volume for B over mineral site A. The chemical reaction rate is a function of the interfacial concentrations ($\tilde{C}_i$), but does not depend on the bulk concentrations C_i. The constants γ_{ij} represent the stochiometric mass ratio of species i to j. In this application this is the ratio of the molecular weight of species i to species j times the stochiometric coefficient.

Since $\tilde{C}_i$ appears in Equation (8), the R_i term in Equation (5) cannot, in general, be expressed solely as a function of the bulk solution phase concentrations C_i. The three independent equations represented by Equation (8) must therefore be solved simultaneously with the mass balances, Equations (4) and (5), to determine surface concentrations and global rates.

The choice of a kinetic expression for uranium from those in the literature was very much dictated by the fact that the process to be modeled is a continuous one. In a batch leaching operation, sufficient amounts of oxidants such as sodium chlorate and manganese dioxide can be added to the leach solution to maintain all of the iron present in the +3 oxidation state. This is virtually impossible in an *in situ* operation. Even if large amounts of these oxidants are added, they would eventually become depleted as the leach solution progresses through the ore body, and ferrous iron would begin to appear. Thus to correctly model an *in situ* leach, the presence of ferrous iron must be recognized and its effect on the reaction rate included in the model.

Only one of the studies discussed above included the effect of ferrous ion on the oxidation rate of uranium. That was the study conducted by Nicol et al.,[11] and their results, as presented in Figure 1, were used to quantify this effect for use in the model. The rate expression for the the three regimes are,

$$R_1 = k_1' \bar{C}_1 (\tilde{C}_{31})^{0.73} \quad , \quad \text{Regime 1} \tag{9a}$$

$$R_1 = k_1'' \bar{C}_1 (\tilde{C}_{31} / \tilde{C}_{21})^{0.5} , \quad \text{Regime 2} \tag{9b}$$

$$R_1 = k_1''' \bar{C}_1 (\tilde{C}_{31} / \tilde{C}_{21})^{1.1} , \quad \text{Regime 3} \tag{9c}$$

In Equations (9) the uranium leach rate is assumed first-order in surface uranium concentration. The second subscript of the surface concentrations $\tilde{C}_{ij}$ indicate the surface concentration of species i above mineral j (see Fig. 2). The form of Equation (9), and all subsequent such equations, implies that all solid-fluid reactions are assumed irreversible. The details on how the regimes were chosen in the numerical model are given elsewhere.[4] The relationships between the kinetic constants for each regime - k_1', k_1'', and k_1''' - can be determined from Figure 1. Clearly at a transition point, the rates predicted by the two appropriate kinetic expressions must be equal, therefore, $k_1' = 4800k_1''$, and $k_1'' = 0.7k_1'''$. We use k_1'' as a fitting parameter below.

The rate expression for pyrite (component 4) is taken as[18]

$$r_4 = k_4 \bar{C}_4 (\tilde{C}_{34} / C_{3I}) \tag{10}$$

where the subscript I denotes injected concentration of Fe(III). Equation (10) reflects the fact that in this system, the pyrite concentration is variable. A similar expression for the molybdenum is also used

$$r_5 = k_5 \bar{C}_5 (\tilde{C}_{35} / C_{3I}) \tag{11}$$

Mass transfer coefficients are correlated by Fedkiw[24] for packed bed flow wherein two distinct trends are observed

$$\frac{\phi h}{D_i} = 0.07 \frac{u}{aD_i} \quad \text{for} \quad \frac{u}{aD_i} \leq 10 \tag{12a}$$

$$\frac{\phi h d_p}{D_i} = 1.09 \left(\frac{u d_p}{D_i}\right)^{1/3} \quad \text{for} \quad \frac{u}{aD_i} \geq 10 \tag{12b}$$

Note that the mass transfer coefficients h in these equations are on a unit surface area basis rather than unit pore volume as were the h_i in Equation (8). The conditions of our experiments generally are satisfied by Equation (12a); therefore, Equation (12c)

$$h = 0.07 \left(\frac{u}{\phi}\right) \tag{12c}$$

will be used to describe the mass transfer coefficient. The h given by Equation (12c) is an effective mass transfer coefficient; no attempt was made to calculate a local, or film coefficient (see below). We determine the coefficient of the u/ϕ term in Equation (12c) by history match below.

The mass transfer coefficient in Equation (12c) is per unit surface area. Since concentrations and reaction rates in the mathematical model are expressed per unit pore volume, it is necessary to convert the mass transfer coefficient to a similar basis. This can be done by multiplying the h in Equation (12c) by the ratio of the surface area of mineral i to the total pore volume. This factor can, in turn, be decomposed into a product of the ratio of total surface area to total pore volume as determined from the BET technique, times the ratio of the mineral i surface area to the total surface area. No information with which to estimate the latter ratio is available, so it was assumed to be proportional to the fractional weight of each mineral

$$h_i = \frac{h a_v}{\rho_r} \left(\frac{\phi}{1-\phi}\right) \bar{C}_i = h a_m \bar{C}_i , \quad i = 1,4,5 \tag{13}$$

where a_v and a_m are the specific surface areas on a pore volume and mass of solid phase basis respectively. Equation (13) implies that h is the same for each species, an assumption whose accuracy can be judged only by comparing the results of the mathematical model to experiment.

Equilibrium Processes - The ferric (C_3) and ferrous (C_2) cations can chemically bond to a negatively charged site in the rock matrix. These sites are preferentially localized on the clay minerals, and the exchange of C_2 and C_3 with adsorbed cations -- predominantly hydrogen, sodium and calcium -- originally present is called cation exchange.[25] The exchange rate of cation exchange is quite fast; hence, the concentrations of cations bound to the stationary phase by this process ($\bar{C}_2$ and $\bar{C}_3$) are in local equilibrium with the bulk concentrations (Fig. 2). The relationship $\bar{C}_2 = \bar{C}_2(C_2)$ or $\bar{C}_3 = \bar{C}_3(C_3)$ are isotherms where the clay-bound ferric and ferrous concentrations are assumed functions of their respective bulk phase concentrations only. The solid and flowing phase mass balance for these components may, therefore, be combined by adding equations (5) and (6). Based on experimental work not reported here[4] we determined the cation exchange of ferric ion to be described by a Langmuir-type relation

$$\bar{C}_3 = \frac{(CEC) K_1 C_3}{1 + K_2 C_3} \tag{14}$$

where the experimental best values of the constants are $K_1 = 1300$ cc/meq and $K_2 = 0.33\ell$/mg. In the history matching phase of this work we determined the CEC (cation exchange capacity) separately for each ore by history match.

It is recognized that the Langmuir isotherm expressed by Equation (14) is an approximation which is valid only for systems containing relatively large concentrations of hydrogen ions (pH 1.5-2.0) and if the hydrogen ion solution concentration is held fixed. The plateau value of $\bar{C}_3 = (CEC)K_1/K_2$ may be interpreted as resulting from adsorbed hydrogen ions which, at the solution pH, tend not to exchange for ferric ion. This adsorption isotherm is important since those ferric ions adsorbed are not available to oxidize uranium. The isotherm given by this equation is, however, not a general one. It can be expected to depend on pH and temperature and to vary considerably from ore to ore.

An exchange isotherm was also obtained for the Fe(II) - H^+ pair; however, the results were inconclusive. Since the exchange of Fe(II) did not seem to impact on the leach experiments, those results are not given here.

The above equations represent 15 independent equations in as many unknowns which must be simultaneously solved. Table 1 summarizes the equations and unknowns based on combinations of Equations (5), (6), (8), (13), and (14). The r_i terms are regarded as given by Equations (9) - (11). The dimensionless mass transfer coefficient $h_D = h\phi/u$ expresses the ratio of the rates of mass transfer to bulk fluid flow, and is analogous to a Stanton number for permeable media flow. From Equation (12c) h_D should be a constant. Note that the surface concentration of urnaium above a UO_2 site never appears since it does not affect the kinetic expression, Equation (9). The parameters to be determined in the history matching below are h_D, k_1'', k_4 and k_5.

The numerical procedure used to solve these equations was a backwards in time and space finite difference procedure with an implicit reaction rate term. This semi-implicit procedure is similar to the procedure used by MacDonald and Coats.[26] The dispersion term was approximated by the numerical truncation error based on the formulas of Lantz.[27] Readers interested in more details of the numerical procedure are referred to the original work.[4]

Experimental

Leach tests - The important variables in this study were fluid velocity and oxidant concentration. Formation fluid velocities are usually limited by fracturing considerations; thus, the velocities in the laboratory leach tests did not exceed 0.5 cm/sec although higher velocities may be possible for close well spacing or high permeability formations. The oxidant concentration was varied to determine its effect and to attempt to define an optimum.

Other variables such as pressure and pH were considered only to the extent that they affected the experimental results. The experiments were conducted at atmospheric rather than formation pressures as the leach process is not believed to be sensitive to pressure. The pH was maintained at 1.8 throughout a run to prevent the precipitation of iron as ferric hydroxide $Fe(OH)_3$ and to enhance the dissolution rate of UO_2.

The ore used was from the Felder uranium deposit, which is located in the South Texas Coastal Plain. The uranium in this ore occurs as uraninite UO_2 and coffinite $U(SiO_4)_{(1-x)}(OH)_{4x}$. These are present in approximately equal proportions, the total uranium concentration being about 0.25%. Other minerals occurring in significant quantities are pyrite (7%) and molybdenite (.07-.08%). The ore was also rich in calcium carbonate. A description of the geology of the Felder deposit is given by Klohn and Pickens.[28]

Leach test procedure - Each of the six acid leach tests reported here consisted of the following steps:

1) preparing the ore for packing and packing the column;
2) injecting an acid solution into the packed column in order to reduce the liquid phase pH;
3) injecting an acid solution containing ferric iron in order to extract the uranium; and
4) measuring the dispersion coefficient and pore volume of the packed column by the use of a step change in the input concentration of chloride ion.

Results

Preliminary calculations - Quantitative characterizations of the physical effects of dispersion, and mass transfer are required so that these may be included in the model described above.

Dispersion coefficients were determined at the end of each run under the conditions of the leach experiments. For the six laboratory leaches described here, the dispersion coefficients ranged from 1.1 to 8.8 x 10^{-3} cm^2/sec (see Table 2) with Peclet numbers from 25 to 100. These values are in agreement with published correlations[29] for unconsolidated porous media.

The summary of low Reynolds' number mass transfer data given above was used to estimate mass transfer coefficients for the various runs. Based on a measured BET surface area of 2.5 m^2/gram, a average particle diameter of 326 μ, and an assumed diffusion coefficient of 10^{-5} cm^2/sec, we determined that Equation (12a) was appropriate for the leach tests. From Equation (12c), therefore, we may calculate the mass transfer coefficient only from the interstitial velocity and the porosity. These values are in the fifth column of Table 2.

Preliminary acid flush - The purpose of this part of the leach was to acidify the column by dissolving acid-consuming materials. Figures 3 to 5 show effluent concentration histories for uranium and calcium, as well as effluent pH histories, for run 7. Run 7 is the run used for illustration throughout this paper because it had the smallest amount of pre-oxidant uranium production (column 7, Table 2), and because its results were typical of all the runs.

The uranium histories (Fig. 3) indicate an initial small production of uranium which is thought to represent the uranium already solubilized in the groundwater. Once the groundwater is displaced, no more uranium is produced until the pH drops to 2. At this point, considerably more uranium was produced as oxidized uranium dissolves in the acid solution. This pre-oxidant uranium production was a feature in all the runs and indicated some uranium oxidation prior to the acid flushes, probably in the ore handling processes. The percent pre-oxidation ranged (Table 2) from 23 (run 7) to 84 percent (run 11). Note that the "initial" uranium concentrations given in Table 2 and used in the history matching are after the preliminary acid flush.

The calcium histories (Fig. 4) indicate a high level of calcium in the effluent which drops off rapidly as calcium carbonate is dissolved. The drop occurs at approximately the same time as the drop in pH. However, high levels of calcium in the effluent, and therefore significant amounts of carbonate in the ore, still remain even after the effluent pH has dropped. For this reason, injection of ferric iron cannot begin until the calcium concentration has reached acceptable levels, which as seen in Figure 4, requires about 14 liters preflush.

Leach runs - Table 2 gives the experimental conditions for each of the seven runs. The variables were initial ferric concentration and fluid velocity. Unfortunately, due to the considerable pre-oxidation of the ore, uranium concentrations at the start of a leach were also variable. These were determined from material balance by integrating under the uranium production curves and given in Table 2. Very little iron was produced in this step, indicating insignificant pre-oxidation of pyrite. Thus, its concentration at the beginning of a leach was taken to be the originally measured concentration.

From each leach experiment, concentration histories for uranium, reported as U_3O_8, molybdenum, and ferric and ferrous iron were obtained. The computer model was then used to generate concentration histories which were matched to the experimental ones by manipulating the parameters which affected their behavior. These were cation exchange capacity, mass transfer coefficients, and kinetic constants. The dispersion coefficients were set by the results of the dispersion runs given in Table 2.

Figure 6 to 11 are a series of plots showing the experimental data (points) and the best fit obtained (by inspection) with the model for run 7. These are concentration histories for uranium as U_3O_8 (Figure 6), ferric (Figure 7), and ferrous (Figure 8) iron, molybdenum (Figure 9), and total iron (Figure 10), as well as cumulative percent recovery of uranium (Figure 11 - experimental points not shown) with time. The manner in which the parameters corresponding to a "best fit" were obtained is described below.

Parameter fit - It was determined from manipulation of the parameters in the computer model that the data could be matched reasonably well whether mass transfer control (k_1'' large) or kinetic control (h large) was assumed. Simply matching the experimental data did not ensure that the mass transfer coefficients and kinetic constants so obtained were reasonable or even represent a unique solution. Thus another criterion was needed - consistency among the runs. Such a requirement dictated that the kinetic constants obtained from matching all the runs be equal (recall that temperature and pressure is constant for all runs). Moreover, the mass transfer coefficients obtained must be consistent, i.e., they must have been calculated in the same manner. It can be seen from Equation (12c) that the group h_D should be constant for all runs.

Figure 12 demonstrates schematically the possible relationships between the best fit (designated by the subscript BF) values of the kinetic constant k_{BF} and mass transfer group $(h\,\phi/u)_{BF} = (h_D)_{BF}$ for a single arbitrary reaction. These relationships exist because, as discussed above, no one pair of these parameters used in the computer model uniquely matches the data from a given run. Thus, for each run an infinite set of pairs exist which adequately fit the data. These pairs are represented by the curves shown in Figure 12. Figure 12a represents the most general case where only one point occurs for which the kinetic constants and mass transfer coefficients are consistent amoung all the runs. The kinetic constant-mass transfer coefficient pair at the intersection point are the overall best fit values.

Figure 12b represents a system in which the reaction rate is so slow that the resistance to mass transfer has a negligible effect on the overall reaction rate. Thus, even though a satisfactory match of the experimental data for a given run can be obtained with an appropriate $(k, \phi\,h/u)_{BF}$ pair, no single such pair can be selected to fit all of the runs unless $(h_D)_{BF}$ approaches infinity. Only at this point do the curves for all runs converge and at this point the kinetic constants for all the runs are equal.

Figure 12c represents the opposite system, in which the kinetic reaction rate is so fast that only the rate of mass transfer controls the overall reaction rate. Again, any appropriate pair of these parameters can be used to acceptably match a given run, but only one pair is consistent among all the runs. This is the pair in which the kinetic constant approaches infinity and $(h_D)_{BF}$ is the same for all the runs.

Figure 13 shows the best fit (by inspection) relationship between mass transfer and kinetics obtained for uranium. The results shown here clearly follow the trends depicted in Figure 12c. We therefore conclude that the reaction is mass transfer controlled and the mass transfer coefficient group h_D is0.091, excluding run 6, with a mean deviation of 14%. No trend is evident with degree of oxidation or velocity. Thus, fairly consistent values were obtained for this case, indicating that the rate of production of uranium is entirely controlled by resistance to mass transfer. It follows that no information concerning the kinetics of the reaction can be determined from these or other similar flow tests.

The average value of h_D determined from the data is slightly higher than, but still in good agreement with, that corresponding to the previously estimated value of 0.07, Equation (12c). This difference probably arises from the assumption that the fractional surface coverage of a mineral is proportional to its fractional weight; however, the overall consistancy of the mass transfer group seems to generally support the correctness of this assumption.

Kinetic studies which consider heterogeneous reactions are usually designed such that the surface coverage is known and very nearly constant over the range of the experiment. It is then incorporated into the kinetic constant. In this system however, the fraction of the exposed surface covered by uranium is not known, nor is it constant. It was assumed in this study that the mass available for reaction (or surface area) of uranium was proportional to the total mass of uranium in the system. Since this assumption is to some extent justified by the excellent agreement between the mass transfer values which fit the data and those obtained from packed bed mass transfer studies, it would appear that the surface area of the uranium mineral does indeed decrease as the uranium is leached from the system. This would lead one to reject the hypothesis that the uranium is deposited in thick layers on the outside surface of the grains in the ore, as the surface area would then remain virtually constant as layer after layer was oxidized and dissolved until the uranium is depleted.

Figure 14 shows the curves of $(h\phi/u)_{BF}$ versus $(k)_{BF}$ generated for molybdenum. Although it is clear that mass transfer is of importance in this reaction, the degree of importance cannot be determined from this figure. No consistent best fit pair exists for all the runs. This difficulty could be caused by either or both of the following:

1) As for uranium, an inherent assumption in this analysis is that the fractional surface coverage is proportional to the fractional weight of a mineral, and that the rate of change of the former is equal to the rate of change of the latter. This may not be a valid assumption for molybdenum. Tests conducted to determine the way in which molybdenum is deposited on the ore could be used to modify or refine this assumption and determine its effect on the results presented in Figure 14.

2) The kinetic expression used for molybdenum did not originate from a fundamental study of the reaction rate of this mineral with ferric ions but was based on an equation which paralleled that used for pyrite.

Whatever the cause, it is obvious from looking at the concentration histories (Figure 9) for molybdenum that these are the most poorly matched of all the concentration histories. Thus, it is not surprising that very little is gained from an analysis of the results which concern molybdenum.

The course of the leaching of pyrite cannot be directly observed as with uranium and molybdenum, since the reaction yields no unique and distinguishable product. The only means available for determining the oxidative behavior of this mineral is to look at end-of-run ferric and ferrous iron concentrations. As uranium and molybdenum minerals are depleted, ferric iron consumption decreases, and the ferric iron concentration gradually increases until it reaches a plateau (Figure 7). This plateau corresponds to the consumption of ferric iron when only pyrite is left in the column since pyrite is present in a large excess. A similar plateau is reached by the ferrous iron concentration (Figure 8). The kinetic constants for pyrite were obtained by matching the model generated curves to these end-of-run plateaus in the ferric and ferrous iron concentration histories.

End-of-run ferric and ferrous concentrations change very slowly for two reasons. As previously pointed out, the oxidation of pyrite is very slow compared to the oxidation of the other minerals. Moreover, pyrite is present in the ore in much greater quantitities than uranium and molybdenum. The reaction rate is thus very nearly constant. On the time scale of uranium and molybdenum production the process of pyrite leaching can essentially be considered a steady state process. A steady-state version of Equation (6) for ferric ion (C_3) assuming no mass transfer resistance, neglecting dispersion, and including only the pyrite reaction is

$$\frac{\partial C_3}{\partial x_D} = -\gamma_{43}\left(\frac{L\phi}{u}\right)k_4\bar{C}_4\left(\frac{C_3}{C_{3I}}\right) \tag{20}$$

from Equation (5) and (10). If the dimensionless pyrite concentration is taken to be constant because of the large excess, the solution to Equation (20) is

$$C_3(x_D) = C_{3I}\, e^{-\gamma_{43}\left(\frac{L\phi}{u}\right)k_4\bar{C}_4 x_D/C_{3I}} \tag{21}$$

from which the effluent concentration may be found by letting $x_D = 1$. With the plateau ferric ion concentration determined from the experimental effluents, the value of k_4 may be determined directly.

No effect of velocity is evident in the rate constant (k_4) obtained for pyrite determined in this manner. They are all within 15% of the average value of 7.5×10^{-8} sec^{-1} except for run no. 11 which was 13×10^{-8} sec^{-1}. This independence of velocity means that the oxidation of pyrite is a very slow process and is, therefore, not affected by mass transfer resistances.

The need for consideration of cation exchange is shown best in Figure 10 which shows a simulated concentration history for total iron for run no. 7 with and without cation exchange. The criterion for determining the cation exchange capacity for a particular run (column 10, Table 2) was that the Fe(III) and total iron break-through curves match the experimental data as closely as possible. In the absence of cation exchange the model correctly predicts that under these circumstances the total iron front will move through the column unchanged except by dispersion, and will break through at one pore volume. However, this is not what occurs experimentally. The total iron front is retarded in the column as shown by the experimental data (the effect appears small on the figures because of the compressed horizontal scale). Since the ferric iron which is consumed by the various minerals is merely reduced to ferrous iron, the total iron concentration should not change. In fact, a quantity of iron is actually produced in the pyrite oxidation reaction. Thus the retardation of the total iron front can only be due to cation exchange. As seen by comparing the two calculated curves in Figure 10, the discrepancy between experimental and model curves is decreased once cation exchange is included in the model. The time of occurrence of the U_3O_8, Fe(II), and molybdenum peaks is slightly better simulated when cation exchange is included in the model (see Figs. 6 and 9). The time at which Fe(III) breaks through is also more accurately simulated (Fig. 8).

Conclusions

The rate of production of uranium is a function of a number of different variables. The global reaction rate of uranium being leached by Fe(III) in acidic solutions has been determined to be controlled by mass transfer rather than kinetics. Although knowledge of uranium kinetics is therefore not necessary in order to model an in situ leach, considerable knowledge as to the physical positions of the deposited mineral is required in order to correctly reflect the change in the amount of uranium (as well as other minerals) available for reaction initially, and as the leach proceeds. Recoveries of 96% to 99% were achieved. Pre-oxidation of the ore was a considerable experimental problem although it is not believed to have affected the conclusions of this work. It was determined that for the ore studied here, the amount of uranium available for reaction was proportional to the total uranium.

Because the global reaction rate for uranium is mass transfer controlled, the maximum oxidant concentration and velocity possible should be used. Increases in each of these enhance the mass transfer rate. The maximum allowable velocity will be limited by fracturing considerations, and the optimum oxidant concentration will be dictated by economic considerations. Temperature effects will be less important than in a kinetically controlled system.

The global reaction rate of pyrite was found to be kinetically controlled, and that of molybdenite was undetermined. The latter may have been because of an inaccurate kinetic expression for molybdenite oxidation.

Competing minerals such as molybdenite and pyrite affect uranium production in two ways. If, as with molybdenite, the oxidation rate of the mineral is of the same order of magnitude as that of uranium, the mineral will actively compete for the oxidant and thus prolong the time required for recovery. If, as with pyrite, the oxidation rate of the mineral is much slower than that of uranium, the mineral will act chiefly in limiting the well spacing.[4]

The presence of acid-consuming minerals such as calcium carbonate in a given ore will severely limit the applicability of acid leaching to that ore. The effect of acidifying the formation and of cation exchange on the restoration of the ore bodies subjected to acid leaching has yet to be explored.

Acknowledgement - We acknowledge Gulf Research and Development for supporting this work. Drs. Peter Lindley, Sandy Riese, Gary Harloff and Tom Hardwick of Gulf have all contributed to the progress of the research. R. S. Schechter is the Senior Cockrell Chaired Professor of Engineering.

Nomenclature

Arabic Characters

a_v	Specific surface area, L^2/L^3- pore volume
a_m	Specific surface area, L^2/mass solid phase
C_i	Bulk concentration of species i, mass i/L^3-pore volume
$\bar{C}_i$	Solid phase concentration of species i, mass i/L^3-pore volume
C_{ij}	Concentration of species i at solid-fluid interface over mineral site j, mass i/L^3-pore volume
CEC	Cation exchange capacity, Mass/mass solid phase
D_i	Molecular diffusion coefficient, L^2/t

d_p	Particle diameter, L
h	Mass transfer coefficient per unit surface area, L/t
h_i	Mass transfer coefficient per unit pore volume, 1/t
k_i	Kinetic reaction constant
K_1, K_2	Cation exchange isotherm constants
K	Dispersion coefficient, L^2/t
L	Total system length, L
N_{Pe}	Peclet number
r_i	Chemical reaction rate of species i, mass i/t - L^3-pore volume
R_i	Global reaction rate, Mass i/t - L^3- pore volume
t	Time, t
u	Superficial velocity, L/t
x	Position, L

Greek Characters

γ_{ij}	Stochiometric mass ratio
ϕ	Porosity, fraction

Subscripts

BF	Designates best-fit quantity
D	Denotes dimensionless quantity
i,j	Species index (1 = uranium, 2 = ferrous, 3 = ferric, 4 = pyrite, 5 = molybdenum)
I	Denotes injected concentrations

References

1. Barnes, F. Q., "Uranium - Where are the Reserves?," Mining Congress Journal (1976) 62, No. 2, 28-33.
2. Ellis, T. R., "Estimation of Minimum Specifications for Economically Explorable Sandstone-Type Uranium Deposits," Colorado School of Mines Quarterly (1979) 74, No. 1.
3. Bommer, P. M. and Schechter, R. S., "Mathematical Modeling of In-Situ Uranium Leaching," Soc. Pet. Eng. J., (December 1979) 393-400.
4. Taton, Therese Aline, "Factors Influencing Production Rates in Acid Leaching of Uranium Ores," M. S. Thesis, The University of Texas at Austin, 1981.
5. Rhees, R. C., "E.M.F. Values - What They Are - What They Mean," American Potash and Chem. Corp. Symp., Henderson, Nevada (1962) 115-152.
6. Betts, R. H., "Kinetics of the Oxidation of Uranium (IV) by Iron (III) in Aqueous Solutions of Perchloric Acid," Canadian Journal of Chemistry (1955) 33, No. 12, 1780-1791.
7. Minc, S., Sobkowski, J., and Stok, M., "Kinetics of the Reaction of Oxidation of U(IV) Ions by Fe(III) Ions in Sulfuric Acid Solutions," Nukleonika (1965) 10, No. 12, 7-11.
8. Kanevskii, E. A. and Filippov, A. P., "Influence of the Ionic Composition of Solutions of Fe(III) on the Solution of Uranium Dioxide," Soviet Radiochemistry (1963) 5, No. 5, 558-562.
9. Kanevskii, E. A. and Filippov, A. P., "Kinetics of the Interaction of UO_2 and Fe(III) in Perchloric Acid Solutions," Soviet Radiochemistry (1965) 7, No. 2, 205-211.
10. Laxen, P. A., "The Dissolution of UO_2 as an Electron Transfer Reaction," The Recovery of Uranium, International Atomic Energy Agency, Vienna, 1971.
11. Nicol, M. J., Needes, C. R. S., and Finkerstein, N. P., "Electrochemical Model for the Leaching of Uranium Dioxide - Part 1 Acid Media," National Institute for Metallurgy, Report No. 1380, Jan. 10, 1972.
12. Amell, A. R. and Langmuir, D., "Factors Influencing the Solution Rate of Uranium Dioxide Under Conditions Applicable to In Situ Leaching," Bureau of Mines, Final Report No. HO272019.
13. Celada, J. and Lara, J. G., "The Kinetics of the Dissolution of Uranous-Uranic Oxide in Sulfuric Acid in Presence of Ferric Ion," J. Inorg. Nucl. Chem. (1965) 27, 2561-2572.
14. Warren, I. H., "The Generation of Sulfuric Acid from Pyrite by Pressure Leaching," Australian Journal of Applied Science (1956) 7, 346-357.
15. McKay, D. R. and Halpern, J., "A Kinetic Study of the Oxidation of Pyrite in Aqueous Suspension," Trans. of the Metallurgical Society of AIME (1958) 301, 301-308.
16. King, W. E., Jr., and Perlmutter, P. D., "Pyrite Oxidation in Aqueous Ferric Chloride," AIChE Journal (1977) 23, No. 5, 679-685.
17. Meyers, R. A., "Desulfurize Coal Chemically," Hydrocarbon Processing (1975) 54, 93-95.
18. Garrels, R. M., and Thompson, M. E., "Oxidation of Pyrite by Iron Sulfate Solutions," American Journal of Science (1960) Bradley Volume 258A, 57-67.
19. Bhappu, R. B., Reynolds, D. H., and Roman, R. J., "Molybdenum Recovery from Sulfide and Oxide Ores," Journal of Metals, N.Y. (1965) 17, 1199-1205.
20. Usataya, E. S., "Oxidation of Molybdenite by Water Solutions," Zap. Vses. Miner. Obsch. (1951) 8, 298-303.
21. Bryner, L. C. and Anderson, R., "Microorganisms in Leaching Sulfide Minerals," Ind. Eng. Chem. (1957) 49, No. 10, 1721-1724.
22. Lake, Larry W., Pope, Gary A., Carey, Graham, F., and Sepehrnoori, Kamy., Isothermal, Multiphase, Multicomponent Fluid-Flow in Permeable Media., Center for Enhanced Oil and Gas Recovery, The University of Texas at Austin, 1981.
23. Smith, J. M., Chemical Engineering Kinetics, 2nd. Ed., McGraw-Hill, N. Y., 1970.
24. Fedkiw, P., "Mass Transfer Controlled Reactions in Packed Beds at Low Reynolds' Numbers," PhD dissertation, University of Berkeley, (1979).
25. Malcolm, R. L., and Kennedy, V. C., "Rate of Cation Exchange on Clay Minerals as Determined by Specific Ion Electrode Techniques," Soil Science Amer. Proc., 33, 247 (1969).
26. MacDonald, R. C., and Coats, K. M. "Methods for Numerical Simulation of Water and Gas Coning," Soc. Pet. Eng. J., 10, 425-436.
27. Lantz, R. B., "Quantitative Evaluation of Numerical Diffusion (Truncation Error)," Soc. Pet. Eng. J., 11, 315-320.
28. Klohn, M. L, and Pickens, W. R.,"Geology of the Felder Uranium Deposit," Soc. of Min. Eng. of AIME (1970). Reprint.
29. Perkins, T. K, and Johnston, O. C., "A Review of Diffusion and Dispersion in Porous Media," Soc. of Pet. Eng. J. (1973), 70-84.

Table 1 - Summary of Equations

	Equations	Dependent Variable: Identity	Dependent Variable: Number
URANIUM	$\frac{\partial C_1}{\partial t_D} + \frac{\partial C_1}{\partial x_D} - \frac{1}{N_{Pe}} \frac{\partial^2 C_1}{\partial x_D^2} = \frac{L\phi}{u} r_1$	C_1	1
	$\frac{\partial \bar{C}_1}{\partial t_D} = -\frac{L\phi}{u} r_1$	$\bar{C}_1$	1
FERROUS	$\frac{\partial C_2}{\partial t_D} + \frac{\partial C_2}{\partial x_D} - \frac{1}{N_{Pe}} \frac{\partial^2 C_2}{\partial x_D^2} = La_m h_D \cdot$ $\{\bar{C}_1(C_2 - \tilde{C}_{21}) + \bar{C}_4(C_2 - \tilde{C}_{24}) + \bar{C}_5(C_2 - \tilde{C}_{25})\}$	$C_{21}\ \tilde{C}_{21}$ $\bar{C}_4, \tilde{C}_{24}, \bar{C}_5, \tilde{C}_{25}$	6
	$r_i = \gamma_{i2} a_m h_D u \bar{C}_i (C_2 - \tilde{C}_{2i})/\phi, \quad i = 1,4,5$	-	-
FERRIC	$\frac{\partial C_3}{\partial t_D} + \frac{\partial \bar{C}_3}{\partial t_D} + \frac{\partial C_3}{\partial x_D} - \frac{1}{N_{Pe}} \frac{\partial^2 C_3}{\partial x_D^2} = -La_m h_D \cdot$ $\{\bar{C}_1(C_3 - \tilde{C}_{31}) + \bar{C}_4(C_3 - \tilde{C}_{34}) + \bar{C}_5(C_3 - \tilde{C}_{35})\}$	$C_3, \bar{C}_3, \tilde{C}_{31}$ $\tilde{C}_{34}, \tilde{C}_{35}$	5
	$\bar{C}_3 = (CEC) K_1 C_3/(1 + K_2 C_3)$	-	-
	$r_i = \gamma_{i3} a_m h_D u \bar{C}_i (C_3 - \tilde{C}_{3i})/\phi, \quad i = 1,4,5$	-	-
PYRITE	$\frac{\partial \bar{C}_4}{\partial t_D} = -\frac{L\phi}{u} r_4$	-	-
MOLYBDENUM	$\frac{\partial C_5}{\partial t_D} + \frac{\partial C_5}{\partial x_D} - \frac{1}{N_{Pe}} \frac{\partial^2 C_5}{\partial x_D^2} = La_m h_D \bar{C}_5 (C_5 - C_{55})$	$C_5, \tilde{C}_{55}$	2
	$r_5 = a_m h_D u \bar{C}_5 (C_5 - \tilde{C}_{55})/\phi$	-	-
	$\frac{\partial \bar{C}_5}{\partial t_D} = -La_m h_D \bar{C}_5 (C_5 - \tilde{C}_{55})$	-	-

Total Equations = Total Unknowns = 15

Table 2. Summary of Initial and Preliminary Data for Leach Tests.

Run #	Porosity	Dispersion Coefficient, K_ℓ $\times 10^3 cm^2/sec$	Interstitial velocity, cm/sec	Mass transfer, coefficient, h $\times 10^4 cm/sec$	Initial uranium ppm	Percent pre-oxidized uranium	Injected concentrations, ppm Fe^{+3}	Injected concentrations, ppm Fe^{+2}	CEC meq/100g
6	0.49	6.3	0.43	4.99	2190	25	1990	24	7.7
7	0.51	8.0	0.43	4.97	2120	23	994	12	23.1
8	0.46	2.5	0.21	2.50	1350	42	1970	23	11.5
9	0.46	5.3	0.33	3.87	1330	42	1970	23	15.4
10	0.46	1.1	0.14	1.63	723	70	1970	23	15.4
11	0.46	8.8	0.39	4.51	463	84	490	6	15.4
13	0.47	2.4	0.47	5.48	437	83	1970	25	7.7

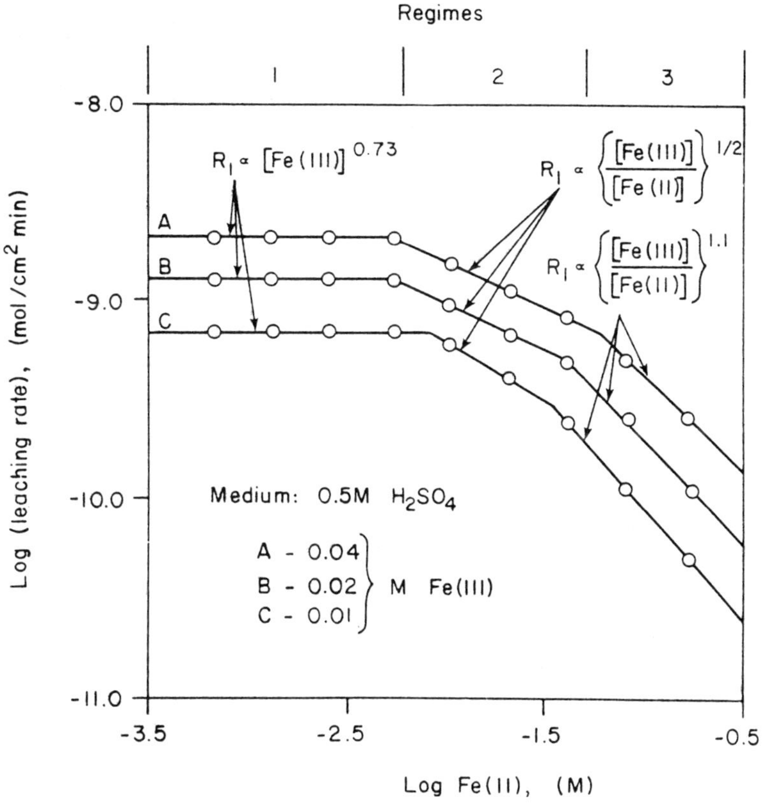

Figure 1 - Uranium leaching rate from Nichol et al.

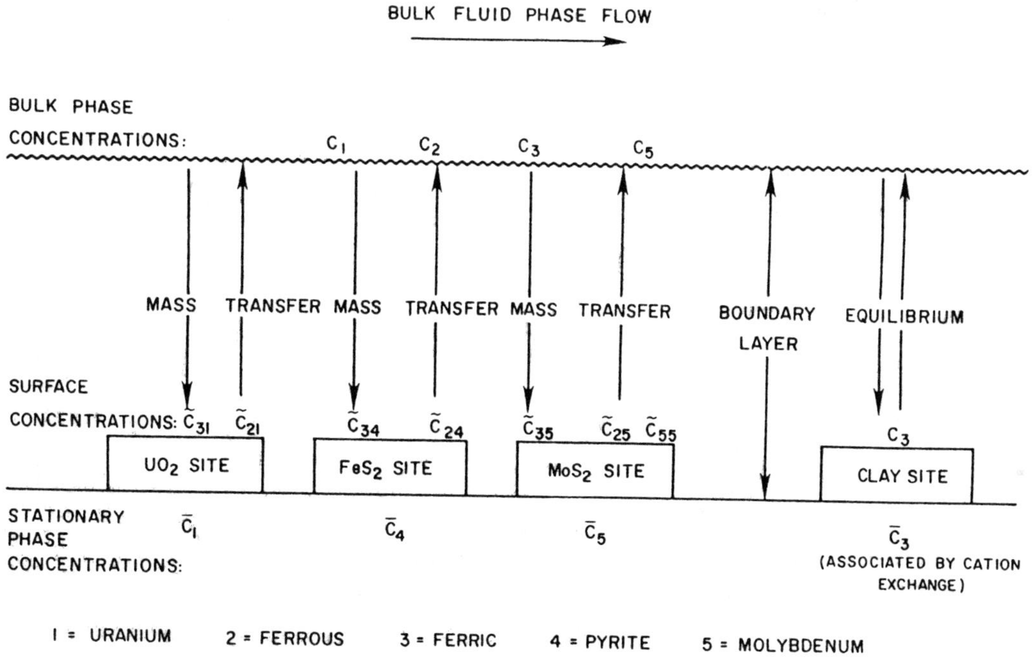

Figure 2 - Schematic of uranium leaching process in the presence of competitive minerals.

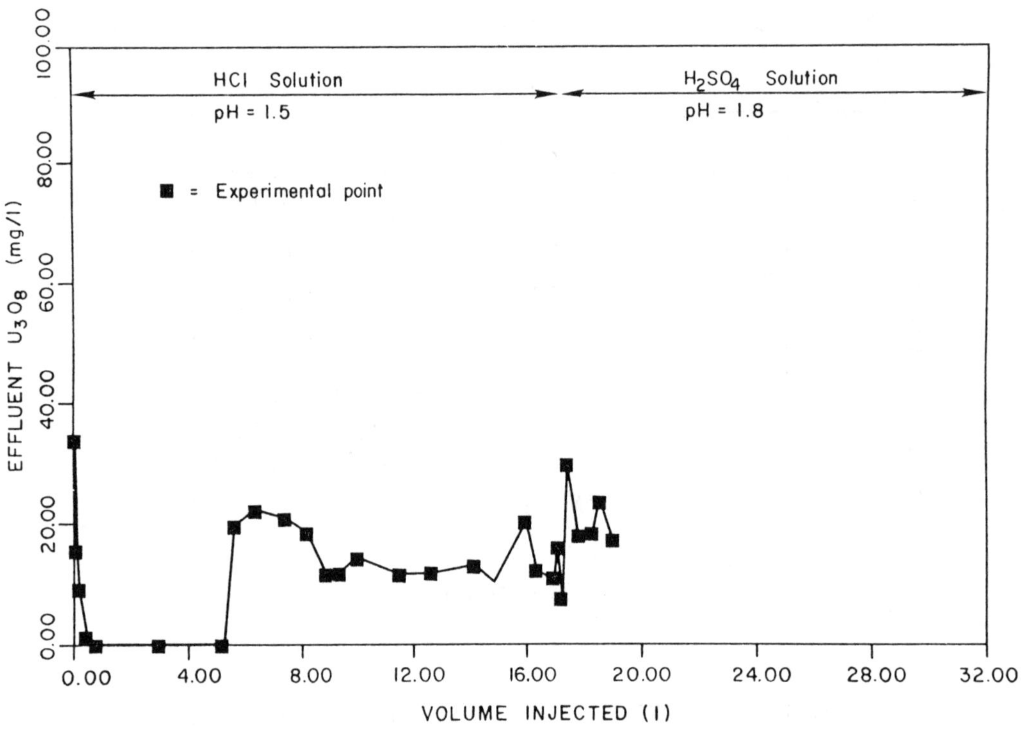

Figure 3 - Uranium effluent history during preliminary acid flush.

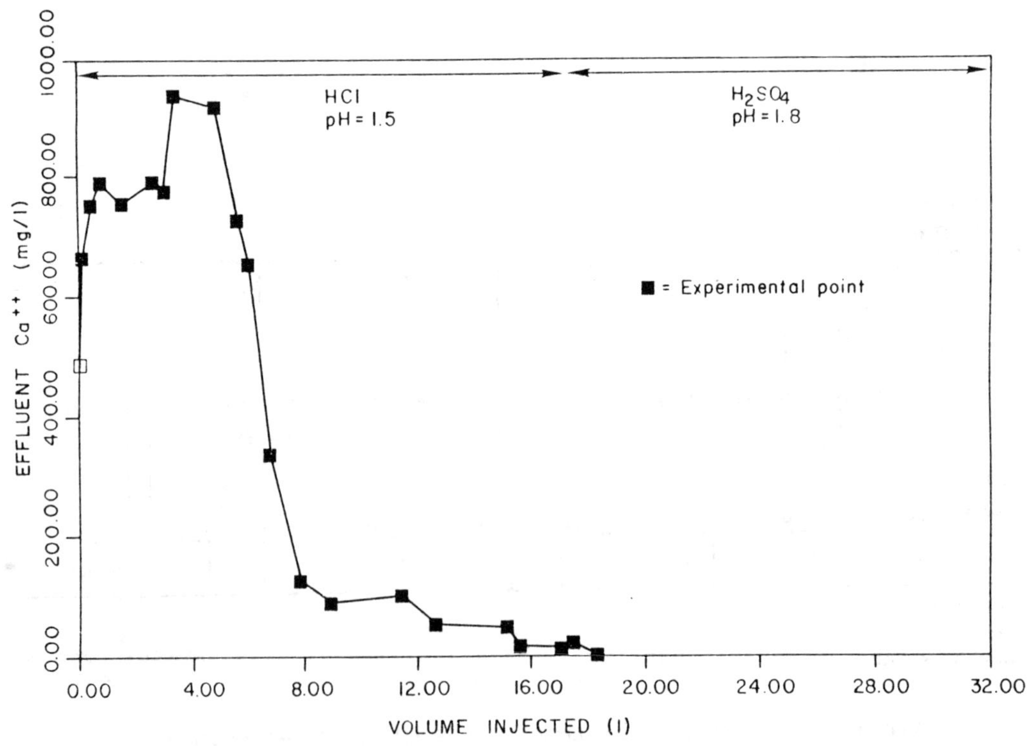

Figure 4 - Calcium effluent history during preliminary acid flush.

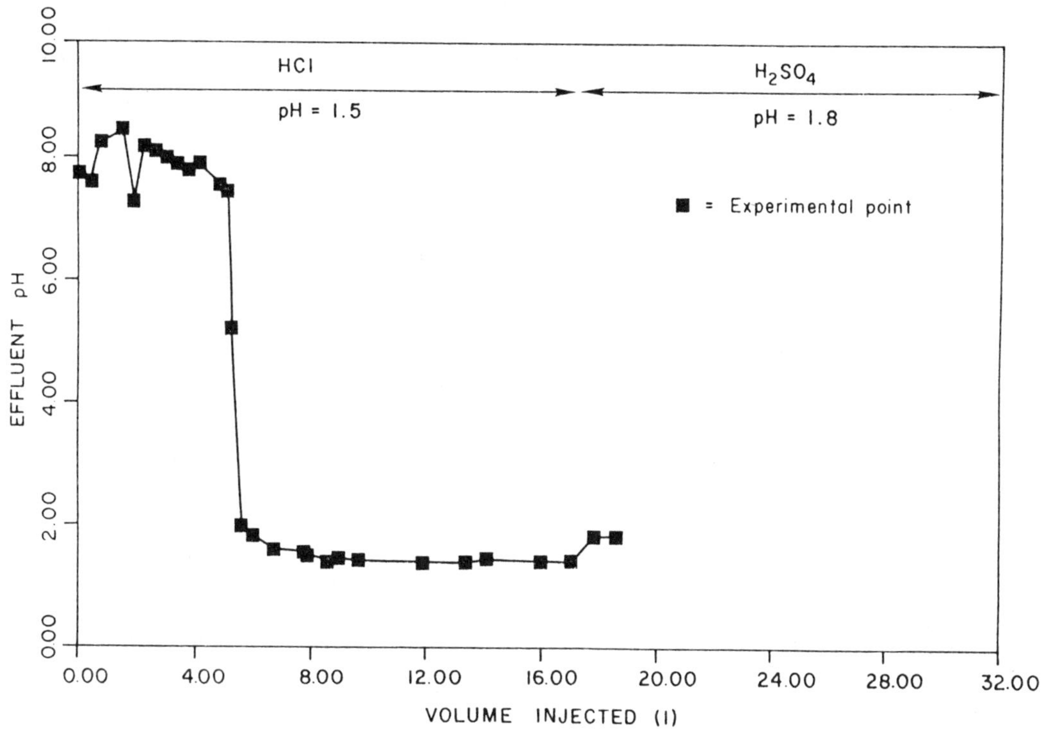

Figure 5 - Hydrogen effluent history during preliminary acid flush.

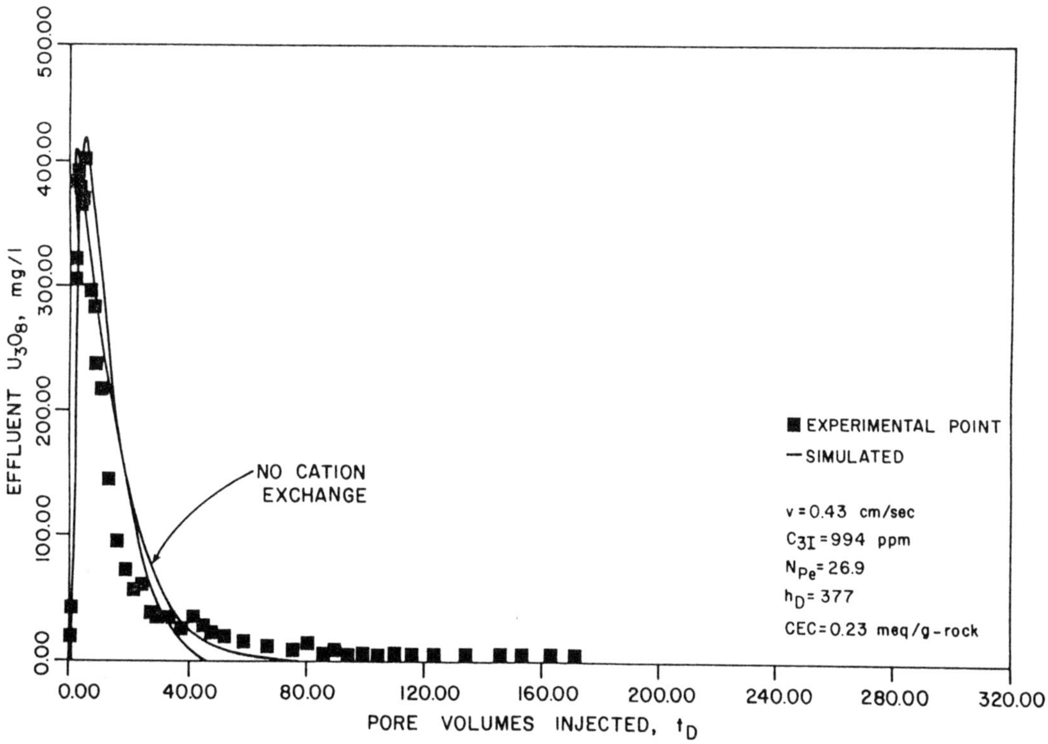

Figure 6 - Uranium effluent history during ferric leach.

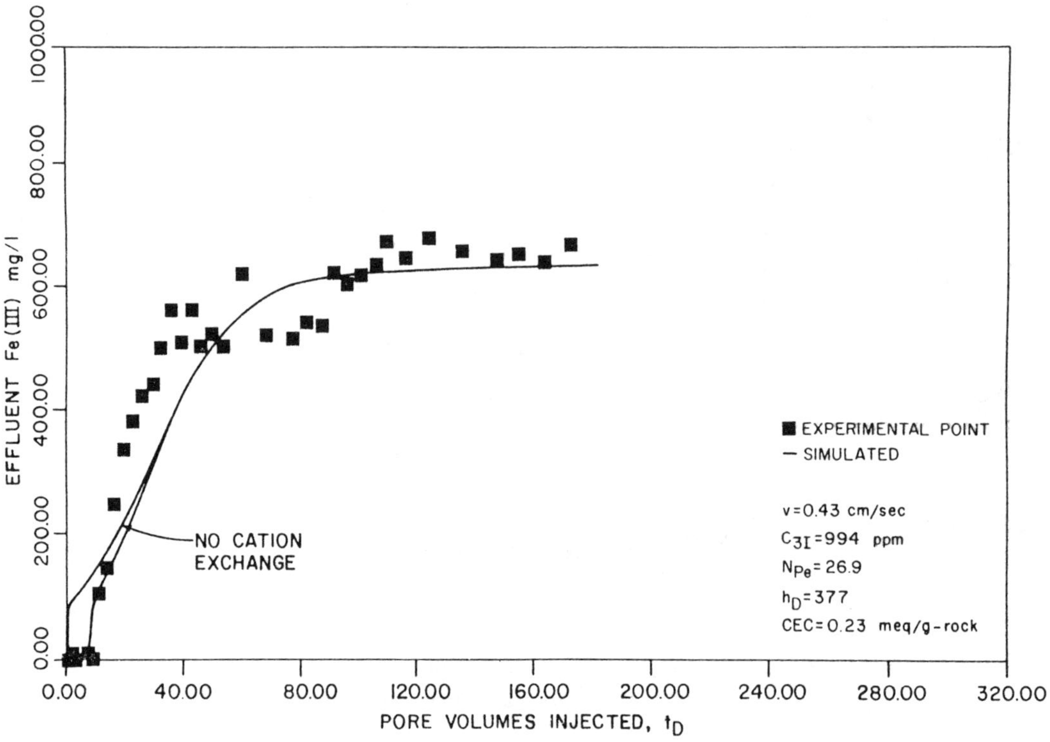

Figure 7. Ferric ion history effluent history during ferric leach.

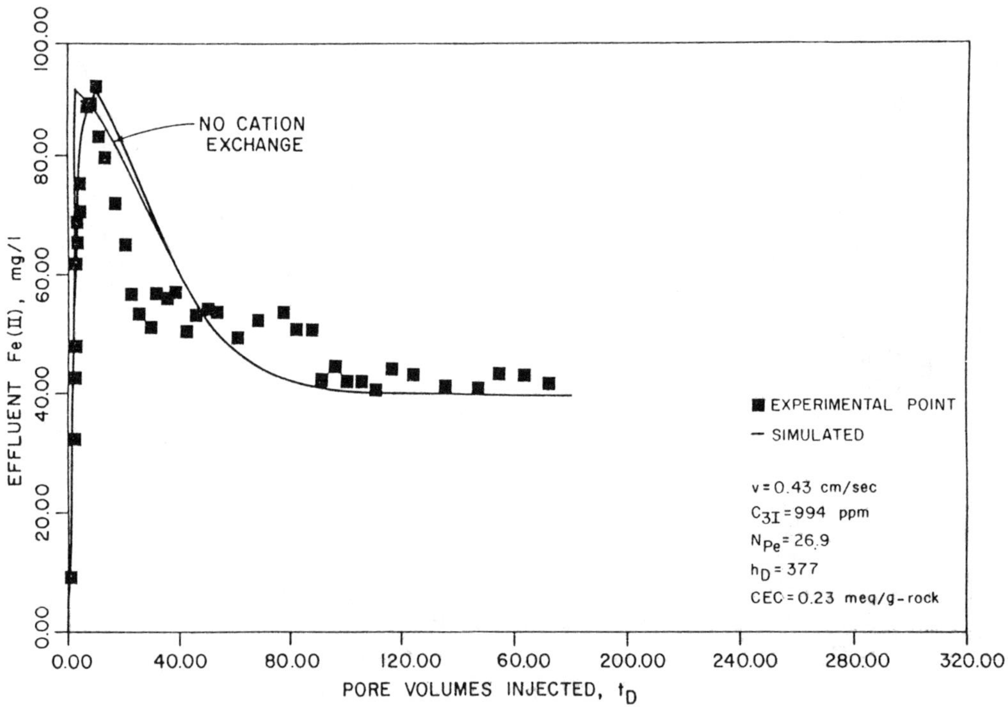

Figure 8. Ferrous ion effluent history during ferric leach.

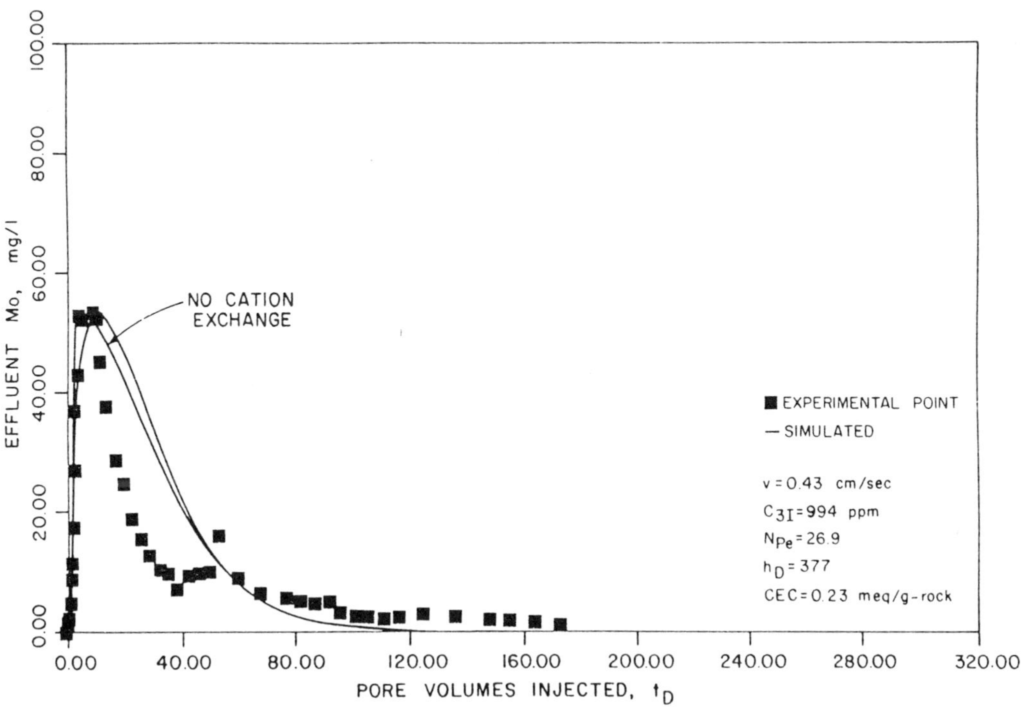

Figure 9 - Molybdenum effluent history during ferric leach.

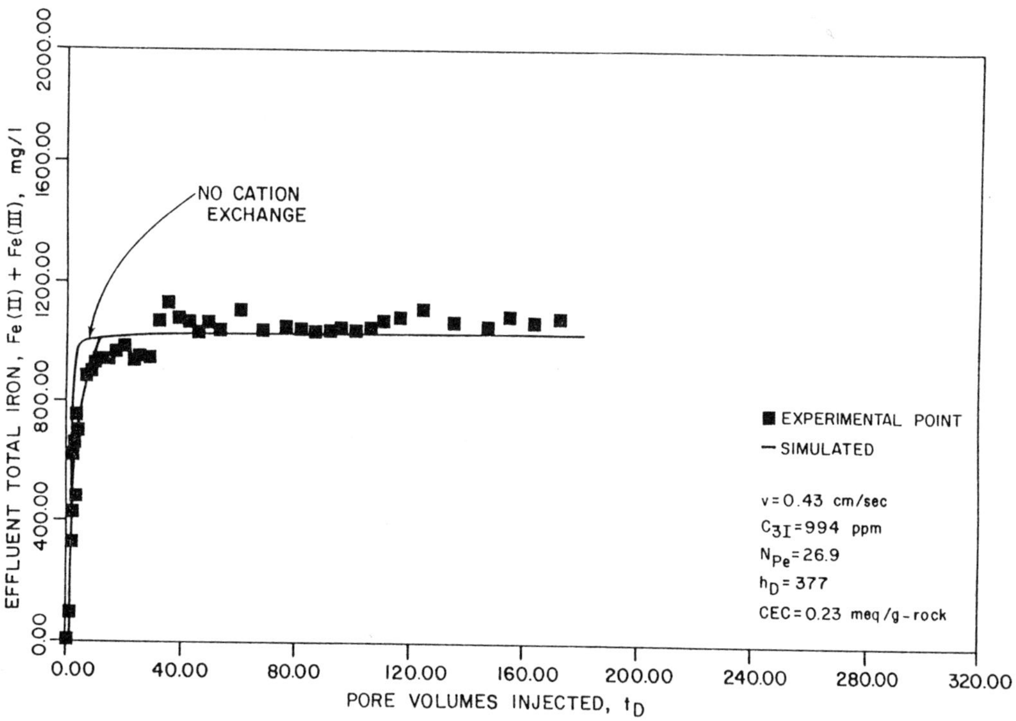

Figure 10 - Total iron effluent history during ferric leach.

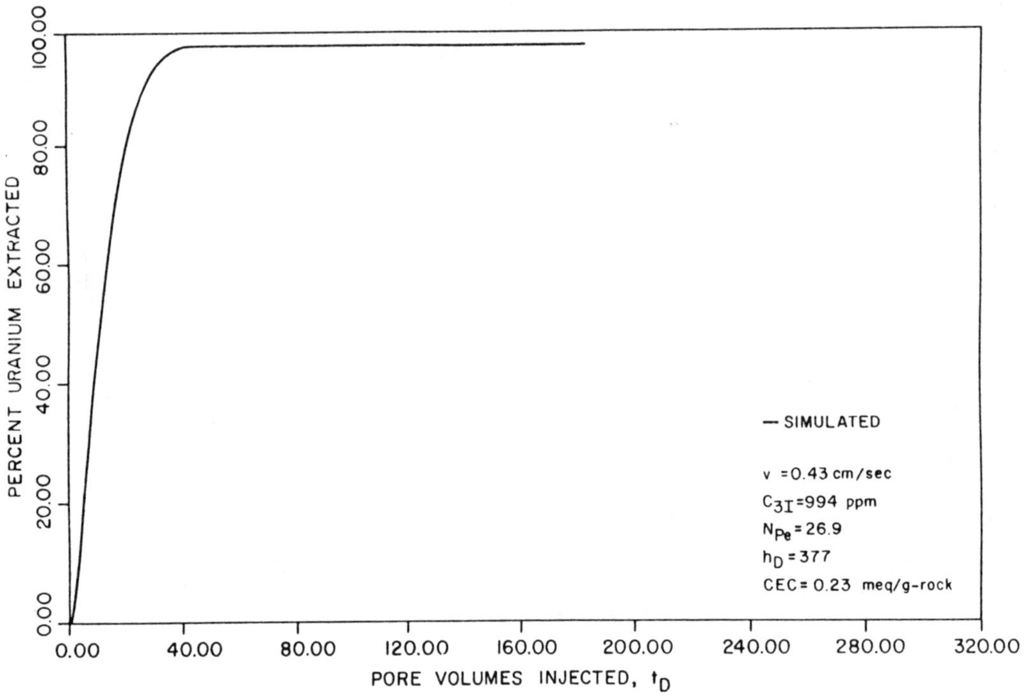

Figure 11 - Cumulative percent uranium recovery during ferric leach.

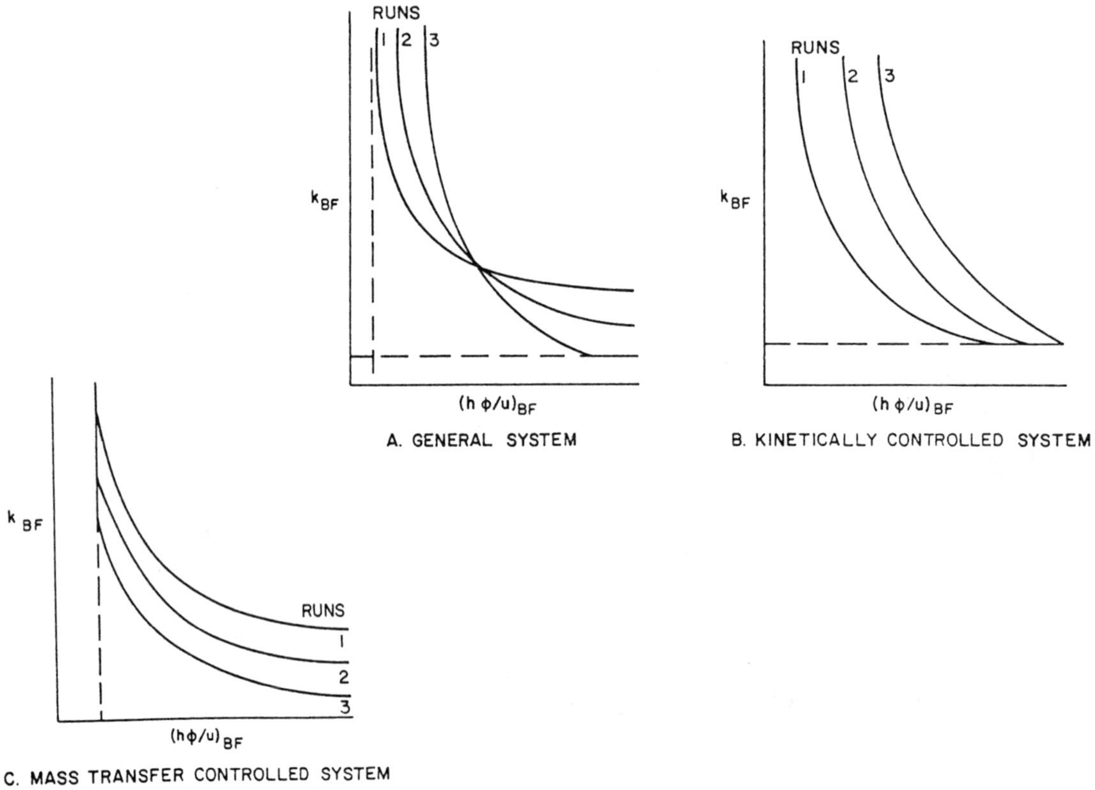

Figure 12 - Schematic illustration of parameter estimation process.

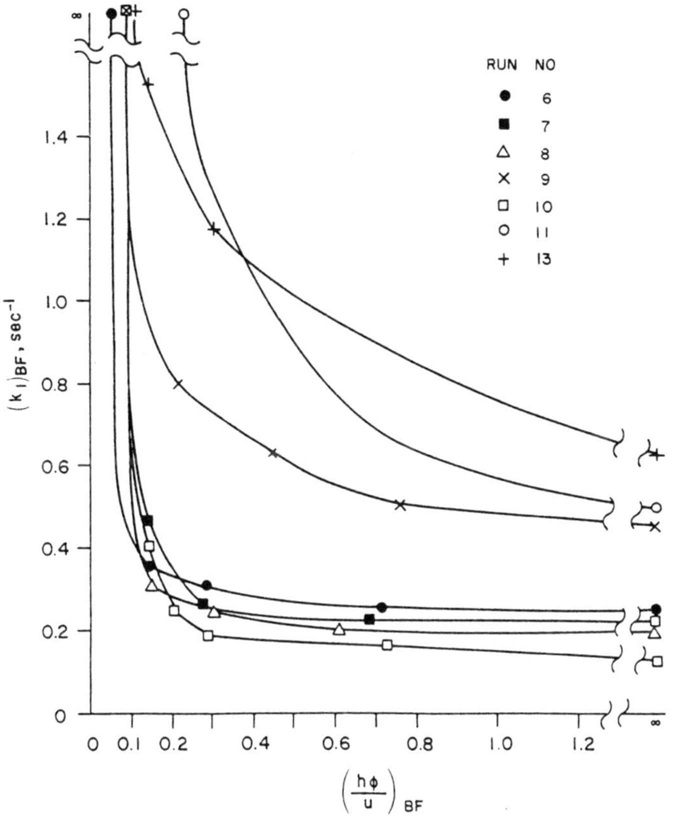

Figure 13 - Parameter estimation for uranium.

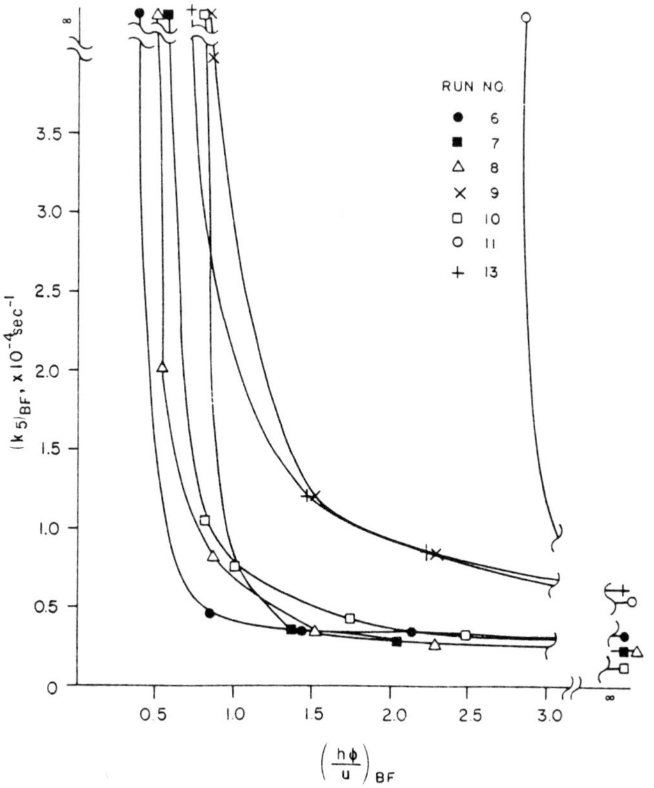

Figure 14 - Parameter estimation for molybdenum.

Chapter 13

A KINETIC RATE EXPRESSION FOR CARBONATE LEACHING OF URANIUM

David A. Shumbera
John W. Sibert
Research & Development Department
ARCO Oil and Gas Company
Dallas, Texas 75221

ABSTRACT

This paper presents a new kinetic rate expression for the oxidation-dissolution of U_3O_8 by oxygen and bicarbonate leachate in the presence of oxygen-consuming gangue constituents. Experimental U_3O_8 effluent concentrations under a given set of oxygen and bicarbonate conditions were obtained from differential reactor runs. These experimental data were successfully fit using a reaction model that assumed two uranium species reacting simultaneously at different rates.

The power law rate expression obtained in this study indicated a first order dependency for each of the uranium species, a 1/2 order dependency on oxygen, and essentially no (zero order) dependency on bicarbonate. This corresponds to an oxidation-dissolution mechanism in which the oxidation of UO_2 is rate controlling in the presence of excess bicarbonate. The 1/2 order oxygen depencency agrees with previous determinations of oxygen dependency, whereas the zero order bicarbonate dependency from this study differs from the 1/2 order bicarbonate dependency found previously.

It is recommended that further experimentation be done to verify the rate expression obtained in this report.

Introduction

The predictive capability of an in-situ leaching simulator depends on how accurately the assumed reaction kinetics describe the actual reaction mechanism of uranium oxidation-dissolution and gangue oxidation. The kinetic expressions herein presented are based on an unpublished process variable study of leaching parameters done by ARCO Oil and Gas Company (1976) and the work of Schortmann and DeSesa (1958). Values for the rate constants for uranium oxidation and dissolution and gangue oxidation rate expressions reported in the above-mentioned work did not appear to adequately describe the uranium and gangue leaching kinetics observed by recent experimental work.

History matching with a leaching simulator of long-core carbonate leaching experiments was done in an attempt to obtain better estimates of the rate constants. However, uncertainties in the gangue characteristics and variability in the feed concentrations of oxygen and bicarbonate led to wide variances in the estimates of the kinetic rate constants for various experimental long core studies. Moreover, history matching can only result in rate constant estimates that are correct relative to one another. A study was initiated to obtain a more accurate rate expression to describe uranium ore oxidation and dissolution by carbonate leaching in the presence of oxygen-consuming gangue constituents.

Experimental

A laboratory procedure was devised to experimentally obtain more accurate rate data for the uranium oxidation-dissolution and gangue oxidation reactions.

A reactor was packed with a small amount of ore with a known U_3O_8 grade. A high leachate flow rate was used so the oxygen and bicarbonate concentrations were approximately constant throughout the reactor at all times, thus providing a "differential" reactor system. Point effluent samples from the reactor were taken and analyzed for U_3O_8 and sulfate concentrations. In this way, U_3O_8 production could be measured at constant oxygen and bicarbonate leachate conditions for a run. Different oxygen and bicarbonate concentrations were used for the various experimental runs.

Data Evaluation

Table 1 summarizes the experimental conditions for the differential reactor runs. Only four runs with two oxygen concentrations and two bicarbonate concentrations were made due to time constraints. Two replicate runs were also made. More experimental runs at different bicarbonate and oxygen concentrations are essential for verification of the form of the kinetic equations and the rate constants.

Point effluent concentrations of U_3O_8 and SO_4 at their corresponding sampling times were successfully obtained. A familiar problem of preoxidation of some of the uranium (probably due to exposure to air) was seen in the first one or two analyzed samples for each of the runs. In long core studies, this problem has been eliminated by preleaching the core with oxygen-free leachate to dissolve and remove any preoxidized uranium. No preleaching was done to the ore in the differential reactors since it was assumed such a small amount of ore would probably produce little preoxidized uranium. For the most part, the preoxidized uranium concentration was small and, for this reason, was included in the effluent U_3O_8 concentrations.

The rate parameters used in modeling the differential reactor runs were the same as in the previous rate expressions incorporated in the in-situ leaching simulator. See Table 2 for the definitions of these paramters.

The reactor for each run had been packed with ore of the same ore grade. Thus, each reactor contained 360 mg U_3O_8 at the start of the run. Uranium production during any time increment, Δt, was calculated by averaging the U_3O_8 concentrations at the beginning and end of the increment and assuming a constant flow rate of 0.3 liters/hr. In this way, total U_3O_8 production at any time, t, could be calculated along with the amount of U_3O_8 remaining in the ore phase.

From the raw experimental data, the following values were calculated:

- U308P - U_3O_8 production for each time interval in mg.
- U308R - The amount of U_3O_8 remaining in the ore in mg.
- FRACR - The fraction of U_3O_8 remaining in the ore based on the original U_3O_8 content of the ore.

Tables 3 through 6 summarize these calculated values for each experimental run.

Experimental data showed that gangue constituents were not in sufficient quantities to obtain accurate point effluent sulfate concentrations throughout the runs. Therefore, calculation of a gangue oxidation rate expression was not possible.

Modeling of Differential Reactor Kinetic Data

All modeling work in this report was done using the Statistical Analysis System (SAS).

It was hoped that the previous rate expressions incorporated into the existing in-situ leaching simulator would provide a reasonably good fit and that only adjustments in the rate constants or exponents in the power law rate expressions would be needed.

Since bicarbonate concentration and oxygen concentration were held constant for a run, the U_3O_8 remaining in the ore, ω, was the only variable to be fit for a run. The approach was to obtain a power law relationship to describe ω versus time for all four runs, with differences in ω due to different bicarbonate and oxygen concentrations among the runs adequately accounted for with only rate constant differences for each run. These rate constants for each of the runs could then be modeled as a power law rate expression in terms of the different bicarbonate and oxygen concentrations. In this way, an overall rate expression could be obtained which would be a function of ω; oxygen concentration, f; and bicarbonate concentration, b. That is,

$$\frac{d\omega}{dt} = k_1\omega^n \qquad (1)$$

n = order of the reaction with respect to uranium remaining in the ore.

This relationship would hold for all four runs with only the rate constant k_1 changing to account for different b_i and f_i levels in the runs. The rate constants, k's, obtained for the four runs could be modeled as a function of b_i and f_i.

$$k_1 = k' b_i^{n'} f_i^{n''} \qquad (2)$$

k' = rate constant (same for all four runs).

n' = order of reaction with respect to HCO_3^-.

n" = order of reaction with respect to O_2.

By combining the two fits, Equations 1 and 2, the overall power law rate expression:

$$\frac{d\omega}{dt} = k' b_i^{n'} f_i^{n''} \omega^n \qquad (3)$$

could be obtained and be incorporated into the model.

However, no suitable k_1 or n could be obtained for Equation 1 to describe all four runs simultaneously.

Modeling Using an Unleachable Uranium Fraction

From the U_3O_8 point production concentrations versus time, it appeared as if the U_3O_8 production were reaching a limit fairly far from complete dissolution of the uranium in the ore. An unleachable fraction was assumed for the uranium in the ore phase. That is,

$$\frac{d(\omega-b)}{dt} = -k_1(\omega-b)^n \qquad (4)$$

where,

b = unleachable uranium fraction.

A plot of ω versus inverse time (1/t) as shown in Figure 1 provided estimates of the unleachable uranium fraction when extrapolated to zero time. The results are shown in Table 7.

This unleachable fraction was fit as a function of the oxygen leachate fraction f_i to serve as a predictive method of removing unleachable uranium from the original uranium content obtained from ore grade measurements for the in-situ leaching model. The following simple equation was obtained.

$$b = (663.8)f_i \qquad (5)$$

where,

b = unleachable uranium fraction.

For all subsequent modeling, the unleachable uranium fraction was subtracted from ω, and ω was based on the <u>leachable</u> amount of uranium.

Modeling of Two Uranium Species Reacting in Parallel

From inspection of the experimental data, it was noticed that two straight line segments would fit the data well for all runs. This could be interpreted as a parallel reaction scheme of two uranium species, one species fast-reacting, the other slower, that is,

$$\text{Uran.}_1 \xrightarrow{k_1} \text{UTC}$$

$$\text{Uran.}_2 \xrightarrow{k_2} \text{UTC} \qquad (6)$$

The variables ω_1 and ω_2 were defined as the fraction of the two uranium species based on the total amount of original leachable uranium. The assumed overall rate expression then became:

$$\frac{d\omega}{dt} = -(k_1\omega_1 + k_2\omega_2). \qquad (7)$$

Using SAS, the plots of ω versus time for each of the four runs were fit as two separate first order with respect to ω_1 or ω_2 rate expressions. That is,

$$\frac{d\omega_1}{dt} = -k_1\omega_1 \qquad (8)$$

$$\frac{d\omega_2}{dt} = -k_2\omega_2 \qquad (9)$$

Since the infinity value for ω_1 was not known, ω_2 versus time was modeled first and the fit for ω_1 versus time was obtained from the fit for ω_2 versus time.
Integrating Equations 8 and 9,

$$\ln \omega_1 = -k_1 t + c_1 \quad (10)$$

and

$$\ln \omega_2 = -k_2 t + c_2 \quad (11)$$

The curve of ln ω versus time for each of the four runs was divided into two separate segments with the experimental data from each segment used separately. See Figure 2.

Using GLM (General Linear Method) of SAS and the second part of the ln ω vs time curves for each of the four runs, fits were obtained for:

$$\ln \omega_2 = -k_2 t + c_2 \quad (12)$$

Run	k_2	c_2
1	0.028 ± 0.00018	- 0.23215 ± 0.0076
2	0.01557 ± 0.000395	- 1.37809 ± 0.0174
3	0.0248 ± 0.00039	- 0.775378 ± 0.0148
4	0.0305 ± 0.0008	- 1.66252 ± 0.0307

To incorporate the contribution of ω_1 to the total ω versus time curve, the following procedure was used:

$$\ln \omega_1 = -k_1 t + c_1 \quad (13)$$

$$\ln(\omega - \omega_2) = -k_1 t + c_1 \quad (14)$$

Since ω, the overall uranium fraction remaining, is known from the experimental data and ω_2 over the entire time interval is calculable from Equation (12) for each of the four differential reactor runs, Equation (14) can be fit in the same manner as Equation (12) to obtain k_1 values for each of the runs. Since the fit for ω_2 put a constraint on c_1 (Equation (15)), c_1 was determined by Equation (15) so that only k_1 in Equation (14) remained to be fit.

$$\ln\left(\omega - e^{-k_2 t} e^{c_2}\right) = -k_1 t + c_1 . \quad (15)$$

At $t = 0$, $\omega = 1$.

Therefore,

$$\ln\left(1 - e^{c_2}\right) = c_1 . \quad (16)$$

The following fits were obtained for Equation (14):

Run	k_1	c_1
1	+ 0.10513 ± 0.00229	- 1.5742
2	+ 0.11490 ± 0.00191	- 0.2904
3	+ 0.14280 ± 0.0030	- 0.6172
4	+ 0.18087 ± 0.00203	-0.2103

Combining the two separate rate equations, the following equations were obtained,

Run	Equation
1	$\omega = e^{-0.10513t}[0.2072] + e^{-0.028t}[0.79283]$
2	$\omega = e^{-0.11490t}[0.748] + e^{-0.01557t}[0.25206]$
3	$\omega = e^{-0.1428t}[0.5395] + e^{-0.0248t}[0.4605]$
4	$\omega = e^{-0.18087t}[0.8103] + e^{-0.0305t}[0.1897]$

To find a single equation to describe all four runs simultaneously, the k_1's, k_2's, and $\omega_2{}^o$'s had to be fit as functions of oxygen f_i and bicarbonate b_i. ($\omega_2{}^o$ is the constant term multiplying the exponential term for the ω_2 vs time fits).

Table 8 summarizes the data that were fitted as a function of f_i and b_i.

$$k_1 = k_1' f_i^{a} b_i^{b} \quad (17)$$

$$k_2 = k_2' f_i^{a'} b_i^{b'} \quad (18)$$

$$w_2^{o} = k_2^{o} f_i^{a''} b_i^{b''} \quad (19)$$

The following results were obtained:

Equation (17)

k_1' = 23.02 ± 6.78
a = 0.5482 ± 0.1064
b = 0.0907 ± 0.0411

Equation (18)

k_2' = 0.31556 ± 0.1058
a' = 0.3961 ± 0.1211
b' = - 0.1055 ± 0.0320

Equation (19)

k_2^{o} = $5.738 \times 10^{-5} \pm 9.548 \times 10^{-6}$
a" = - 0.5969 ± 0.1869
b" = - 0.5673 ± 0.0723

Therefore, the final equation to describe ω versus time was found to be:

$$\omega = e^{-23.02 f_i^{0.5482} b_i^{0.0907} t}$$

$$[1 - (5.738\text{x}10^{-5}) f_i^{-0.5969} b_i^{-0.5673}]$$

$$+ e^{-0.3156 f_i^{0.3961} b_i^{-0.1055} t}$$

$$[(5.738\text{x}10^{-5}) f_i^{-0.5969} b_i^{-0.5673}] \qquad (20)$$

From this final equation, the following rate equations can be written:

$$\left(\frac{1}{\omega_1{}^o}\right) \frac{d\omega_1}{dt} = -23.02 f_i^{0.5482} b_i^{0.0907} \omega_1 \qquad (21)$$

$$\left(\frac{1}{\omega_2{}^o}\right) \frac{d\omega_2}{dt} = -0.3156 f_i^{0.3961} b_i^{-0.1055} \omega_2 \qquad (22)$$

Equation (20) was tested to determine its accuracy in predicting ω.

Equation (20) was found to model the experimental data for all runs very well.

Conclusions and Recommendations

A satisfactory rate expression to describe the fraction of leachable uranium remaining in the ore with time, bicarbonate concentration, and oxygen concentration was found. The rate expression used nine parameters to totally describe the reaction mechanism. More experimentation with different oxygen and bicarbonate levels are necessary to verify the validity of the rate expression over the entire range of leaching concentrations for oxygen and bicarbonate.

The final rate expression, Equation (20), obtained in this study indicated that oxygen (f_i) was approximately 1/2 order, while there was essentially no bicarbonate (b_i) dependency (zero order with respect to bicarbonate) as long as excess bicarbonate was present. This type of dependency would occur if the uranium oxidation step was the slow step (controlling step) in the mechanism of the oxidation/dissolution of UO_2.

References

1. Schortmann, W. E. and M. A. DeSesa, U.S. Peaceful Uses of Atomic Energy Proceedings, 2nd International Conference, Geneva, September 1958, Vol. 11, pp. 333-341.

Table 1. Differential Reactor Runs 1 Through 4 Experimental Conditions

Run	$[HCO_3^-]$, ppm (mg/ℓ)	$[O_2]$, ppm (mg/ℓ)
1	500	177
2	3000	177
3	500	354
4	3000	354

Table 2. List of Variables for Rate Expressions for Differential Reactor Runs

$$\omega_i = \frac{\text{gm U in ore at time t}}{\text{gm U in virgin ore}}$$

$$b_i = \frac{\text{gm } HCO_3^- \text{ in solution}}{\text{gm of solution}}$$

$$f_i = \frac{\text{gm } O_2 \text{ in solution}}{\text{gm of solution}}$$

Table 3. Calculated Variable Values from Experimental Data Run #1

OBS	TIME (hrs)	U3O8P (mg)	U3O8R (mg)	FRACR
1	0	0.00	360.00	1.00000
2	1	12.40	347.60	0.96556
3	2	11.80	335.80	0.93278
4	3	11.40	324.40	0.90111
5	4	10.35	314.05	0.87236
6	5	9.75	304.30	0.84528
7	6	9.30	295.00	0.81944
8	7	9.00	286.00	0.79444
9	8	8.70	277.30	0.77028
10	9	8.10	269.20	0.74778
11	10	7.80	261.40	0.72611
12	11	7.80	253.60	0.70444
13	12	7.65	245.95	0.68319
14	13	7.50	238.45	0.66236
15	14	7.20	231.25	0.64236
16	15	6.75	224.50	0.62361
17	16	6.75	217.75	0.60486
18	17	6.60	211.15	0.58653
19	18	6.15	205.00	0.56944
20	19	5.70	199.30	0.55361
21	20	5.25	194.05	0.53903
22	21	4.95	189.10	0.52528
23	22	4.65	184.45	0.51236
24	23	4.50	179.95	0.49986
25	24	4.50	175.45	0.48736
26	25	4.50	170.95	0.47486
27	26	4.42	166.52	0.46256
28	27	4.07	162.46	0.45128
29	28	3.69	158.77	0.44103
30	29	3.53	155.25	0.43125
31	30	3.38	151.87	0.42186
32	31	3.22	148.65	0.41292
33	32	3.09	145.56	0.40433
34	33	3.02	142.55	0.39597
35	34	2.96	139.60	0.38778
36	35	2.88	136.72	0.37978
37	36	2.80	133.92	0.37200
38	37	2.73	131.19	0.36442
39	38	2.62	128.56	0.35711
40	39	2.51	126.06	0.35017
41	40	2.37	123.69	0.34358
42	41	2.19	121.50	0.33750
43	42	2.10	119.39	0.33164
44	43	2.10	117.29	0.32581
45	44	2.03	115.27	0.32019
46	45	1.89	113.38	0.31494
47	46	1.82	111.56	0.30989
48	47	1.76	109.81	0.30503
49	48	1.70	108.11	0.30031
50	49	1.67	106.45	0.29569
51	50	1.65	104.80	0.29111
52	51	1.62	103.18	0.28661
53	52	1.59	101.59	0.28219
54	53	1.55	100.05	0.27792
55	54	1.50	98.55	0.27375
56	55	1.47	97.08	0.26967

Table 4. Calculated Variable Values from Experimental Data Run #2

OBS	TIME (hrs)	U3O8P (mg)	U3O8R (mg)	FRACR
1	0	0.00	360.00	1.00000
2	1	20.00	340.00	0.94444
3	2	18.50	321.50	0.89306
4	3	17.50	304.00	0.84444
5	4	16.20	287.80	0.79944
6	5	15.60	272.20	0.75611
7	6	15.15	257.05	0.71403
8	7	15.00	242.05	0.67236
9	8	14.25	227.80	0.63278
10	9	13.20	214.60	0.59611
11	10	12.60	202.00	0.56111
12	11	11.40	190.60	0.52944
13	12	10.20	180.40	0.50111
14	13	9.45	170.95	0.47486
15	14	9.00	161.95	0.44986
16	15	8.25	153.70	0.42694
17	16	7.05	146.65	0.40736
18	17	6.30	140.35	0.38986
19	18	5.85	134.50	0.37361
20	19	5.40	129.10	0.35861
21	20	4.68	124.42	0.34561
22	21	4.05	120.37	0.33436
23	22	3.68	116.69	0.32414
24	23	3.41	113.29	0.31469
25	24	3.18	110.11	0.30586
26	25	2.94	107.17	0.29769
27	26	2.70	104.47	0.29019
28	27	2.45	102.02	0.28339
29	28	2.15	99.88	0.27744
30	29	1.92	97.96	0.27211
31	30	1.77	96.19	0.26719
32	31	1.59	94.60	0.26278
33	32	1.50	93.10	0.25861
34	33	1.38	91.72	0.25478
35	34	1.20	90.52	0.25144
36	35	1.11	89.41	0.24836
37	36	1.04	88.37	0.24547
38	37	0.98	87.40	0.24278
39	38	0.93	86.47	0.24019
40	39	0.87	85.60	0.23778
41	40	0.78	84.82	0.23561
42	41	0.69	84.13	0.23369
43	42	0.66	83.47	0.23186
44	43	0.63	82.84	0.23011
45	44	0.57	82.27	0.22853
46	45	0.53	81.74	0.22706
47	46	0.51	81.23	0.22564
48	47	0.51	80.72	0.22422
49	48	0.53	80.20	0.22278
50	49	0.51	79.69	0.22136
51	50	0.48	79.21	0.22003
52	51	0.47	78.74	0.21872
53	52	0.44	78.31	0.21753
54	53	0.42	77.89	0.21636

Table 5. Calculated Variable Values from Experimental Data Run #3

OBS	TIME (hrs)	U308P (mg)	U308R (mg)	FRACR
1	0	0.00	360.00	1.00000
2	1	20.00	340.00	0.94444
3	2	17.50	322.50	0.89583
4	3	15.60	306.90	0.85250
5	4	14.12	292.79	0.81331
6	5	13.29	279.50	0.77639
7	6	12.60	266.90	0.74139
8	7	11.55	255.35	0.70931
9	8	10.05	245.30	0.68139
10	9	9.00	236.30	0.65639
11	10	8.40	227.90	0.63306
12	11	8.03	219.87	0.61075
13	12	7.50	212.37	0.58992
14	13	6.78	205.59	0.57108
15	14	6.29	199.31	0.55364
16	15	5.73	193.58	0.53772
17	16	5.19	188.39	0.52331
18	17	4.77	183.62	0.51006
19	18	4.38	179.24	0.49789
20	19	4.02	175.22	0.48672
21	20	3.77	171.45	0.47625
22	21	3.59	167.87	0.46631
23	22	3.42	164.45	0.45681
24	23	3.21	161.24	0.44789
25	24	3.00	158.24	0.43956
26	25	2.79	155.45	0.43181
27	26	2.52	152.93	0.42481
28	27	2.28	150.65	0.41847
29	28	2.06	148.59	0.41275
30	29	1.92	146.67	0.40742
31	30	1.82	144.86	0.40239
32	34	6.18	138.68	0.38522
33	35	1.34	137.34	0.38150
34	36	1.29	136.05	0.37792
35	37	1.25	134.81	0.37447
36	38	1.20	133.61	0.37114
37	39	1.19	132.42	0.36783
38	40	1.16	131.27	0.36464
39	41	1.07	130.20	0.36167
40	42	0.99	129.21	0.35892
41	43	0.95	128.27	0.35631
42	45	1.80	126.47	0.35131
43	46	0.86	125.61	0.34892
44	47	0.84	124.77	0.34658

Table 6. Calculated Variable Values from Experimental Data Run #4

OBS	TIME (hrs)	U308P (mg)	U308R (mg)	FRACR
1	0.0	0.00	360.00	1.00000
2	0.5	27.00	333.00	0.92500
3	1.5	25.30	307.70	0.85472
4	2.5	22.98	284.72	0.79089
5	3.5	21.30	263.42	0.73172
6	4.5	19.65	243.77	0.67714
7	5.5	18.15	225.62	0.62672
8	6.5	16.87	208.75	0.57986
9	7.5	14.93	193.82	0.53839
10	8.5	12.65	181.18	0.50328
11	9.5	10.55	170.63	0.47397
12	10.5	8.93	161.71	0.44919
13	12.5	15.03	146.68	0.40744
14	13.5	5.97	140.71	0.39086
15	14.5	4.80	135.91	0.37753
16	15.5	4.10	131.81	0.36614
17	16.5	3.50	128.32	0.35644
18	17.5	3.05	125.27	0.34797
19	18.5	2.67	122.60	0.34056
20	19.5	2.33	120.28	0.33411
21	22.5	5.76	114.52	0.31811
22	23.5	1.56	112.96	0.31378
23	24.5	1.46	111.50	0.30972
24	25.5	1.41	110.09	0.30581
25	26.5	1.27	108.82	0.30228
26	27.5	1.13	107.69	0.29914
27	28.5	0.99	106.70	0.29639
28	29.5	0.87	105.83	0.29397
29	31.5	1.50	104.33	0.28981
30	32.5	0.66	103.67	0.28797
31	33.5	0.62	103.06	0.28628
32	34.5	0.59	102.47	0.28464
33	35.5	0.55	101.93	0.28314
34	36.5	0.50	101.43	0.28175
35	37.5	0.47	100.96	0.28044
36	38.5	0.44	100.53	0.27925
37	39.5	0.41	100.12	0.27811
38	40.5	0.39	99.73	0.27703
39	42.5	0.77	98.97	0.27492
40	44.5	0.71	98.26	0.27294
41	46.5	0.57	97.69	0.27136

Table 7. Unleachable Uranium Fraction as Determined by ω Versus $Time^{-1}$ Curve Extrapolation to Zero Time

Run No.	Intercept ($\omega_{unleachable}$)
1	0.115
2	0.125
3	0.240
4	0.220

Table 8. Parameter Values to be Fit as Functions of f_i and b_i

Run	k_1	k_2	$\omega_2{}^o$	f_i	b_i
1	0.10513	0.028	0.79283	177×10^{-6}	0.0005
2	0.11490	0.01557	0.25206	177×10^{-6}	0.0030
3	0.1428	0.0248	0.4605	354×10^{-6}	0.0005
4	0.18087	0.0305	0.1897	354×10^{-6}	0.0030

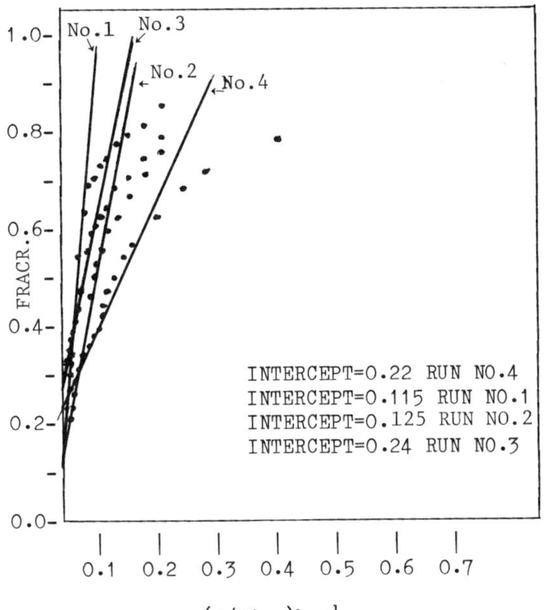

Figure 1. Determination of Unleachable Fractions for Runs 1 to 4

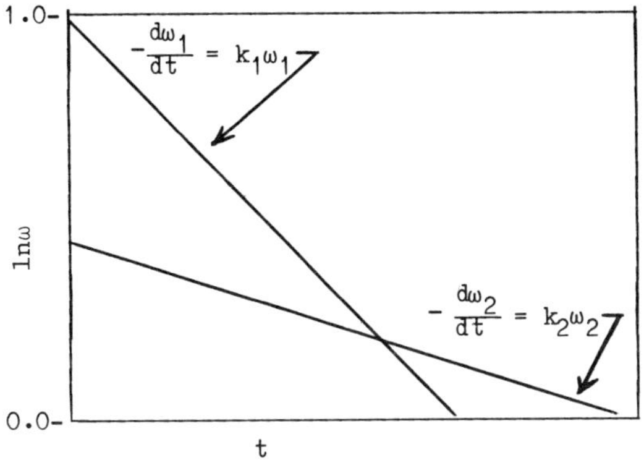

Figure 2. Segmented ln ω versus Time Curve

Chapter 14

THE EFFECT OF pH AND Eh ON THE CHEMICAL AND BIOLOGICAL LEACHING OF A PYRITIC URANIUM ORE

A. Bruynesteyn
A. Vizsolyi
Division of Extractive Metallurgy
B. C. Research
Vancouver, B. C., Canada V6S 2L2

ABSTRACT

Results are presented from a series of tests designed to determine the beneficial effects of the microorganism ***Thiobacillus ferrooxidans*** upon the leaching of uranium from a pyritic uranium ore.

Minus 1-inch (–25mm), 0.0711 percent uranium-containing, pyritic ore from the Rio Algom Mine at Elliot Lake, Canada was leached with lixiviants of various Eh and pH as well as being subjected to a biologically-assisted ferric iron leach. In all cases, the biological leach gave superior extraction. At pH 1.6, the biological leach gave 100.6 percent extraction and the ferric iron leach, 69.8 percent. This improvement in extraction is associated with a relatively rapid biological breakdown of the pyritic material, thus liberating uranium mineralization associated with these pyritic materials and regenerating the ferric sulphate lixiviant in situ.

INTRODUCTION

In many cases, uranium mining operations process relatively high grade ores, utilizing today's technology involving mining, crushing, milling and chemical leaching in vats or tanks, followed by uranium recovery by solvent extraction or ion exchange. However, vast quantities of low-grade mineralization are not amenable to this technology, mostly for economic reasons. Some of these low-grade mineralizations are associated with current mining operations, others are in self-contained orebodies. Development of an economic process for extraction of these low-grade materials will enhance the resource estimate of most countries significantly, as well as contribute to the profitability of associated mining operations.

Presently, chemical leaching of uranium, particularly from an acidic gangue, involves acid ferric iron solutions. The 6-valent uranium is soluble at the acidic pH values involved, whereas solubilization of 4-valent uranium involves oxidations to the 6-valent species with ferric iron as the primary oxidant. Under most commercial vat leaching conditions, the pH is modestly below the minimum level at which microbiological leaching occurs. In heap or *in situ* leaching, microbiological leaching is responsible for a substantial proportion (in mine drainage situations for all) of the uranium solubilized. The bacterium, *Thiobacillus ferrooxidans,* oxidizes the sulfides of most metals to the corresponding sulfates, thus generating acidic solutions. Where iron sulfides are involved the iron moiety is solubilized and concomitantly oxidized to the ferric form. The acidic ferric solutions, thus generated, solubilize both the 4- and 6-valent forms of uranium. The pH values over which this microbiological action proceeds are considered to be in the approximate range of 1.8 to 2.6. As the 4-valent uranium is oxidized to the 6-valent form the iron which is reduced to the ferrous form is quickly re-oxidized to the ferric state by *T. ferrooxidans*, thus enabling leaching to continue. Microbiological oxidation of ferrous to ferric iron is in no way dependent on the original source of the ferrous iron. Thus microbiological oxidation can interact with chemical leaching processes, where pH and other conditions permit.

The chemistry and biology of the microbiological leaching of copper from its ores are relatively well understood. Considerable progress has been made in utilizing this basic information in maximizing the rate and extent of copper release from dumps of low-grade ore and from *in situ* leaching of ores. However, comparable studies to produce the same pragmatic information for uranium are virtually non-existent. A thorough understanding of the actual operating variables which control performance of commercial dump, heap and *in situ* leaching,of both low and high grade uranium ores,undoubtedly will have a significantly beneficial effect on those existing and future mining operations leaching underground stopes of broken ore or surface heaps or dumps of run-of-mine and crushed ore.

This paper discusses the results of a series of tests designed to evaluate the effect of pH and Eh upon the chemical and biologically-assisted leaching of a low grade ore.

EXPERIMENTAL PROCEDURES

Equipment

A series of 9 small PVC columns, 7.6 cm diameter and 58 cm high were used; each containing approximately 6 kg of run-of-mine ore, crushed to -25 mm.

The 9 columns were grouped into three sets of three columns each. Within each set, all three columns were maintained at the same Eh, whereas individual columns were maintained at pH values of 1.2, 1.6 and 2.0. The Eh of the first set was that which pertained naturally without any control. With the second set hydrazine sulfate was used to keep the dissolved iron in a reduced form and thus maintain a low Eh value. The Eh of the third set was kept at approximately 580 mV by oxidizing all the dissolved iron to the oxidized form with hydrogen peroxide. All non biological columns were maintained sterile through the addition of dichlorophene (Parabis).

Column operation involved trickling 4,000 ml of water of appropriate pH through the ore bed and collecting the leachate in a barren reservoir. The solution feed rate onto the top of each column was 46.4 ml/h. Water was added to the contents of the barren reservoir to restore the volume to 4,000 ml and the pH and Eh adjusted as required. The leachate, with adjusted Eh and pH, was fed back to the ore column. Thus, only the initial 4,000 ml feed solution had no Eh control. Iron was not added to the leach solutions.

Ore Samples

The tests were performed on minus 1-inch (-25 mm) pyritic ore from the Rio Algom Mine at Elliott Lake, Ontario, Canada.

Analytical Procedures

Screen assays were performed dry on a Gilson vibratory screen. Head assays involved HNO_3 and H_2SO_4 digestion of 1.00 g portions of minus 60 mesh pulverized sample, followed by oxidation with hydrofluoric acid and perchloric acid.

Soluble uranium analyses were performed with a Scintrex UA-3 uranium analyzer. pH was measured with a Chemtrix type 45 AR pH meter.

Eh was measured, relative to the calomel electrode, with a Radiometer type 28 Eh-pH meter.

RESULTS

Screen assays performed on the sample of the rod mill feed sized ore showed that the minus 35 mesh particles were significantly enriched in uranium values and accounted for 14.1 percent bw of the total sample (Table 1). 35.5 percent of the sample was +1.5 inch (+13 mm) and 44.7 percent was +3/8 inch (+9.5 mm). The head assay, calculated as the mean value for the 9 different size fractions (Table 1), showed the sample to contain 0.0711 percent uranium and 3.96 percent iron.

Non-oxidative leach

The first three test columns were operated with a barren solution containing ferrous iron only. Any ferric iron produced while the solution percolated through the orebed was reduced before the solution was recycled.

The characteristics of the leach solutions for the three different test conditions show that, in the non-oxidative leach (Figures 1, 2, 3), ferric iron concentrations were less than 500 mg/l, total iron concentrations were highest at pH 1.2 (1,750 mg/l) and lowest at pH 2.0 (1,000 mg/l). Dissolved uranium concentrations reached 700 mg/l at pH 1.2. The extractions for the non-oxidative leach, calculated from solution assay data, were highest at pH 1.2 (Fig. 4). At pH 1.2, 40 percent extraction, representing readily acid soluble uranium, was obtained within 3 days whereas at pH 1.6 it took 26 days and at pH 2.0 as much as 73 days to obtain 40 percent extraction.

Oxidative ferric iron leach

In this test series, all the dissolved iron was maintained in the ferric form. The solution assays for the oxidative leach tests (Figs. 5, 6, 7) show that the highest total iron concentration (2.5 g/l) developed at pH 1.2. The lowest total and ferric iron concentrations developed in the pH 2.0 test conditions. Dissolved uranium concentration varied from 650 mg/l at pH 2.0 to 920 mg/l at pH 1.2. Extractions calculated from solution assays (Fig. 8) were very much affected by leach solution pH. Although the initial 40 percent extraction was obtained more rapidly than in the similar tests under non-oxidative conditions, the significant difference in the oxidative leach is in the rate the extraction takes place between 40 percent and 70 percent. The extraction profiles (Fig. 8) subsequent to day 140 suggest precipitation of uranium from solutions at the lower two pH values. The levelling off of the total iron concentration curves in Figures 5 and 6 suggest that iron salt precipitation may be associated with the uranium removal.

Biological leach

The commencement of biological activity is identified by the start of ferrous iron removal from solution (Figs. 9, 10, 11). At pH 1.6 and 2.0 this happens after only 14 days but at pH 1.2 it takes as long as 60 days probably because of the inhibitory effect of high hydrogen ion concentrations. As a result of the rapid biological pyrite oxidation, compared with the ferric iron leach, ferric iron concentrations are very high, reaching over 12 g/l at pH 1.6. Due to the lower solubility of jarosite at pH 2.0, ferric iron concentrations at pH 2.0 are less than 5 g/l. This does not mean that less pyrite was oxidized at pH 2.0 than at pH 1.6, but that a significant portion of the ferric iron produced did precipitate from solution. The effect of pH in the biological leaching process (Fig. 12) is less pronounced than in the ferric iron leach (Fig. 8). The limiting effect of low pH upon the bacterial activity is demonstrated in the fact that extraction at pH 1.2 is lower than at pH 1.6. This is in contrast to the extractions from the ferric iron leach.

Effect of pH

A summary of extractions obtained at pH 1.2 (Fig. 13) shows that the ferric iron leach initially produced the best results, that after 120 days the biological leach had caught up and that, subsequently, precipitation problems reduced the extraction (based on solution assays) for the ferric iron leach.

At pH 1.6, (Fig. 14) the biological leach was superior in all aspects, as was the case at pH 2.0 (Fig. 15).

Upon completion of the leach, the ore charges in the columns were washed twice with washwater of appropriate pH. The amount of uranium removed from the ore in the two wash cycles represents a significant portion of the overall uranium extraction (Table 2) and varied between 11.8 and 7.6 percent.

A comparison of the final extractions, calculated from both leach solution assays and wash solution assays (Fig. 16) clearly demonstrates the superiority of the biological leach at all three pH values, with the most significant difference in extraction occurring at pH 1.6 where the ferric iron leach gave 69.8 percent extraction and the biological leach 100.6 percent.

Discussion

The results, as indicated by Figure 16, strongly suggest that at pH 1.6, the biological leach gives maximum extraction and the ferric iron leach provides only 70 percent of maximum extraction. Since the difference between the two leaching processes most probably involves the relatively rapid bacterial breakdown of pyrite, it can be concluded that approximately

30 percent of the uranium values in the ore are associated with pyritic materials. Therefore, a bacterial leach is required for heap or dump leaching this type of ore and the construction of any such heap or dump should be done in such a way that conditions limiting bacterial activity will not develop, otherwise the leach will revert back to the less efficient ferric iron leach.

The bacterial activity at pH 1.2 was slow to develop but eventually, extraction at pH 1.2 increased to well above those obtained in the non-biological leaches. It is possible that maximum extraction has not yet been reached in the pH 1.2 biological leach at the time the test was terminated although the uranium concentrations in the leach solution (Fig. 9) suggests that uranium extraction had reached maximum values after 120 days. However iron solubulization continued to occur up to termination at day 190 which could indicate that additional uranium could still be released.

The washing operations following termination of the leach test produced significant amounts of uranium, indicating that the residual solution in the orebed did contain high uranium concentrations. If the proportion of the uranium removed by washing is assumed to be the uranium content of the residual leach solution in the drained ore, then the ore would have to contain from 10.9 to 14.0 percent of the original leach solution. The weight percentage of this amount of solution then suggests that the drained ore contains between 7.3 and 8.8 percent of water by weight.

This amount of hold up water is quite possible for fine material. Any heap or dump leach operation using finely sized material will therefore require some kind of counter current washing operation to extract the 10 to 15 percent of uranium contained in the residual waters in the heap or dump.

The decrease in solution concentrations of uranium, observed during leaching, does suggest that some part of the uranium extracted during the washing operations may represent a readily acid soluble uranium containing salt, or other precipitate from which the uranium can be readily extracted.

ACKNOWLEDGEMENT

The results presented in this paper represent part of the work undertaken by B.C. Research, under a multicompany sponsored, multiyear investigation, on the basic and applied aspects of the biological and chemical leaching of uranium ores. Thanks are extended by the authors to the sponsors, for their support and permission to publish this data.

TYPE OF LEACH

	BIOLOGICAL		FERRIC IRON	
pH	1.2	1.6	1.2	1.6
Leach Solution (ml)	4100	4250	4250	4450
Uranium Content (g)	3.010	2.987	3.842	4.150
Uranium Content Wash				
Solution (g)	0.367	0.419	0.526	0.451
Ratio %	12.1	14.0	13.7	10.9
Equivalent Hold Up (ml)	496	595	582	485
Ore Charge (g)	6,777	6,772	6,592	6,430
Hold Up, % of Ore Weight	7.3	8.8	8.8	7.5

TABLE 1

SCREEN ASSAY RIO ALGOM ORE

SIZE	NET WEIGHT (lb.)	%	ASSAY U (%)	ASSAY Fe (%)
+ 3/4"	14.8	12.4	0.0582	3.71
-3/4"+ 1/2"	27.6	23.1	0.0664	4.16
-1/2"+ 3/8"	11.0	9.2	0.0799	4.12
-3/8"+ 4#	20.3	17.0	0.0655	4.39
- 4# + 8#	11.7	9.8	0.0614	3.74
- 8# + 18#	10.5	8.8	0.0580	2.82
-18# + 35#	6.5	5.5	0.0655	2.57
-35# + 50#	5.3	4.5	0.0975	4.22
-50#	11.4	9.6	0.1143	4.81
Calculated head assay			0.0711	3.96

TABLE 2

EXTRACTIONS - Eh-pH TEST

LEACH CONDITION	RETENTION (TERMINATION) DAY	FINAL SOLUTION			WASH (1)	WASH (2)	U % EXTRACTION			
		EmV	pH	U g/l	U g/l	U g/l	Solution	Wash (1)	Wash (2)	Total
Biological pH 1.2	189	560	1.4	.922	.108	.023	76.6	9.0	1.9	87.5
1.6	189	705	1.6	1.104	.092	.015	91.7	7.6	1.3	100.6
2.0	189	650	2.0	.777	.098	.022	65.5	8.2	1.8	75.5
EmV < 390 pH 1.2	189	400	1.3	.717	.105	.035	60.7	8.9	2.9	72.5
1.6	189	400	1.7	.607	.074	.017	51.1	6.3	1.3	58.7
2.0	189	395	2.2	.511	.072	.035	44.0	6.1	3.0	53.1
EmV > 580 pH 1.2	189	470	1.2	.691	.066	.026	59.1	5.7	2.2	67.0
1.6	189	465	1.7	.693	.074	.031	60.7	6.4	2.7	69.3
2.0	189	450	2.1	.585	.078	.031	52.9	7.0	2.7	62.6

Figure 1: COMPOSITION OF SOLUTION Eh<390 mV pH 1.2

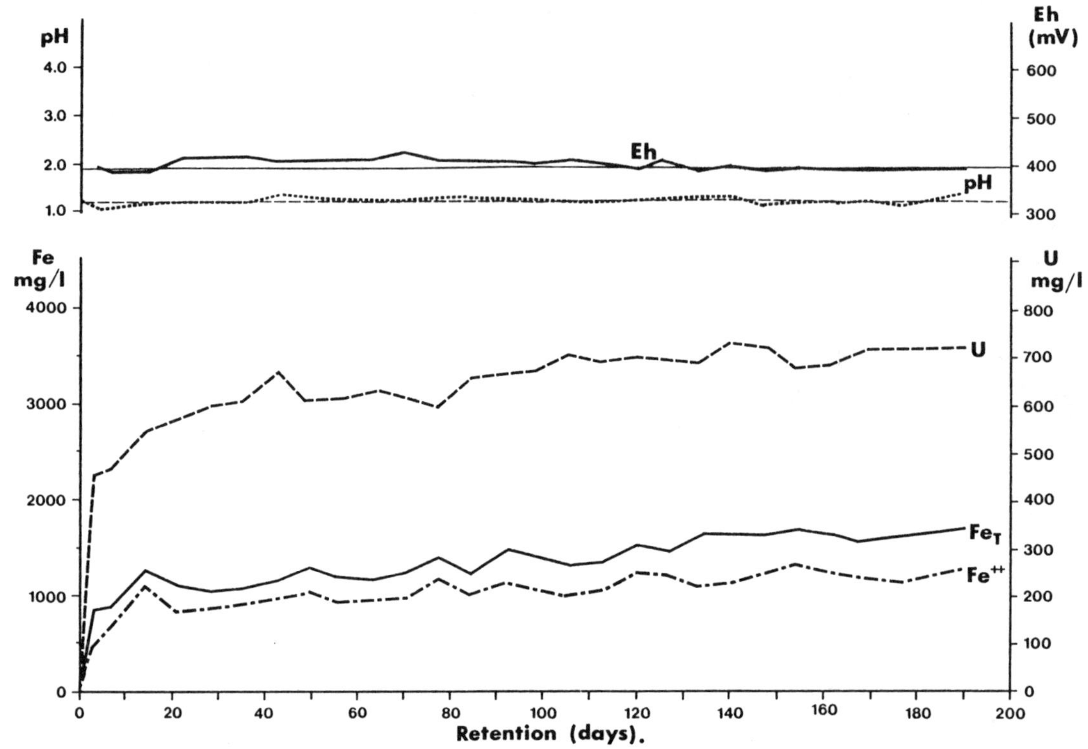

Figure 2: COMPOSITION OF SOLUTION Eh<390 mV pH 1.6

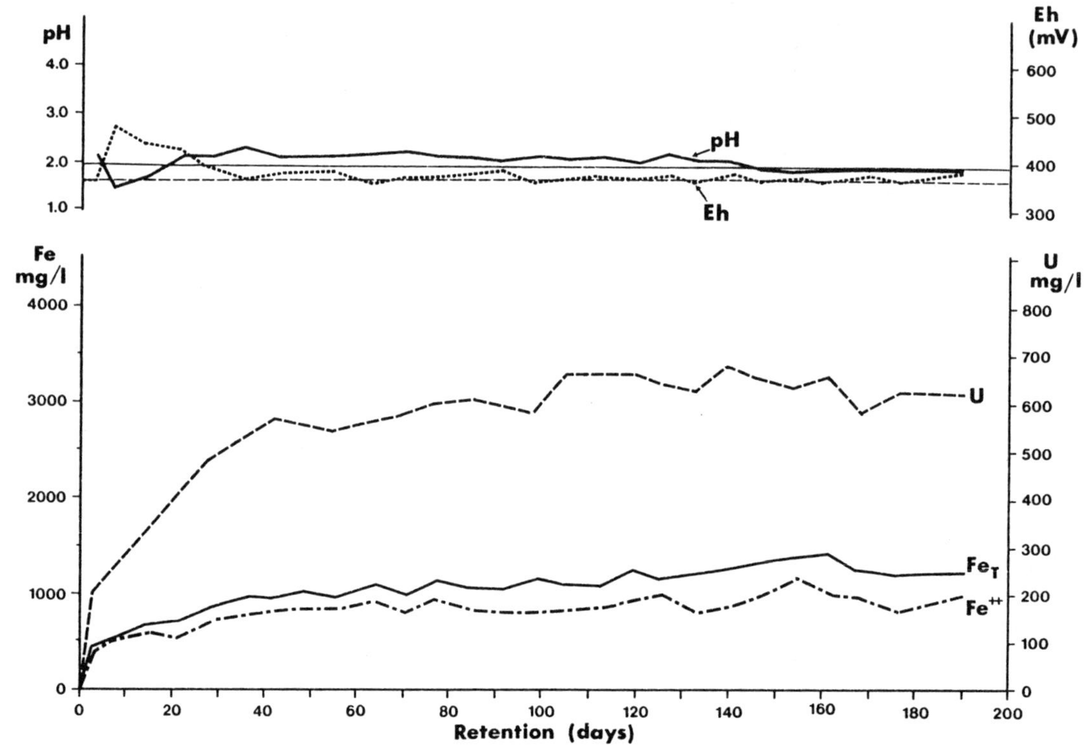

Figure 3: COMPOSITION OF SOLUTION Eh<390 mV pH 2.0

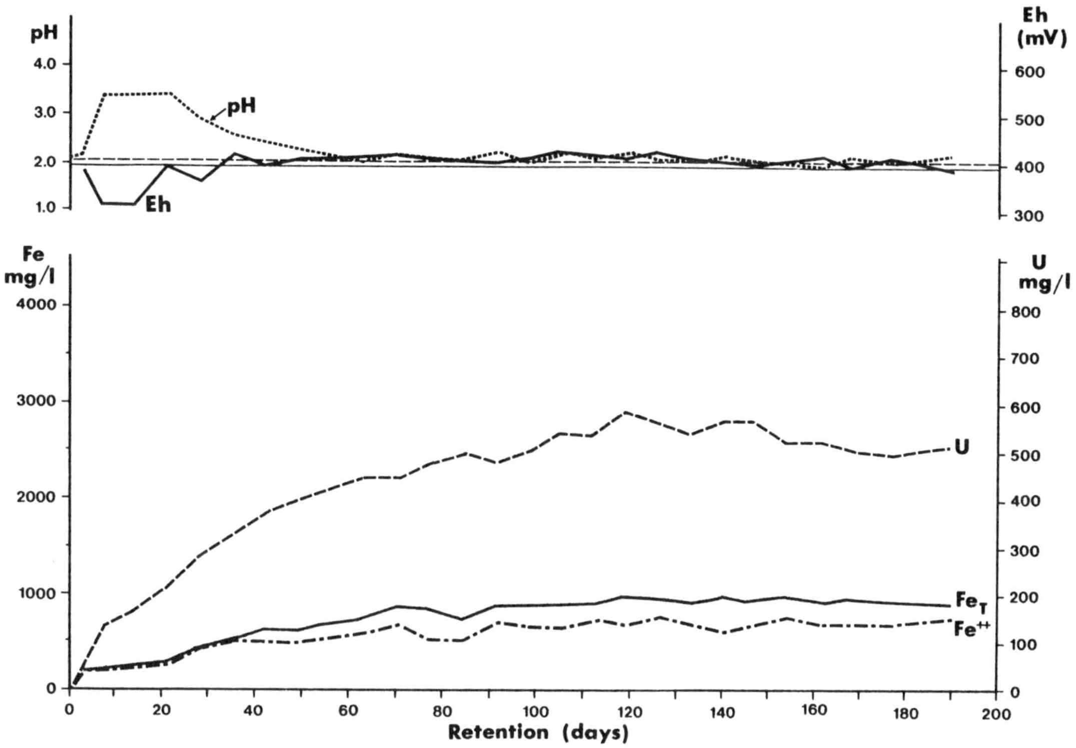

Figure 4: Eh – pH COLUMNS – EXTRACTION AT < 390 mV (~100 % Fe^{++})

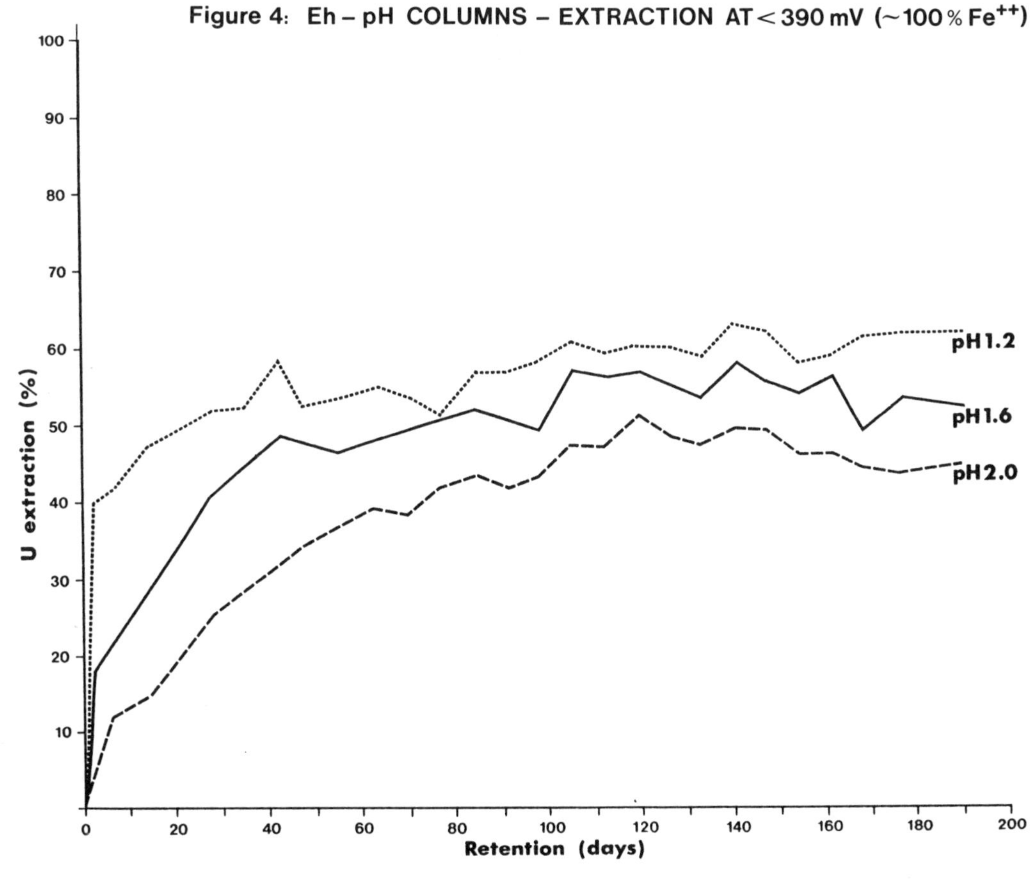

Figure 5: COMPOSITION OF SOLUTION Eh>580 pH 1.2

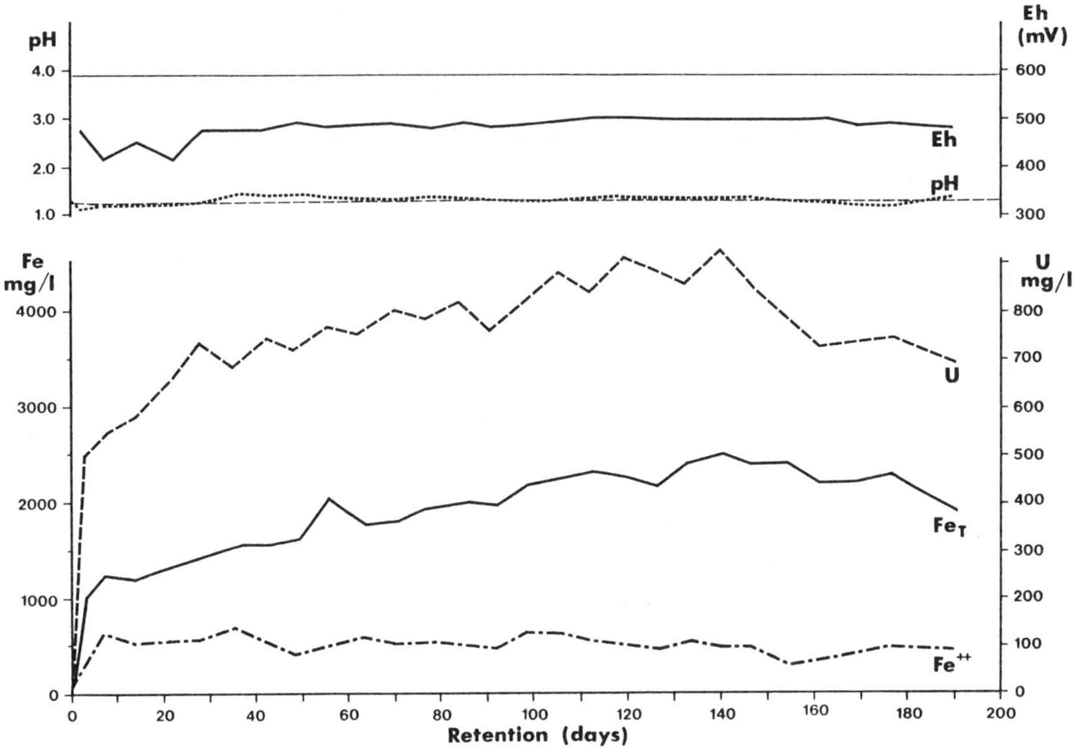

Figure 6: COMPOSITION OF SOLUTION Eh>580 pH 1.6

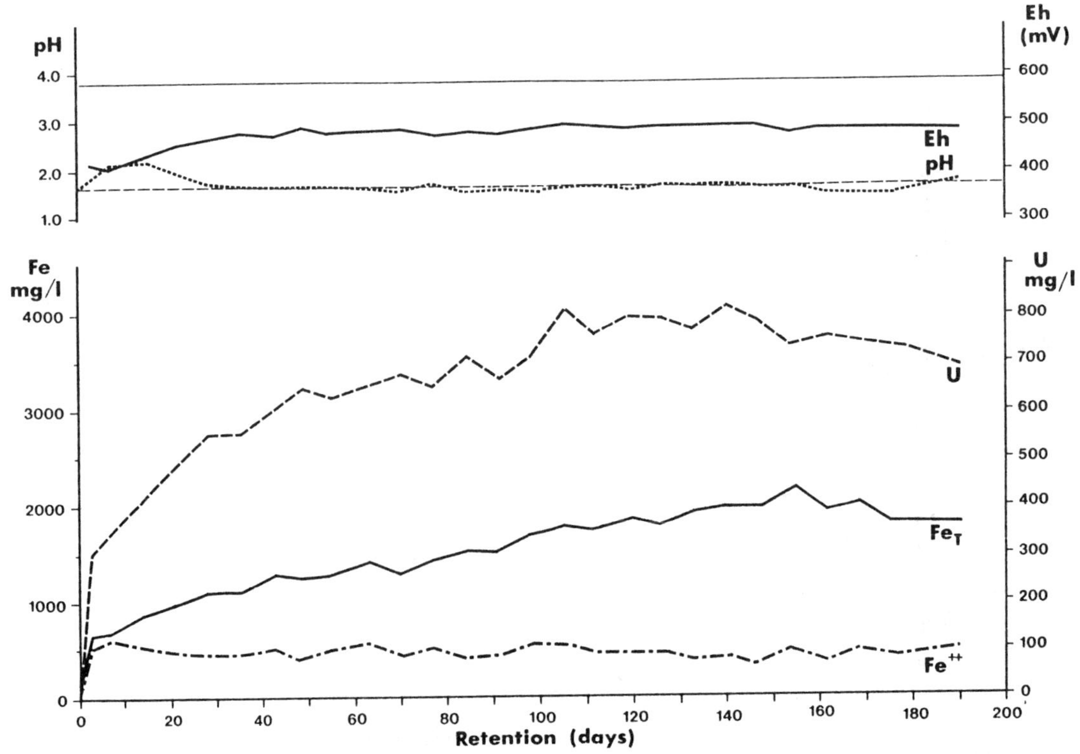

Figure 7: COMPOSITION OF SOLUTION Eh>580 mV pH 2.0

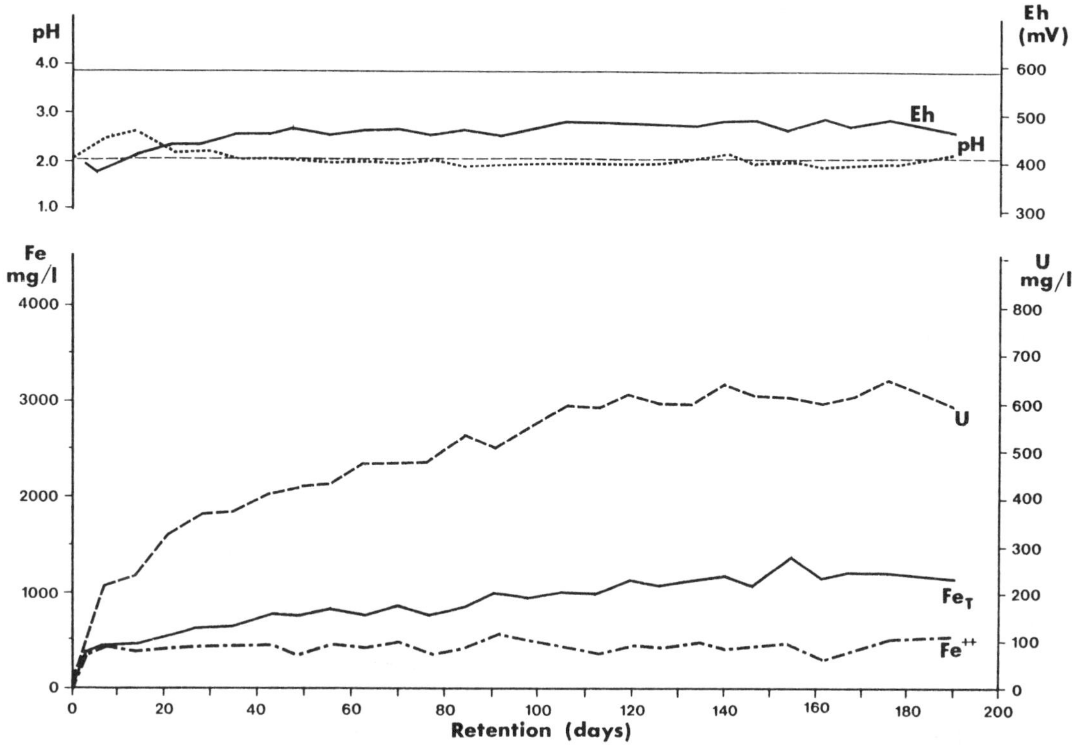

Figure 8: Eh – pH COLUMNS – EXTRACTION AT >580 mV (~100% Fe^{+++})

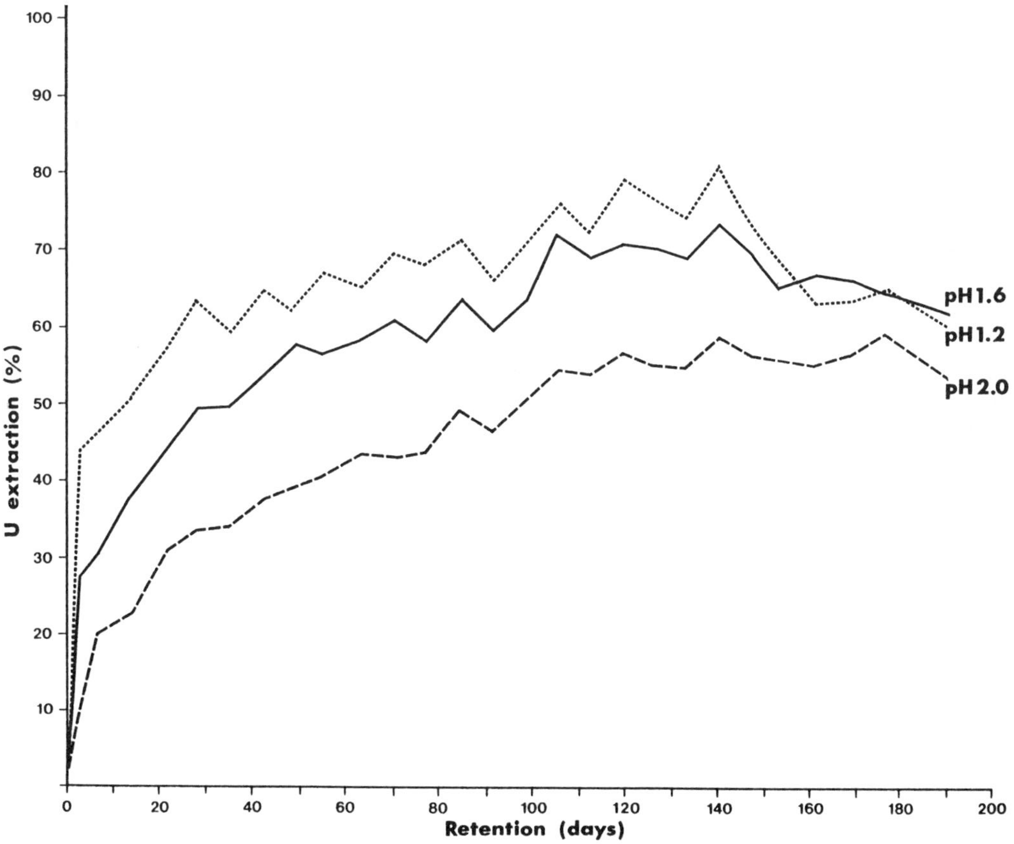

Figure 9: COMPOSITION OF SOLUTION BIOLOGICAL LEACH pH 1.2

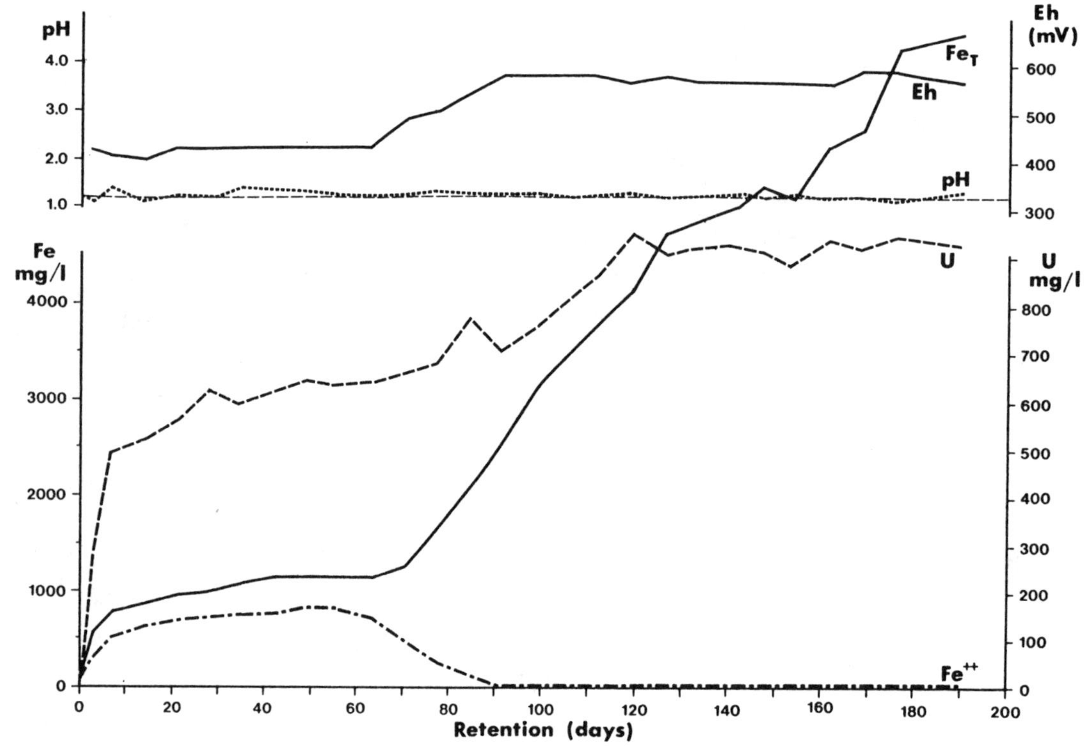

Figure 10: COMPOSITION OF SOLUTION BIOLOGICAL LEACH pH 1.6

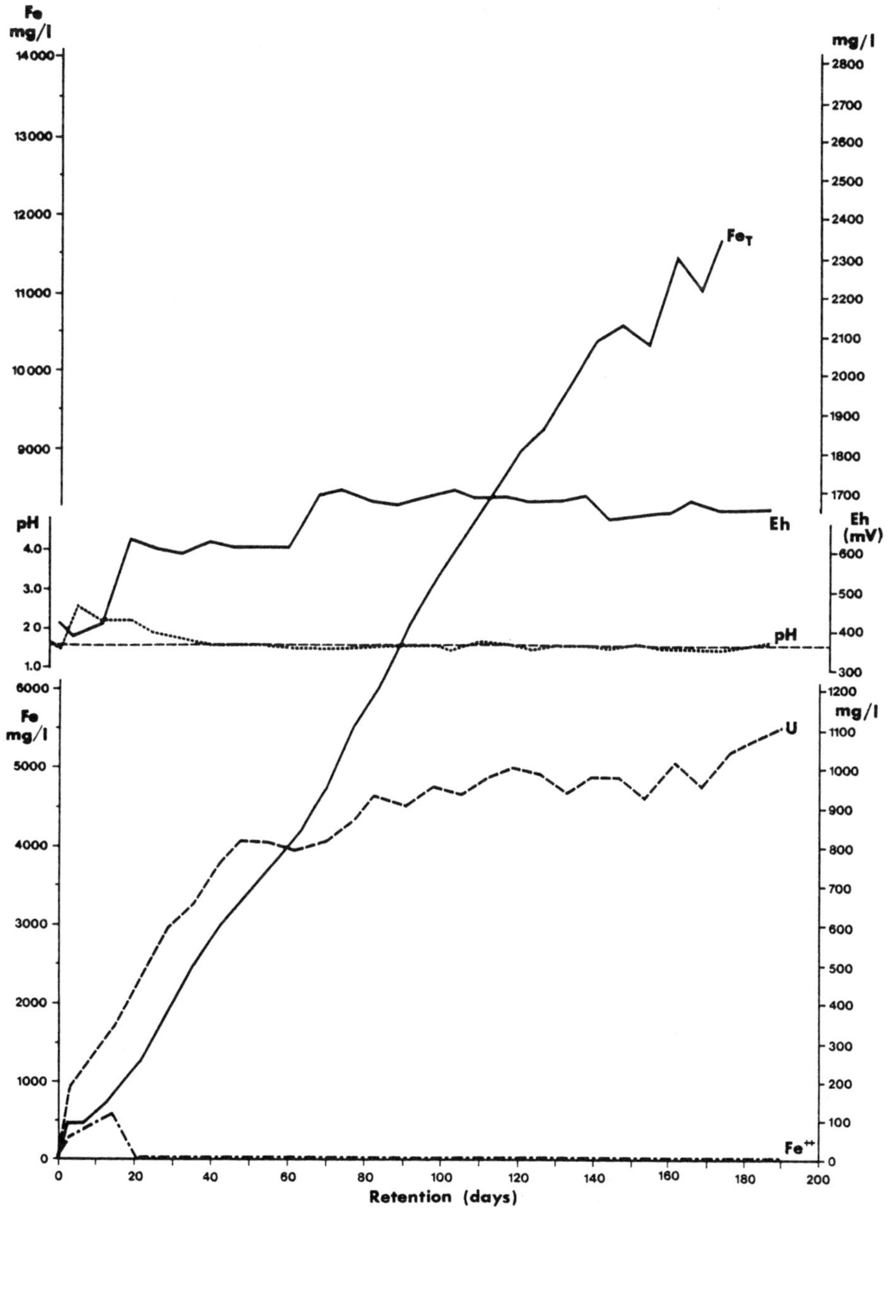

Figure 11: COMPOSITION OF SOLUTION BIOLOGICAL LEACH pH 2.0

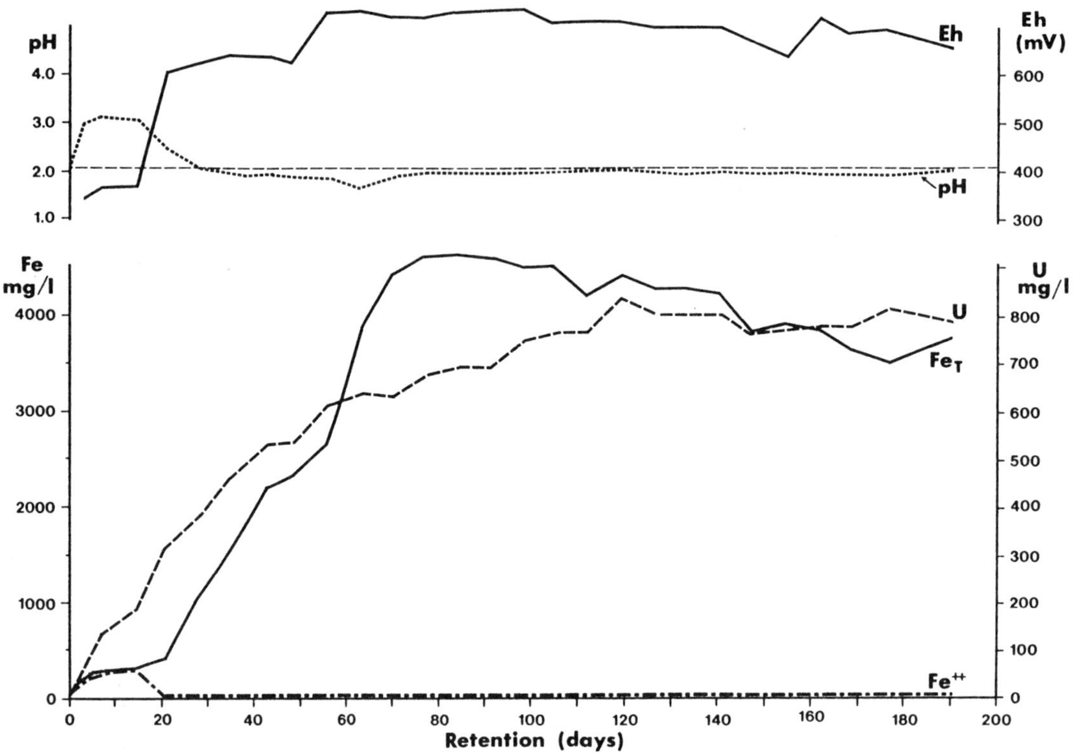

Figure 12: Eh – pH COLUMNS – EXTRACTION OF CONTROL – BIOLOGICAL LEACH

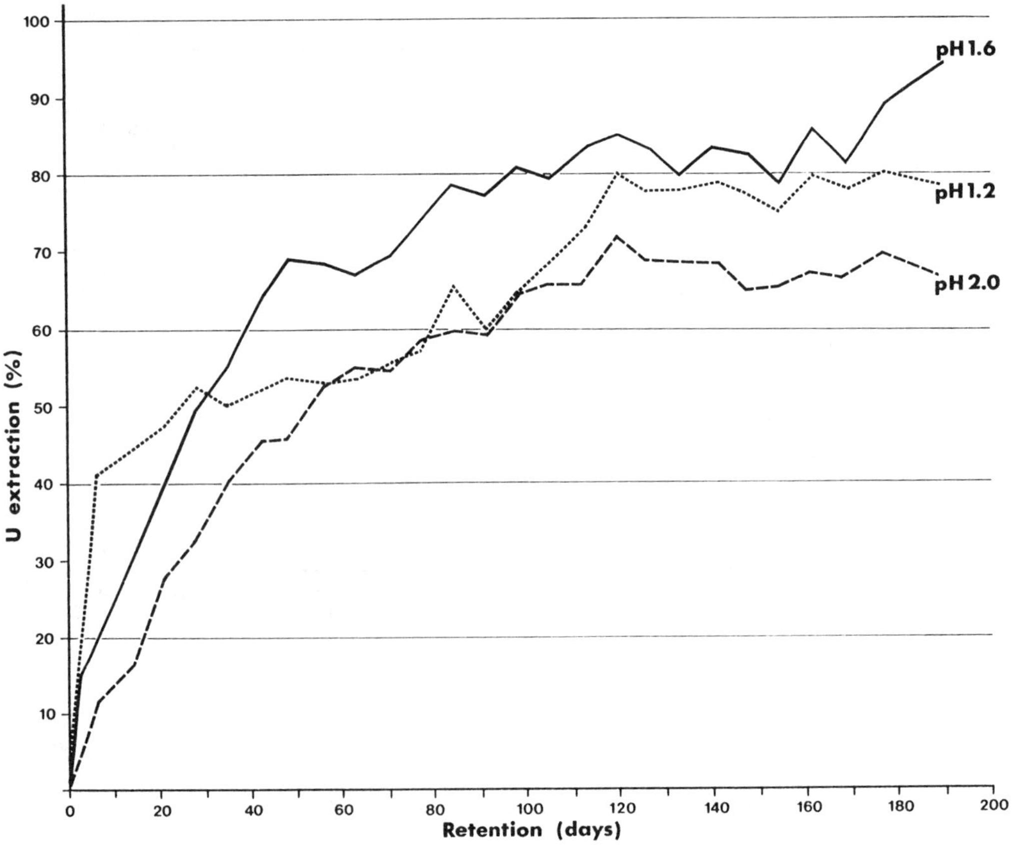

Figure 13: Eh – pH COLUMNS – EXTRACTION AT pH 1.2

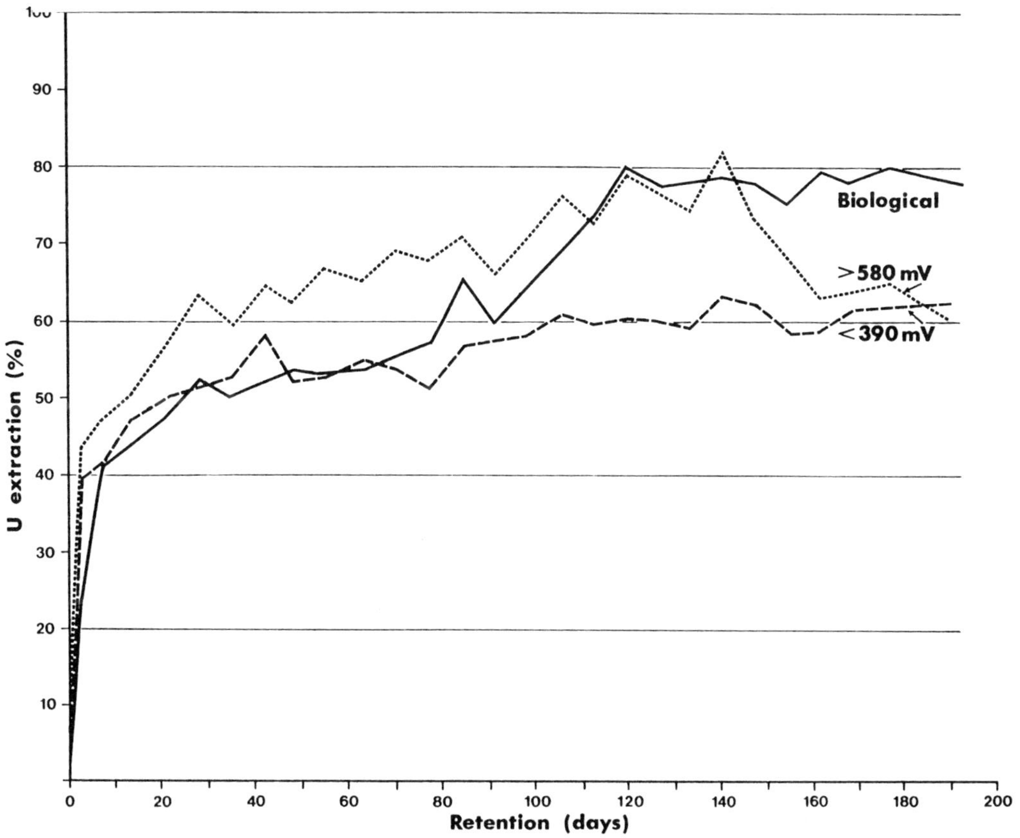

Figure 14: Eh – pH COLUMNS – EXTRACTION AT pH 1.6

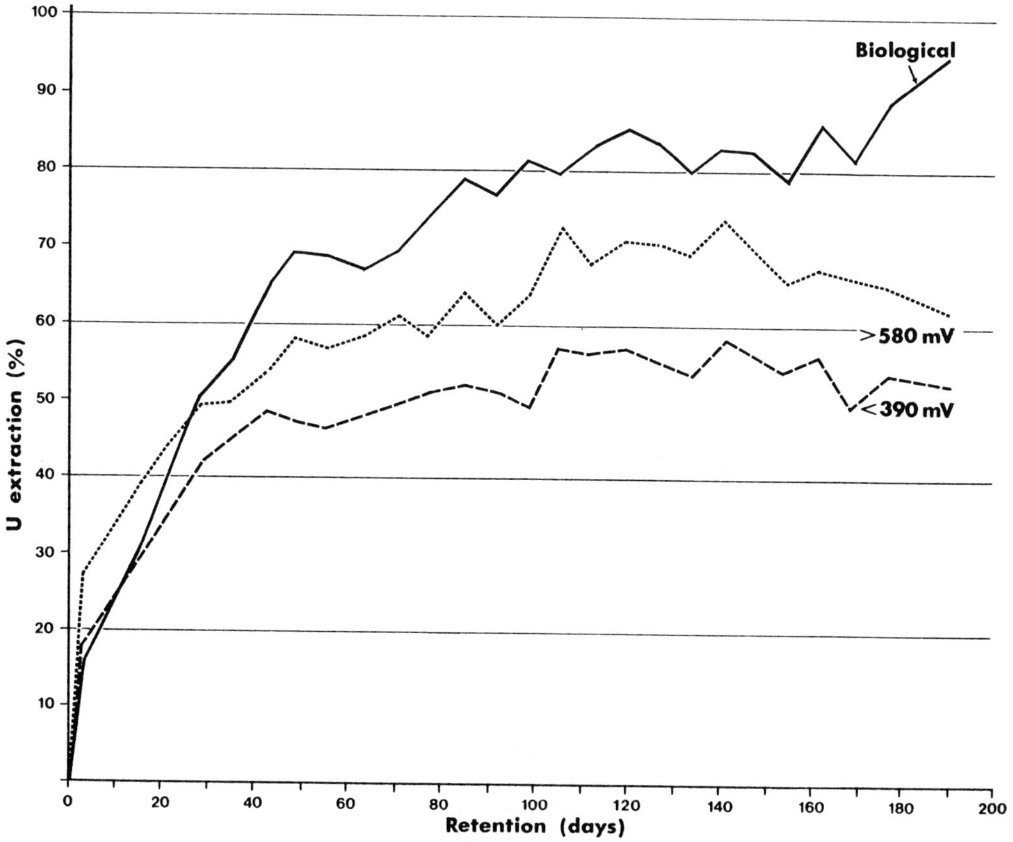

Figure 15: Eh – pH COLUMNS – EXTRACTION AT pH 2.0

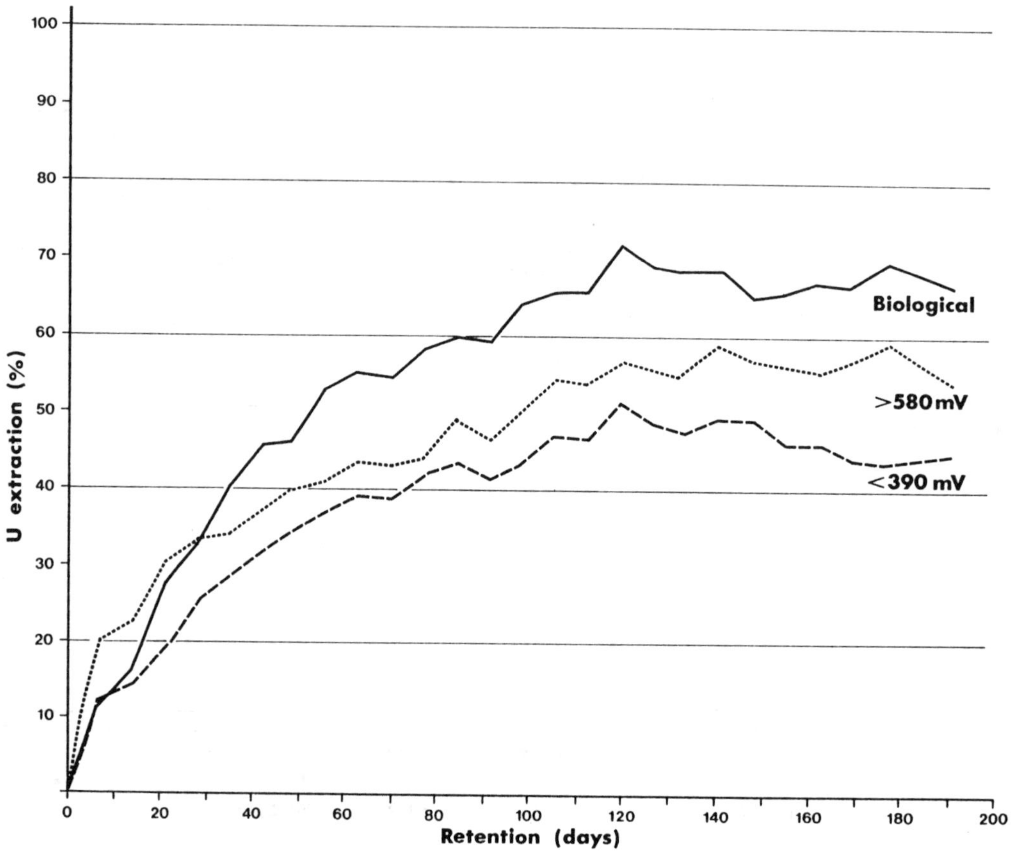

Figure 16: Eh – pH COLUMNS – RIO ALGOM ORE
FINAL EXTRACTIONS AFTER WASHING (0.0711 % U Headassays)

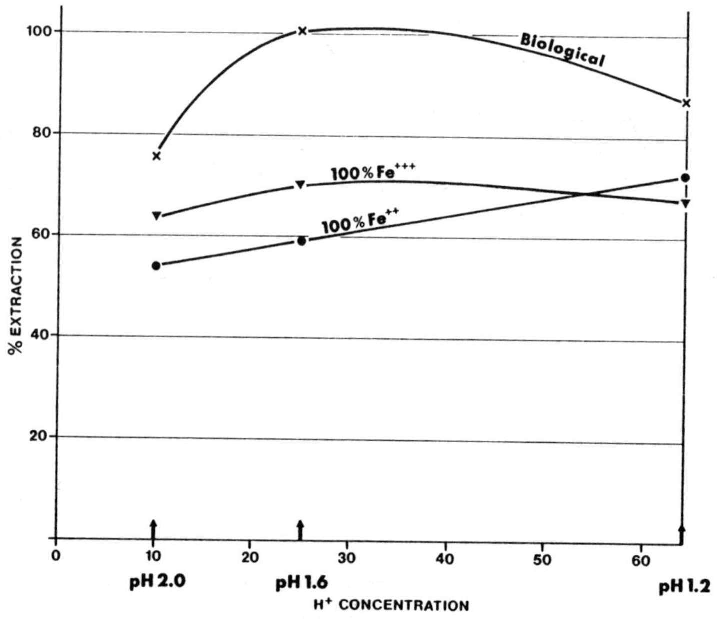

Chapter 15

IN SITU LEACHING OF CROWNPOINT URANIUM ORE: III LABORATORY STUDY OF STRONG LEACHING SYSTEMS – SODIUM HYPOCHLORITE

T. C. Vogt
E. T. Strom
W. F. Johnson
P. B. Venuto
Research Department
Mobil Research and Development Corporation
Dallas, Texas 75221

ABSTRACT

The organic carbonaceous matter found intimately associated with the uranium mineral in some Crownpoint ore trends appears to shield some of the uranium from contact with a chemically-mild leaching system. Sodium hypochlorite (NaOCl) is a strong enough oxidant to attack the carbonaceous matter and expose trapped uranium mineral species. Batch and pack laboratory tests show that alkaline bicarbonate solutions containing 0.1–5.0 wt.% NaOCl are effective for rapid recovery of uranium from Crownpoint refractory ore. Recoveries of 90% and greater were obtained in the laboratory tests. However, NaOCl is such a strong oxidant that it also reacts extensively with pyrite and other gangue minerals present in the ore. Thus, chemical consumption is high. Since oxidation with NaOCl generates sodium chloride, high chloride levels tend to build up in the leaching circuit. Leaching with NaOCl is also associated with high levels of dissolved organic carbon. Electron microprobe studies of ore samples after leaching with NaOCl showed the presence of holes and cracks in the residual carbonaceous matrix as well as the presence of chloride. Since NaOCl is highly corrosive, plastic and fiberglass must be used in parts of the leaching apparatus.

INTRODUCTION

The past history of uranium production has been dominated by strip and shaft mining methods, but recently in-situ leaching has emerged as a viable alternative for uranium recovery. (1-8) Mobil was the first company to start in-situ uranium leaching in New Mexico. Mobil's Crownpoint project areas are in McKinley County and are part of the Grants Mineral Belt. The major uranium deposits in this area occur in the Westwater Canyon Member of the Jurassic Morrison Formation. (9-11) The ore depth at the Crownpoint project is about 2000 ft.

In an earlier paper we discussed the uranium mineralogy of an areally broad sampling of Crownpoint ore. (12) The organic carbonaceous material often associated with the coffinite ore was characterized in some detail. The organic material appeared to be a very mature, coaly-type kerogen derived from plant components. This result was consistent with other studies on such carbonaceous material. (13-15)

In a second paper we explored the utility of high pressure oxygen (O_2) or hydrogen peroxide (H_2O_2) in sodium bicarbonate ($NaHCO_3$) solution as a leachate for recovering uranium (expressed as U_3O_8) from a broad sampling of Crownpoint ore. (16) The leaching rates were found to correlate with ultimate U_3O_8 recovery. A wide range of leaching rates and U_3O_8 recoveries were observed. Coffinite ores from some areas leached very slowly and gave low U_3O_8 recovery. Such ores were called refractory. The uranium in these slow leaching, refractory ores was intimately associated with organic matter. We postulated that the carbonaceous material was shielding the uranium mineral from contact by the leachate. A similar effect of leaching inhibited by carbonaceous material had been observed by Rhett on ore samples taken about 15 miles to the northeast of the Crownpoint leases. (17) If carbonaceous material indeed inhibits leachate attack, then a chemically severe leaching system must be used to disrupt the carbonaceous barrier. This paper deals with one such system using alkaline sodium hypochlorite (NaOCl) as the oxidant. In an accompanying paper we report studies with another chemically severe leaching system, H_2SO_4 and O_2. (18)

NaOCl is a powerful and readily available oxidant. The use of hypochlorite (OCl^-) in removing U_3O_8 from its ores has been studied previously (19-22) In an early evaluation of a group of chemical oxidants, DeSesa et al concluded that OCl^- was by far the most economical of the reagents per equivalent of oxidizing capacity. (20) As regards in-situ leaching, OCl^- has been considered for NH_3 removal after NH_4HCO_3 leaching. (23-25)

The aqueous chemistry of OCl^- is quite complex. (25,26) Hypochlorous acid (HOCl) is a weak acid (the equilibrium constant = 3.4×10^{-8}), so OCl^- at pH lower than 8 to 9 should be present in the form of HOCl. In acid solution HOCl can in principle be generated as follows:

$$Cl_2 + HOH \rightarrow HCl + HOCl \qquad (1)$$

however, the equilibrium lies to the left. In a 0.09 M chlorine water solution, two-thirds of the Cl_2 is undissociated, and only one-third is present as HOCl. (26) Hypochlorous acid can decompose directly to O_2 as follows:

$$HOCl \rightarrow HCl + 1/2\ O_2 \qquad (2)$$

This reaction is catalyzed by ultraviolet light.

Hypochlorite can be formed more efficiently under basic conditions by the reaction:

$$Cl_2 + 2OH^- \rightarrow Cl^- + OCl^- + H_2O\ . \qquad (3)$$

The equilibrium constant for this reaction is quite favorable, 7.5×10^{15}. Even here, however, a decomposition can occur to generate chlorate ion, i.e.,

$$3\ OCl^- \rightarrow 2Cl^- + ClO_3^-\ . \qquad (4)$$

Fortunately, this disproportionation is slow at normal temperatures.

These considerations suggest that alkaline NaOCl leaching should provide the most efficient use of the OCl^- reagent. An alkaline leach has the further advantage of minimizing the mobilization of both calcium and iron. The alkalinity is provided by $NaHCO_3$, which also serves as a complexing agent for the uranyl ion. The U_3O_8 is leached as the extremely stable uranyl tricarbonate ion, $UO_2(CO_3)_3^{4-}$. (27) The formation constant for this complex, recently identified by carbon-13 nuclear magnetic resonance, (28,29) is in the range of 10^{18} to 10^{23}, as shown in recent compilations. (30,31)

EXPERIMENTAL

Batch and packed column leach tests were used to evaluate the leach rate and recovery of U_3O_8 from refractory Crownpoint ore with NaOCl. In the tables and figures in this paper the same sample designation is used as in the previous papers. (12,16) The letter represents a general location within the project areas, while the numerals represent a well number followed by the depth in parenthesis where multiple samples were taken from a single well.

Batch Tests

These provided rapid assessment of the utility of NaOCl in leaching of Crownpoint ores and utilized polyethylene bottles mounted on a shaker apparatus (10 gm ore, 50 cc leachate). Duplicate experiments were carried out for three and 24 hr so that pseudo first-order rate constants could be derived. The rate constants, $\underline{k}$, were determined in the following manner: The U_3O_8 concentration in the leachate contacting the ore was measured after three and 24 hr reaction times by analyzing one cc aliquots. The initial U_3O_8 level in the ore had been previously measured; subtraction of the amount of U_3O_8 in solution gave the amount of U_3O_8remaining in the ore. With this data, the pseudo first-order rate constant, $\underline{k}$, was calculated as

$$\underline{k} = \frac{1}{t}\,(\ln c_0 - \ln c)\ , \qquad (5)$$

where c_0 = initial concentration of U_3O_8 in ore (wt % or other units), c = concentration in consistent units of U_3O_8 remaining in the ore at time, t, in hours or days. The $\underline{k}$ values have units of reciprocal time.

Earlier (16), we had studied the batch test response of an areally-broad sampling of Crownpoint ore using a mild alkaline leaching system. In that study ores that leached with rate constants, $\underline{k}$, greater than 10 x 10^{-3} $hour^{-1}$ were classified as easy-to-leach; those with $\underline{k}$ between 2 and 10 x 10^{-3} $hour^{-1}$ were designated as intermediate; those with $\underline{k}$ less than 2 x 10^{-3} $hour^{-1}$ were designated slow-leaching or refractory ores.

Packed Column Leach Tests

Four packed column tests were conducted in polyvinylchloride reactors of varied lengths and diameters, ranging from 9 to 26 in. long and 3/4 to 1 1/2 in. in diameter. Small plugs of sand and glass wool were placed in the bottom of each pack prior to packing with ore. The pore volume (PV) of a pack was measured experimentally by imbibing distilled water into an evacuated pack, measuring the volume of water imbibed, and correcting for the volume of the end plugs. Occasionally, the PV was determined indirectly by weighing the ore, measuring the volume of the pack, correcting for the volume of the end plugs, and determining the volume of ore from its weight and specific gravity. For Pack 4, both methods were used and were found to agree within 5%.

Stainless steel fittings used in early runs (Packs 1 and 3) were attacked by NaOCl and were replaced by plastic and brass fittings. Subsequent runs (Packs 2 and 4) used only fittings inert to NaOCl.

Packed column leach tests in general were performed by pumping leachate downflow through the ore-packed reactors. Injection rates were varied by timer-programmed pumps. Preceding the leaching tests, the packed columns were flushed with $NaHCO_3$ solution to remove preoxidized U_3O_8. Samples were taken daily during the early part of the runs (less often, later) and were analyzed for U_3O_8 by the BROMOPDAP method. (32)

Ore Analyses

Batch tests were carried out on an areally-broad selection of ore from Crownpoint. The analytical data for most of these ores were reported earlier. (16) Column tests were performed on a representative refractory ore sample from Crownpoint. This was a dried composite of ore from Area B, Well 360. The ore grade averaged 0.091 % U_3O_8. The analyses for other key components of representative Area B ores are shown in Table 1.

TABLE 1

SUMMARY OF ANALYSIS OF SELECTED COMPONENTS FOR KEY WELLS FROM AREA B

Component	Area B, Wt, %	
	Average[a]	Range
Uranium (as U_3O_8)	0.080	0.005 - .227
Selenium	0.0012	0.0005- .0047
Vanadium (as V_2O_5)	0.036	0.010 - .093
Organic Carbon	0.184	0.040 -3.33
Calcium Carbonate ($CaCO_3$)	1.43	0.04 -3.33
Iron	-	0.70 - .95
Copper	-	0.002 - .005
Titanium	-	0.0075- .0525

(a) Based on 16 samples.

Analyses for U_3O_8 in the ore sand before and after leaching were carried out by Core Laboratories.

Calculation of U_3O_8 Recoveries and Material Balances

For packed column tests it is possible to calculate U_3O_8 recovery in two different ways. One method entails analyzing the U_3O_8 in the original ore pack before leaching and then measuring the amount produced. The ratio of U_3O_8 produced to the U_3O_8 content of the pack before leaching gives the fractional recovery. Alternatively, the residual U_3O_8 in the leached ore pack can be determined. This amount added to that produced in the leachate gives the total U_3O_8present in the ore pack before leaching. The fractional U_3O_8

recovery then is the ratio of produced U_3O_8 to the sum of residual and produced U_3O_8.

The U_3O_8 recoveries given herein are based on the U_3O_8 content of the ore pack before leaching, because the %-error in this number is less than the %-error for very small amounts of residual U_3O_8 in a leached pack. Nevertheless, recoveries based on residual U_3O_8 in the leached packs were also calculated to check the U_3O_8 material balance.

RESULTS

Batch Tests

Batch tests were used in scoping studies to test different oxidants (or even solvents) that might dissolve carbonaceous material to allow leachate to contact the U_3O_8-mineral. The pseudo first-order rate constants derived for 18 different leachate compositions are shown in Table 2. The refractory ore used the composite from Area B, Well 360, described above. Oxidants tested were NaOCl, H_2O_2, sodium chlorate ($NaClO_3$), and potassium permanganate ($KMnO_4$) with varying pH's and amounts of $NaHCO_3$. Some systems utilized water-soluble organic solvents (methanol and acetone) or additives (citric acid and NH_4HCO_3). The tabulated rate constants clearly indicated the effectiveness of NaOCl as an oxidant for recovering U_3O_8 from the refractory ore. Figure 1 shows a comparison of alkaline hypochlorite with the milder alkaline peroxide leachate using batch tests.

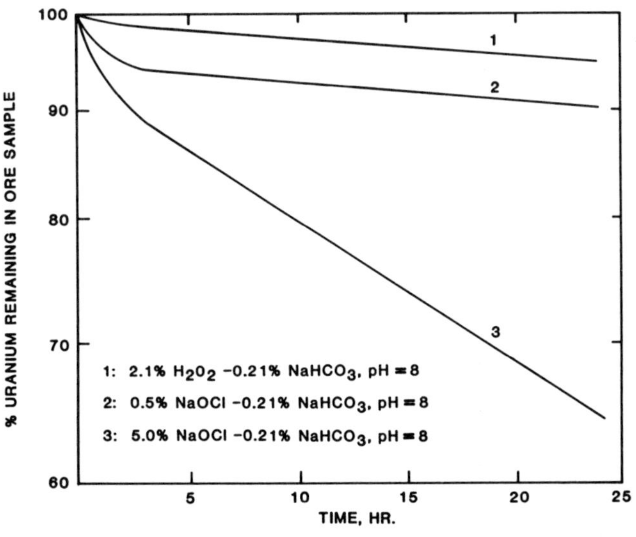

FIGURE 1. COMPARISON OF BATCH LEACH TEST RESULTS FOR ALKALINE NaOCl AND H_2O_2 LEACHATES IN CONTACT WITH SAMPLE B-360

TABLE 2

SCOPING BATCH LEACH TEST RESULTS[(a)]

Leachate Composition	pH	Rate Constant, k 10^{-3} x hr^{-1}
1) 50 gpl NaOCl 2.1 gpl $naHCO_3$	8.3 to 9.6	12 to 15
2) 10 gpl NaOCl 2.1 gpl $NaHCO_3$	8.3 to 9.3	10
3) 5 gpl NaOCl 2.1 gpl $NaHCO_3$	8.3	2.3
4) 2.1 gpl H_2O_2 H_2SO_4 to pH 1.2	1.2	2.3
5) 21 gpl H_2O_2 2.1 gpl $NaHCO_3$	8.3	2.3
6) 2.1 gpl H_2O_2 2.1 gpl $NaHCO_3$	8.0 to 8.3	2.0
7) 2.1 gpl H_2O_2 H_2SO_4 to pH 2	2.0	1.5
8) 2.1 gpl H_2O_2 HCl to pH 2	2.0	1.5
9) 2.1 gpl H_2O_2 2.1 gpl $NaHCO_3$ 50% Acetone	8.3	1.3
10) 2.1 gpℓ H_2O_2 2.1 gpl $NaHCO_3$ 50% Methanol	8.3	< 1
11) 2 gpl $NaClO_3$ 2.1 gpl $NaHCO_3$	8.3	< 1
12) 10 gpl $NaClO_3$ 4.2 gpl $NaHCO_3$	6.5	< 1
13) 10 gpl $NaClO_3$ 10 gpl $NaHCO_3$	5.0	< 1
14) 10 gpl $NaClO_3$ H_2SO_4 to pH 2	2.0	< 1
15) 2.1 gpl H_2O_2 2.5 gpl Na_2CO_3	11.3 to 12.0	< 1
16) 5 gpl $KMnO_4$ 21 gpl $NaHCO_3$	8.3	< 1
17) 4 gpl H_2O_2 2.3 gpl Citric Acid HCl to pH 2	2.0	< 1
18) 21 gpl H_2O_2 Saturated with NH_4HCO_3	8.3	< 1

(a) Composite ore from Area B, Well 360.

TABLE 3

COMPARISON OF RATE CONSTANTS FOR BATCH LEACH TESTS WITH HYDROGEN PEROXIDE AND SODIUM HYPOCHLORITE LEACHATES ON AN AREALLY-BROAD SAMPLING OF CROWNPOINT URANIUM ORE

Sample Description			Pseudo First Order Rate Constant ($10^3 \times hr^{-1}$)		Ratio, $\underline{k}_{(NaOCl)}$ to $\underline{k}_{(H_2O_2)}$
Area-Well No.	Depth(a)	%U_3O_8	$\underline{k}_{(H_2O_2)}$(b)	$\underline{k}_{(NaOCl)}$(c)	
L-31	A	0.046	1.9	9.1	5
	B	0.017	1.6	3.3	2
	C	0.108	0.9	8.2	9
A-88	A	0.041	1.4	5.4	4
B-358	A	0.047	1.3	3.4	2
B-359	A	0.029	2.9	24.7	8
B-507	A	0.027	2.4	6.8	3
	B	0.061	1.8	15.1	9
	C	0.054	1.2	10.1	9
	D	0.143	1.3	34.8	26
	E	0.147	2.1	32.1	15
B-512	A	0.071	1.1	15.0	13
	B	0.242	1.3	24.2	18
	C	0.030	0.9	5.8	7
	D	0.046	1.3	3.3	3
	E	0.026	1.2	4.8	4
M-45	A	0.142	1.1	11.3	10
C-195	A	0.296	2.5	15.1	6
C-196	A	0.087	1.6	8.0	5
	B	0.047	1.4	4.2	3
E-781	A	0.150	8.4	37.3	4
J-128	A	0.030	0.9	1.9	2
	B	0.260	0.3	1.9	7
D-174	A	0.205	2.7	26.2	10
D-175	A	0.187	1.0	11.4	11
N-78	A	0.067	0.7	6.8	9
	B	0.150	1.1	9.5	8
G-515	A	0.026	1.2	2.0	2
	B	0.153	3.0	16.8	6
H-191	A	0.042	2.2	12.2	6
H-192	A	0.082	2.0	14.9	8
I-74	A	0.063	10.0	111.5	11
K-206	A	0.020	16.8	98.8	6

(a) Letters describe ore sands in order of increasing depth, i.e., A = shallowest footage, B = next shallowest footage, etc.

(b) 4.6 gpl H_2O_2, 2.1 gpl $NaHCO_3$, pH = 8.5.

(c) 10 gpl NaOCl, 2.1 gpl $NaHCO_3$, pH = 8.5.

Batch test studies were also carried out comparing the capabilities of H_2O_2 and NaOCl in leaching U_3O_8 from an areally-broad sampling of Crownpoint ores (Table 3). These samples encompassed easy-to-leach and intermediate leaching ores as well as refractory ores. The derived rate constants, <u>k</u>, for both NaOCl and H_2O_2 and their ratio are given in Table 3. In every case the NaOCl-$NaHCO_3$ system gave faster leaching rates, with the ratios varying from two to 26.

Column Pack Tests

Pack tests provide a reasonable method of measuring ultimate U_3O_8 recovery in the laboratory. Four tests were conducted to explore the effect of NaOCl concentration on U_3O_8 leaching rate and recovery. The NaOCl concentrations varied from 0.1 to 5 wt %. Results are presented in order of increasing NaOCl concentration. Packs were leached to maximum feasible U_3O_8 recovery.

For all pack tests, the U_3O_8 level in the effluent was measured as a function of PV throughput. The point at which NaOCl broke through in the effluent was also determined. In one test (see Pack 3 below) organic carbon production was also monitored during leaching. The number of PV's and average U_3O_8 level to the 80% U_3O_8 recovery point - an optimistic recovery target - was noted in each case. A plot of %-U_3O_8 recovery and U_3O_8 level as a function of PV for a characteristic pack (Pack 3, 1.0% NaOCl-0.4% $NaHCO_3$) is shown in Figure 2.

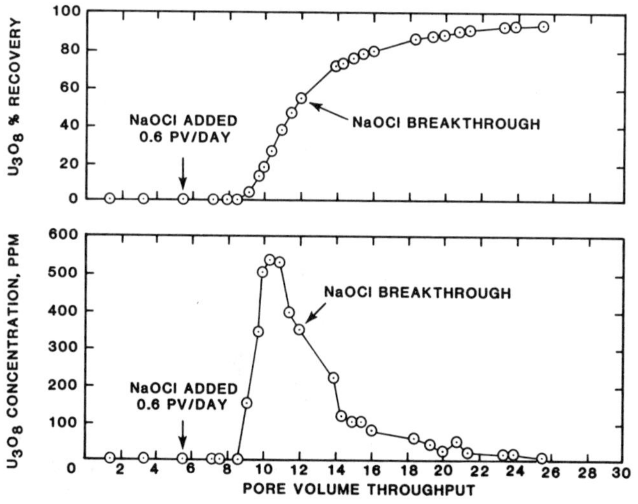

FIGURE 2. <u>U_3O_3 PRODUCTION LEVEL AND RECOVERY VERSUS PORE VOLUME THROUGHPUT FOR 1% NaCOl-0.4% $NaHCO_3$ LEACHATE AND SAMPLE B-360</u>

<u>Pack 1 - Leachate: 0.1% NaOCl, 0.2% $NaHCO_3$</u>. This dilute solution performed very well in recovering U_3O_8, reaching 80% recovery in 94.3 PV. The ultimate recovery was 87.3% after 116.4 PV. The highest U_3O_8 level was 75 ppm at 17.7 PV. The average U_3O_8 level to 80% recovery was 34 ppm.

The initial injection rate was 0.6 PV per day. This rate was increased to 2.0 PV per day after 11 PV. The rate was then reduced to 1 PV per day after passage of 30 PV. After 92 PV throughput, the rate was further reduced to 0.2 PV per day. Breakthrough of NaOCl occurred after 59 PV.

<u>Pack 2 - Leachate: 0.5% NaOCl, 0.4% $NaHCO_3$</u>. This leachate effectively recovered 80% of the U_3O_8 after 40.8 PV. The U_3O_8 level peaked at 153 ppm, 7.9 PV after NaOCl injection began. The average U_3O_8 concentration to 80% recovery was 77 ppm. The ultimate recovery was 90.2% after 60.7 PV.

Initial injection rate was 2.0 PV per day. The NaOCl breakthrough occurred at 7.9 PV, the same point at which the U_3O_8 level was maximum. The flow rate was then reduced to 1 PV per day, and NaOCl ceased to break through. The breakthrough, however, reoccurred after 18.8 PV. The flow rate was further reduced to 0.3 PV per day after 38.9 PV. The U_3O_8 level increased for a short time before declining again.

<u>Pack 3 - Leachate: 1% NaOCl, 0.4% $NaHCO_3$</u>. This pack gave an ultimate recovery of 92.5% after passage of 18.4 PV. The leachate injection rate was a constant 0.6 PV per day. The U_3O_8 level reached a maximum value of 540 ppm after 5 PV throughout and then steadily decreased. The average U_3O_8 level to 80% recovery was 187 ppm. NaOCl breakthrough occurred after 6 PV had been injected.

<u>Pack 4 - Leachate: 5% NaOCl, 1% $NaHCO_3$</u>. This pack was run at a constant rate of 0.2 PV/day. The ultimate recovery was 92.7% after 21.6 PV. The maximum U_3O_8 level was 1007 ppm after 2.1 PV. The average U_3O_8 level to 80% recovery was 390 ppm. The breakthrough of NaOCl occurred very late, after 11.6 PV and 81.6% recovery. After 1.3 PV a red solid material (probably ferric hydroxide) was observed in the pregnant lixiviant. Its appearance was accompanied by a drop in pH from 8.4 to 6.5. The pH rose again quickly when the solid ceased to form.

<u>Residual U_3O_8 in Leached Packs and Uranium Material Balance</u>. To determine U_3O_8 recoveries by the residual U_3O_8 method, we sectioned each pack measuring the U_3O_8 residue in each segment. These data for the four packs are summarized in Table 4. For the exhaustively leached packs (Packs 2, 3, and 4), there was no discernable trend in U_3O_8 abundance with position in the pack. For Pack 1, however, a minor trend did appear. The amount of U_3O_8 was least near the inlet and greatest near the exit, as might be expected if the oxidant was rapidly consumed.

TABLE 4

RESIDUAL URANIUM IN LEACHED ORE AND MATERIAL BALANCE IN COLUMN PACK STUDIES

Pack No.	Vertical Section[(a)] (in.)	Residual U_3O_8 (wt %)	Average Residual U_3O_8 (wt %)	U_3O_8 % Recovery Based on Residual U_3O_8	U_3O_8 % Recovery Based on Original U_3O_8
1	0 - 2 3/4	.0151			
	2 3/4 - 5 1/2	.0175			
	5 1/2 - 8 1/4	.0190			
			.0172	81.3	87.3
2	0 - 3	.0119			
	3 - 6	.0079			
	6 - 9	.0080			
	9 - 12	.0158			
	12 - 15 1/2	.0101			
			.0109	88.0	90.2
3	0 - 3	.015			
	3 - 6	.015			
	6 - 9	.015			
	9 - 12	.018			
	12 - 14	.015			
			.016	83.0	92.5
4	0 - 5	.0128			
	5 - 10	.0086			
	10 - 15	.0070			
	15 - 20	.0083			
	20 - 26	.0090			
			.0091	90.0	92.7

(a) The zero point is the inlet.

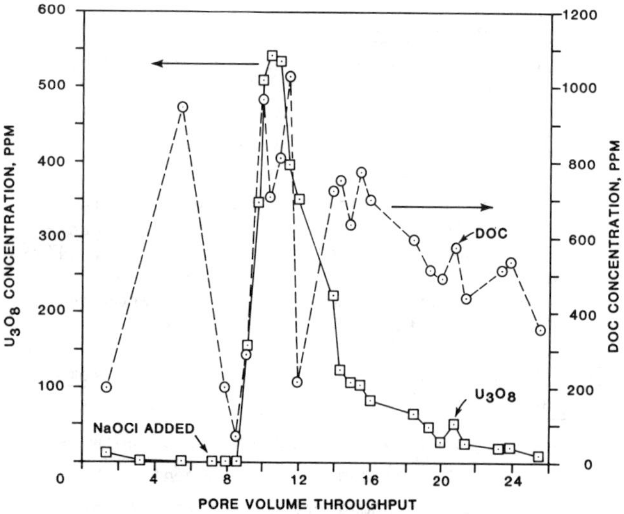

FIGURE 3. U_3O_8 AND DOC PRODUCTION LEVELS VERSUS PORE VOLUME THROUGHPUT DURING LEACH TEST OF SAMPLE B-360 WITH 1.0% NaOCl

Organic Carbon and Other Species in Pregnant Leachate. For the leach test of Pack 3, the amount of total organic carbon was measured in the pregnant lixiviant as a function of PV. Figure 3 shows the concentrations of organic carbon and U_3O_8 present in the effluent during the leach test.

The figure shows that copious amounts of dissolved organic carbon (DOC) ranging up to 1000 ppm were present in the effluent from the pack test. There was also a large amount of DOC produced in the pre-flush. When NaOCl was added, the concentration of organic matter correlated fairly well with the concentration of U_3O_8. The DOC level fluctuates more than the U_3O_8 concentration, however.

Post-Leach Water Flush Studies. After leaching of the packs was complete, a waterflush quickly lowered U_3O_8 levels in the effluent. This data is summarized in Table 5.

Microprobe Examination of Carbonaceous Uranium Grains Exposed to Hypochlorite Leachate

Carbonaceous grains containing coffinite from Well 174 in Area D of Crownpoint were exposed to 50 gpl NaOCl containing $NaHCO_3$ for 72 hours. X-ray diffraction estimates of coffinite originally in place and U_3O_8 analyses of the leachate

TABLE 5

POST-LEACH WATER FLUSH
OF SODIUM HYPOCHLORITE LEACHED PACKED COLUMNS[a]

Pack No.	Leach Solution	Flushing Fluid	Conc., ppm U_3O_8 Start	End	PV
1	0.1% NaOCl, 0.2% $NaHCO_3$	0.2% $NaHCO_3$, 0.1% NaCl	5.8	1.2	3.3
2	0.5% NaOCl, 0.4% $NaHCO_3$	0.2% $NaHCO_3$, 0.1% NaCl	6.7	1.1	5.1
3	1.0% NaOCl, 0.4% $NaHCO_3$	Tap Water	10.0	1.2	4.5
4	5.0% NaOCl, 1.0% $NaHCO_3$	1.0% $NaHCO_3$, 2.5% NaCl	4.5	< 0.2	3.6

(a) Columns contained Crownpoint B-360 Ore.

showed about half of the U_3O_8 was removed from the leached ore.

Electron microprobe examination of polished thin sections of the grains exposed to the leachate demonstrated that coffinite, pyrite, and carbonaceous material were still present in the grains. Analysis of the carbonaceous material showed that chlorine was present in diffuse homogeneous distribution on the exposed surface. The carbonaceous matrix was mutilated with holes and cracks.

DISCUSSION

Utility of Alkaline NaOCl for Leaching Refractory Uranium Ore

The pack tests demonstrate that alkaline NaOCl from 0.1 to 5.0 wt % gave high ultimate recovery of U_3O_8, on the order of 90%, from the B-360 ore stock. The rates of recovery, however, were a function of NaOCl concentration, with the more concentrated solutions giving more rapid rates. Comparisons between the various concentrations are not completely straightforward because certain runs were carried out at very high pump rates (1-2 PV per day). Ideally, one would want to study all concentrations at the same (preferably slow) injection rate to model field conditions. The one NaOCl leachate that was studied at a slow injection rate throughout was 5% NaOCl. The idea was, that by using a very slow injection rate, this extremely concentrated oxidant might remove all the U_3O_8 in one or two PV. Actually, the leach rate with 5% NaOCl was only slightly faster than that with 1% NaOCl, although higher U_3O_8 levels were observed with 5% NaOCl. Perhaps the slow injection rate allowed more reaction of NaOCl with oxidizable sulfide, e.g. jordisite or molybdenite (33) and other gangue material.

One way of measuring effectiveness of the various NaOCl concentrations is to compare the amount of NaOCl in each formulation necessary to produce one lb of U_3O_8. We chose the 80% recovery point as an optimistic target recovery. For a completely efficient process it should take 0.266 lb of NaOCl to produce one lb of U_3O_8. By dividing 0.266 by the experimentally determined ratio of lb oxidant per lb of U_3O_8, one obtains the efficiency at each concentration. The ultimate recoveries, average U_3O_8 production levels to 80% recovery, and efficiencies to 80% recovery for the four different NaOCl formulations are given in Table 6.

Table 6 shows that the more concentrated NaOCl systems were the least efficient. These more concentrated solutions have the advantage of maintaining a high average ppm of U_3O_8 to achieve faster recovery. The efficiency value for the 0.5% NaOCl run seems lower than expected. These results may be due to the fact that the injection rate was neither constant nor identical for the four leachates. Since one would like to maintain an average concentration in a field operation of at least 50 ppm U_3O_8 as a feed to the ion exchange columns during surface processing to achieve good column efficiency, it appears that a NaOCl level of 0.2-0.3% would give the best trade-off between efficiency and reasonable ppm of U_3O_8.

Comparison of NaOCl, $NaHCO_3$ and O_2 or H_2O_2, $NaHCO_3$ Leaching Systems. Earlier, we found that large leaching rate constants resulted in high U_3O_8 recoveries from Crownpoint ore. (16) By this criterion, the much larger rate constant values shown in Table 3 for the NaOCl, $NaHCO_3$ leachate system, as compared to the H_2O_2, $NaHCO_3$ system, imply that alkaline NaOCl would give superior recoveries in leaching Crownpoint ores. A direct comparison is available for the B-360 ore used in our pack tests. Earlier we had measured U_3O_8 recoveries of 60% in a pack test and 66% in a core test with B-360 ore, using a high pressure O_2, $NaHCO_3$ leachate. (16) The 90% recoveries achieved with alkaline NaOCl for this ore indicate that the NaOCl-$NaHCO_3$ leachate gives about 50% greater recovery than the O_2, $NaHCO_3$ system.

The recovery advantage of the NaOCl system is shown graphically in Figure 4. Here the U_3O_8 recoveries and concentrations for the most dilute NaOCl leachate, the 0.1% NaOCl, 0.2% $NaHCO_3$ system (Pack 1), are superimposed on a high pressure O_2,

TABLE 6

SUMMARY OF PACK COLUMN URANIUM LEACHING TESTS USING HYPOCHLORITE CROWNPOINT B-360 COMPOSITE LOT 2 ORE

		Ultimate (Final) Value		Values to 80% Recovery			
Leach Solution	Pack No.	Recovery	PV	PV	Av. Conc., ppm	lbs Oxidant/ lb U_3O_8	Efficiency,[a] %
0.1% NaOCl, 0.2% $NaHCO_3$	1	87.3	116.4	94.3	34	29	0.92
0.5% NaOCl, 0.4% $NaHCO_3$	2	90.2	60.7	40.8	77	65	0.41
1% NaOCl, 0.4% $NaHCO_3$	3	92.5	18.4	10.5	187	53	0.50
5% NaOCl, 1% $NaHCO_3$	4	92.7	21.6	9.7	390	128	0.21

(a) Moles of U_3O_8/mole of NaOCl x 100%.

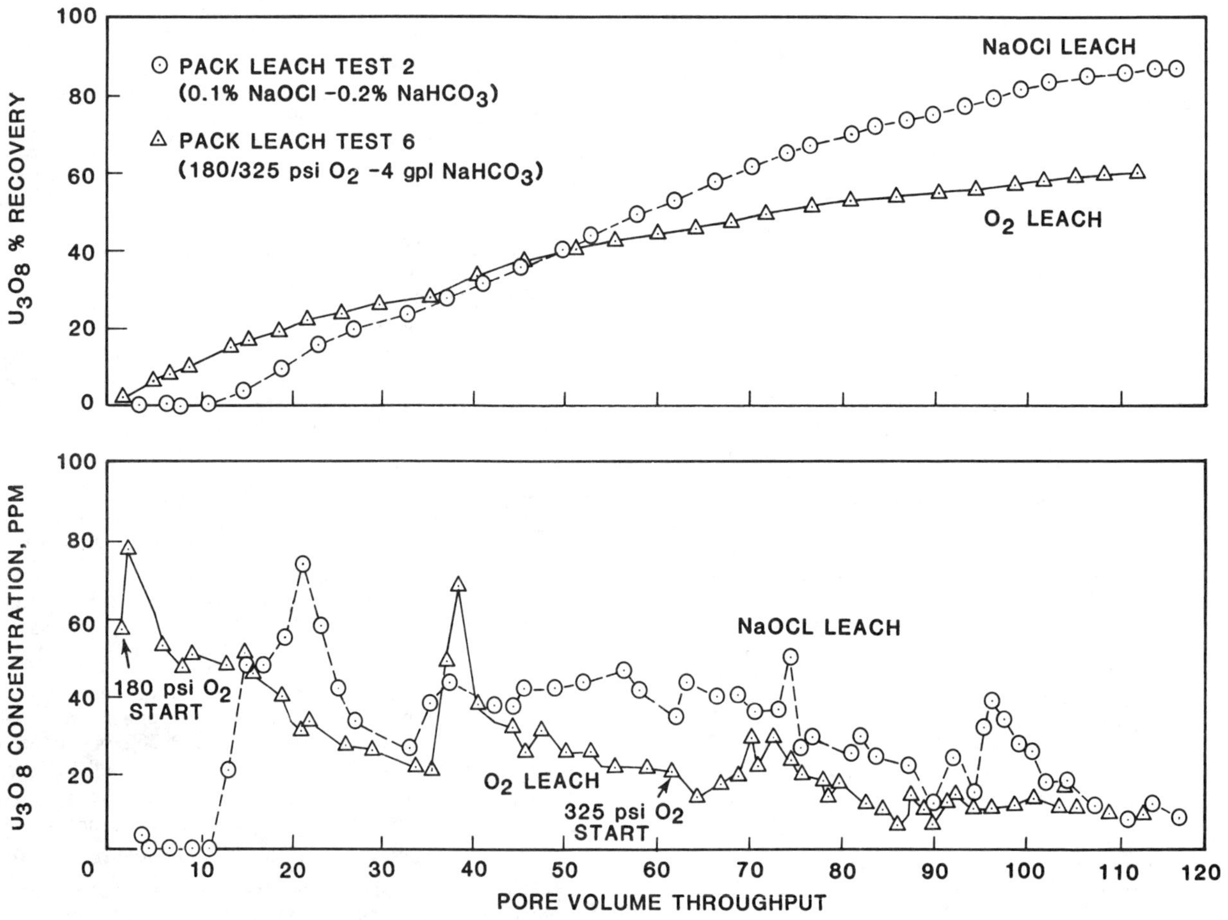

FIGURE 4. COMPARISON OF U_3O_8 PRODUCTION LEVEL AND RECOVERY FOR SAMPLE B-360 FOR TWO LEACHATE FORMULATIONS

$NaHCO_3$ leach test of the same ore, described in our earlier work. (16)

Another example shows how alkaline NaOCl leaches U_3O_8 unavailable to the O_2, $NaHCO_3$ system. We had recovered 87% of the U_3O_8 in a core from Crownpoint Area D, Well 175, with an 850 psi O_2, 2 gpl $NaHCO_3$ leachate. The maximum U_3O_8 production level achieved during the leach was 400 ppm. The ore contained organic carbonaceous material but did not exhibit a refractory ore leach rate. This leached core was flushed with a 1% NaOCl, 4 gpl $NaHCO_3$ leachate. The U_3O_8 concentration rose to 304 ppm in just 1.2 PV. At this point the experiment had to be terminated because of a line rupture.

Kinetics. In Table 3 we compared the effectiveness of alkaline NaOCl and H_2O_2 in leaching various Crownpoint ores using pseudo first-order rate constants, k. There are certain limitations to these k values. The differences in leach rate demonstrated in the batch tests signify that the U_3O_8 in the ores must occur in a variety of environments, each with its characteristic rate constant. However, different microchemical environments can occur not only in ore samples from different areas, but also with ore depth in a single well. In a given ore sample, the most accessible uranium crystallites are leached first at a particular rate. As the conversion progresses, the less accessible environments react at a slower rate. We have tried to narrow the range of these uncertainties by measuring the k's over a 24-hour period, but even then there may be several different leaching regimes as leaching continues. Therefore, the k's should perhaps be considered leaching indices rather than first-order constants.

The limitations discussed above became apparent for certain pack tests when the log of % U_3O_8 remaining in the four packs is plotted on semi-log paper versus leachate throughput (Figure 5). A straight line was drawn through the early experimental points for data from each of the four packs. The deviations of the experimental data from these straight lines reflect the deviation form first order behavior. All four pack runs can be characterized by pseudo first-order rate constants derived from the lines. The measured k values (in PV^{-1}) are 0.018, 0.055, 0.301, 0.16 for the 0.1%, 0.5%, 1% and 5% NaOCl runs (Packs 1, 2, 3, and 4), respectively. However, these k values should not be directly compared because of the differences in pump rates for the four pack runs.

Figure 5 shows that Packs 3 and 4 appear to follow first-order kinetics until 70% of the U_3O_8 is recovered. Pack 2 follows first-order kinetics until 55% of the U_3O_8 is recovered. However, Pack 1 follows first-order kinetics for only 20% of the recovery curve. This sequence is plausible because the more dilute NaOCl leachates are least likely to have oxidant in large excess of that required to oxidize the U_3O_8, a necessary condition for pseudo first-order kinetics to prevail. (34) Unfortunately, Packs 1 and 2 were leached part of the time at faster pump rates than Packs 3 and 4, so definitive kinetic interpretation is hazardous.

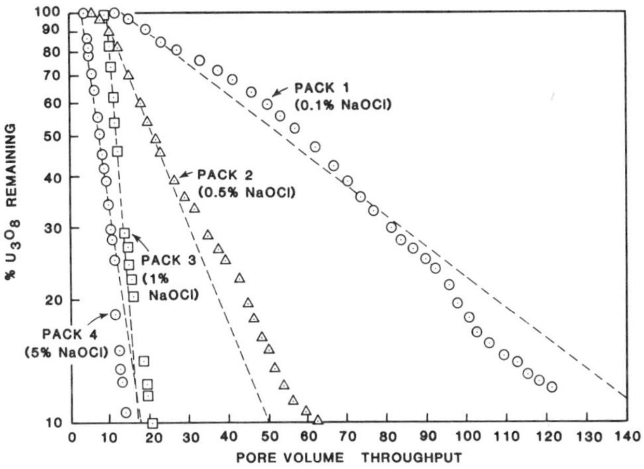

FIGURE 5. U_3O_8 RECOVERY VERSUS PORE VOLUME THROUGHPUT FOR PACKED COLUMN LEACH TEST USING B-360 ORE SAMPLES AND VARIOUS NaOCl CONCENTRATIONS IN THE LEACHATE

Nevertheless, it is clear that Pack 4 (5% NaOCl), although leachate was injected at the lowest pump rate, has a smaller rate constant than Pack 3 (1% NaOCl). The 5% NaOCl leachate is also much less efficient than the 1% NaOCl leachate. The more concentrated NaOCl solution may react more extensively with the gangue material, while the more dilute solutions appear to have greater selectivity for U_3O_8. In a large volume batch test conducted on B-360 ore, we found about 200 ppm $SO_4^=$ present in the produced leachate. This possibly reflects oxidation by NaOCl of metal sulfides such as pyrite, molybdenite and/or jordisite in the gangue material. (25,33)

Carbonaceous Material. In our earlier mineralogical study of Crownpoint ores, we had determined that the organic material associated with the uranium was a very mature, coaly-type kerogen derived from plant material. (12) In our study of the leachability of Crownpoint ore with mild leaching agents, we speculated that encapsulation of U_3O_8 ore with this carbonaceous material was the cause of certain ores being refractory. (16) In this regard, the data of Figure 3 is quite significant. The leaching of U_3O_8 was accompanied by the presence of large amounts of DOC, and the level of organic material production in the leachate correlated with the U_3O_8 production level. The electron microprobe results also demonstrated that NaOCl attacked the carbonaceous material. These results strongly suggest that some degree of degradation of carbonaceous material is necessary for the leaching of U_3O_8 from these Crownpoint refractory ore samples.

The Use of NaOCl for In-Situ Leaching - Process Implications. These laboratory studies show that alkaline NaOCl is a powerful and technically effective system for in-situ leaching of refractory U_3O_8 ores. The NaOCl-$NaHCO_3$ system gave extremely high recoveries at a rapid rate. The most dilute and consequently least corrosive formulation showed the greatest efficiency.

NaOCl is, of course, a corrosive system. Piping and tanking for in-situ U_3O_8 leaching field operations are generally made of fiberglass or PVC, and no change would need to be made. However, special pumps with all exposed parts made of corrosion resistant material would be required, since unreacted NaOCl might reach the producing wells. Also, any unreacted NaOCl in the produced leachate stream would have to be destroyed before contacting the ion exchange resin, since resin beads cannot tolerate prolonged exposures to oxidant of any kind.

Buildup of Cl^- in the system must also be handled. If NaOCl is produced on-site by the reaction

$$Cl_2 + 2\ NaOH \rightarrow NaOCl + NaCl + H_2O\ , \qquad (6)$$

then NaCl is introduced immediately. That problem could be circumvented by producing NaOCl electrolytically from NaCl solution. However, the oxidative action of NaOCl produces NaCl, i.e.

$$UO_2 + NaOCl \rightarrow UO_3 + NaCl \qquad (7)$$

or

$$S^{=} + 4\ NaOCl \rightarrow SO_4^{=} + 4\ NaCl\ . \qquad (8)$$

This buildup of Cl^- would make loading of $UO_2(CO_3)_3^{4-}$ on ion exchange resins difficult because of competition of Cl^- for the resin cation sites.

The DOC generated by oxidation of the carbonaceous material might also require removal, i.e. by adsorption on activated carbon, prior to the ion exchange column. Finally, the loss of much of the NaOCl in oxidation of gangue material such as FeS_2 would have to be carefully evaluated from the standpoint of economics.

CONCLUSIONS

- Alkaline NaOCl is a powerful and technically effective leaching system for recovering uranium from refractory Crownpoint ores in the laboratory batch and column pack tests.
- The NaOCl-$NaHCO_3$ system gives much greater leaching rates than H_2O_2-$NaHCO_3$ in comparable batch tests on an areally-broad sampling of Crownpoint ores, with $k_{NaOCl}/k_{H_2O_2}$ varying from two to 26, depending on the sample.
- High ultimate recoveries (~ 90%) of U_3O_8 were observed in column pack tests with 0.1, 0.5, 1.0 and 5.0% NaOCl leachates; greatest efficiency was observed with 0.1% NaOCl, but higher NaOCl levels gave faster U_3O_8 recovery rates (higher ppm of U_3O_8); ~ 0.2-0.3% NaOCl appears to represent the best tradeoff between efficiency and rate.
- The strong oxidant OCl^- appears to attack the organic carbonaceous material that encapsulates a significant fraction of the coffinite in these samples, thereby rendering the uranium mineral accessible to the leachate; presence of DOC in column effluents and electron microprobe observations support this conclusion.
- Significant amounts of sulfate in the produced leachate from a large volume batch test infer oxidation by NaOCl of metal sulfides (i.e., pyrite) also present in the ore.
- First order kinetics were observed in the early stages of the leach runs, but major deviations from first order occurred as recovery (i.e., conversion) increased, particularly with dilute NaOCl systems where oxidant levels were low.
- In any process application, corrosion-resistant materials would be required in certain equipment, and NaOCl must be prevented from contacting ion exchange resin beads. Chloride ion produced in the oxidation reaction must be handled, and the economic significance of NaOCl consumption from oxidation of gangue minerals assessed.

ACKNOWLEDGEMENT

We acknowledge the excellent support given by E. T. Habib in conducting and analyzing the laboratory leaching studies; D. P. Nichols (packed column leach tests); G. M. Williams (uranium solution analyses); J. Crofford (sample selection and preparation) and S. A. Dixon (electron microprobe analyses).

REFERENCES

1. Merritt, R. C., The Extractive Metallurgy of Uranium, Colorado School of Mines Research Institute, Golden, 1971, pp. 108-110.
2. Livingston, C. W., "Method of Mining Ore In-Situ by Leaching", U. S. Patent 2,818,240, Dec. 31, 1957.
3. MacGregor, R. A., "Recovery of U_3O_8 by Underground Leaching", Canadian Mining and Metallurgy Bulletin, Vol. 59, No. 649, May 1966, pp. 583-587.

4. Gardner, J. and Richie, M. I., "Leaching of Uranium Ore In-Situ", U. S. Patent 3,309,140, Mar. 14, 1967.
5. Anderson, J. S., and Richie, M. I., "Solution Mining of Uranium", Mining Congress Journal, Vol. 54, No. 1, Jan. 1968, pp. 20-23, 26.
6. Erickson, J. W., "Underground In-Situ Mining - A New Mining Method", Mining Engineering, Vol. 30, No. 11, Nov. 1978, pp. 1532-1534.
7. Chase, C. K., Winters, H. J., and Bhappu, R. B., "Economic Evaluation of In-Situ Extraction of Uranium", In Situ, Vol. 2, No. 1, 1978, pp. 49-63.
8. Conine, W. D., "Uranium Solution Mining: A Comparison of New Mexico with South Texas", Third Annual Uranium Seminar, SME/AIME, Casper, Wyoming, Sept. 9-12, 1979, pp. 126-128.
9. Granger, H. C., "Localization and Control of Uranium Deposits in the Southern San Juan Basin Mineral Belt, New Mexico: An Hypothesis", U. S. Geological Survey Professional Paper 600-B, 1968, pp. 60-70.
10. Granger, H. C., "Clays in the Morrison Formation and their Spatial Relation to the Uranium Deposits at Ambrosia Lake, New Mexico", U. S. Geological Survey Professional Paper 450-D, 1962, pp. 15-20.
11. Hilpert, L. S., "Uranium Resources of Northwestern New Mexico", U. S. Geological Survey Professional Paper 603, 1969, 166 pp.
12. Vogt, T. C., Dixon, S. A., Strom, E. T., Johnson, W. F., and Venuto, P. B., "In-Situ Leaching of Crownpoint, New Mexico, Uranium Ore, I. Mineralogical Frame of Reference", presented at the Fifth Annual Uranium Seminar, SME/AIME, Albuquerque, New Mexico, Sept. 21-23, 1981.
13. Breger, I. A., and Deul, M., "Association of Uranium with Carbonaceous Materials, with Special Reference to Temple Mountain Region", U. S. Geological Survey Professional Paper 320, 1959, pp. 139-149.
14. Jacobs, M., "Elucidation of the Organic Matrix from a Uranium Ore", Ph.D. Thesis, Colorado State University, 1970, 170 pp.
15. Haji-Vassiliou, A., "The Association of Uranium with Naturally Occurring Organic Materials in the Colorado Plateau and Other Areas", Ph.D. Thesis, Columbia University, 1969, 263 pp.
16. Vogt, T. C., Strom, E. T., Dixon, S. A., Johnson, W. F., and Venuto, P. B., "In-Situ Leaching of Crownpoint, New Mexico Ore. II. Laboratory Study of a Mild Leaching System", presented at the 56th Annual Fall Technical Conference and Exhibition of the Soecity of Petroleum Engineers of AIME, San Antonio, Texas, Oct. 5-7, 1981.
17. Rhett, D. W., "Mechanism of Uranium Retention in Intractable Uranium Ores from Northwestern New Mexico", Journal of Metals, Vol. 31, No. 10, Oct. 1979, pp. 45-50.
18. Vogt, T. C., Strom, E. T., Johnson, W. F., and Venuto, P. B., "In-Situ Leaching of Crownpoint, New Mexico Ore. IV. Laboratory Study of Strong Leaching Systems: Oxidant-Sulfuric Acid", this volume, pp. 191-198.
Mining Symposium, Denver, November 18-20, 1981.
19. Magno, P. J., and DeSesa, "Oxidants in Carbonate Leaching of Uraniferous Ores", USAEC Report WIN-86, Aug. 23, 1957, National Lead Co., Inc., Winchester, Mass., 44 pp.
20. DeSesa, M. A., Magno, P. J., Gardner, H. E., and Dickerman, E., "Cupric Ammonium Reagent as an Oxidant in the Carbonate Leaching of Uraniferous Ores", Proceedings of the Second United Nations International Conference, Geneva, Vol. 3, 1958, pp. 342-345.
21. Merritt, op. cit., pp. 104-108.
22. Merritt, R. C., Peterson, H. D., Wentz, C. N., "Heap Leaching Studies on Uranium Ore. Phase IV", Bureau of Mines Open File Report No. 141-77, Mar. 18, 1977, 162 pp.
23. Yan, T. Y., "A Method for Removing Ammonium Ions from a Subterranean Formation after In-Situ Uranium Leaching", Journal of Petroleum Technology, Vol. 33, No. 5, May 1981, pp. 921-926.
24. Humenick, M. J. and Garwacka, K., "Oxidative Destruction of Ammonia", Technical Report EHE-79-79, Dec. 1980, The University of Texas at Austin, Austin, Texas.
25. Paul, J. M., Johnson, W. F., Fletcher, A., and Venuto, P. B., "In-Situ Leaching of South Texas Uranium Ores. II. Oxidative Removal of Adsorbed Ammonium Ions with Sodium Hypochlorite", presented at the 56th Annual Fall Technical Conference and Exhibition of the Society of Petroleum Engineers of AIME, San Antonio, Texas, Oct. 5-7, 1981.
26. Cotton, F. A., and Wilkinson, G., Advanced Inorganic Chemistry, 4th Edition, Wiley, New York, 1980, pp. 556-558.
27. Merritt, op. cit., pp. 84-86.
28. Strom, E. T., Woessner, D. E., and Smith, W. B., "^{13}C NMR Spectra of the Uranyl Tricarbonate-Bicarbonate System", Journal of the American Chemical Society, Vol. 103, No. 5, March 11, 1981, pp. 1255-1256.
29. Ciavatta, L., Ferri, D., Grenthe, I., and Salvatore, F., "The First Acidification Step of the Tris (carbonato) dioxouranate (VI) Ion, $UO_2(CO_3)_3^{4-}$", Inorganic Chemistry, Vol. 20, No. 2, Feb. 1981, pp. 463-467.
30. Cinneide, S. O., Scanlan, J. P. and Hynes, M. J., "Equilibria in Uranyl Carbonate Systems I, The Overall Stability Constant of $UO_2(CO_3)_3^{4-}$", Journal of Inorganic and Nuclear Chemistry, Vol. 37, No. 4, April 1975, pp. 1013-1018.
31. Ciavatta, L., Ferri, D., Grimaldi, H., Palombari, R., and Salvatore, F., "Dioxouranium (VI) Carbonate Complexes in Acid Solution", Journal of Inorganic and Nuclear Chemistry, Vol. 41, No. 8, 1979, pp. 1175-1182.
32. Johnson, D. A., and Florence, T. H., "Spectrophotometric Determination of Uranium (VI) with 2-5-Bromo-2-pyridylazo-5-diethylaminophenol", Analytica Chimica Acta, Vol. 53, No. 1, Jan. 1971, pp. 73-79.
33. Bhappu, R. B., Reynolds, D. H., Stahmann, W. S., "Studies on Hypochlorite Leaching of Molybdenite", Unit Processes in Hydrometallurgy, Eds. M. E. Wadsworth, and F. T. Davis, Gordon and Breach, New York, 1964, pp. 95-113.
34. Moelyn-Hughes, E. A., Physical Chemistry, Pergamon, New York, 1961, p. 1215.

Chapter 16

IN SITU LEACHING OF CROWNPOINT URANIUM ORE: IV LABORATORY STUDY OF STRONG LEACHING SYSTEMS – OXIDANT/SULFURIC ACID

T. C. Vogt
E. T. Strom
W. F. Johnson
P. B. Venuto
Research Department
Mobil Research and Development Corporation
Dallas, Texas 75221

ABSTRACT

This paper presents laboratory data on the effectiveness of a fairly severe leaching system—sulfuric acid (H_2SO_4) and an oxidant – in recovery of uranium from refractory Crownpoint uranium ore. In combination with a strong mineral acid such as H_2SO_4, an oxidant such as oxygen or hydrogen peroxide (H_2O_2) should be able to degrade the kerogen intimately associated with uranium in slow-leaching uranium ores, thus exposing the uranium mineral to contact by the leachate. Scoping batch leach tests show that a leachate such as H_2SO_4 with ferric ion gives good recoveries at a fast rate from Crownpoint refractory ore. For detailed study of the H_2SO_4–oxygen system, a composite core was fabricated with ore segments from several wells in an area in which ore leaches slowly with mild leachates. With this system, comprising 0.5% H_2SO_4, 24.5 g/l Na_2SO_4, 1 g/l NaCl and 0.2 g/l CO_2 with 800 psi O_2 65% recovery of uranium was observed rapidly in about 30 pore volumes. This is almost double the recovery observed with a mild leaching system (O_2–bicarbonate) in the same number of pore volumes. Plugging that occurred twice during the leach run appears related to movement of feldspar and quartz fines rather than gypsum. Chemical consumption in these H_2SO_4 experiments was fairly high.

INTRODUCTION

Earlier we described the uranium mineralogy of an areally broad sampling of Crownpoint ore, including the characterization of organic carbonaceous material associated with the uranium mineral. (1) We also reported on the leaching behavior of Crownpoint ores with a mild alkaline leaching system, oxygen (O_2)-sodium bicarbonate ($NaHCO_3$). (2) There was a wide range of leaching responses, with samples from some areas leaching at a slow rate with low uranium recovery. Evidence indicates that in these samples the organic material is intimately associated with the uranium mineral and shields it to some extent from contact by the leachate. (1,2)

One method to obtain faster leach rates and higher recoveries is to use a chemically severe leaching system such as sodium hypochlorite (NaOCl) as an oxidant. (3) The use of sulfuric acid (H_2SO_4) and an oxidant for leaching ore in surface milling operations is widely practiced to effect high recovery of uranium (U_3O_8) from the ore. (4,5,6) The U_3O_8 is mobilized as the disulfate complex, $UO_2(SO_4)_2{}^{2-}$, or possibly, the trisulfate complex, $UO_2(SO_4)_3{}^{4-}$, is the soluble species. (7) Sulfuric acid leachate formulations were used in the first in-situ leaching operation in the Shirley Basin in Wyoming. (8) More recently H_2SO_4 formulations have been used in the Rocky Mountain Energy field tests in the Powder River Basin (Bear Creek) and in the Teapot formation at Nine Mile Lake site near Casper. (9,10) Laboratory studies related to the use of H_2SO_4 for in-situ leaching have been reported (11-17) and Yan has reported packed column leach studies on Crownpoint ore using H_2SO_4 and oxidant. (18)

In this paper we report laboratory studies on in-situ leaching behavior of refractory Crownpoint ore with H_2SO_4 and O_2, hydrogen peroxide (H_2O_2) and sodium chlorate ($NaClO_3$) using consolidated core samples in continuous-flow tests, as well as crushed composite samples in batch tests.

TABLE 1. ANALYSIS OF SELECTED PARAMETERS FROM AREA B ORE SAMPLES

WELL NUMBER	B-514	B-516	B-520	B-360
Depth Interval, ft	2135-2156	2027-2157	2136-2158	2021-2166
U_3O_8, Wt %				
Average	0.036	0.117	0.03	0.08
Range	0.007-0.062	0.043-0.2	0.02-0.052	0.005-0.23
Fe^{++}, Wt %				
Average	1.00	1.21	0.92	0.09
Range	0.88-1.12	0.82-3.58	.58-1.68	.05-.12
$CaCO_3$, Wt %				
Average	5.5	4.1	2.5	1.41
Range	0.1-12.2	0.2-23.2	0.17-10.8	0.04-3.33
Organic Carbon, Wt %				
Average	0.09	0.24	0.11	0.18
Range	0.07-0.14	0.11-.32	0.08-0.13	0.04-0.6

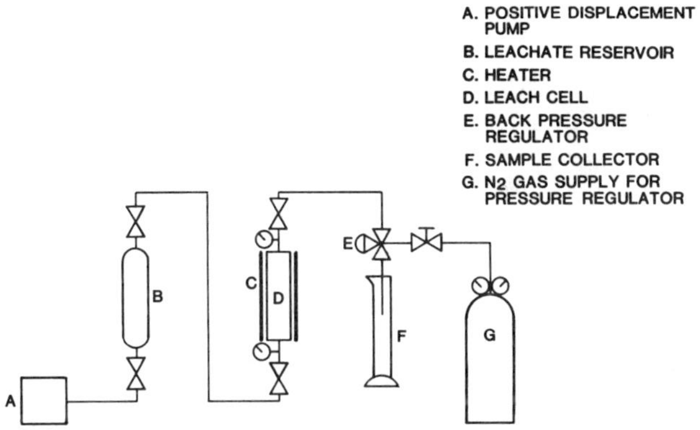

FIGURE 1. CORE LEACHING APPARATUS

EXPERIMENTAL

The refractory ore samples used were from Area B at Crownpoint, referred to in our earlier paper. (1) Ore samples from the area exhibited intermediate to slow leachability in batch and packed column tests as reported earlier. (2) Core analyses are shown in Table I.

The core leach test was conducted on composite samples from four wells distributed within a 200 x 200 ft area. The eight core samples were nominally three inches in diameter and from one to three inches in length for a total length of 14.8 inches. The method of assembly and assaying core samples is described in detail in an earlier report. (2) The leaching apparatus is shown in Figure 1. A heater assembly around the core cell held the temperature at 90°F.

The leachate used in the core test contained 1 gpl NaCl, 15 gpl of sulfate ($SO_4{}^{2-}$) as Na_2SO_4, 5 gpl H_2SO_4, 0.2 gpl CO_2, and O_2 gas dissolved at 800 psi. Before the oxidative leach began, approximately six pore volumes (PV) of a preflush solution were flowed through the core to remove pre-oxidized U_3O_8. The pre-oxidation occurred from exposure of the core during handling and sample preparation. The pre-flush solution contained 1 gpl NaCl and 15 gpl of $SO_4{}^{2-}$ as Na_2SO_4 adjusted to pH = 2 with H_2SO_4. The effluent solution was analyzed during pre-flush and oxidative leaching phases for 17 parameters, including U_3O_8. The leachate analyses are given in Table 2. The bulk of the total dissolved solids is from the Na_2SO_4 dissolved in the leachate to simulate recycled $SO_4{}^{2-}$ levels in a field operation.

Batch tests were of two types. The first type was conducted in 4 and 6 ounce polyethylene bottles using composite ore samples from a given well. This type of test has been described earlier. (2) Another type of batch test was conducted under contract by Hazen Research, Inc. (Golden, Colorado). These tests were conducted on a semi-continuous batch leach basis on the same composite ore sample as in the batch test described above. Enough sample was placed in a

TABLE 2. ANALYSIS OF SELECTED PARAMETERS FROM AREA B ORE SAMPLES

PARAMETER, ppm	PORE VOLUME THROUGHPUT: PREFLUSH				OXIDATIVE LEACHING							
	0.25	1.94	4.06	5.96	6.99	7.84	8.68	10.14	12.04	14.13	18	31
Total Dissolved Solids	26775	24460	25780	26340	27860	27245	26440	26790	26625	26955	30125	31380
Chloride as Cl	698	621	617	614	648	613	614	617	614	620		
Sulfate as SO_4	17994	15500	16620	17052	18128	14984	16558	16401	16480	16978	19243	20455
pH @ 24°C	2.17	8.13	4.42	4.04	4.30	7.57	8.16	8.16	8.38	7.68		
Carbonate as CO_3	0.0	0.0	0.0	0.0	0.0	0.0	0.0	0.0	8.2	0.0		
Bicarbonate as HCO_3	0.0	463	0.0	0.0	0.0	1489	726	1389	913	696		
Sodium as Na	6700	6300	6700	6925	7240	7100	6925	6975	6875	6875		
Potassium as K	4.4	155	83.0	41.3	31.0	37.8	31.0	27.8	22.4	19.8		
Calcium as Ca	66	171	152	169	160	174	135	177	159	172	209	54.8
Magnesium as Mg	1.8	38.2	30.0	53.5	34.5	66.5	90.5	97	101	100		
Silicon as Si	6.8	19.9	10.6	16.2	16.0	26.0	28.9	29.6	27.1	25.5		
Iron as Fe	416	<0.05	49.3	121.5	45.0	<0.05	<0.05	<0.05	<0.05	<0.05		
Molybdenum as Mo	0.30	10.4	0.19	0.23	0.27	38.4	18.3	5.50	1.02	1.92	0.034	0.045
Arsenic as As	0.454	0.323	0.392	0.315	0.381	0.312	0.419	0.369	0.381	0.396	0.107	0.258
Selenium as Se	0.489	0.543	0.577	0.562	0.581	0.596	0.664	0.639	0.644	0.620	0.121	0.280
Uranium as U_3O_8	3.77	19.1	2.24	0.59	1.30	116.7	250.3	173.5	136.8	152.4	528	19.2
Dissolved Organic Carbon (DOC)	21.3	16.8	6.2	12.9	16.5	40.1	35.2	42.9	35.0	24.6	15.6	11.5

five-pint reagent bottle to give 25% solids. The bottle was agitated by placing it on a roller apparatus. After 24 hours the bottle was removed and the solids allowed to settle. One-half of the solution was removed as clear supernatant and replaced with fresh leachate. The bottle was then returned to the roller for another 24 hours. The procedure was repeated for five leaching cycles.

RESULTS

Core Leach Tests

The U_3O_8 recovery and production level for the core leach test are shown in Figure 2. After preflush, oxidative leaching was begun and continued for a total of about 29 PV throughput. At that point about 65% of the U_3O_8 had been recovered. Plugging of the leach cell occurred on two occasions during oxidative leaching as can be seen in Figure 3 by the very abrupt increases in the pressure differential across the core. In the first instance at about 9 PV throughput, disconnecting the inlet line and momentarily allowing the fluid under pressure in the core to flow back eliminated the problem. In the second instance of plugging, at about 15 PV cumulative throughput, the discharge line had to be drilled out. X-ray diffraction showed plugging particles to be quartz and feldspar.

The maximum U_3O_8 level in the production stream was slightly over 600 ppm. In addition to analysis for U_3O_8, the production level of selected species was monitored during the leaching test. See Table 2. The total dissolved solids and pH levels of the produced fluid were determined as a function of leachate throughput.

Batch Leach Tests

This series of batch leach tests (Table 3) was conducted to compare the effect on leaching of ferric ion (Fe^{3+}) as ferric sulfate ($Fe_2(SO_4)_3$) and Na_2SO_4, both in H_2SO_4 solution. The use of different oxidants, i.e., H_2O_2 and $NaClO_3$, was also tested. The $Fe_2(SO_4)_3$ and Na_2SO_4 systems were compared at three concentration levels at approximately equivalent $SO_4^{=}$ concentrations of 10, 20 and 50 gpl. The oxidant level was always 0.5 gpl. The acid was initially introduced at 5 gpl in each test. The U_3O_8 concentration was measured at 3 and 24 hr., and a pseudo-first-order rate constant was calculated as described in a previous report. (2) In all cases when Fe^{3+} ion was used, the reaction rate increased by a factor of about 10 on the average. Increasing the SO_4^{2-} concentration also caused the rate to increase.

A leachate consisting of 5 gpl Fe^{3+} ion, 5 gpl H_2SO_4 and 0.5 gpl of either H_2O_2 or $NaClO_3$ was used in the semi-continuous batch tests (Table 4). In one experiment with $NaClO_3$, the concentration of H_2SO_4 was raised to 15 gpl. In the semi-continuous batch tests, results showed that U_3O_8 recovery with Fe^{3+} ion was 79 to 84%, whether $NaClO_3$ or H_2O_2 was used as the oxidant.

A comparison of leach rates and U_3O_8 levels for mild and chemically severe systems is given in Table 5. With refractory ore B-360, the 1% NaOCl batch test gave a 5.3 fold increase in leach rate over a mildly oxidizing leachate. U_3O_8 recovery values are based on packed column leach tests. Using 800 psi O_2 and 5 gpl H_2SO_4, a leach rate and U_3O_8 recovery were obtained which were similar to those with NaOCl. The ore sample used in the core leach test (Area B composite) in Figure 1 was shown to be more refractory than sample B-360. The leach rate and recovery was slightly more than half that for the B-360. Yet in the acid leach test of the core, the recovery improved to 65%,-- a factor of about 1.8.

DISCUSSION

Leaching of refractory ore from the Crownpoint area is relatively slow when using O_2 or H_2O_2 in mild alkaline solution as reported earlier. (2) The use of NaOCl in alkaline $NaHCO_3$ solution, a more chemically severe oxidizing system, gave U_3O_8 recoveries of 90% and greater in laboratory leach studies. (3)

In this study we found that 5 gpl of H_2SO_4 containing dissolved O_2 at 800 psi is also sufficiently severe to effect leaching of the Crownpoint refractory ore. The U_3O_8 recovery in the core leach test described here was 65% after about 30 PV of the acid leachate had passed through the core (Figure 2). With mild alkaline leaching of refractory ore, U_3O_8 recovery was often little more than 35%. (2) Yan also showed the effectiveness of H_2SO_4-O_2 systems in leaching packed columns of refractory Crownpoint ore. (18) The data indicated that O_2 pressure as low as 400 psi were adequate for leaching the refractory ore.

TABLE 3. BATCH LEACHING TESTS USING H_2SO_4 WITH $NaClO_3$ AND H_2O_2; EFFECT OF ADDING SULFATE AND FERRIC IONS

Experiment Number	Leachate	Wt % U_3O_8 Original Sample	mg U_3O_8 per Liter 3 hr	mg U_3O_8 per Liter 24 hr	pH Initial	pH Final	first order rate constant k, $hr^{-1} \times 10^{-3}$
1	5 gpl H_2SO_4 14.7 gpl Na_2SO_4 [a] 0.5 gpl $NaClO_3$	.091	20.64	28.77	1.71	2.99	2.46
2	5 gpl H_2SO_4 29.4 gpl Na_2SO_4 [b] 0.5 gpl $NaClO_3$	.091	20.40	32.55	1.90	3.16	3.72
3	5 gpl H_2SO_4 73.5 gpl Na_2SO_4 [c] 0.5 gpl $NaClO_3$	.091	21.82	44.57	2.00	3.05	7.29
4	5 gpl H_2SO_4 17.8 gpl $Fe_2(SO_4)_3$ [d] 0.5 gpl $NaClO_3$	.091	69.57	143.86	1.55	1.94	51.48
5	5 gpl H_2SO_4 35.6 gpl $Fe_2(SO_4)_3$ [e] 0.5 gpl $NaClO_3$	.091	84.90	153.30	1.55	1.76	58.04
6	5 gpl H_2SO_4 88.9 gpl $Fe_2(SO_4)_3$ [f] 0.5 gpl $NaClO_3$	.091	107.31	161.55	1.40	1.54	61.68
7	5 gpl H_2SO_4 29.4 gpl Na_2SO_4 [b] 0.5 gpl H_2O_2	.091	29.36	56.37	1.81	2.97	9.27
8	5 gpl H_2SO_4 35.6 gpl $Fe_2(SO_4)_3$ [e] 0.5 gpl H_2O_2	.091	93.16	153.30	1.38	1.76	53.81

a) 10 gpl SO_4^{2-} b) 20 gpl SO_4^{2-} c) 50 gpl SO_4^{2-}
d) 12.8 gpl SO_4^{2-} e) 25.6 gpl SO_4^{2-} f) 64.1 gpl SO_4^{2-}

TABLE 4. SEMI-CONTINUOUS BATCH LEACHING TEST OF CROWNPOINT REFRACTORY ORE: ACID (H_2SO_4/Fe^{3+}) LEACHING SYSTEM

Leachate, gpl				Residual U_3O_8 Wt %	Uranium Recovery, %
Fe^{3+}	H_2SO_4	H_2O_2	$NaClO_3$		
5	5	0.5	0	.018	84
5	15	0	0.5	.017	84
5	5	0	0.5	.022	79

Also we see from Figure 2 that a small fraction of the ore is easily leached. The preflush with no oxidant but only SO_4^{2-} to complex the pre-oxidized U_3O_8 produced about 8% of the U_3O_8. We interpret this as meaning that some of the U_3O_8-mineral was accessible to contact with air during core handling and sample preparation and was very easily oxidized.

In the earlier reports, we identified the intermediate and hard-to-leach Crownpoint ores as having coffinite mineral intimately associated with organic carbonaceous material. (1,2) Similar results were reported by Rhett. (19) A sample with "intermediate" leachability is postulated as having a smaller fraction of the U_3O_8-mineral crystallites encapsulated or shielded from contact by the mild leachate than a sample classified as "refractory" from a leach test. We see chemically severe leachate systems like NaOCl or H_2SO_4-O_2 appear to be capable of attacking the organic barriers allowing some additional fraction of the U_3O_8 to be contacted. Yet some fraction of the coffinite in the ore of intermediate leachability and refractory ore still appears to be shielded by the organic material. This was observed in the "before" and "after" scanning electron microscope images in our earlier work (2,3) and in the study by Rhett. (19)

The production of dissolved organic carbon (DOC) shown in Table 3 is also indicative of the disaggregation of the organic matter to expose U_3O_8 for improved leaching. Similar results were observed in NaOCl leaching of refractory ore. (3) In leaching the refractory Crownpoint ore with mild alkaline leachate, the DOC levels are usually less than 1 ppm during leaching, indicating only minor breakdown of the carbonaceous encapsulation.

The plugging of the discharge tube with quartz and feldspar fines described in the results can be graphically seen in Figure 3 as an abrupt increase in differential pressure across the leach cell. Packed column leach tests showed little change in the permeability of the column as leaching progressed. (16) The action of lower pH on sandstone cores has been shown to cause migration of fines. (20) In addition, modification of mineral assemblage by H_2SO_4 has been discussed by Riese. (14,21)

Kaolinite, feldspar and iron-rich chlorite, observed in Crownpoint ore, can all be affected by contact with H_2SO_4 to give fragmented products capable of producing fines. (17)

The use of H_2SO_4 with dissolved Fe^{3+} ion as a co-oxidant has been commonly used in surface mill uranium leaching operations. (4,16,22) Batch tests in Table 3 indicate a significant improvement in the leaching rate with Fe^{3+} ion. These values are similar to those for fast leaching Crownpoint ore samples with mild leachate reported earlier. (1)

Vanadium (V) also serves as an effective co-oxidant in acid leaching. (4,9) The Rocky Mountain Energy in-situ pilot at Nine Mile Lake demonstrated the use of V as a co-oxidant in the H_2SO_4 leaching process. The V was present as a naturally occurring mineral, was solubilized by the H_2O_2-acid leachate, and was recycled during leaching.

As with NaOCl leaching systems, the H_2SO_4-oxidant system is capable of attacking the organic

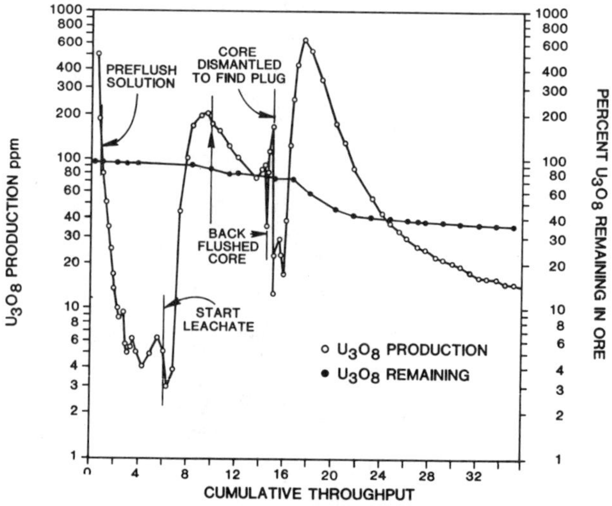

FIGURE 2. URANIUM RECOVERY AND PRODUCTION LEVEL VERSUS ACID LEACHATE THROUGHPUT

TABLE 5. COMPARATIVE LEACHING RATE AND U_3O_8 RECOVERY FOR VARIOUS LEACHATE SYSTEMS WITH CROWNPOINT REFRACTORY ORE

Leachate Formulation,	Sample B-360: Relative Leach Rate[a]	Sample B-360: U_3O_8 Recovery[b] (wt %)	Area B-Composite: Relative Leach Rate[a]	Area B-Composite: U_3O_8 Recovery (wt %)
1. 2 gpl H_2O_2, 2 gpl $NaHCO_3$	1	60	.6	35[b]
2. 2 gpl NaOCl, 2 gpl $NaCO_3$	5.3	92	-	-
3. 800 psi O_2, 5 gpl H_2SO_4	5.1	94[c]	-	-
4. 0.5 gpl H_2O_2, 5 gpl H_2SO_4	-	-	-	65[d]

a) Batch test result relative to leach rate of Sample B-360 using leachate formulation No. 1.
b) Pack test result
c) T. Y. Yan pack data, reference (18)
d) Core test result

matter in the Crownpoint ore as evidenced by the dissolved organic material shown in the analysis of the produced fluid from the core leach test (see Table 2). The highest level of dissolved organic carbon in the H_2SO_4 tests was only about 40 ppm, and the produced fluid had a brownish color. With NaOCl as oxidant, the dissolved organic levels were usually much higher as might be expected from a more powerful oxidant. (3) The observation that improved U_3O_8 production from refractory ore was accompanied by an increase in the amount of dissolved organic carbon is indicative of the shielding effect of the U_3O_8-mineral by carbonaceous material.

Molybdenum (Mo) production was maximized at 38 ppm after 7.84 PV, long before the U_3O_8 production declined. Mo in its reduced state is more easily oxidized than the reduced U mineral and Mo production was completed by the time U_3O_8 levels had reached their maximum value of 528 ppm at 18 PV. There was some Fe^{3+} production during preflush. The acidic preflush probably attacked pyrite to give the maximum production of 121 ppm Fe^{3+} observed just before oxidative leaching began at 6 PV. Although arsenic and selenium were mobilized during the preflush and oxidative leaching, they never exceeded 1 ppm. Calcium (Ca^{2+}) levels are consistent with the solubility product for $CaSO_4$ since SO_4^{2-} ion is in excess. The acid was substantially consumed by the gangue minerals such as $CaCO_3$ in the leach test as indicated by the high pH values and bicarbonate levels in the effluent from the core. Complete consumption of the $CaCO_3$ is calculated to occur at some point greater than 33 PV of injection, provided the acid can contact all the carbonate material. We speculate that $CaSO_4$ is probably precipitated in the core and may coat the surface of $CaCO_3$ moieties preventing further contact by acidic leachate.

H_2SO_4-oxidant leaching formulations give faster leaching rates and higher U_3O_8 recoveries for refractory Crownpoint ore than the mild alkaline carbonate leaching system. Consumption of H_2SO_4 by gangue minerals, primarily $CaCO_3$, must be factored into the economics. High SO_4^{2-} levels may possibly cause $CaSO_4$ precipitation in processing equipment and in the formation. Fines mobilization appears possible, as the acid interacts with formation materials releasing feldspar and quartz fragments. DOC must be prevented from fouling ion exchange beads in the surface process complex. These are among factors that would need to be addressed in planning and design of an in-situ leaching operation in the field.

CONCLUSIONS

- Crownpoint refractory ore can be leached with good recovery of U_3O_8 using a chemically severe leachate system such as H_2SO_4-O_2 as demonstrated in laboratory tests.

- The recovery level using the acid-oxidant systems ranges from about 65% for tests on consolidated core without added Fe^{3+} ion to about 80% U_3O_8 recovery from crusted ore samples with acid containing Fe^{3+} ion.

- Fines movement initiated or caused by acid leaching caused plugging in laboratory flow systems.

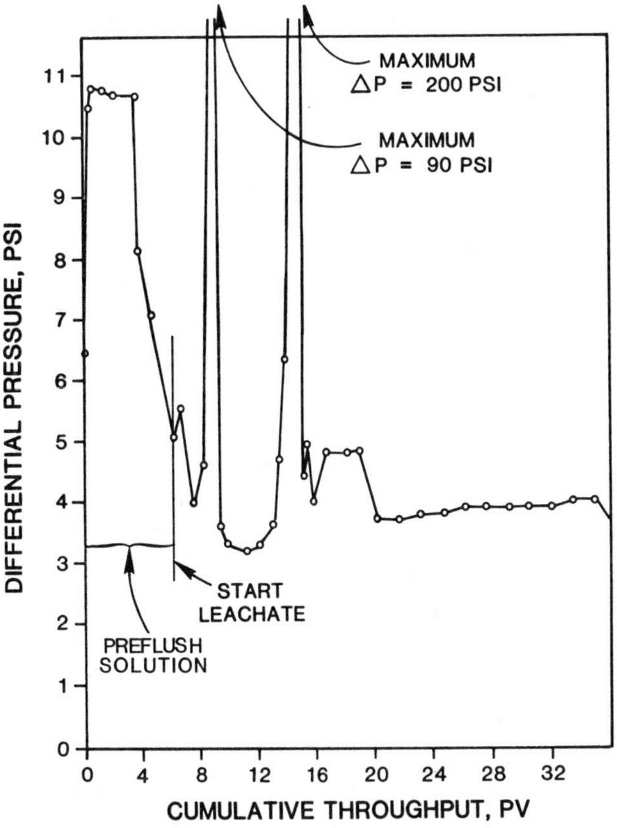

FIGURE 3. DIFFERENTIAL PRESSURE ACROSS CORE DURING ACID LEACHING

ACKNOWLEDGEMENT

We are indebted to M. L. Mathis for conducting the core leaching tests; J. Crofford (sample acquisition and preparation); G. M. Williams (uranium solution analyses); and, J. T. Edwards (X-ray diffraction studies). The semi-continuous batch tests were performed at Hazen, Inc., Golden, Colorado.

REFERENCES

1. Vogt, T. C., Dixon, S. A., Strom, E. T., Johnson, W. F., Venuto, P. B., "In-Situ Leaching of Crownpoint New Mexico Uranium Ore. 1. Mineralogical Frame of Reference", presented at the Fifth Annual Uranium Seminar, Albuquerque, Sept. 20-23, 1981.

2. Vogt, T. C., Strom, E. T., Dixon, S. A., Johnson, W. F., Venuto, P. B., "In-Situ Leaching of Crownpoint New Mexico Uranium Ore. II. Laboratory Study of a Mild Leaching System", presented at the 56th Annual Fall Technical Conference and Exhibition of the SPE of the AIME, San Antonio, Oct. 4-7, 1981.

3. Vogt, T. C., Strom, E. T., Johnson, W. F., and Venuto, P. B., "In-Situ Leaching of Crownpoint New Mexico Uranium Ore. III. Laboratory Study of Strong Leaching Systems: Sodium Hypochlorite", this volume, pp. 177-189.

4. Merritt, R. C., The Extractive Metallurgy of Uranium, Colorado School of Mines Research Institute, Golden, 1971, pp. 59-83.

5. Clegg, J. W., and Foley, D. D., Uranium Ore Processing, Addison-Wesly Publishing Co., Inc., Reading, Mass., 1958.

6. International Atomic Energy Agency, Uranium Ore Processing, Proceedings of an Advisory Group Meeting, Washington, D. C., Nov. 24-26, 1975, Vienna, 1976.

7. Merritt, R. C., op. cit., p. 60.

8. Anderson, J. S., and Ritchie, M. I., "Solution Mining Uranium", Mining Congress Journal, Vol. 54, No. 1, 1968, pp. 20-23, 26.

9. Phillips, P. E., "Rocky Mountain Energy's Experiences with In-Situ Leaching of Uranium with Acid", South Texas Uranium Seminar, South Texas Minerals Section of AIME, Corpus Christi, Texas, Sept. 10-13, 1978.

10. Tweeton, Daryl R., Anderson, G. R., Ahlness, J. K., Peterson, O. M., Engelmann, W. H., "Geochemical Changes during In-Situ Uranium Leaching with Acid" presented at the SME-AIME Annual Meeting, New Orleans, February 10-22, 1979.

11. Hardwick, T. J., "Acid Leach of Uranium Ore: In-Situ Conditions", presented at the 54th Annual Fall Technical Conference and Exhibition of the SPE of the AIME, Las Vegas, Sept. 23-26, 1979.

12. Merritt, R. C., Peterson, H. D., and Wentz, C. N., "Heap Leaching Studies on Uranium Ore Phase III", U. S. Bureau of Mines Open File Report 118-76, July 6, 1976, 62 pp.

13. Merritt, R. C., Peterson, H. D., and Wentz, C. N., "Heap Leaching Studies on Uranium Ore, Phase IV, U. S. Bureau of Mines Open File Report 141-77, Mar. 18, 1977, 71 pp.

14. Riese, A. C., and Popp, C. J., 1980, "Application of Solution-Mineral Equilibrium Chemistry to Solution Mining of Uranium Ores" in Geology and Technology of the Grants Uranium Region 1979, Christopher A. Rautman, compiler: New Mexico Bureau of Mines and Mineral Resources, Memoir 38, pp. 344-354.

15. Sundar, P. S., "In-Situ Leaching Studies of Uranium Ores - Phases 1 through III", U. S. Bureau of Mines Open File Report 140-77, 1977, 392 pp.

16. Grant, D. A. "In Situ Leaching Studies of Uranium Ores -- Phase IV", U. S. Bureau of Mines Open File Report No. 52-79, Jun. 11, 1979, 497 pp.

17. Harloff, G. J., and Riese, A. C., "Experimental and Theoretical Study of Uranium In-Situ Solution Mining with H_2SO_4", presented at the 55th Annual Fall Technical Conference and Exhibition of the SPE of the AIME, Dallas, Sept. 21-22, 1980.

18. Yan, T. Y., "In-Situ Leaching of Uranium Using Dilute Sulfuric Acid and Molecular Oxygen", in press.

19. Rhett, D. W., "The Mechanism of Uranium Retention in Intractable Uranium Ores from Northwestern New Mexico", Journal of Metals, Vol. 31, No. 10, Oct. 1979, pp. 45-50.

20. Mungan, N., "Permeability Reduction through Changes in pH and Salinity", Journal of Petroleum Technology, Vol. 17, Dec. 1965, p. 1449.

21. Riese, A. C., "An Application of Solution Mineral Equilibrium Chemistry to Uranium Leaching", M. S. Thesis, New Mexico Inst. of Mining and Technology, Secorro, New Mexico, 1978.

22. Dutrizac, J. E., MacDonald, J. C., "Ferric ion as a Leaching Medium", Minerals Service Engineering, Vol. 6, No. 2, April 1974, pp. 59-100.

Chapter 17

A MATHEMATICAL MODEL FOR THE IN SITU LEACHING OF PRIMARY COPPER ORE

H. W. Gao
H. Y. Sohn
M. E. Wadsworth
Department of Metallurgy and Metallurgical Engineering
University of Utah
Salt Lake City, Utah 84112

ABSTRACT

This paper discusses an unsteady-state model which simulates the in situ leaching of a deeply buried rubblized copper ore body. Leaching is accomplished by pumping water saturated with oxygen into the bottom of the flooded rubble chimney. The physical processes incorporated in the present model include the axial convective transport of both mass and heat, axial dispersion of mass, mass transfer between the bulk fluid and solid surface, and pore diffusion within the ore fragments. The coupled model equations are solved numerically by an implicit finite-difference method. The paper discusses important factors which affect the in situ leaching process.

INTRODUCTION

During the last decade, the feasibility of applying the in situ leaching technology to the recovery of copper from deeply buried deposits containing chalcopyrite has received much attention.[1-4] The major incentive for using this technology is that under certain circumstances, copper in deep underground deposits, which cannot be mined economically by conventional means, can be recovered economically. Basically, this technology involves rubblization of the deeply buried copper ore body followed by in-place leaching of the rubblized copper ores. Rubblization of the ore can be achieved by some suitable method such as the use of block caving mining. The rubblized chimney is then filled with water until the original water level is reached. Leaching occurs when the water saturated with oxygen is introduced into the chimney and flows through the fractured rock masses and within pores of rocks. As the dissolved oxygen oxidizes the primary sulfides, heat is generated. Because of low conductivity of the surrounding rocks, the generated heat is able to raise the temperature of the chimney. The increase in the temperature of the chimney and the relatively high content of dissolved oxygen in the leach solution under high hydrostatic pressure will result in an enhanced dissolution of the primary sulfide minerals.

To ensure a successful application of the in situ leaching technology, a complete understanding of the effects of the ore particle size distribution, ore grade mineralogy, temperature, pressure, lixiviant concentration, pH, etc. on the rate of copper recovery is necessary so that copper can be recovered efficiently and the amount of copper extracted can be predicted. These can be achieved through mathematical modeling of the in situ leaching process and conducting laboratory tests. In this paper we will present a mathematical model that simulates the in situ leaching process as depicted in Figure 1. To model this process properly, not only the kinetic and thermodynamic aspects of leaching chemistry must be known, but the effects of heat generation, the physical aspect of fluid flow and other related multicomponent transport phenomena must be considered. Furthermore, since most factors involved in the leaching process change with time, a leaching system must be described using unsteady state equations.

Laboratory studies of the leaching of primary copper sulfide ores conducted by Braun et al.[2,5] have provided the necessary leaching chemistry and kinetics for the in situ copper leaching model development. Their results indicate that leaching chemistry strongly depends on the types of ores. For an ore having a pyrite to chalcopyrite molar ratio (Py/Cp) of 2, the net reaction is

$$CuFeS_2 + 2FeS_2 + \frac{45}{4}O_2 + \frac{43}{6}H_2O = Cu^{+2} + Fe_3(SO_4)_2(OH)_5 \cdot 2H_2O + \frac{1}{3}S + \frac{11}{3}SO_4^{-2} + \frac{16}{3}H^+ \qquad (1)$$

For an ore with Py/Cp = 1, the net reaction is

$$CuFeS_2 + FeS_2 + \frac{15}{2}O_2 + \frac{13}{3}H_2O = Cu^{+2} + \frac{2}{3}Fe_3(SO_4)_2(OH)_5 \cdot 2H_2O + \frac{1}{3}S + \frac{7}{3}SO_4^{-2} + \frac{8}{3}H^+ \qquad (2)$$

In either case, sulfuric acid is formed and the leach solution becomes acidified. The hydrogen ions produced from these reactions are partly consumed by reacting with the gangue minerals producing equivalent amount of cations such as Ca^{+2}, Mg^{+2}, Al^{+3}, Na^+, K^+, etc.

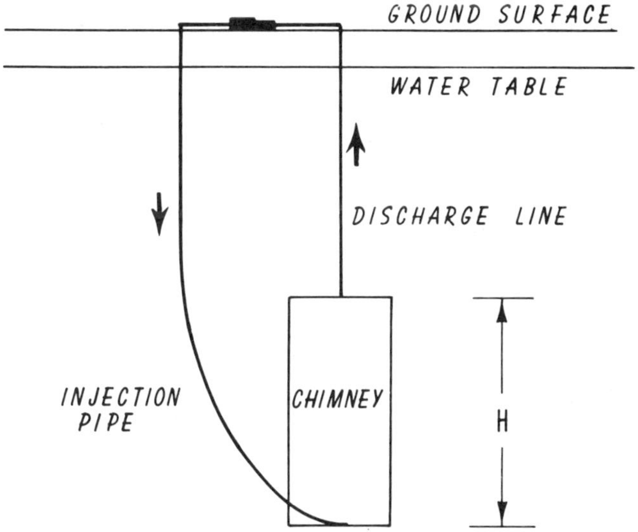

Figure 1. Configuration of Rubble Chimney

In experiments at 90°C, the pH of the leach solution is observed to decrease from its initial value near neutral to a steady-state value of about 2, indicating that the system is buffered. This buffering effect represents a steady-state balance among the H^+-related reactions and is an important feature of the leaching system.

The stoichiometry number which relates the oxygen consumption and copper production is different for different ores. Reaction (1) indicates that 11.25 moles of oxygen are required for each mole of copper to be dissolved, while in reaction (2), only 7.5 moles of oxygen are required to dissolve one mole of chalcopyrite. It can also be shown that for an ore which has a Py/Cp mole ratio of equal to 0.5, the stoichiometry number is 5.625. Furthermore, if the ore contains some biotite,[5] the stoichiometry number will be slightly greater.

Regarding the leaching kinetic model, Bartlett[6] has incorporated the diffusion of oxygen within pores of rocks and the kinetics of the dissolution of chalcopyrite and pyrite to develop a computer model. He applied the continuity equation for oxygen to spherical coordinates to describe the oxidative leaching of copper sulfide ores in a system as described by Lewis and Braun.[2] This model results in an unsteady-state concentration gradient in the ore particle. Braun, Lewis and Wadsworth[5] have successfully used a moving reaction zone model to describe their laboratory leaching data for San Manuel ores over a wide range of ore particle sizes. Their model takes into account two stages of leaching. First, the enhanced initial leaching characteristics is ascribed to the higher grade of $CuFeS_2$ in an exterior shell of the ore. Second, enhanced leaching rates in the later stage are due to the physical changes of the ore, resulting in the increase of the effective interfacial reaction-zone area. This model has also been used by Braun and Mallon(7) in their in-situ copper leaching model calculations. More recently a modified steady-state approximation of the continuity equation has been applied to the leaching of enriched copper sulfide ores by Madsen and Wadsworth.(8) In this paper, the model developed by Braun, et al.,(5) is used as the basis for our model calculations.

MATHEMATICAL FORMULATION

In contrast to the combined leach-circulation model as proposed by Braun and Mallon(7) in which the fluid is circulated by the injected oxygen bubbles and the temperature is assumed to be uniform throughout the chimney, we propose an unsteady-state, one-dimensional, one-phase fluid flow model. This model is suitable for the case when the fluid introduced into the bottom of the chimney consists only of a liquid phase. In another paper, we will present a two-phase fluid flow model for an in stiu leaching system where the fluid injected into the chimney contains both liquid phase and undissolved oxygen bubbles.

In the present model, the assumption that the fluid is considered as one phase is based on the following considerations. Since the fluid introduced into the bottom of the chimney does not contain oxygen bubbles, any oxygen bubbles which may form in the chimney are due to the decrease of oxygen solubility in the leach solution with a decrease in the hydrostatic pressure and the increase in the temperature of the chimney.(9) Bubbles formed through this process are believed to be very small and have very small velocities. Therefore we assumed that they move at the same velocity as the liquid. Furthermore, the volume of the bubbles occupy a very small fraction of the fluid; the average velocity of the liquid is not affected. Hence, one-phase fluid flow description of the system should be adequate.

Species Continuity Equations

The equation of continuity for each species present in the liquid is

$$\frac{\partial}{\partial t}(\varepsilon C_i) + V_z \frac{\partial C_i}{\partial z} - \frac{\partial}{\partial z}\left(D_{ai}\frac{\partial C_i}{\partial z}\right) = \dot{R}_i \qquad (3)$$

where ε is the porosity of the rubble chimney; C_i is the concentration of species i in the liquid; V_z is the superficial velocity; D_{ai} is the axial dispersion coefficient; t is time; z is the vertical position measured from the bottom of the chimney; and $\dot{R}_i$ is the net rate of production per unit volume of chimney for species i.

The axial dispersion coefficient D_{ai} can be calculated from the following equations:(10)

$$D_a = 0.7\,D_o + 1.75 \times \frac{V_z d_p}{\varepsilon} \qquad \text{for } \frac{V_z d_p}{\varepsilon D_o} < 50 \qquad (4)$$

and

$$D_a = 1.75\left(\frac{V_z d_p}{\varepsilon}\right)^{1.2} / D_o^{0.2} \qquad \text{for } \frac{V_z d_p}{\varepsilon D_o} > 50 \qquad (5)$$

D_o is the molecular diffusion coefficient; and d_p is the weighted average particle diameter.

By incorporating the external mass transfer process of oxygen from the bulk liquid to the external surfaces of the ore particles into the rate equation developed by Braun, et al.,(5) the rate of production per unit volume of the chimney for oxygen can be expressed as

$$\dot{R}_{O_2} = -\sum_{j=1}^{N} q_j \sigma \frac{4\pi r_j^2}{\phi_j} \times \left(\frac{C_{O_2}}{\left(\frac{r_j}{R_j}\right)^2 \frac{\sigma}{k_c} + \frac{1}{G\beta} + \frac{\sigma}{D_e}\frac{r_j}{R_j}(R_j - r_j)} \right) \qquad (6)$$

$$\text{where } \beta = \frac{3\rho_r \delta k}{r_p \rho_p} \qquad (7)$$

C_{O_2} is the oxygen concentration in the bulk liquid; R_j is the initial spherical radius for the jth size fraction; r_j is the radius of unreacted core for the jth size fraction; q_j is the number of particles in the jth size fraction per unit volume of the chimney; ϕ_j is the geometry factor for jth size fraction including both boundary roughness and sphericity factors; σ is the stoichiometry number; D_e is the effective diffusion coefficient; G is the ore grade; ρ_r is the bulk rock density; δ is the thickness of reaction zone; k is the rate constant; and k_c is the effective external mass transfer coefficient. The equation for the rate of change of the unreacted core radius is

$$\frac{dr_j}{dt} = -\frac{M_w}{32\, x\phi_j \rho_r G} \left(\frac{C_{O_2}}{\left(\frac{r_j}{R_j}\right)^2 \frac{\sigma}{k_c} + \frac{1}{G\beta} + \frac{\sigma}{D_e}\frac{r_j}{R_j}(R_j - r_j)} \right) \qquad (8)$$

M_w is the molecular weight of chalcopyrite.

In Equations (6) and (8), the ore grade G varies with the radial position of the ore particle according to Equations (27)-(29) of reference 6 and the variation of the shape factor ϕ_j with r_j is described by their equation of (31).

The effective external mass transfer resistance considered is a sum of two separate resistances, $\frac{1}{k_f}$ and $\frac{1}{k_d}$.(11) k_f is the external mass transfer coefficient and is expressed as(12)

$$k_f = \frac{1.1068\, V_z\, N_{Re}^{-0.72}}{Sc^{2/3}} \quad \text{for } N_{Re} < 10 \qquad (9)$$

and

$$k_f = \frac{0.4548\, V_z\, N_{Re}^{-0.4069}}{Sc^{2/3}} \quad \text{for } N_{Re} > 10 \qquad (10)$$

N_{Re} is the Reynolds number based on the particle diameter, $(d_p \rho_\ell V_z/\mu)$; Sc is the Schmidt number, $(\mu/\rho_\ell D_o)$; μ is viscosity and ρ_ℓ is liquid density.

K_d describes the effects due to longitudinal dispersion and molecular diffusion and is expressed as[11]

$$k_d = \frac{V_z}{a_p \left(\frac{d_p}{Pe_d} + \frac{D_o}{2V_z}\right)} \qquad (11)$$

in which a_p is the external surface area of ore particles per unit volume of chimney; and Pe_d is the Peclet number for axial dispersion, $(d_p V_z/D_e)$.

The rate equation for copper extracted per unit volume of chimney is related to the rate of oxygen consumed by the equation

$$\dot{R}_{Cu} = -\frac{1.986}{\sigma}\dot{R}_{O_2} \qquad (12)$$

Two species, oxygen and copper, are considered in this paper. The following initial and boundary conditions are used for Eq. (3):

I.C.

$$C_{O_2}(z,0) = 0 \qquad (13)$$

$$C_{Cu}(z,0) = 0 \qquad (14)$$

B.C.

$$C_{O_2}(0,t) = C_s(T,P) \qquad (15)$$

$$C_{Cu}(0,t) = C_b(t) \qquad (16)$$

$$\left.\frac{\partial C_{O_2}}{\partial z}\right|_{(H,t)} = 0 \qquad (17)$$

and

$$\left.\frac{\partial C_{Cu}}{\partial z}\right|_{(H,t)} = 0 \quad \text{for } t > 0 \qquad (18)$$

where C_s is the solubility of oxygen in water; T is temperature; P is pressure; and H is the height of chimney.

The dependence of the solubility of oxygen in water on temperature and pressure has been measured[9] and summarized by the following Henry's constant (kg/m^3/psi) for T in °C[5]

$$k_h = 3.418 \times 10^{-3} - 3.738 \times 10^{-5} T + 1.95 \times 10^{-7} T^2 \qquad (19)$$

Therefore C_s in Eq. (15) can be expressed as

$$C_s = k_h P \qquad (20)$$

The copper concentration, C_b in Eq. (16), depends on how the process is operated. Due to the presence of other ions which have been dissolved by sulfuric acid into leach solution, there will exist a maximum allowable copper concentration in the solution. Whenever the copper concentration in the pregnant solution reaches this level, some of the copper in the solution must be removed prior to reinjecting the solution to the chimney. The maximum allowable copper concentration level has not been established and this requires more study on the solution chemistry of a leaching system. We have arbitrarily set it at a reasonable value. Under the condition that no processing of the pregnant solution is made before the copper concentration reaches this value,

$$C_b(t) = C_{Cu}(H, t-\tau) \qquad (21)$$

where τ is the residence time for the leach solution flowing from the top of the chimney to the bottom of the chimney.

Energy Conservation Equation

By assuming that the axial conduction of heat is negligible and that the temperatures of fluid and the ore particle are in local equilibrium, the energy conservation equation can be written as

$$\left((1-\varepsilon)\rho_s c^s + \varepsilon\rho_\ell c^\ell\right)\frac{\partial T}{\partial t} + \rho_\ell c^\ell V_z \frac{\partial T}{\partial z} = \dot{Q} \qquad (22)$$

where ρ_ℓ and ρ_s are the densities of liquid and ore particles, respectively; and $\dot{Q}$ is the rate of heat generated per unit volume of chimney. The above assumptions should be valid since the reaction is not fast and the conductivity of rocks surrounding the chimney is low. The initial and boundary conditions used are

$$T(z,0) = T_0 \quad \text{for all } z \qquad (23)$$

and

$$T(0,t) = T_b(t) \quad \text{for } t > 0 \qquad (24)$$

The heat generated during leaching mainly comes from the oxidation of pyrite and chalcopyrite. The net amount of heat generation as calculated from standard heats of formation is 1.05 megacal per mole of copper for Eq. (1) and 0.7 megacal per mole of copper for Eq. (2). For an ore having Py/Cp = 0.5, the calculated heat of generation is 0.526 megacal per mole of copper.

The temperature T_b in Eq. (24) depends on the temperature of the pregnant solution withdrawn from the top of the chimney and the amount and temperature of oxygen supplied from the oxygen plant. The generated heat for the dissolution of oxygen in water at 1 atm and 18°C is 3297 cal per mole of oxygen.[13] In addition, in order to facilitate the processing of the pregnant solution above the ground and to avoid unsatisfactory recovery rate,[3] the maximum value of T_b is restricted to 90°C.

Pressure Drop Equation

The total pressure drop per unit height of chimney due to the frictional flow of the leach solution between the ore particles is calculated from the Ergun equation,[14]

$$\Delta P = (150 \times \frac{(1-\varepsilon)\mu}{d_p \rho_\ell V_z} + 1.75)\ (\frac{1-\varepsilon}{\varepsilon^3})\ (\frac{1}{d_p})\ (\rho_\ell V_z^2) \quad (25)$$

NUMERICAL METHOD

The above coupled model equations are solved by an implicit-finite difference method.[15] If subscripts k and n represent vertical position and time increments, respectively, the derivatives of the species concentrations C, temperature T and unreated core radii r_j are approximated as:

$$\frac{\partial C}{\partial z} = \frac{C_{n+1,k+1} - C_{n+1,k-1}}{2\Delta z} \quad (26)$$

$$\frac{\partial^2 C}{\partial z^2} = \frac{C_{n+1,k+1} - 2C_{n+1,k} + C_{n+1,k-1}}{(\Delta Z)^2} \quad (27)$$

$$\frac{\partial T}{\partial z} = \frac{T_{n+1,k} - T_{n+1,k-1}}{\Delta z} \quad (28)$$

$$\frac{\partial C}{\partial t} = \frac{C_{n+1,k} - C_{n,k}}{\Delta t} \quad (29)$$

$$\frac{\partial T}{\partial t} = \frac{T_{n+1,k} - T_{n,k}}{\Delta t} \quad (30)$$

and

$$\frac{dr_j}{dt} = \frac{r_{n+1,k,j} - r_{n,k,j}}{\Delta t} \quad (31)$$

for k = 2,3,...M and j = 1,2,...N; M is the total number of position increments and N is the number of different sizes of the ore particles. The first derivative of C at the top of the chimney is approximated by the second-order backward difference equation:

$$\left.\frac{\partial C}{\partial z}\right|_{z=H} = \frac{3C_{n+1,M+1} - 4C_{n+1,M} + C_{n+1,M-1}}{2\Delta z} \quad (32)$$

Model calculations were made for two modes of operation: (1) constant flow rate and (2) variable flow rate of the leach solution. In the constant flow mode, if the flow rate of the leach solution is not high enough, not only the rate of dissolution of the primary sulfide minerals will be lowered, but also some undesirable side effects such as the depletion of copper in solution to form secondary copper sulfides[3,5] may occur. If the flow rate of the leach solution is too high, due to the decrease of the leaching rate with time, excessive oxygen may result and form gas bubbles in the later stages of leaching. To compromise between these cases, one might vary the flow rate of the leach solution such that no oxygen bubbles will appear and yet the content of the dissolved oxygen is close to the saturation concentration during the period of leaching. This concept results in a variable flow mode.

The procedure for solving the linealized finite-difference equations depends on the mode of operation. After all the quantities at time step t = n are known, the oxygen concentration profile at time step t = n + 1 is first calculated. The leaching rate is calculated from the dissolved oxygen concentration at t = n. The next step in the numerical procedure is to advance temperature profile. In doing so, the average of the heat generated at t = n and t = n + 1 with the temperature dependent variables evaluated at the time step t = n is used. The pressure and solubility of oxygen at each grid point are then calculated and the comparison between the oxygen concentration and the solubility is made. For the variable flow mode if the calculated oxygen concentration at any grid point exceeds the solubility at that grid point, the velocity of the liquid for this time step is allowed to decrease and the calculation is repeated until the oxygen concentration at each grid point is less than or equal to the solubility at that grid point. The copper concentration profile and the rate of copper extracted are then computed. In addition, in order to have the oxygen concentration close to their saturation solubilities in the variable flow mode, the velocity of the liquid is also allowed to increase.

To investigate the effects of grid-spacing and time step on the numerical solutions, a series of preliminary tests were made. Results for grid spacings ranging from 50 points to 300 points and a time step of Δt = 0.1 day have always agreed to one another within 1%. Results for copper concentration profile are more sensitive to the size of the time step than those for other quantities. It was also found that appropriate time step size depends on the liquid velocity and the porosity of the chimney. Nevertheless, the size of the time step used is always below 0.2 day and the results are accurate to within about 2% except for the copper concentration profile which is accurate to within about 5%.

RESULTS AND DISCUSSIONS

In all calculations, the chimney is assumed to have a height of 360 m and a distance of 360 m below the water table. The particle size distribution used ranges from 0.635 cm to 30.48 cm in diameter (with two exceptions) and is shown in table 1. The initial temperatures of the chimney, the leach solution and the temperature of oxygen supplied from the oxygen plant are assumed to be 60, 60 and 20°C, respectively.

Table 1

Ore Particle Size Distribution

Size Range (inches)	(cm.)	Wt.%
-12 + 10	-30.48 + 25.4	9.5
-10 + 8	-25.4 + 20.32	10.6
- 8 + 6.5	-20.32 + 16.51	8.9
-6.5 + 5.0	-16.51 + 12.7	10.2
-5.0 + 3.5	-12.7 + 8.89	12.1
-3.5 + 2.5	-8.89 + 6.35	9.8
-2.5 + 1.5	-6.35 + 3.81	12.4
-1.5 + 1.0	-3.81 + 2.54	8.0
-1.0 + 0.5	-2.54 + 1.27	10.7
-0.5 + .25	-1.27 + 0.635	8.0

Typical results for the fraction of copper extracted as a function of time are shown in Fig. 2 for constant flow rates of 10, 15, 20 and 30 m^3 of leach liquid per unit cross-sectional area of the chimney. Although the fraction of copper extracted increases with flow rate for the period of leaching shown, the leaching rate does not always increase with flow rate. Figure 3 shows how the total fraction of copper extracted varies with time and position for the simulated conditions shown. These results indicate that those primary sulfide minerals near the bottom of the chimney are dissolved faster than those near the top of the chimney as expected since the content of dissolved oxygen decreases with the vertical distance from the bottom of the chimney.

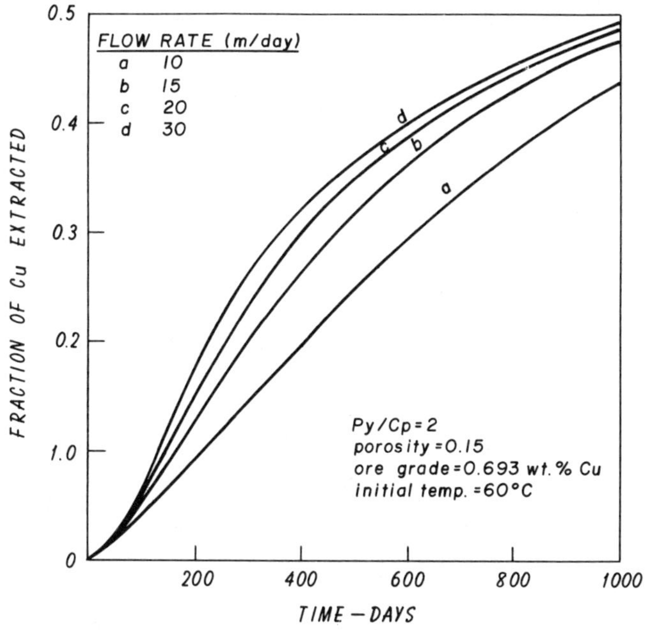

Figure 2. Effect of flow rate on the extraction of copper from primary sulfide ore.

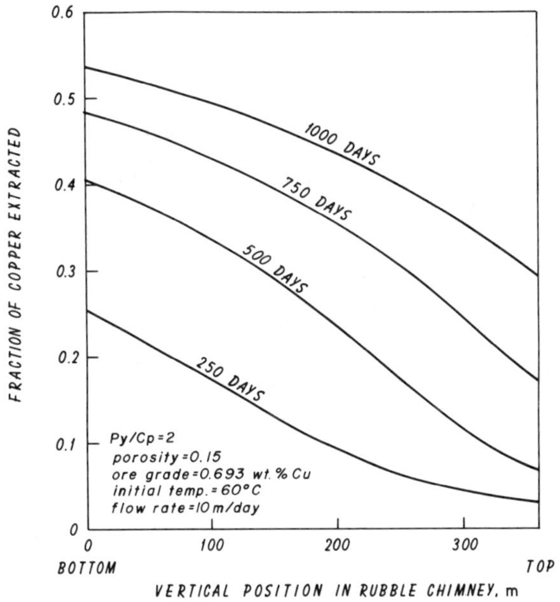

Figure 3. Weight fraction of copper extracted as a function of position at 250, 500, 750 and 1000 days (case 1a).

Figure 4 shows the temperature profiles in the chimney at 50, 100, 150, 200 and 1000 days for the same simulated conditions as in Figure 3. At any day of the first 160 days of leaching, the variation in the temperature along the chimney is less than 2°C. While at 200 days, the variation increases to about 5°C, it is dropped to about 3° at 1000 days. This is due to the decreases in the leaching rate and hence the heat generated.

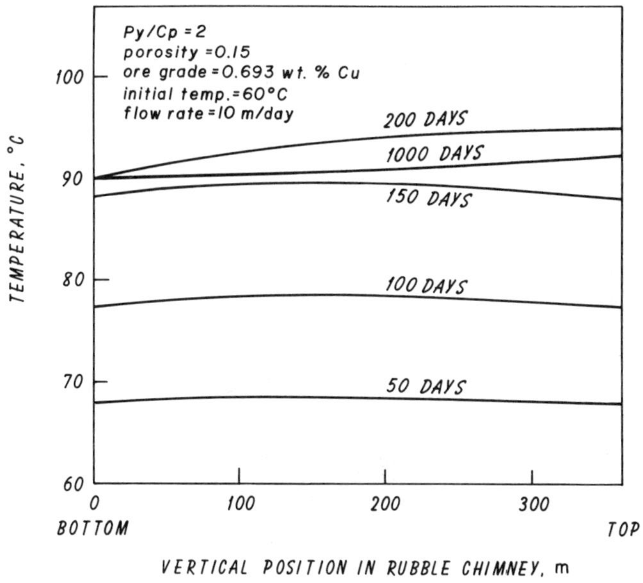

Figure 4. Temperature vs. position relationship for case 1a at 50, 100, 150, 200 and 1000 days.

Typical oxygen concentration profiles for a constant flow mode and a variable flow mode are shown in Figs. 5 and 6, respectively. For the chimney properties and the simulated conditins shown in Fig. 5, oxygen bubbles can be formed during the first 50 days and after 500 days of leaching, resulting from the slow consumption of the oxygen supplied. Figure 6 shows that the dissolved oxygen concentration can be maintained close to or equal to the solubility in any part of the chimney by varying the flow rate of the leach solution. The variation of the solution velocity with time for the chimney properties and simulated conditions of Fig. 6 is shown in Fig. 7 for a leaching period from 200 to 1000 days. During this period, the required superficial velocity of leach solution is decreased from 50 m/day to 11 m/day due to the decrease in the leaching rate with time.

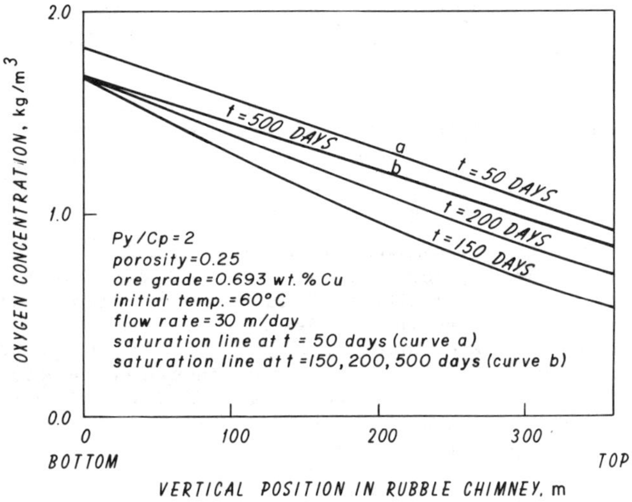

Figure 5. Dissolved oxygen concentration profiles at 50,150,200 and 500 days (case 2c).

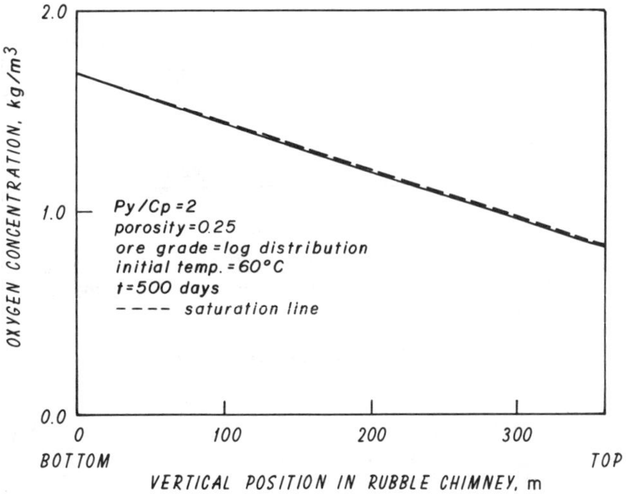

Figure 6. Dissolved oxygen concentration profile at 500 days (case 9a).

Table 2 summarizes the results of all calculations. As shown in cases 1a to 1d and cases 2a to 2c, the total fraction of copper extracted increases with the flow rate of the leach solution for the first 1000 days of leaching. However, in either chimney, the fractions of copper extracted for flow rates of 20 and 30 m^3/m^2-day do not differ much from each other at 1000 days, indicating that a flow rate of 30 m^3/m^2-day for further leaching is too high. Indeed, after 1000 days, a flow rate at a level higher than 10 m^3/m^2-day is still too high for both chimneys since the results of case 1a and case 2a have indicated that the dissolved oxygen concentrations have already reached the respective solubilities. The calculated results from the variable flow mode for both chimneys (case 1e and 2d) show that at 1000 days, the flow rates for these cases have dropped to 6.406 m^3/m^2-day and 5.85 m^3/m^2-day, respectively. Yet, the dissolved oxygen concentrations are still within 97% of the corresponding solubilities. The number of days required for the leach solution to reach a temperature of 90°C decreases with increasing flow rate.

The results of cases 1a to 1d and 2a to 2c also indicate that the fraction of copper extracted is not very sensitive to the chimney porosity. However, the total copper extracted is higher from the chimney having the lower porosity. This is not surprising because we have a larger amount of ore in a chimney packed to a lower porosity.

Calculated results also show that the fraction of copper extracted is sensitive to the ore grade. At a flow rate of 20 m^3/m^2-day, 50.7% of Cu have been extracted in 1000 days for an ore grade of 1.5 wt% in $CuFeS_2$ as shown in case 3a, while only 48.7%, 43.7% and 36.7% have been extracted in the same period of leaching for ore grades of 2%, 3.3% and 5.43%, respectively. Results from variable flow mode also show the same trend. Although the percentage of Cu extracted decreases with increasing ore grade, the total copper extracted is increased. The number of days required for the solution to reach 90°C also decreases with increasing ore grade.

The major effect of Py/Cp mole ratio in the in situ leaching process is probably on the amount of oxygen required from the oxygen plant. During the first 1000 days of leaching, while only 228 kg-moles of oxygen are required for recovering 2540 kg of Cu for an ore having a Py/Cp mole ratio of 0.5, 298 kg-moles of oxygen are required for extracting 2499 kg of Cu from an ore having Py/Cp mole ratio of 1 and 422 kg-moles of oxygen are required for extracting only 2371 kg of Cu from an ore having Py/Cp mole ratio of 2. The higher the Py/Cp mole ratio, the more the oxygen required to extract the same amount of copper.

Calculated results for cases 7a and 7b were obtained for a uniform particle size distribution. The diameter of the ore particles is 0.012 m, which is the weighted average particle diameter of the ore particle size distribution shown in Table. 1. In 1000 days, 98% and 94.5% of copper have been extracted in cases 7a and 7b, respectively. These results, compared with cases 2b and 2d, respectively, indicate that the weighted average particle size cannot be used to represent a very wide particle size distribution.

The decrease in the shape factor is equivalent to the increase in the roughness of the ore particle and thus the interfacial area available for reaction. Therefore, more copper is expected to be extracted.

Table 2. Summary of In Situ Leaching Model Calculations

Case No.	Ore grade wt% $CuFeS_2$	Py/Cp mole ratio	Shape factor	Chimney porosity	Initial chimney temp. °C	Flow rate (m^3/m^2-day)	Days to reach 90°C	Percentage of Cu extracted at 250 Days	500	750	1000	Total copper extracted at 1000 days (Kg)	Days for copper loading in exit stream to leach 10 Kg/m^3	Maximum Pressure Drop (psi)	Total oxygen supplied for 1000 days (Kg-mole/square meter of chimney)
1a	2	2	0.65	0.15	60°C	10	159	12.1	24.9	35.8	43.8	2415	214	0.68	429
1b						15	126	16.5	31.8	41.5	47.5	2621	168	1.04	464
1c						20	112	18.8	34.8	43.2	48.7	2685	144	1.41	476
1d						30	103	22.6	36.7	44.3	49.4	2726	128	2.19	483
1e						variable	103	23.9	37.4	44.8	49.9	2752	125	3.6	487
2a	2	2	0.65	0.25	60°C	10	182	12.7	26.7	37.7	45.1	2195	356	0.12	391
2b						20	138	19.3	35.0	43.2	48.7	2371	252	0.24	422
2c						30	131	21.7	36.3	44.0	49.3	2396	230	0.37	427
2d						variable	133	22.1	36.5	44.3	49.5	2409	217	0.7	429
3a	1.5	2	0.65	0.25	60°C	20	186	18.6	36.1	45.0	50.7	1852	329	0.24	331
3b						variable	187	19.1	36.5	45.3	51.1	1865	315	0.38	333
4a	3.3	2	0.65	0.25	60°C	20	95	16.0	29.7	38.3	43.7	3509	186	0.24	623
4b						variable	73	23.5	34.8	41.4	45.9	3685	117	1.20	655
5a	2	1	0.65	0.25	60°C	20	212	17.3	36.1	45.4	51.4	2499	269	0.24	298
5b						variable	217	17.0	36.2	45.5	51.6	2511	267	0.32	299
6a	2	0.5	0.65	0.25	50°C	20	296	12.3	34.7	45.4	52.2	2540	316	0.24	228
6b						variable	308	11.3	33.9	45.2	52.2	2540	326	0.19	228
7a(p)	2	2	0.65	0.25	60°C	20	76	30.3	60.4	87.1	98.0	4769	181	0.24	845
7b(p)						variable	69	47.9	75.0	88.8	94.5	4598	106	0.54	813
8a	2	2	0.35	0.25	60°C	20	95	26.1	44.1	52.9	58.5	2848	186	0.24	506
8b						variable	72	35.1	48.2	55.4	60.3	2934	118	1.06	522
9a(q)	2	2	0.65	0.25	60°C	20	83	10.7	20.6	28.1	33.1	4377	169	0.24	776
9b(q)						variable	50	19.1	27.6	32.7	36.4	4809	80	1.9	851
10a	5.43	2	0.65	0.25	60°C	20	79	10.9	21.6	30.3	36.7	4849	164	0.24	861

(p) Uniform particle size distribution (d_p = 0.012 m)

(q) Ore grade follows a log distribution: $\log G = -0.21 \log d_p - 1.096$ (d_p in centimeters, G = % $CuFeS_2$/100)

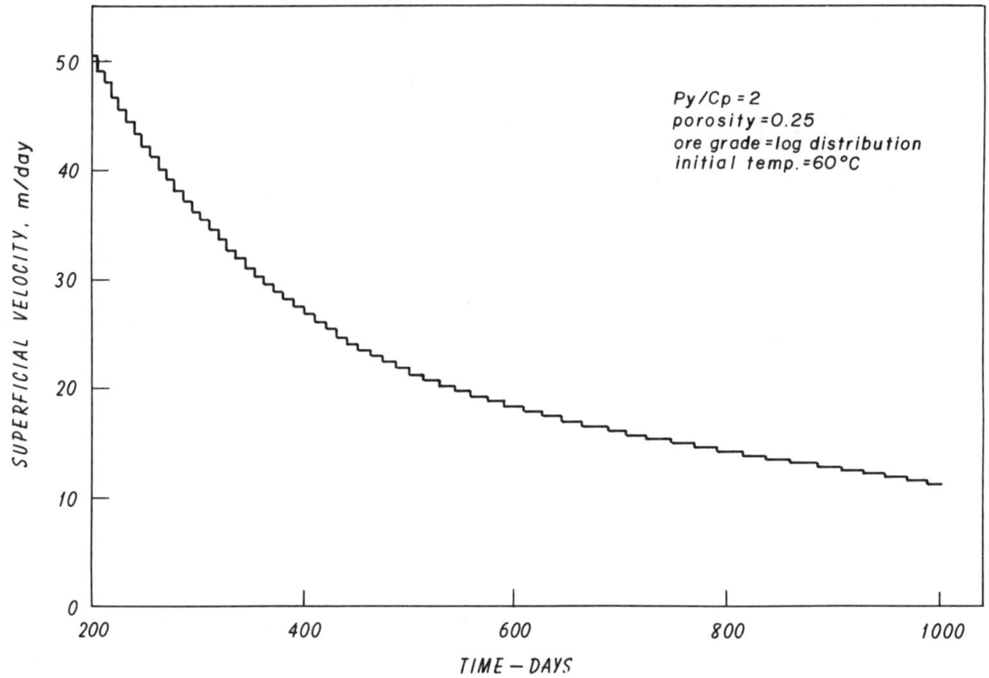

Figure 7. Solution velocity vs. time relationship (case 9a).

Results of cases 8a and 8b show such an effect. A decrease of the shape factor from 0.65 to 0.35 has resulted in an increase of the copper extracted by 10% in 1000 days of leaching.

In addition to considering all the ores to have the same ore grade, model calculations can also take into account the variations of the ore grade with the particle size distribution. Results of cases 9a and 9b were obtained for an ore grade following a log distribution. Results of case 10a were obtained for an ore grade of 5.43 wt% in $CuFeS_2$, which is the average ore grade of the log distribution used in cases 9a and 9b. These calculated results indicate that if the variation of the ore grade with the particle size distribution is not properly considered, the predicted fraction of copper extracted can be different.

Table 2 also lists the days required for the copper loading in the exit stream of the chimney to reach 10 kg/m^3 and the total amount of copper extracted in 1000 days of leaching for each case. Here, we have assumed that with the presence of other cations, copper loading in the leach solution can be built up to a level of 10 kg/m^3 without causing unfavorable side effects. Indeed, this level of copper loading has been built up in an in-situ leaching experiment performed by Braun et al.[5] Typical results for copper loading in the exit stream of the chimney as a function of time and flow rate are shown in Figure 8. Calculated results for each case show that copper concentration in the leach solution can vary along the chimney during early days of leaching. However, at the time when the copper loading in the exit stream of the chimney has reached a level of 10 kg/m^3, the concentration of copper in the leach solution along the chimney is uniform and has a value of 10 kg/m^3.

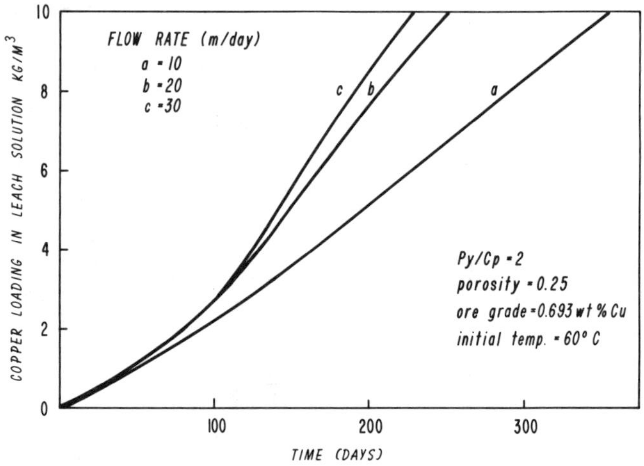

Figure 8. Typical copper concentration in the exit stream as a function of time and flow rate.

The calculated total amount of oxygen required from the oxygen plant in the first 1000 days of leaching as shown in Table 2 for each case is based on the condition that no oxygen is lost during leaching.

Finally, the maximum pressure drops for the leach solution flowing through the chimney for all cases are listed in the last column of table 2. In the case of constant flow mode, the maximum pressure drop usually occurs within the first few days, when the solution has the lowest temperature and highest viscosity. In the case of variable flow mode, it occurs at around the time when the temperature of the system just reaches 90°C. At this time, the flow rate of leach solution is highest. As shown, the maximum pressure drops are rather small, indicating that the leakage of the leaching solution might not present a serious problem.

CONCLUSION

The model proposed is capable of predicting the rate of leaching of a rubblized copper deposit incorporating basic kinetic data, fluid flow, pore diffusion and heat and mass transfer. The model provides a basis to check laboratory results and to extend these results to field application.

ACKNOWLEDGEMENTS

The authors wish to acknowledge many helpful technical discussions with Dr. K.J. Richards, Dr. David Davies, Mr. W. Henderson, Dr. J.B. Hiskey of Kennecott Minerals Company, Dr. R.L. Braun, Lawrence Livermore National Laboratory and Dr. H.-J. Sohn of the University of Utah.

This work has been supported by the U.S. Department of the Interior under Grant Nos. G 5195047 and G 5105088.

NOMENCLATURE

a_p = external surface area of ore particles per unit volume of chimney

C = concentration

C_i = concentration of species i in the bulk liquid phase

C_s = solubility of oxygen in water

c^{ℓ} = heat capacity of liquid phase

c^s = heat capacity of solid phase

D_a = axial dispersion coefficient

D_{ai} = axial dispersion coefficient for species i

D_e = effective diffusivity

D_o = molecular diffusivity

d_p = weighted average particle diameter

G = ore grade

H = height of chimney

k = rate constant

k_c = effective external mass transfer coefficient

k_d = apparent conductance due to axial dispersion

k_f = fluid-phase mass transfer coefficient

k_h = Henry's constant

M = total number of position increments

M_w = molecular weight of chalcopyrite

N = number of size fraction of the ore particles

N_{Re} = Reynolds number

P = pressure

Pe_d = Peclet number for axial dispersion

$\dot{Q}$ = rate of heat generation per unit volume of chimney

q_j = number of ore particles in the jth size fraction per unit volume of chimney

$\dot{R}_i$ = rate of production per unit volume of chimney for species i

R_j = initial spherical radius for the jth size fraction

r_j = radius of unreacted core for the jth size fraction

r_p = average spherical radius of copper sulfide particle

Sc = Schmidt number

T = Temperature

t = time

V_z = superficial velocity

z = vertical position measured from the bottom of chimney

Greek Letters

δ = thickness of leaching zone

ε = porosity of rubble chimney

μ = viscosity

ρ_{ℓ} = density of liquid phase

ρ_p = density of copper sulfide

ρ_r = bulk rock density

ρ_s = density of solid phase

σ = stoichiometry number

τ = residence time

ϕ_j = geometry factor for the jth size fraction

Superscripts

ℓ = liquid phase

s = solid phase

Subscripts

b = bottom of chimney

i = species i

j = jth size fraction of ore particles

k = position increment

ℓ = liquid phase

n = time increment

s = solid phase

REFERENCES

1. U.S. Energy Research and Development Administration, Kennecott Copper Corporation and Lawrence Livermore Laboratory, "In Situ Leaching of a Nuclear Rubblized Ore Body," Vol. 1 and Vol. 2, 1975.

2. A.E. Lewis and R.L. Braun, "Nuclear Chemical Mining of Primary Copper Sulfides," Trans. SME-AIME, Vol. 254, pp. 217-224, 1973.

3. D.L. Leah and R.L. Braun, "Leaching of Primary Sulfide Ores in Sulfuric Acid Solutions at Elevated Temperatures and Pressures," Trans. Soc. Ming. Eng., AIME, Vol. 260, pp. 41-48, March 1976.

4. F.F. Aplan, W.A. McKinney, and A.D. Pernichele, Solution Mining Symposium 1974, Proc. of a Symposium, 103rd AIME Annual Meeting, Dallas, Texas, February 25-27, 1974.

5. R.L. Braun, A.E. Lewis and M.E. Wadsworth, "In-Place Leaching of Primary Sulfide Ores: Laboratory Leaching Data and Kinetics Model," Metallurgical Transactions, Vol. 5, pp. 1717-1729, August 1974.

6. R.W. Bartlett, "A Combined Pore Diffusion and Chalcopyrite Dissolution Kinetics Model for In-Situ Leaching of a Fragmented Copper Porphyry," in Proc. Intl. Syp. Hydrometallurgy, AIME, pp. 331-372, 1973.

7. R.L. Braun and M.G. Mallon, "Combined Leach-Circulation Calculation for Predicting In-Situ Copper Leaching of Primary Sulfide Ore," Trans. Soc. Ming. Eng., AIME Vol. 258, No. 2, pp. 103-110, June 1975.

8. B.W. Madsen and M.E. Wadsworth, "The Application of a Mixed Kinetics Model to an Ore Containing an Assemblage of Different Copper Minerals," Conf-770440, pp. 619-629, April, 1977.

9. L.M. Zoss, S.N. Suciu and W.L. Sibbitt, "The Solubility of Oxygen in Water," Trans. ASME, Vol. 76, pp. 69-71, 1954.

10. T.K. Perkins and O.C. Johnston, "A Review of Diffusion and Dispersion in Porous Media," Soc. Pet. Eng. Jour. pp. 70-84, March, 1963.

11. R.H. Perry and C.H. Chilton, Perry's Chemical Engineers' Handbook, 5th Ed., McGraw-Hill, New York, N.Y. pp. 16-20, -21, 1973.

12. P.N. Dwivedl and S.N. Upadhyay, "Particle-Fluid Mass Transfer in Fixed and Fluidized Beds," Ind Eng. Chem., Process Des. Dev. Vol. 16, No. 2, pp. 157-165, 1977.

13. "International Critical Tables," Vol. 5, p. 176, McGraw-Hill, 1928.

14. R.B. Bird, W.E. Stewart and E.N. Lightfoot, "Transport Phenomena," John Wiley and Sons, Inc., New York, N.Y., 1960.

15. B. Carnahan, H.A. Luther and J.O. Wilkes, "Applied Numerical Methods," John Wiley and Sons, Inc., New York, N.Y., 1969.

Chapter 18

HEAP LEACHING PRACTICES AT ORTIZ GOLD MINE, SANTA FE COUNTY, NEW MEXICO

Robin J. Hickson
Gold Fields Operating Co. — Ortiz
Gold Fields Mining Corporation
Cerrillos, New Mexico 87010

ABSTRACT

The Ortiz Mine is operated by Gold Fields Mining Corporation, a subsidiary of Consolidated Gold Fields Ltd., one of the world's major gold producers. The Ortiz Mine is Gold Fields only heap leaching facility as well as their first operating mine in the U. S. A.

The hydrometallurgical processes utilized for recovery of gold and silver at Ortiz are discussed along with the several practical changes in operation adopted since start-up. Included are heap leaching by cyanidation, concentration of values on activated carbon, desorption, and the final electrowinning steps prior to shipment.

Construction costs and pertinent technical data are shown.

Location

The Ortiz Mine, located in the center of an original Spanish Mine Grant, is one of the earliest gold producing areas in the United States, having produced gold values exceeding $2 million since early in the nineteenth century.

The Mine Grant's location is shown in Figure 1. It is in north central New Mexico, midway between the two cities of Santa Fe and Albuquerque, and about 40 kilometers (25 miles) east of the Rio Grande.

The Ortiz Grant totally encompasses the Ortiz Mountains, an igneous intrusive mass rising to 2700 meters (8850 feet) above sea level and towering some 900 meters over the surrounding sedimentary plains.

Current operations of Gold Fields Mining Corporation (Gold Fields) are sited on the northeastern slopes of the Ortiz Mountains at the 2125 meter (6970 foot) elevation mark.

Gold Fields has a 99 year mineral lease covering the eastern two-thirds of the Mine Grant, approximately 145.7 km^2 (36,000 acres), and owns 17.4 km^2 (4,300 acres) of the surface lands immediately surrounding the mine site.

History

The first gold nugget within the Ortiz Mine Grant was found in the summer of 1828 by overland freighters transporting goods from Mississippi to the west coast. Their discovery triggered the western United States' first gold rush and from that date to the present mining operations have taken place in the area on a more or less continuous basis.

The Spanish government bestowed a Mine Grant of 245.5 km^2 (60,659 acres) surrounding the site of the discovery to Jose Francisco Ortiz in 1833.

During the first few years of activity mining operations were all on surface placers but, as those deposits quickly diminished, underground extraction was initiated by miners following either quartz veins in solid rock or nuggets along the bedrock underlying the placer gravels. Gold separation was by placer techniques; either dry, with fan forced separation of the lighter gravels, or wet, using melted snow. In all probability, the housewives selling water for 20 cents a bottle from the few hand dug wells in the area made more money than the majority of miners!

The first technically trained mining engineer in the West, a Spaniard named Don Cano, continued the initial boom sinking the first underground shaft in 1835.

Through the mid 1800's the area saw little activity as richer strikes further west in California and Nevada attracted the prospectors.

At the turn of the century, escalating metal prices not only encouraged the development of the district's deepest underground mine, serviced by a 152m (500 ft) shaft, but it brought America's foremost scientific mind, Thomas A. Edison, to the area. In 1897 Mr. Edison erected a large plant at Dolores (one kilometer east of the present operation) in an attempt to extract gold via static electricity. Mr. Edison gave the following statement to the "St. Louis Republic" on September 10, 1899: "This process for extracting gold from sand is certainly the biggest thing I ever invented. Near Santa Fe, New Mexico, is a region of 10 square miles (25.9 km^2) containing gold worth $800 million." Alas, Mr. Edison's mining ability did not match his other scientific talents and his patented process for the separation of gold from gravel without the use of water was a dismal failure. He finally closed his plant down in 1903, just four years after the announcement.

Gold mining activity decreased as the present century progressed. The underground operations closed down in the late 1920's, due predominantly to declining grades with depth, and the placer mines closed at the end of the Depression as people found easier ways to make a living than mining.

Although at least 51 separate investigations of the area were conducted from the end of second World War until the early 1970's none found a successful method of mining the area's remaining low grade deposits until Gold Fields brought their modern gold extraction technology to the district in 1973.

Gold Fields became aware of the district in May 1972, took a long term lease in May 1973 and started exploration on June 4, 1973. Exploration was completed by late 1976. Feasibility studies were then conducted both in-house and by outside consultants with the decision to open an open-pit mine being made by the Board of Directors in March 1978.

Preliminary plant designs were formulated in July 1978, site work commenced in October of that year, construction started in February 1979 and the operation came on stream on February 1, 1980. Full production of 2722 tonnes (3,000 st) per day was achieved in early 1981.

A total of less than $15,000,000 has been spent to bring the new Ortiz Mine into operation; see Table 1. Capital Cost Summary. Engineering design and construction management were performed by Bovay Engineers, Inc., of Albuquerque, New Mexico.

Geology

The Ortiz Mine is located near the center of a 65 km (40 mile) long, 10 km (6 mile) wide, north-south trending porphyry belt, locally known as the San Pedro-Cerrillos Porphyry Belt. The belt stretches from South Mountain, about 40 km (25 miles) east of the city of Albuquerque, then continues north, paralleling the Rio Grande through the San Pedro Mountains and Ortiz Mountains, ending in the Cerrillos Hills about 32 km (20 miles) south of the capital city of Santa Fe; see Figure 1. Location Map of Ortiz Mine Grant.

All of the aforementioned intrusive upland areas have been scenes of mining operations in the past and both the Ortiz and San Pedro Mountains support gold operations today.

Mesa Verde sedimentary sandstones and shales of the Cretaceous Age form the principal rocks of this high desert region though the Paleozoic rocks of the Galisteo formations frequently lie unconformably on the Mesa Verde horizons in the immediate Ortiz area.

In subsequent Eocene and Oligocene time these original sedimentary rocks were intruded with great masses of volcanic rocks, predominantly monzonitic stocks, which pushed the sediments up into domes but did not, initially, break through to the surface to form lava flows. This volcanism was the first intrusive phase within the Ortiz Mountains, forming what is known as the main Ortiz Basin Breccia Stock and resulting, at the time, in the local fragmentation and collapse of overlying brittle Mesa Verde formations. This action increased the porosity and permeability of the sandstones and thus the sedimentary horizons were prepared as a host rock for subsequent mineralization.

A second intrusive phase quickly followed the initial volcanic activity and emplaced latite sills as well as numerous dikes along the faults and fractures that had been earlier formed in the region by the PreCambrian Tijeras Fault System. The Tijeras Fault System cuts Northeast - Southwest through the Ortiz Mountains, actually forming the Southwestern edge of the Ortiz Basin Breccia Stock.

One of these second stage intrusive dikes, while upwelling toward the surface, came into contact with the intensely brecciated Mesa Verde sandstone formations above. The fluid magma quickly boiled to the surface in this low pressure zone and further jumbled the fragments of the sandstones, now predominantly metamorphosed to quartzites. This semi-explosive event allowed the gases to dissipate, leaving the solid mineral fractions to be deposited in the fractures and pore spaces of the newly formed quartzite breccias. The present Ortiz Mine ore body has such an origin, it being an intensely brecciated mineralized quartzite containing disseminated particles of native gold.

Throughout the San Pedro - Cerrillos Porphyry Belt mineralization is associated with the second intrusive phase. In the Ortiz Mountains mineralization appears to be restricted to the edges of the main Ortiz Basin Breccia Stock, with the southern edge yielding the most favorable returns to date (site of original underground Ortiz Mine, Benton Mine as well as the current open pit Ortiz Mine).

Unfortunately much of the original mineralization at the present mine site has been eroded away by the local, seasonally intermittent, streams of the Las Norias and Cunningham Gulches. These two streams have randomly deposited most of the gold minerals in the pediment gravels of the Espinosa formations and the Galisteo River Basin to the east leaving but a small portion of the original mineral deposit exposed as an outcrop on the formed ridge, known as Cunningham Hill, between the two stream beds.

Mineralization

The Ortiz ore body has an upper oxidized zone which varies in thickness from 6 to 43 m (20 to 140 ft) and averages 27 m (90 ft). In this zone pyrite, the predominant (and often the only mineral with the exception of gold), has been completely broken down to limonite.

The semi-oxidized zone, in which pyrite and limonite are present together, is below the oxidized zone, averaging 18 m (60 ft) thickness in the center of the ore body but thinning around the margins.

Underneath this is the sulphide, or unoxidized, zone containing pyrite but no limonite.

The principal minerals within the ore body are pyrite, hematite, magnetite, tungsten, gold and silver. Trace amounts of pyrrhotite, chalcopyrite and galena are occasionally seen as microscopic filling of microfractures within the pyrite.

Pyrite is by far the dominant mineral and generally occurs as a molding on earlier formed iron oxides of hematite and/or magnetite. (The hematites and magnetites are intimately intergrown, with the hematite always occurring as martitic replacement of the magnetite.) The bulk of the pyrite is deposited in fractures and voids of the host rocks, and occurs either as a complete healing of the fractures or as surface encrustations in the voids.

The gold occurs essentially as free particles of native gold throughout the deposit. Gold content appears to be related directly to the amount of pyrite and iron oxide minerals present which in turn is dependent upon the abundance and volume of voids in the host rock. In fresh unoxidized samples gold occurs as granular inclusions, either in the microfractures of the pyrite, in introduced carbonate gangue or along the grain boundaries of pyrite and hematite. Granular gold is also present in oxidized material but most of the gold there occurs either as thin films associated with limonite boxworks or, rarely, as thin coatings on tarnished pyrite.

Gold particle size ranges from 2 to 600 microns with 85% of the gold occurring in the 40 to 250 micron size range. The gold grain size is slightly larger in the oxidized zone than elsewhere.

Silver is found in close association with the gold but, unfortunately, only in trace or negligible amounts. A gold to silver ratio of 6:1 seems to be fairly typical.

Exploration

The Ortiz gold deposit has been extensively explored and sampled by Gold Fields as well as by numerous prior operators and owners. The exploration sampling methods are summarized in Table 2. Gold Fields utilized diamond drilling, underground drifting, cross-cutting and slabbing of pre-existing tunnels and, to a lesser extent, surface trenching. Figure 2. illustrates these various sampling methods in diagrammatic form. Gold Fields' drilling amounted to only one half of the total footage drilled on the deposit and, as such, reappraisal of the old data was an important step in the evaluation of the ore body.

Excellent correlation between our own data and the earlier ventures was found. Variograms were prepared for each sampling method and for each prior exploration program.

Reserves were estimated by a plan polygonal method, a crossectional method, a rolling mean method, and by a block kriging method. Results were computed for a range of cut-offs and the methods are generally in good agreement (± 10% for both tonnage and grade).

Block kriging is the method currently utilized by the Ortiz Engineering staff. Basic block dimensions are 15.24 m by 15.24 m (50 ft x 50 ft) horizontally by 7.62 m (25 ft) vertically, equivalent to approximately 4536 t (5000 st) per block. Block sizes were chosen to correspond to proposed mine benches and thus are readily used for detailed pit planning. Although the relative standard deviation (R.S.D.) for ore grade in any single block ranges from 17% upwards, the R.S.D. for the entire deposit grade is only 2% and for the deposit tonnage is less than 3%.

Exploration, prior to commencement of mining operations yielded an ore reserve of 6,841,000 tonnes at 1.82 g/t (0.053 oz. Au/st) using a 0.68 g/t (0.02 oz. Au/st) cut-off, and 1,580,000 tonnes of low grade ("Protore") material averaging 0.51 g/t (.015 oz. Au/st). Since start-up a little over 320,000 t of material has been added to reserves at similar grades. On-going exploration activity continues to expand this number even further.

Mining

The Ortiz Mine ore body outcrops on the south side and crest of Cunningham Hill, from an elevation of 2200 m (7220 ft) at the ridge line down to an elevation of 2135 m (7000 ft) in the floor of Cunningham Gulch. The ore body then disappears underground as it plunges southwest to its ultimate depth of about 2012 m (6600 ft) in elevation, 123 m below surface.

The present mining operation will eventually develop into an open pit 335 m (1100 ft) in length along Cunningham Hill and 275 m (900 ft) wide. The pit is programmed to stop at the ground water table, some 40 m (130 ft) below the surface at the 2195 m (6875 ft) elevation. Additional permits will be required to mine the ore below the 2195 m elevation. The economic evaluations for extracting this lower ore are currently underway and a decision on possibly increasing the mine life should be made by early 1982. The current operation has a little more than a 7 year life utilizing a 0.68 g/t grade cut-off, ore production rate of 680,000 tonnes (750,000 st) per year and stripping ratio waste: ore of 1.24:1.

If the rest of the known, lower, ore was to be extracted it would extend the mine life from 7 years to either 10.1 years for a 0.68 g/t cut-off or 12.6 years for a 0.34 g/t cut-off assuming the same 680,000 t per year production rate as at present. (All mineral grading between 0.68 and 0.34 g/t (0.01 and 0.02 oz. Au/st)) is being stockpiled in protore stockpiles, separate to waste and ore, and the present intention is to process this marginal material at the end of the current mine life or at whatever other time period is deemed appropriate by the economics of the day). If the ore below the water table is eventually extracted then a little over 18.4 million tonnes of material will be moved, yielding a stripping ratio of 1.67:1 for the 0.68 g/t cut-off or 1.15:1 for a 0.34 g/t grade cut-off. Figure 3 shows a section through the center of the Ortiz Mine.

All mining operations are being carried out by a local Albuquerque, New Mexico contractor, Mountain States Constructors, Inc., under Gold Fields' direction. The Contractor is paid a fixed amount for every ton of ore, or waste, he drills, blasts, hauls and dumps. This payment covers equipment maintenance, road maintenance, explosives costs and supervision as well as profit to the Contractor. The Contractor's equipment fleet is listed in Table 3.

Gold Fields provides engineering, surveying and ore control services. Gold Fields also controls the scheduling of ore blocks, benches, and ore/waste ratios on a week by week basis.

The mine is being excavated in 7.62 m (25 ft) benches which are designated by their elevation in feet above sea level. Mine development is now at the 7050 bench having progressed 45 m (150 ft) down from the crest of Cunningham Hill. Broken rock is hauled by four 45 t (50 st) trucks on roads with average grades of 1%, maximum grades of 12%, and minimum widths of 15 m (50 ft). The roads are built with run-of-mine waste rock being initially developed by crawler tractors and then final-graded with a Caterpillar 12E grader.

Waste is currently drilled by a single Gardner-Denver GD-25C track mounted rotary drill with 11,340 kg (25,000 lb) pull down. 165 mm (6½ in) holes are drilled on a 5.5 m (18 ft) square pattern 8.84 m (29 ft) deep at a single pass yielding a 1.22 m(4 ft) sub-grade. Ore is drilled with the same rig but on 3 m by 3 m (10 ft x 10 ft) centers to provide better fragmentation. A single 40% strength booster is placed in the bottom of each hole with a millisecond delay electric cap and the hole is then filled to within 3 m of its collar with ANFO. Approximately 40-50 holes are blasted at one time with blasting taking place generally three times a week.

The Contractor's crews work a 10 hour shift with half an hour out for lunch, 5 days per week and adequately keep pace with the maximum daily mine requirements of 5265 t (5800 st) of waste and 3175 t (3,500 st) of ore. One waste drill hole currently yields about 590 tonnes of rock and an ore hole yields 175 tonnes. Thus about 27 holes per day (9 waste and 18 ore) is the mine drill's quota. Depending upon ground conditions within the quartzite breccia, drilling of a single 8.84 meter hole takes from 12 to 60 minutes. Shortfalls are made up by a single LeRoi hydraulic track drill hauling a 0.087 m^3/s (185 CFM) air compressor and utilizing 3.65 m (12 ft) extension steels with 115 mm (4½ in) screw-on cruciform tungsten carbide bits. This drill trims up benches, does all the hillside 'edge' drilling as well as secondary blasting. Both drills are manned by single operators.

Blasting is done by the supervisors and air-track operator.

Ore is not visually discernible. Thus, all drill holes have their cuttings sampled and analysed by the Gold Fields laboratory personnel to ensure that mine grades are as predicted and that the ore-waste boundary is adhered to. The ore, low grade and waste zones are color-flagged post blasting so that the equipment operators can be directed to the desired material.

Two front end loaders with 5.7 m^3 (7½ yd^3) buckets load the 45 tonne trucks. Most of the broken material is pushed by two crawler tractors to the loaders, this being the contractor's preference.

The waste rock is disposed of in terraced "dumps" built predominantly to the northwest of the pit. The dumps will be graded, contoured and revegetated to hopefully blend with the surroundings when mining is complete.

Protore is currently separated in two dumps to the west of the pit.

The ore rock is transported by the haul trucks to the Primary Crusher. Preferably the trucks dump directly into the 100 tonne crusher feed hopper, but, if the crusher is not ready to receive ore, then the trucks dump at an adjacent ore stockpile, leaving the rock for later rehandling by a front end loader.

Ore Processing

General

Heap leach gold recovery techniques are employed at Ortiz. Caustic cyanide solutions are sprayed onto crushed rock piles, dissolving the metallic gold from within the rocks and carrying it out as a 'pregnant' gold containing solution. The solution is then passed over carbon particles which the gold preferentially precipitates upon. The gold loaded carbon is removed and placed into a desorption circuit, where other caustic solutions redissolve the gold, thus effectively stripping the precious metal values from the carbon. These strip solutions subsequently flow through electrowinning cells where the metal values are precipitated out onto steel wool cathodes for further concentration by more electrolysis and then final sale to an outside refinery.

Crushing and Ore Stackout

The crushed ore flow sheet is shown in Figure 4. Three stage crushing is employed at Ortiz. An average of 2722 tonnes (3000 st) of ore rock from the mine is dumped daily into an Allis-Chalmers 1016 mm by 1067 mm (40 in x 42 in) primary jaw crusher where reduction to about 140 mm (5½ in) takes place. The crushed ore is transported and elevated via inclined conveyor to form a 29,000 tonne fine ore storage pile. This stockpile permits independent operation of the primary and secondary crushers. Ore is drawn as necessary out of the fine ore storage pile through two Eriez vibratory feeders at the bottom of the pile. Conveyors then transport this ore through two cone crushers where final reduction to about 9.5 mm (3/8 in) occurs.

Johnson-March dust suppression systems utilizing chemical wetting agents are used to effectively reduce fugitive dust in both crushing plants as well as the reclaim tunnel containing the vibratory feeders.

The cone crushing plant is an open circuit. It currently yields 60% passing 9.5 mm (3/8 in), see Table 4, Screen Analysis of Feed to Pad. Material from the fine ore stockpile is fed onto a 1.83 m by 3.66 m (6 ft x 12 ft) double-deck Allis-Chalmers vibrating screen utilizing 12.7 mm (½ in) square screen cloth (9.5 mm effective opening) on the bottom deck. Undersize rock bypasses the cone crusher and feeds straight to the leach pad. Oversize from the double-deck screen feeds through the secondary crusher, a 1.7 m (5½ ft) Symons Standard Cone Crusher equipped with coarse liners set at 25.4 mm (1 in) opening, and then over a 1.83 m by 3.66 m single deck Allis-Chalmers vibrating screen also utilizing a 12.7 mm square screen cloth and into the tertiary crusher. Undersize from the second screen bypasses the tertiary crusher, a 1.7 m Symons Short Head Cone Crusher with fine liners set at 9.5 mm opening, and passes straight to the leach pad.

The combined crushing plant product is then conveyed across an electronic weightometer to a Thomas Systems Incorporated custom built travelling gantry and stacked 4.5 m (15 ft) high on a reusable impervious leach slab. Current slab capacity is about 136,000 tonnes (150,000 st). The gantry is a unique, fully automated, 61 m (200 ft) span moveable bridge conveyor with tripper and it lays down a flat topped, rectangular, uncompacted rock pile to any desired height from 0 to 4.5 meters. The gantry distributes the fine ore material uniformly through the heap, thus avoiding the segregation (fines core) problem of radial stacker formed heaps, the layering of mobile equipment formed heaps and the compaction problems of dozed flat-top heaps. Besides greatly enhancing recovery rates due to its pile formation characteristics, the gantry has the further advantage of operating at less than a third of comparable mobile equipment costs. Total pad loading operating and maintenance costs are below 15 cents/st.

At initial plant start-up a 32 mm (1¼ in) open circuit crushed product was conveyed to the pad and coarse liners were present in both cone crushers. However, with over 42% of the primary product found to be finer than 32 mm it was quite obvious the cone crushing plant was being under utilized. Hence bowl liners were replaced, finer screen cloths were installed and the smaller 9.5 mm crusher product is now produced.

Leaching

The Leach Slab is constructed of high grade asphalt, 178 mm (7 in) thick. The bottom 51 mm (2 in) is an impervious hydraulic asphalt while sandwiched in the center of the upper asphalt is a further impervious protective rubber membrane. The slab is subdivided by 230 mm (9 in) surface curbs into eight 61 m (200 ft) long by 49 m (160 ft) wide sections or "piles". A pad, comprised of eight 17,000 tonne (18,750 st) sections; is loaded on average every 10 weeks. The tops of the piles are contiguous but the curbs separate their bases.

Chemical solutions bearing a sodium cyanide (NaCN) concentration averaging 0.07% (varies from 0.05 to 1.5 mg/L depending on sulphide content of ore) and pH adjusted to 10.2 (±1) via lime additions are sprayed over the top of the crushed rock piles at a rate of 0.0018 L/m^2 (0.005 gallons/ft^2) by Rainbird Model P-55 all plastic low angle, impulse sprinklers.

The impulse sprinklers have adjustable trajectory angles and are spaced at 7.6 m (25 ft) centers, this distance adequately overlapping their 12.8 m (42 ft) radius spray circles. Eight separate 102 mm (4 in) pipe headers, fed by the main incoming 254 mm (10 in) barren solution pipe line, feed the individual pad sections. The flow through each section's individual 102 mm pipe header is pressure regulated at a minimum of 13 atm (20 psi) and a maximum of 24 atm (35 psi) depending on wind conditions. Each header has an individual flow totalizer, which allows daily calculation of gold recovery for each section, as well as a continuous flow read-out meter in gallons per minute. This latter instrument allows the operator to equalize spray distribution to all sections. Each 102 mm header then feeds into eight separate 51 mm (2 in) hose lines spaced at 7.6 m intervals containing six sprinklers also spaced at 7.6 m intervals. The sprinklers actually attach to the 51 mm feed hoses via 19 mm (3/4 in) diameter tail hoses to allow maximum local adjustment, as well as ensuring complete spray coverage with zero wind loss. Although Bagdad wigglers were initially utilized at Ortiz their erratic performance and high labor maintenance cost precipitated the change over to impulse sprinklers, which are now much preferred. However, the Rainbird sprinklers have not been as successful spraying the pile sides where the lime in solution frequently causes the arc segment limitation trigger of the sprinkler to stick (This problem does not occur on the pile top where the sprinklers rotate 360°). Substitution of Toro Model AG-1 agricultural sprinklers for the impulse sprinklers on the pile sides in recent days has been found to eliminate this sticking problem completely.

No lime plugging problems have occurred with any of the sprinkler or wiggler types utilized. The addition of 12.7 mm (½ in) crushed pebble lime (CaO) onto the stacker ore feed stream rather than milk of lime through a slaking plant into the barren solution is felt to be the reason for the success in this area. The pebble lime is fed onto the ore stream leaving the cone crushing plant at an average rate of 3.0 kg/t (6 lbs/st) of ore; feed ratio being maintained via electronic connection to the belt weightometer. A slaking plant is available as back-up to the pebble lime plant if necessary. Lime plays a dual role by maintaining a high pH and inhibiting other metals from attack by the cyanide.

The solution percolates down through the heaps dissolving the gold it encounters and ultimately contacting the impermeable barrier of the leach slab. The now 'pregnant' solution, containing gold, migrates with the 4% slope of the pad to a collection ditch on the northern edge of the pad.

Gold dissolution in the heaps is denoted by the following chemical equation:

$$4Au + 8Na\,CN + O_2 + 2H_2O = 4NaAu\,(CN)_2 + 4NaOH$$

The pregnant solutions flow via gravity out of the collection ditch, actually an integral part of the leach slab, through trash screens, settling sumps and into concrete storage sumps. Pipelines used on route are 305 mm (12 in) Schedule 20 steel enclosed within 1067 mm(42 in) reinforced concrete tunnels.

The ore piles are subject to the leaching process for a minimum period of 50 twenty-four hour days. After this period application of cyanide solutions cease and pure water is applied intermittently over a further period of three days to effectively remove any residual cyanides from the leached residue.

The pile is then allowed to drain for four additional days, auger samples are taken of the pile at strategic locations to ensure thorough washing took place and, finally, the material is loaded onto trucks and removed to the residue area for permanent storage.

It should be noted, that, although the winters in New Mexico are quite mild and temperatures generally are above 0°C at least for a portion of every day of the year the high altitude of Ortiz means night time temperatures from November to March are predominantly below freezing. Hence sprays do not operate for 24 hours per day during winter, generally being turned off whenever temperatures drop below -3°C (26°F) to prevent ice sheets covering the piles. To maintain the minimum 1,200 hours desired spraying time (50 days x 24 hours) the rock is left on the pads for more than 50 calendar days in the winter. Consequently, tonnage to the pads is less than 2722 t/day (3000 st/day) in the winter, dropping to as low as 1900 t/day in February. Throughputs of 3175/day are thus scheduled in the summer to makeup these winter short-falls. Evaluation continues of the cold weather distribution system. Recovery results were better than anticipated in Ortiz's first winter. Perhaps continual 24 hour spraying is not really necessary?

The residue removal is conducted by the contract miner, generally taking 3 days to remove a single 17,000 tonne pile. The contract miner is paid a fixed additional amount per ton for this service. Normally two of the four mine trucks are utilized along with the rehandle front end loader from the primary crusher. However, the toothed bucket utilized in rehandle has to be exchanged for a flat spade-front bucket for all removal work. This ensures integrity of the slab.

The leaching and washing solutions are then pumped from the 132.5 kL (35,000 gallon) pregnant storage sumps to the Gold Recovery Plant where the gold is removed by adsorption onto activated carbon columns which the solutions up-filter through. The solution, now barren of gold, returns to the barren storage sump where additional makeup chemicals are added as necessary. Current NaCN reagent consumption 0.5 kg/t (1 lb/st) of ore. The barren solutions are then pumped, through 254 mm steel pipes, back to the ore piles to again leach gold and continue the cycle once more.

Worthington deep well turbine pumps with high density polymer bearings are used to move solutions from the storage ponds to the Au. plant and back to the leaching piles, with flow meters measuring the rate of flow both ways.

A 5678 kL (1½ million gallon) C.P.E. lined holding pond sited at the side of the drain system can store all the solutions in circulation in the event of a 100 year frequency rain storm and/or a concurrent 24 hour power outage.

Sixteen monitoring wells, nine surrounding the leach slab, four surrounding the storage pond, one below the residue area and two deep wells downstream from the entire plant area, assure that no cyanide has escaped the processing system.

Leaching and Recovery Flow Sheets are shown in Figure 5.

Gold Recovery

Solutions from the pregnant storage sumps moving at 2840 L/min (750 g.p.m.), pass through five countercurrent tower adsorption tanks in series. The 2.44 m (8 ft) diameter x 1.83 m (6 ft) high tanks each contain 1,360 kg (3,000 lbs) of activated carbon granules (actually burnt coconut shells), 12 x 30 mesh in size. The carbon is maintained in an expanded bed state by the pregnant solutions upflowing through the tanks at a rate of 692 L/min per m^2. The gold adsorbes onto the activated carbon, starting at the highest #1 tank and diminishing in concentration as the solutions gravity drop through the five towers.

Activated carbon has the ability to adsorb the $NaAu(CN)_2$ ion onto its surface. With typical pregnant gold values being 1.0 g/t (0.03 oz/st) and barren solution gold values running between 0.14 and 0.34 g/t (0.004 to 0.010 oz/st) it means 75% of the gold is normally adsorbed by the carbon on a single 7 minute journey through the five towers. With the solutions coming back time and time again for further adsorption this performance is considered quite satisfactory.

Approximately every second day 1360 kg of gold-laden carbon averaging 8570 g/t (250 oz Au/Ton) with a range of 4800 to 13,000 g/t is removed from the #1 tank and a corresponding amount is advanced at each of the other adsorption towers. The gold-laden carbon is placed in one of two 2040 kg (4500 lb) insulated, desorption vessels 1.52 m (5 ft) diameter x 2.74 m (9 ft) high. Fresh carbon, or reactivated carbon which has undergone desorption and has subsequently been treated in a custom built Gold Fields propane-fired rotary kiln at 600°C controlled heat, is added back to the lowest #5 Carbon Adsorption Tower to effectively keep the carbon load in balence.

Carbon loses adsorption efficiency after many cycles of adsorption and desorption due to "poisoning" by organics inherent in the process. Reactivation brings the carbon back to an "as new" condition. Carbon losses are confined to small particles formed by attrition in the circuit. This undersized material (less than 7 kg per day) is carefully screened out and sold to a smelter for its gold content.

Soon after operations start-up it was found that in spite of regeneration carbon loading was becoming progressively worse, occasionally as low as 2570 g/t (75 oz Au/st), and hence the gold content of the barren solutions started to rise alarmingly. A lime coating forming on the carbon was causing this problem. Heavy additions of anti-lime precipitant stopped this fouling but at considerable expense. Thus, a carbon acid wash plant has recently been commissioned and since this plant's start-up carbon loading has not dropped below 6850 g/t (200 oz Au/st). Removing 900 kg (2000 lbs) of carbon once per week and acid washing that material with a 208 L (55 gallon) drum of concentrated HNO_3 (suitably diluted of course) is enough to keep the entire carbon circulating load lime-free.

When enough gold-laden carbon has been accumulated in the desorption vessels then a caustic-cyanide 'stripping solution' containing 1% sodium hydroxide and 5% ethanol is filtered up through the carbon at approximately 87°C (near boiling at Ortiz's altitude) and low pressure head.

This hot strong caustic solution reverses the adsorption process, thus redissolving and stripping the gold, plus any silver values, from the loaded carbon. The ethanol alcohol merely acts as a carrier to enable the solution to penetrate the carbon particles. The alcohol reduces the strip process from a 72 hour event to a less than 24 hour process.

The strip solutions are routed through one of two parallel Gold Fields custom made 316 stainless steel electrolytic cells, (each with 18 cathodes and anodes), where the gold is electrically deposited from the solution at 350 to 400 A and 6 V onto steel wool cathodes.

$$2NaAu\ (CN)_2 + 2e = 2Au + 2NaCN$$

The gold is plated onto 0.9 kg (2 lb) cathodes of #3 stainless steel wool (340 Fine) encased in two simple plastic 'bread' trays. The cathode is merely a point of deposit for the gold and takes no part in the reaction. Steel wool is used because of its great surface area and its low cost. Common steel wool originally was utilized but stainless stands up to re-use better. Six wraps of steel wool are utilized within each cathode.

Cathodes load from 600 to 2000 g (20 to 65 oz troy) generally averaging a little less than 50% Au.

Both cells were originally utilized in parallel for the stripping process but it was found that using one cell with a higher voltage (Ortiz initially used 2.5 V) resulted in a higher gold content per cathode. The steel wool cathodes are 685 mm long by 585 mm wide by 76 mm thick, the holders being purchased from Westates of Los Angeles.

In recent days 0.1% NaCN has been added to the strip solution. This improves the quality of gold forming on the cathode by reducing the amount of foreign material plating out with the gold.

Barren strip solutions are piped back from the electrowinning cells to the carbon in the desorption vessels to remove more gold and then back to the cells in an unending cycle. Chemicals are added as necessary to the solutions to restore them to their original concentration.

Carbon gold loadings can be reduced to the 170 to 350 g/t (5-10 oz/st) range within the 24 hour stripping time by moving a minimum of 13 volumes of solution through the stripping tanks. As the carbon eventually comes back to be stripped again in a later cycle it is not felt to be economic to attempt to go much beyond this 95% strip efficiency.

Gold Fields does not melt the steel wool cathodes to form a dore bullion. Security concerns over this process plus unacceptable capital and operating costs have caused Ortiz to perform, instead, a second electroplating of the gold. The 18 steel wool cathodes are removed periodically from one of the stainless steel electrolytic cells and placed in a similar sized, polypropylene coated, mild steel replating tank custom made for Gold Fields by Americhem Engineering Service of Phoenix Arizona. This second tank contains 18 highly polished 316 stainless steel cathode plates, each held in a woven filter bag. The stainless steel wool cathodes from the first tank become the anodes in this second tank and gold is then replated on the polished steel plate cathodes as a high quality foil, generally over 97% precious metal content (83% + Au; 14% + Ag). The replating cell operates between 2.6 and 3.4 V at 400 A. The filter bags prevent any great amount of loss of foil to the bottom of the replating tank; a considerable problem in our early attempts to make this process work. The replating medium is simply a 5% NaCN solution utilizing softened water. (Calcium carbonate build-up on the gold foil was a large draw-back to early experiments at replating and the addition of a water softener eliminated this problem.)

Polypropylene was used for the replating cell due to dissatisfaction with the capital cost and maintenance repair costs of the initial, first stage stainless steel electrolytic cells. Performance of the new replating cell has been excellent and, whenever the first stage stainless steel electroplating cells are replaced, polypropylene material tanks will be substituted.

Typically, two 1360 kg carbon strips are cycled through a first stage electrolytic cell before the stainless steel cathodes are removed. If carbon fouling causes gold loadings to be low. i.e. less than 5000 g/t (145 oz Au/st), then this is increased to three strips.

Two platings are generally run through the replating tank before the polished steel plates containing the gold foil are removed and hand scraped. Replating takes anywhere from 24 hours to 60 hours dependent upon the gold loading of the first stage steel wool.

Thus, either four or six carbon strips end up as one replating product. Final replating is done at random intervals, mostly on day shift. Hand scraping of the gold foil is done under strict supervision and the product leaves Ortiz by armored courier at random times and days as soon as it is available for shipment. The product is shipped to a precious metals refinery where it is refined and sold on the international bullion market.

Table 5, Ortiz Operating Costs, shows the current costs at the Ortiz operation.

Recovery

Total gold recovery has obviously been of paramount concern from Ortiz's inception through its first one and a half years of operation. Heap leaching is not renowned for its excellent recovery characteristics! Since our first pad of ore was laid down in early 1980, total gold recovery at Ortiz has averaged 78%. The first pad achieved 56% recovery and progress has been steadily improving since that time. The best pad recovery to date has been slightly over 87%. Figure 6 shows the average recovery curve for the last 4 completed pads.

The extremely fine nature of the gold mineralization coupled with the mineral deposition being predominantly in the fractures and pore spaces around the breccia particles makes the Ortiz ore particularly amenable to heap leaching practices. Simple crushing preferentially cracks the rocks along their planes of weakness (the mineral-containing fractures and 'vugs' around the breccia fragments) and thus exposes virtually all the deposited gold to an outside surface. The relatively fine, mineralogically uncomplicated, particles of native gold are then fairly readily dissolved by application of the leach solutions.

Recoveries above the industry average were thus expected from the very start for Ortiz. Achievement of these high recoveries has been aided however by continual operational changes during Ortiz's first 18 months of operation. These changes include, amongst other items: (1) reduction of size of feed material to pad from a 32 mm open circuit product to a 9.5 mm open circuit product; (2) change-over from Bagdad wigglers to impulse sprinklers on pile top with associated more predictable spray patterns as well as elimination of wind loss; (3) improvement of pile edge spray coverage with adoption of Toro sprinklers on pile sides; (4) interlocking of pebble lime feeder with pad feed conveyor belt. This ensures adequate liming as well as eliminates over-liming (over-liming not only wastes money it reduces silver recovery); (5) improved communication from mine department as to when high sulphide ore is about to reach the pad. This allows cyanide make-up to be increased prior to, rather than post, high cyanide consumption; (6) curtailing spraying at -3°C rather than spraying as long as possible in cold winter nights. Ortiz used to spray down to -8°C but the ice sheet that eventually formed prevented later warm weather spray solutions from penetrating throughout the pile; (7) installation of a pre-wetting spray nozzle which wets the ore with leach solution as it travels along the gantry. This not only starts the leaching process prior to hoses being laid out but also does some minor agglomeration work. The cascading motion from the gantry discharge as ore is gently added to the heap further enhances this agglomeration of fines and coarse particles; (8) elimination of fines loss due to installation of Johnson-March dust suppression system; (9) retrofit of pressure regulators and continuous reading flow meters on each pad section to ensure correct spray coverage is fed to all parts of the pad regardless of elevation above pump or operator attendance; (10) raising of initial designed 3 m pile height to current 4.5 m. This has allowed over 50% more spraying time at the same throughput rate. No channelizing or loss of solution coverage has been noted to date from this pile height raising; and (11) retrofitting of curbs along the sides of the pad, allowing the pile to be constructed 2 meters wider yet still not allowing solutions to flow off the pad. This has allowed 3 more days spraying time for every pad section.

Further recovery enhancement techniques may include:

1) Raising the gantry height by a further 2.1 m (7 ft).

2) Reducing product size from the cone crusher to 6.3 mm (¼ in) open circuit.

3) Retrofitting a closed circuit cone crushing system.

4) Adding more pre-wet leach application nozzles from a spray bar on the gantry.

5) Commissioning of a Metallurgical Testing Lab with leach simulation equipment (columns etc). Tests are obviously cheaper in columns than in a heap and such tests are frequently more practical and reliable.

General Support Facilities

Laboratory facilities at Ortiz include an atomic absorption unit, wet lab and fire assaying equipment. These serve exploration programs, pit grade control, leaching, process and residue control programs as well as quantifying the gold -

silver ratio in the marketed product.

A 186 m^2 (2000 square foot) General Office is sited between the first two barriers of a triple barrier chain link security fence. Separate change houses are provided for female, male hourly and staff personnel.

A 2-bay high-bay maintenance shop with 9 tonne overhead crane has been built by Gold Fields and is currently utilized at no charge by the contract miner. A small warehouse facility is located between the changes houses and the maintenance building.

The environmental impact at Ortiz is being very carefully limited by minimizing disturbed areas, by careful dust control in plant and along roads, by landscaping and by proper planning for future reclamation of dump areas. Over $35,000 has already been spent on revegetating disturbed areas with 27 different varieties of native plants and these efforts, which are being conducted under the guidance of a local Santa Fe botanist, are expected to be greatly expanded in future years.

Crushing and leach pad loading are conducted 5 days per week, 2 shifts per day. Leaching and gold recovery operation takes place 3 shifts a day, 365 days per year.

Table 6 lists the current Ortiz Manpower Structure. Average base salary paid to Gold Fields hourly personnel is $8.65 per hour.

In addition to the Gold Fields' personnel there are 16 mine contractor employees permanently assigned to the Ortiz site. These include 2 supervisors, 2 drillers, 4 loader operators, 4 truck drivers, 2 general equipment employees (water truck, dozers, grader etc.) and 2 mechanics.

The Pinkerton agency provides armed security protection around the clock.

Over 77% of the persons employed at Ortiz are from the immediate Santa Fe County local area. Gold Fields is proud of the significant input their presence has had in reducing local unemployment from the 12% range that was prevalent in Santa Fe prior to Ortiz construction start-up. With current payrolls in excess of $1.4 million annually and local taxes paid exceeding $700,000 it is obvious that Gold Fields' impact in the area goes far beyond the goods and services purchased from the local community.

Ortiz is one of the newest and technically most up-to-date gold mines in the country. It has tripled New Mexico's gold production since coming on stream in 1980 and currently produces 4% of the entire U.S.A. gold output.

Acknowledgements

The author is grateful to the staff of the Ortiz Mine for their assistance in helping prepare this paper as well as to the Gold Fields Mining Corporation in granting permission for publication.

Table 1. Capital Cost Summary

	$-000's omitted
Pre Construction (1972 thru'3/1978)	
Exploration/Feasibility etc.	$ 1,920
Access Road	1,152
Mine Development Costs	943
Plant Construction Costs	
Site Preparation (excl. Leach Pad)	179
Buildings - Admin. Offices/Lab	205
- Shop/Warehouse/Change.	481
Primary Crushing Plant	790
Cone Crushing Plant	1,461
Leach Pad Stack-Out System	953
Leach Pads - Base Preparation	562
- Asphalt	535
Leach Solution Piping System	756
Gold Recovery Plant	981
Lime Plants (2)	119
Water Distrib. (incl. Wells and Pumps)	687
Utilities Distrib. (Power, Fuel, Tel.)	119
Monitor Wells	33
Fencing	85
Engineering Design	550
Construction Management	469
Sub-Total:	$ 8,965
Mobile Equipment	50
Other Expenses: Pre-Start-Up	
Land Acquisitions & Lease Payments	962
Admin. Costs (Permits, Planning etc.)	683
Plant "Shake-down" & Training	255
Sub-Total:	$ 1,900
TOTAL:	$14,930

Note: Above costs have been adjusted to include all retrofit modifications necessitated since start-up. However Capital additions since start-up are NOT included.

Table 2. Drilling and Tunnelling For Ortiz Deposit

Operator	Sample Type	No. of Holes	Meters
Previous #1	Vert. Diamond Drill	14	979
Previous #1	Churn Drill	24	1,677
Previous #1	Wagon Drill	12	881
Previous #2	Vert. Diamond Drill	2	428
G.F.M.C.	Vert. & Incl. D. Drill	22	4,032
G.F.M.C.	Tunnel Rounds		410
G.F.M.C.	Slabbed Tunnels		212
Total Drill Footage		74	7,997
Total Underground Dev.			622
TOTAL SAMPLED FOOTAGE:			8,619

Table 3. Contractor Equipment List

1	Gardner Denver	GD-25C	Rotary Drill
1	LeRoi	HDR12E	Hydraulic Crawler Drill
1	Ingersoll-Rand	185-CFM	Air Compressor
3	Caterpillar	988-B	Loaders
1	Caterpillar	988	Loader (Rehandle Unit)
4	Caterpillar	773	45t Trucks
1	Caterpillar	D-8	Crawler Tracter
1	Int. Harvester	TD-25	Crawler Tracter
1	Caterpillar	12-E	Grader
1	Custom Made		15,000L Water Truck
2	Custom Made		Maintenance Vehicles
2	Supervisory		Pickups

Table 4. Screen Analysis of Feed to Leach Pad

Size Range (mm)		Wt. %	Cum. Wt. % Passing
	+19.0 (3/4 in.)	10	100
-19.0	+15.9 (5/8 in.)	8	90
-15.9	+12.7 (1/2 in.)	10	82
-12.7	+11.1 (7/16 in.)	6	72
-11.1	+ 9.5 (3/8 in.)	6	66
- 9.5	+ 6.3 (1/4 in.)	12	60
- 6.3	+20 Mesh	28	48
-20 Mesh	+40 Mesh	14	20
-40 Mesh	+100 Mesh	6	6
-100 Mesh		< 0.3	< 0.3

Table 5. Mid 1981 Operating Costs

	$ Cost/Short Ton Ore
Mining	
- Ore Removal (Contract)	1.09
- Waste Removal (Contract)	1.84
- General Expense	.42
Sub-Total Mining:	3.35
Processing	
- Crushing Plants	1.66
- Leaching	1.06
- Gold Recovery Plant	.82
- Residue Disposal (Contract)	.58
- Shipping and Refining	.18
- General Expense	1.23
Sub-Total Processing:	5.53
Administrative	.96
TOTAL:	9.84

Note: Local and Corporate Taxes, Lease Royalties, Capital Depreciation, Depletion as well as current Exploration costs are all excluded from above figures.

Table 6. Ortiz Manpower

	Salaried	Hourly
Administrative Office	3	1
Mine Department	4	--
Plant Department		
- Supervision	6	--
- Crushing Plant	--	8
- Leaching	--	2
- Gold Recovery Plant	--	4
- General	--	5
- Maintenance	1	7
- Metallurgical Control	1	--
- Assay Lab	3	--
Accounting Department	3	--
TOTAL:	21	27

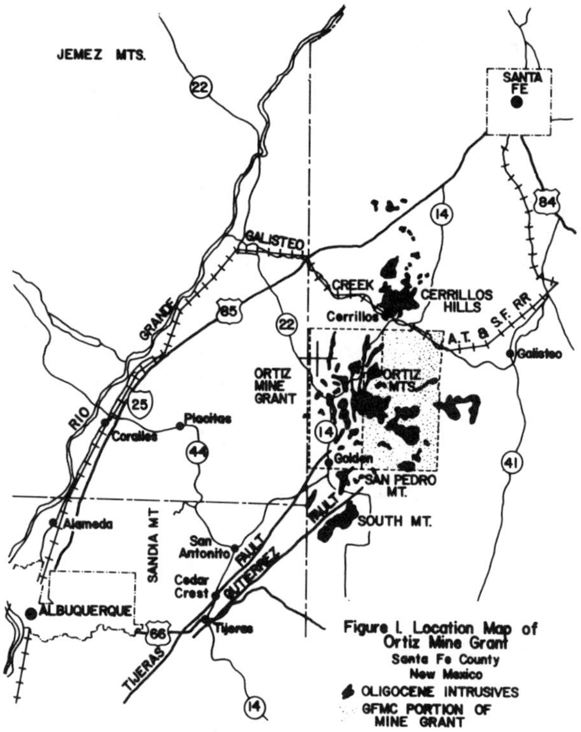

Figure I. Location Map of Ortiz Mine Grant

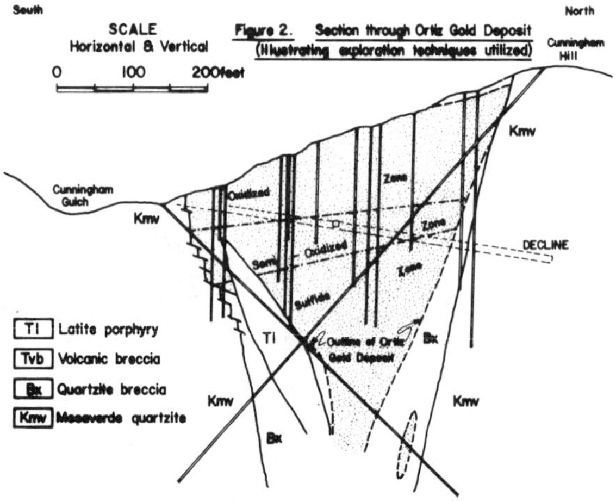

Figure 2. Section through Ortiz Gold Deposit (Illustrating exploration techniques utilized)

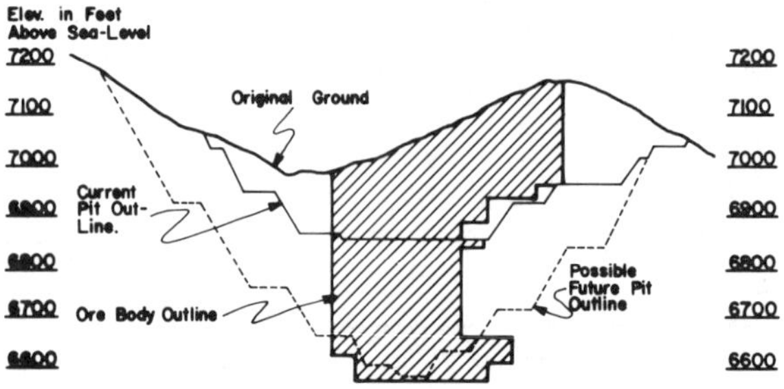

Figure 3. Cross Section through Ortiz Pit

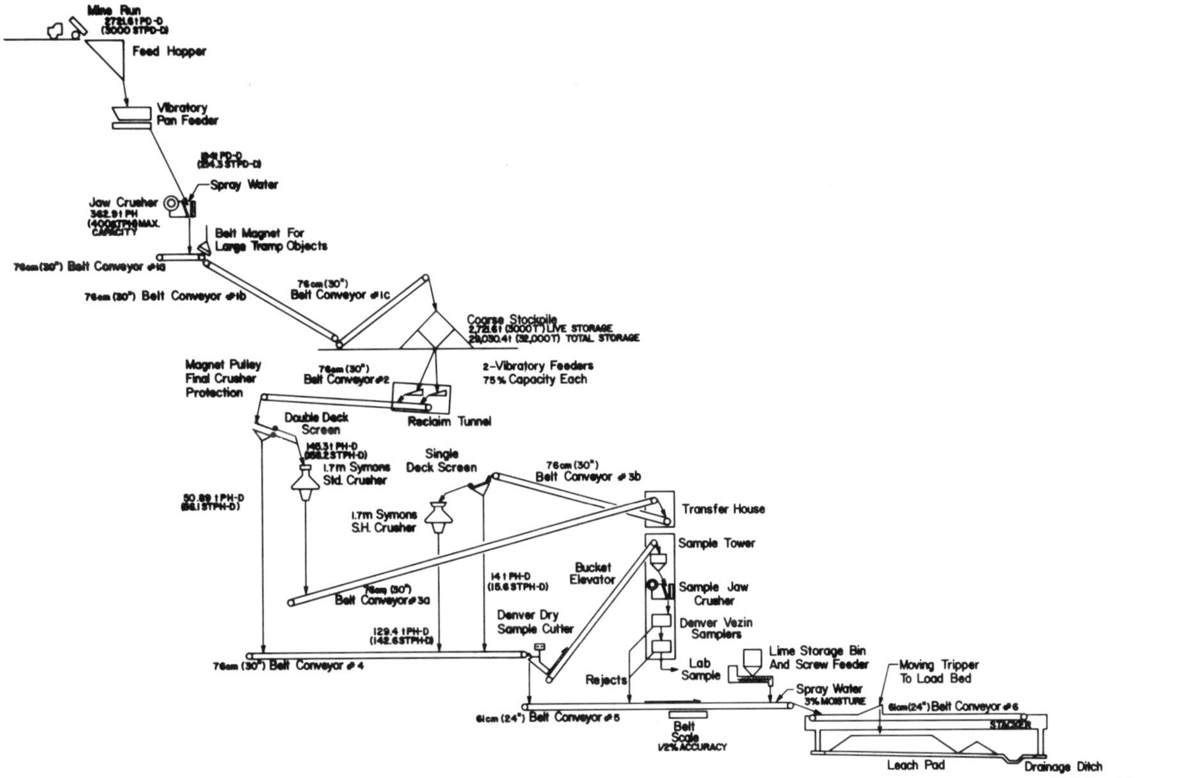

gold fields mining corp.
ORTIZ MINE
CRUSHED ORE FLOW SHEET
Figure 4

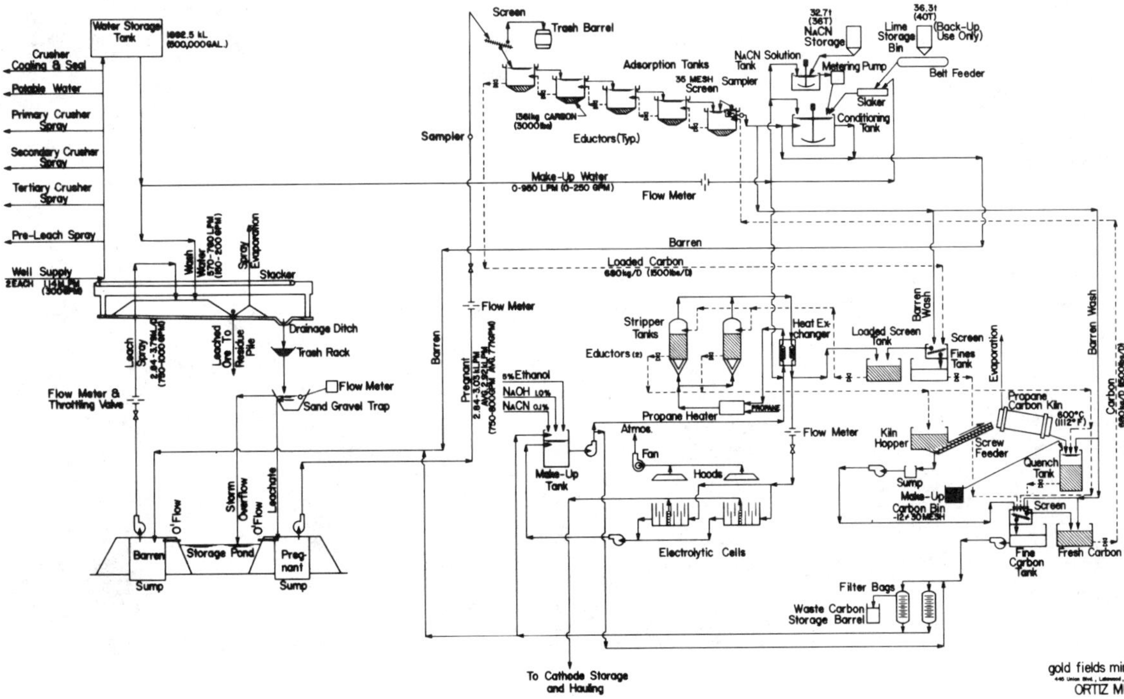

gold fields mining corp.
ORTIZ MINE
LEACH and RECOVERY FLOW SHEET
Figure 5

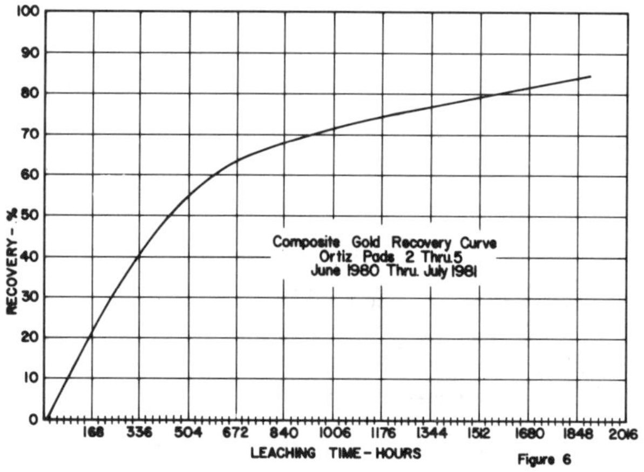

Figure 6

Figure 7. Ortiz Open Pit.

Chapter 19

HEAP LEACHING AND SILVER RECOVERY AT THE STATE OF MAINE

Charles B. Escapule
Louise W. Escapule
C. Bailey Escapule
State of Maine Mining Company
Tombstone, Arizona 85638

David D. Rabb
Tesoro Mining
Tucson, Arizona 85704

ABSTRACT

This paper describes the leaching of silver ores at the State of Maine property near Tombstone, Arizona. It includes discussions of preliminary test work, site selection, and pad preparation; heap construction, reagent preparation, solution application systems, and leaching operations; solution recovery, precipitation, and refining; and the analytical control procedures. The large scale cyanide leaching of several large dumps has been in continuous successful operation at the Tombstone site for more than four years.

INTRODUCTION

The first successful heap leach of silver ores in the State of Arizona has been under continuous operation since early 1976 at the State of Maine Mine, two miles southwest of Tombstone, Arizona. The silver produced from leaching old mine dumps at this property was the first in the Tombstone mining district to achieve a 999+ purity.

While cyanide leaching of silver and gold ores is a time-proven accepted practice, a relatively new method of treatment called "heap" leaching was employed. Heap leaching is a comparatively inexpensive fast way to recover precious metals from low grade material.

SYSTEM DESCRIPTION

Using material from old mine dumps or from current open pit mining, the ore is stacked or heaped on a prepared base and sprayed with dilute cyanide solution. The solution percolates down through the ore dissolving the metal and is collected, filtered to remove all suspended particles and all or nearly all dissolved oxygen is removed before zinc dust is added to precipitate the silver and gold. A simple filter collects the metal precipitates which are then melted and refined and cast into anodes and further refined by electrolysis. (See Fig. 1 for flow sheet.)

HISTORY

In 1878 John Escapule arrived in Tombstone to photograph the activity for the San Francisco Chronicle and stayed to locate the State of Maine Mine (S/MM). Heap leaching requires relatively few workers per ton of ore treated and economy of work force is one of the outstanding features of this method. Being entirely family owned and operated, paperwork and expenses at the State of Maine have been greatly reduced and a closely-knit, efficient work force is the result.

Detailed production figures from the S/MM prior to 1976 are not available but published estimates[(1)] range from two to three hundred thousand dollars, almost all in silver.

By mid 1981 there was an estimated 50,000 tons of ore under leach at three locations on the S/MM property.

Elsewhere in the Tombstone area, there are three other major heap leach operations with a total of over one million (estimated) tons under leach. There are no flotation or gravity operations, but one company is considering the use of hyposulphite as a leaching agent.

GEOLOGY

The country rocks in the Tombstone district range from pre-Cambrian to Quaternary in age. At the S/MM in the western-southwestern part of the district, the important ore bodies, predominantly silver with very little gold (average ratio 200:1), occur in a quartz latite porphyry and in Mesozoic sediments. Geographically, in the Tombstone district, ore with significant gold values seems to be restricted to the northeastern part.

The ore in the State of Maine appears to be in narrow stringers in the crushed sericitized rock in a fault zone, with wall rock alterations (mainly sericitization) extending only a few feet from the vein. All of the rock in the fissured zone weathers to a brownish or reddish color with some rather strong stains of manganese in some areas. Very little copper staining is evident and assay maps indicate that only small concentrations of lead are present in the S/MM area.

Extensive development (seven levels off an inclined shaft) resulted in several large dumps; total - about sixty thousand tons with recoverable values averaging 0.01 ounces of gold per ton and 2.0 ounces of silver. An additional estimated fifty thousand tons of broken ore are available in old stopes and fill material underground.

MINERALOGY

Studies of the mineralogy of the Tombstone district have shown a large number and variety of lead, silver and zinc minerals plus manganese, tellurium, molybdenum, and vanadium.

The silver minerals in the State of Maine ore are predominantly halides: cerargyrite with a little bromyrite and embolite plus some native silver and argentiferous galena. Some manganese, principally polianite and pyrolusite, occur in quantities of economic importance in several areas. The silver is frequently complexed with the manganese and, if so, lower extractions result. The silver-manganese minerals are resistant to cyanide leaching and the bulk of the silver can be recovered only by the Russell or Patera or similar process in which the manganese is first rendered soluble and removed. Difficult settling and filtration problems usually result and the treatment is inordinately expensive. Also, neither flotation or gravity separation tests have been successful on these refractory manganese silver ores.

TEST WORK

In heap leaching the controlling factor as to whether or not a particular batch of ore is to be added to the heap is not the total precious metal content but rather the amount of precious metal which can be made soluble in the leaching solution in a reasonable time. Batch tests as described below cannot only determine reagent consumption and extraction data but also observations can be made as to the character of the ore under leach. Preliminary testing is the key to a successful leach operation, particularly if some of the ore is gathered from various old mine dumps in the district. Each potential lot of ore must be tested and evaluated individually.

Barrel Test

50 pounds ore
3 gallons water
11 grams caustic (1 lb. per ton of ore)
11 grams cyanide (NaCN)

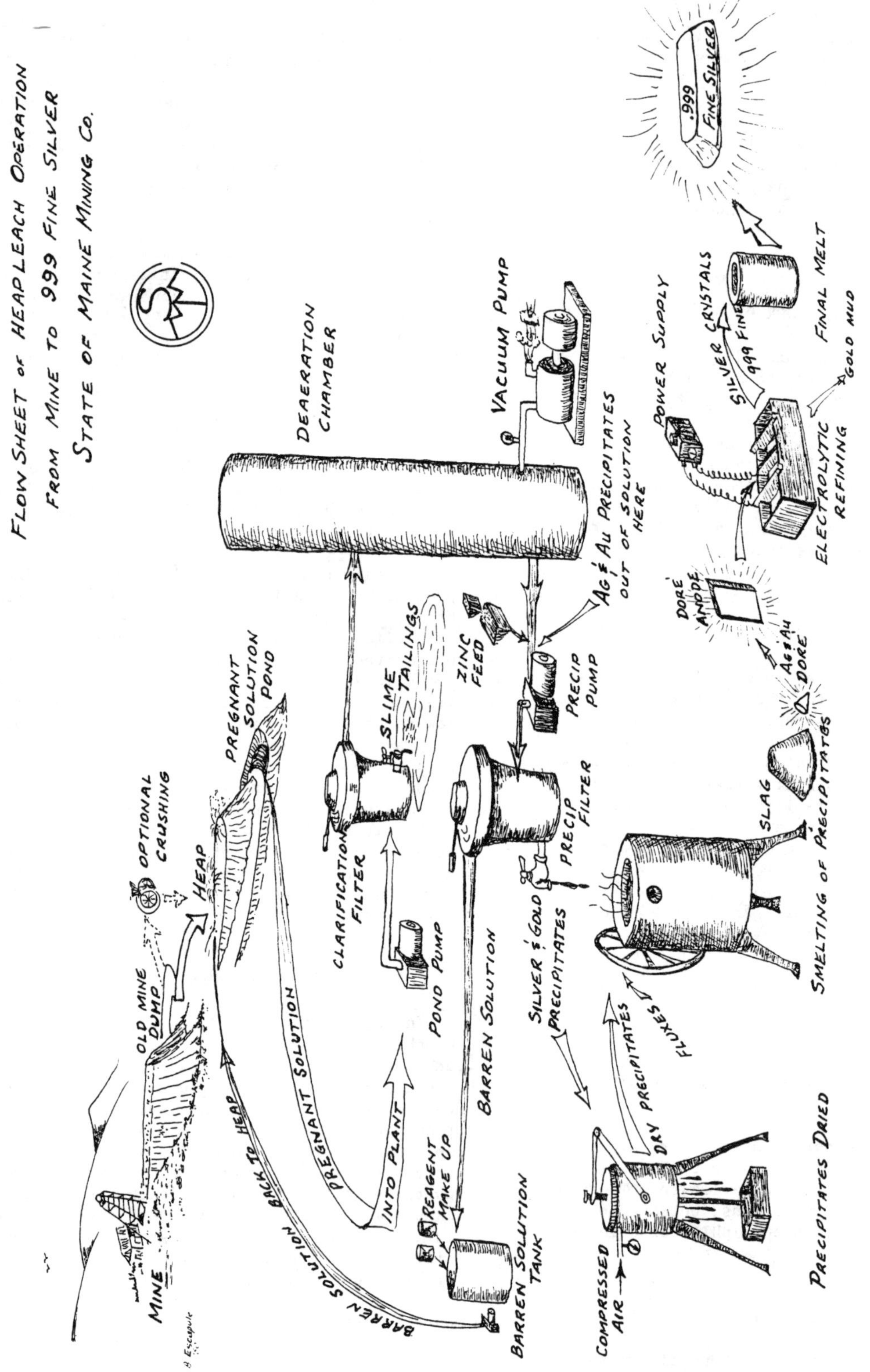

Fig. 1. Schematic Flow Sheet for Silver Heap Leach Operation at the State of Maine Mine.

A 5-gallon container with a perforated bottom sits in a tray (bathinette) with a small circulating pump from an air conditioner.

Add caustic; e.g. pure slacked lime, $Ca(OH)_2$, until solution has a pH of at least 10 or better. Then add cyanide; sodium cyanide, NaCN.

Run percolating solution over the ore in the bucket for 24 hours and take small samples of OFF solution for testing. Add fresh water to system so as to make up for losses from evaporation and sampling.

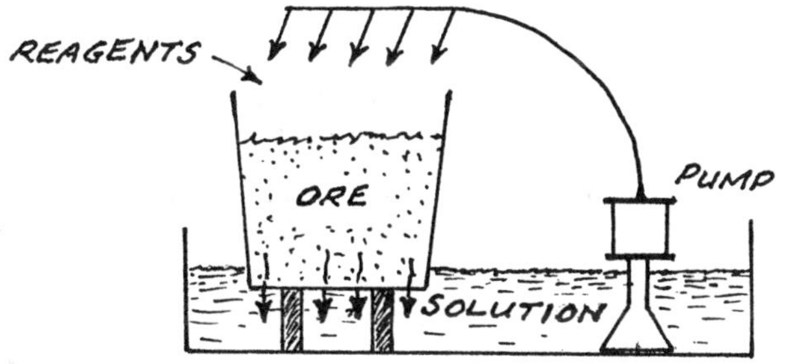

Continue to take samples every 24 hours until precious metal values in solution level off (do not increase). Titrate for free cyanide in the OFF solution and check the pH each time a sample is taken. Adjust pH to 10 or better each time and add more cyanide each time as needed.

This test will tell:

Reagent consumption
Extractions versus time
Percolation rates
Amenability to cyanidation percolation

Shaker Test

42 grams ore (pulverized)
42 ml of cyanide solution (2 pounds of cyanide per ton of solution at pH 10 or better

Add a flake or two of caustic before adding the cyanide solution.

Place in 125 ml flask, shake for 30 to 60 minutes. Filter and assay the solution for soluble precious metals.

This test will tell quickly if values will go into solution.

Agitation Tests

Can be in a variety of ways: a small 1 or 2 pound test in a bottle on rolls or in a cement mixer (about 25 pounds).

Add caustic to a unit pH of 10, then add cyanide.

To determine optimum leaching for that particular ore, tests should be done at different grinds with different concentrations of reagents. An idea of retention time can be determined from the periodic removal and assay of clear preg solution samples.

This type of test should be done only on higher grade ores.

Tests at the S/MM on certain batches of ore which contained clay and fine material indicated that better and faster extraction of metal values was achieved if the material could be pelletized or agglomerated prior to loading on the heap. Barren preg, at a pH of 10 and strong in cyanide, was sprayed on crushed ore on a moving belt. Pellets form as the belt moves, tilted at an angle of about 50 degrees from the horizontal. Leaching time was roughly cut in half to attain the same extraction.

SITE SELECTION AND PREPARATION

Criteria for site selection were:

1. Proximity to the ore to be leached.
2. Natural slope drainage that required minimal devegetation and grading. At S/MM a 8 to 9 percent slope was found to be the most satisfactory.
3. No interference or impoundment of streams or natural run-off.
4. Availability of utilities (especially water) and accessibility to roads.
5. Room for expansion, if desired. Size is determined by the amount of ore to be leached. An area 100 by 100 feet will hold about 4000 tons of ore stacked to 5 feet high.

At the S/MM all criteria were met in a succession of separate pads. All the OFF solutions could be collected in one preg storage pond and processed by one precipitation unit.

Subsurface hydrology and the proximity of Tombstone city water wells demanded that all plant solutions and run-off had to be controlled and contained on the property. Consequently, not one but two earth and rock dams were constructed within (1) 100 yards and (2) 250 yards downstream from the site. Each was lined with an impermeable clay bottom as described later. See pad construction.

Each dam was lined with clay and rock cover and was of sufficient size to contain the volume of run-off water from a 100-year flood plus the total of all solutions in the proposed leaching operation.

This demonstration of good faith and cooperation with the State Water Engineer's Office helped establish excellent rapport with that state agency and the Water Quality Control officials.

A willingness to work with county, state, and federal officials has been helpful in the State of Maine start-up and in subsequent operations.

PAD CONSTRUCTION

After the site was prepared, a pad base consisting of about 15 inches of mill slimes or a fine bentonitic clay from an old nearby tailings pond was laid down. The pad was constructed slightly larger than the dimensions of the heap to go on it; i.e., about

5 feet all around. Subsequent experiments have shown that this base is most effective if it is spread while slightly damp, 2 or 3 inches at a time and rolled or compacted between each application. A multi-stage build-up is the recommended procedure to insure an impermeable base. The final stage of pad construction is the addition of a layer of several inches of ore crushed to minus 1-inch and spread evenly over the clay base at about a 9 percent slope. Some operators favor the use of welded plastic sheeting of about 10 mil thickness, laid on a clay base and covered with gravel and/or fine sand. The most recent pad base now being installed at the S/MM will employ a reinforced plastic on a clay base and a clay and crushed ore cover. In any case, it is essential that an absolutely impervious barrier be emplaced and maintained so as to completely contain the leach solutions from the heap. Pad construction and pond linings were a major point of inspection by governmental agencies both Federal and County.

No solution leakage problems have been encountered at the S/MM due to pad base failure. The integrity of all pads and pond linings has been maintained completely.

LOADING THE HEAP

After the pad is constructed, the ore is placed on the pad with a 1½ yard loader in 5 foot lifts. Loading the heap should be done so that the coarse is at the bottom of the pile and the finer material is at the top. This allows for better solution distribution and decreases the possibility of vertical channels of solution flow through the heap. To do this a ramp is started up to the final level of the lift. Ore is dumped on the edge of the ramp and pushed over. This lets the larger rock roll to the bottom and the finer rock stays at the top.

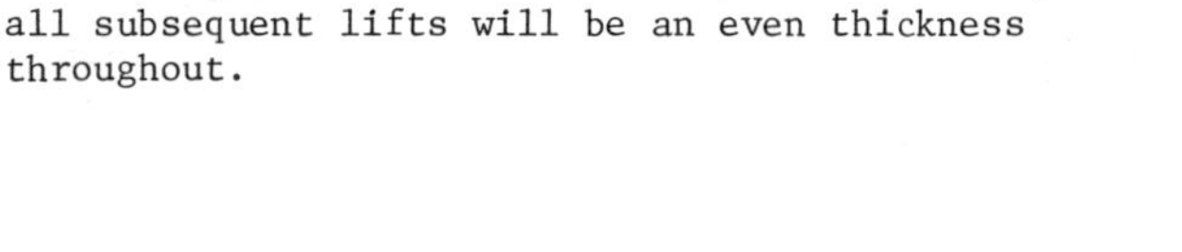

A leveling lift will have to be put down first, then all subsequent lifts will be an even thickness throughout.

After each lift is loaded, the ore is ripped to loosen the areas compacted by machinery. After each lift is leached, the ore is ripped before loading the next lift.

About 800 tons of ore could be loaded in one day, taking 6 working days per lift.

WHY 5 FOOT LIFTS?

The cyanide solution works best if there is an ample supply of oxygen in the solution. Using rainbird plastic sprinklers, the solution is broken down into a spray, allowing oxygen from the atmosphere to be dissolved into the solution.

Frequently, it has been found that the amount of dissolved oxygen is depleted after it has percolated 10 feet into the heap. Beyond this point the cyanide solution will not dissolve the silver or gold effectively. For this reason no lift should be more than 10 feet thick. Also, shallower lifts usually work better for finer material.

PREGNANT SOLUTION PONDS

Pregnant solution moving under the ore heap, along the pad, flows through a small decantation pond. This small pond is used to let some of the solids settle out before entering the main pregnant solution storage pond. Solution is decanted off the top of this small pond through a stand pipe into the main pond. When the solution is very slimey (for example, after heavy rains or a new batch of clayey ore is added to the heap), a mixture of water and lime is added into the decant pipe. This helps to settle some of the slimes and also helps maintain pH control.

The main pregnant solution pond was constructed with a capacity of 400,000 gallons just below the ore heap and the small decantation pond. The bases or liners in these ponds were prepared in the same way as the pad. The pond is designed to hold all solution from the leaching system (72,000 gallons), plus the added volume of water from a 100-year rain over the leach area. A second large dam with a impervious lined pond has been constructed below the main pond to collect and control all cyanide solutions and run-off that could overflow from the primary leach area. This second pond provides additional insurance against accidental overflows in the system.

The values contained in the main pond vary from the time cyanide is first added to the new batch ore until 6 to 8 weeks later, when a new lift of ore is needed for leaching.

Solution from the first days of leaching will have the highest values, ranging from 0.8 to 1.0+ ounces of silver per ton of solution. After 10 to 14 days, values usually drop to about 0.3 to 0.4 ounces of silver per ton of solution. The heap becomes "passive". At this time the heap may be allowed to drain and to aerate. It is then ripped to a depth of 4 or 5 feet with a single scorpion ripper behind a D-8 CAT. Spacing of the ripper pattern is about 6 feet. The pregnant solution then assays 0.6 to 0.8 ounces per ton coming OFF the heap.

After 6 to 8 weeks, depending upon the amount of fines in the heap, values in the solution drop to about 0.3 ounces per ton. The average value in the pregnant solution pond during this 6 to 8 week period is about 0.5 ounces per ton. The plan is always to leach the heap until an economic cut-off point has been reached. Then after ripping again, another 5 foot lift is added and leaching continued in the same manner. While the upper lift is undergoing its initial leach phase, the remaining values in the lower lift(s) are slowly being worked on, if oxygen and free cyanide (in solution) are available.

PLANT OPERATION

Solution from the pregnant storage pond is pumped into a clarifier filter where any remaining slimes are removed from the solution.

After clarification, the solution is run through a vacuum chamber to remove the dissolved oxygen (deaeration).

As the solution leaves the vacuum chamber, zinc dust is introduced and precipitation of the silver and gold takes place almost immediately. At this point the cyanide strength should not be excessive because some of the precipitated precious metal values may redissolve before they can be removed from the circuit in the precipitation filter. The solution with the suspended solids then passes through another filter to remove the silver and gold precipitate. The barren preg solution flows out of the precipitate filter and into a barren solution storage tank. Here samples are taken to determine the amounts of silver and gold remaining in solution.

Cyanide and caustic are added to bring solution back up to the desired pH strength and free cyanide.

Solution from this tank is then pumped back through the sprays and onto the heap.

Titrating for cyanide concentration is done with standard silver nitrate reagent on the pregnant solution coming OFF the heap. This is to determine how much free cyanide is left in solution after leaching. If the solution does not show any cyanide, then all of the available silver and gold may not be taken into solution. When the solution shows that some cyanide is present (usually a few hundredths of a pound per ton), then the available silver and gold is being taken into solution. But, as mentioned before, an excess of cyanide may redissolve silver and gold from the precipitation filter. For this reason, fresh cyanide is added only after precipitation before the barren preg goes back onto the heap.

REFINING

Precipitates mixed with diatomaceous earth from the precipitation filter unit are dried in a filter press and mixed with borax and soda ash. This mixture is then melted in a diesel-fired furnace and cast in a silver doré cone. This cone is then melted in another furnace into an anode 7 inches by 9 inches by ½ inch thick for electrolytic refining.

The silver doré anode is refined to 999+ fine silver crystals using silver nitrate electrolyte and stainless steel cathodes in an electrolytic refining cell. The silver crystals are scraped off the cathodes, rinsed, dried, and melted again into 10 ounce bars or whatever size is on order. Each bar is stamped 999 fine and the S/MM logo added.

The gold is collected with the other impurities from the bottom of the cell and is chemically recovered by using a nitric acid bath to dissolve everything but the gold. This small amount of gold is collected and when enough has been processed, it is melted into 1.5 ounce bars.

TYPICAL OPERATING DATA

Leach 4000 to 5000 tons per lift
Heap size: 180 by 100 feet, 5 feet high
300 tons of solution in 24 hours (50 gpm)
Fresh water make-up 50 tons (worst conditions: dry, hot, windy day and night)
Values in preg solution, ounces of silver per ton of solution:
- First 16 hours - 0.8 to 1.0
- 10 to 14 days - 0.3 to 0.4
- After ripping - 0.6 to 0.8
- At shut down - 0.3
- Average 6 to 8 weeks - 0.5

Pump out decant pond about every two years due to slime build-up.

Precipitates from the Merrill-Crowe precipitation filters average about 50 percent silver. Most of the remainder is diatomaceous earth (silica) used in the filter media.

Use a flux mix of 30 percent borax and 3 percent soda ash (dry weight) when smelting the precipitates. The raw silver anodes measure 7 inches by 9 inches by ½ inch.

Power requirements for the refining cell are 1.8 to 4 volts and 10 amperes D.C. per anode in cell. Power varies with the number of anodes in the cell. Capacity, 6 anodes.

The pH of the barren preg solution ON to the heap is maintained at 10 or slightly better in order to stabilize the cyanide. The free cyanide in the ON solution is dependent upon the determination of how much is coming OFF. Pregnant solutions coming OFF the heap should show between 0.01 and 0.05 pounds of free cyanide per ton of solution. If less, the strength of ON solution is increased by an appropriate amount.

The integrated heap leaching and Merrill-Crowe precipitation circuit required a minimum of operating labor; e.g., one full time worker and one part-time helper per 8 hours, 3 shifts per day, seven days a week.

REFERENCES

1. Butler, B.S., Wilson, E.D., and Rasor, C.A. "Geology and Ore Deposits of the Tombstone District, Arizona", Arizona Bureau of Mines, Bulletin No. 143, 1938, 114 pp.

SKIMMER

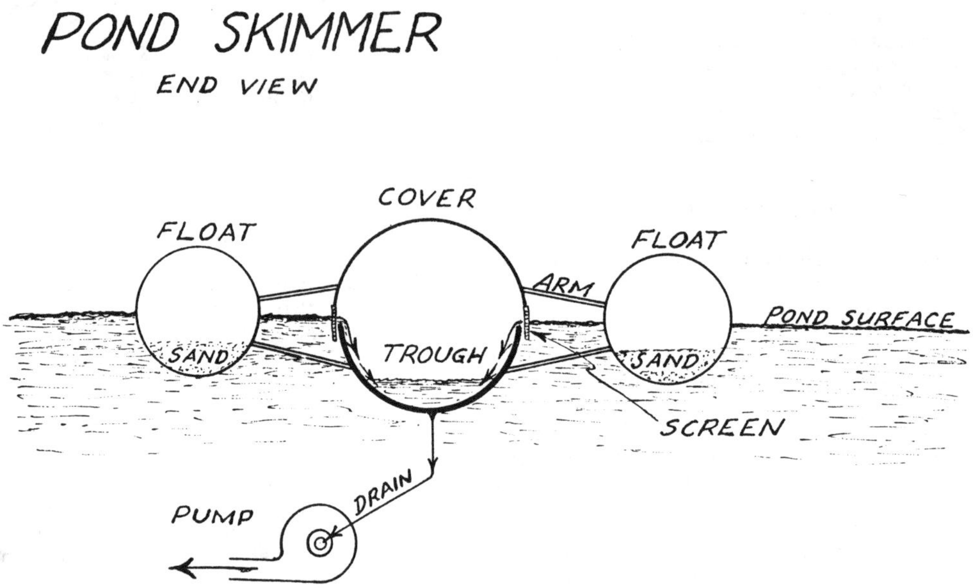

Notes: The trough is 8 to 20 feet long, with closed ends, supported by two outrigger floats, one on each side. The floats are attached to the trough and to the cover with several outrigger arms spaced on each side. The precise levelling and adjustment of trough lip below the pond surface is made by adding sand as a ballast to the two floats.

This skimmer pulls the top 1/8 inch of solution off the surface; therefore, maximum aeration and since it is screened and covered, there are no birds, bugs, or debris in the recovered solution.

Further, there is a minimum of silt or suspended solids trapped and taken off in the out-flow.

When pumping from strong preg storage ponds to the precipitation unit, care should be exercised to avoid any air entrapment by vortices, cavitation, or stirring in of extra air. Any extra air puts an unnecessary load on the deaeration unit.

Chapter 20

NEW DEVELOPMENTS AT HOMESTAKE'S BULLDOG MOUNTAIN CARBON-IN-PULP SILVER PLANT

R. S Kunter
Homestake Mining Company
Golden, Colorado 80401

D. N. Halbe
Western Mining Corporation, Ltd.
Laverton, Western Australia, Australia

ABSTRACT

The Bulldog CIP plant treats the slime portion of current tailings and reclaimed flotation tailings from the adjoining 350 tpd (3.67 kg/s) silver-lead flotation mill. Changes to the dredge reclaimer, pulp storage tank and trash removal system have allowed an increase in tonnage treated to a sustained 460 tpd (4.83 kg/s). CIP feed is 3-4 oz/T Ag (102.9 – 137.1 ppm). Following conventional cyanidation, carbon is added to the slurry.

Silver loaded carbon is stripped in atmospheric-pressure vessels using hot caustic cyanide. Silver recovery from this solution is by Zadra-type cells. Some of the modifications to the plant and operating techniques are discussed.

Introduction

A detailed description of Homestake's Bulldog Mountain CIP plant was presented at the AIME Convention in Las Vegas in 1980. Since the writing of that paper some additional work has been done on the CIP circuit at Creede, and a brief description of that work is included here.

A copy of the original paper is available from the Metallurgical Society as preprint A80-21, "Homestake's Bulldog Mountain Carbon-In-Pulp Silver Plant", by S. Mitchell, L. Trautman and D. Halbe.

Dredge

The original dredge for the Bulldog was built by Al Hack, Inc., of Denver. The dredge cutter head is mounted on a boom made of 6-inch (0.15 m) diameter pipe which is raised and lowered by a cable-pulley system. Maximum cutting depth is 8 feet (2.44 m). Slurry from the dredge flows through a floating 6-inch (0.15 m) diameter pipe supported on 55-gallon (208.2 ℓ) drums which runs from the dredge to the shore and then to a trash screen and pumping system.

The cutter head, suction pump, winches for dredge movement and all other operations are controlled from a small operations dog house located toward the rear of the dredge.

All of the basic dredge equipment is mounted on a chassis about 8' x 16' (2.44 m x 4.88 m) in size. Additional bouyancy was originally provided by sealed 55-gallon (208.2 ℓ) drums supported along the sides of the chassis by a small wooden framework. Additional framework and barrels on the front of the chassis on either side of the cutter boom provided additional stability, particularly when digging in hard material.

The dredge was quite stable, but two problems were apparent: For one, it could have been sunk by someone from the shore firing a rifle into all of the barrels. In prime deer-hunting and sportsman country such as Creede, this is not an idle fear. In addition, with the barrels it was extremely difficult to move the dredge into and out of the water without some damage to these supports.

The first problem could have been eliminated by filling the barrels with a spray-in type foam. The second problem was not as minor as it may initially appear. Winters at Creede are severe, but frequently several days of sub-freezing weather in the fall may be followed by several weeks of Indian Summer. It is necessary to remove the dredge from the water for the winter to prevent ice damage either prior to or immediately after the first freeze. With the original design this was a big job and usually resulted in minor damage to the barrels and framework. Then, even though the weather might then have moderated and made possible several more weeks of dredging, it was either impossible or not worth the effort to put the barge back into the water; this problem would continue to worsen as the level of the pond was lowered with continued dredging increasing the steepness of the banks.

The new design eliminated both of these problems. All barrels were replaced with pontoons running alongside and to the front of the dredge, again on either side of the cutter boom. These pontoons were made of steel plate and were compartmentalized. In addition, the pontoons were angled at each end so that the dredge can be pushed into or pulled out of the water readily. When the dredge was initially tested there was a little more bouyancy in the front than the rear, resulting in a slight rear-ward cant. This was modified by the addition of several sets of barrels attached to a small structure off the stern.

One of the main original operating problems with the dredge was the low percent solids in the dredged product. Shortly after the system was commissioned additional teeth were added to the cutter head to aid in digging the solids. This helped. In addition, the operators soon became more skilled in maintaining higher solids and we moved from an initial 5-6% solids to 15-20% routinely.

(Overflow for the 32' x 35' (9.75 m x 10.67 m) dredge storage tank flows by gravity back to the upper tailings pond. This overflow is normally very low in solids, and this does provide some upgrading in the dredge tank of the percent solids in the system.)

Recently we have installed a larger hydraulic pump system which allows greater pressure on the dredge cutter head with a marked increase in tonnage output. Although we do not measure directly the tonnage from the dredge pond, we do, of course, measure tonnage to the CIP plant. The tonnage of slimes from the mill remains relatively constant, so increases in throughput are due to tonnage from the dredge. In recent experiments we have operated at a sustained rate of 460 tpd (4.83 kg/s) over a 5-day period, as compared with a maximum rate of 360 tpd (3.78 kg/s) the previous year. See Figures 1 and 2.

Pulp Storage Tank Design

As noted in the original report, slurry from the dredge is pumped to a 35' x 32' (9.75 m x 10.67 m) storage tank adjacent to the CIP plant. The original design called for keeping the slurry in suspension by means of three tangential jets located 9½' (2.9 m) up from the flat bottom of the tank. A 6" x 4" (0.15 m x 0.10 m) Galigher horizontal pump circulated slurry from the tank through these jets continuously. A portion of the pumped slurry also went to a small constant head box on the top of the storage tank and flowed from there by gravity to the CIP plant thickener. Any overflow from the box dropped back into the storage tank.

This design would probably have been adequate for -200 mesh (-58×10^{-6} m) material, but once the dredge circuit started it became evident that there was considerable +200 ($+58 \times 10^{-6}$ m) mesh material in the pond. (The mine uses sand fill in the mining operation, and the design assumption was made that almost all of the available sand in the past had been pumped underground and that very little had gone to the pond.)

The original pump suction was located on the side of the tank about one foot (0.3 m) from the bottom. The sand present in the feed fairly quickly built up in the tank in a cone shape, with the top about 3 feet (0.91 m) above the level of the tangential jets. The swirling action imparted by the jets kept the sand from building up at the very edge all the way to the bottom.

This big sand bar reduced the capacity of the tank by about 25%. More seriously, however, sloughing periodically occurred directly into the pump intake. It occasionally plugged the pump, but more frequently just sent surges of sand through the pipeline to the thickener, sanding that line.

Water flushing and constant attention to the line helped somewhat, but the system was not very workable.

After some discussion a proposal was accepted to install a withdrawal system similar to the systems used in the phosphate fields of Florida. These systems use a dual suction line with the suction placed in the center of the bins, and are able to handle surprisingly coarse feed: deslimed phosphate and silica sand sized with cyclones to a nominal 16 x 150 (1600×10^{-6} m x 70×10^{-6} m) mesh.

Since the Bulldog Mountain dredge pumping system is of a relatively low percent solids (15-25%) as compared to the phosphate operations, some modifications were made. (Feed from the phosphate storage systems is approximately 85% solids, and is diluted to 45-50% solids with dilution water line placed near the suction line of the pump.) Since the feed is much finer in the dredging operation than in the phosphate system, a cover was placed over the suction end of the lines to help prevent possible plugging in the event of an extended shutdown. The cover is a 4' x 8' (1.22 m x 2.44 m) sheet of ¼" (0.006 m) plate raised off the bottom 18" (0.46 m). A dilution water line was placed under the cover for use if required. The dilution line is a 2" (0.05 m) line reduced to 1¼" (0.03 m) at the outlet. The outlet is located 6" (0.15 m) away from and aligned with the center of the 6" (0.15 m) suction line.

An additional set of 4 tangential jets was placed 2' (0.61 m) up from the bottom of the tank and located midway between the original jets. A second 6" x 4" (0.15 m x 0.10 m) Galigher pump was installed to provide sufficient volume of pulp flow for operation of both sets of jets when required, as well as to provide a stand-by system for routine operation. The two pumps are manifolded to be used interchangeably or together.

We felt that, particularly after a shutdown, it might be preferable to start circulation with the upper set of jets, then cut in the lower set. The system has operated so well since start-up in May of 1980, that we have not had to shutdown and have not had the opportunity to experiment with the system. We operate routinely, however, using only the lower jets.

The original feed line from the hill tank to the CIP thickener had been a 2" (0.05 m) gravity flow line laid down the hillside with an irregular slope, and crossed one roadway on an elevated trestle. The line was relaid and the trestle raised to increase the minimum slope of the flatter section from 0° to 15°, and to provide sufficient elevation for gravity feed to a DSM trash screen. In addition, as a safety factor a 2" x 2" (0.05 m x 0.05 m) pump was installed at the base of the tank to supplement gravity flow of the slurry to the thickener. Because of the problems we had had previously with the pipeline plugging, the operators used this pump frequently when the system was first started up. Now, however, it is seldom, if ever, used.

All piping is supplied with both high pressure water and air for flushing, and a spare 2" (0.05 m) line was laid from the tank to the thickener in case of serious plugs.

It may sound as if there was a little engineering overkill in the design, but the dredging season is quite short at Creede and we wanted to be able to minimize those startup modifications which might have been required had we designed less liberally. In addition, we had fought the system for several seasons and had all decided, I think, that now we were going to make it work right--period! These "extras" if you wish to call them that, represented, in fact, minor additional cost, and not only provided an additional assurance of trouble-free operation, but would also give some welcome flexibility in operation of the system.

The system started up with no significant problems and continues to operate in a completely satisfactory manner. The amount of sand in the tank when it is drawn down toward the end of the graveyard shift is small. We've been pleased with the operation of the system; and the mill operators, after having expressed some major reservations about whether the design would work, are delighted with its operation. See Figure 3.

DSM Wood Chip Removal Screen

One of the minor but occasionally troublesome problems in the Creede plant is related to the clayey nature of some of the ore from the mine. At times the thickener underflow is so clayey that it forms a rope which can almost be carried around without a bucket.

As described in the previous report: "...at times we have difficulty in maintaining the 48% solids in the thickener underflow. This high clay ore is also difficult to screen through the trash screen, normally requiring more than the regular minimal amount of spray water. Fortunately we are able to operate satisfactorily at lower densities in the carbon adsorption tanks during these periods because the increased viscosity of the slurry allows the carbon to remain in suspension. The addition of small amounts of sand from the dredge circuit is noticeably effective in minimizing the occasional problem of these clays."

The addition of water as a spray to the trash screen did help somewhat, as did additional sand when available, but the problem was not really solved.

A simultaneous solution to this problem and the problem of trash removal from the CIP thickener overflow prior to recycle is under evaluation now.

We have installed a 4' (1.22 m) 2.3 mm opening DSM screen ahead of the CIP thickener. At present either the dredge tank slurry or the tails from the flotation plant can go over this screen. Until maximum tonnages from the dredge circuit are definitely established we will be unable to determine exactly the total area of screen required. However, the screen appears to operate very effectively in removing the wood chips, wire, coarse rock, etc., from the slurry. Since the slurry is fairly dilute at this point the clay presents no problem.

A side benefit is that the wood and other floatable material which would report to the thickener overflow is removed here also. We have periodically checked the wood for silver values but found nothing, so it is sluiced into the tailings sump and goes to the lower tailings pond.

At this point we have experimented with both 0.75 mm and 2.3 mm opening screens, which would be expected to pruduce a +24 mesh (+860 x 10^{-6} m) or +16 mesh (+1600 x 10^{-6} m) product. Both appeared to work satisfactorily in removing trash. The main criterion is that all +16 mesh (+1600 x 10^{-6} m) material (which would otherwise follow the carbon in the CIP circuit) be removed. See Figure 4.

Changes to the Bulldog Mountain Mill Water and Piping System

A number of changes were made to the Bulldog Mountain Mill water and piping system. The intent of these changes was to make better use of existing fresh water consumption to decrease chemical contamination in the sand fraction returned underground, to decrease the volume of solution in the tails line thus releasing capacity, and to return water to the slime thickener for clean-up and re-use.

Fresh water was being used to cool the plant air compressors and the Nash vacuum pump for the concentrate filter. The total water volume from these two uses was estimated at 40 gpm (2.52 x 10^{-3} m^3/sec) and was being discharged to tails via floor sumps. It was decided that a further beneficial use could be made of this fresh water by collecting it in a common sump and pumping it to the sand cells. Repulping the sand fraction with fresh water instead of mill water in the sand flotation cells considerably reduces the amount of any dissolved chemicals in the sand tail slurry. These sand tails are cycloned and repulped again with mine water prior to being discharged underground as backfill. The use of fresh water in the sand cells reduces the possibility of any chemical contamination being introduced underground from the mill circuit, and could be done without increasing fresh water consumption in the mill.

A backup fresh water line was also installed to ensure a sufficient supply of repulping water particularly during mill start-up. Existing mill water lines were left installed as additional backup.

In the past there has been a considerable volume of solution circulating in the mill piping system. In particular, return water from the tailings dam was overflowing the mill surge tank and returning to the tails sump. It was easier from the operator's viewpoint to allow the tailings return water to overflow the surge tank than take the chance of a low level in this tank. During the dredging season, however, the extra capacity in the tails line was needed for the increased amounts of solids being pumped to the tailings pond. It was either install a new tails line or take measures to reduce the volume of needlessly circulating water. It was felt that the percent solids in the tails being pumped (7 to 14%) was sufficiently low that an increase in pulp density could be readily accommodated and in fact might be beneficial in terms of less pipe wear.

Three sets of changes in mill piping and solution flows were proposed to implement the decision to reduce the solution volume in the tails line. Two of these have been completed and the third is to be implemented at a later date.

The first and most significant change was the installation of float valves and a low level alarm on the mill surge tank. The high float was installed on the mill water return line and a lower float was installed on the mine water return line. A low level alarm was installed below the lowest float level; a lighted readout on the alarm was installed in the mill control room. The lines with float valves had manual shut-off valves installed above the float valves.

A third float and system is expected to be installed when the water treatment plant return water line is installed. The float on this line would be at the same level as the mine return float. When this is installed, it is expected that the mine return water valve would normally be closed and not used. There is an unfloated line to the surge tank from the slimes thickener overflow. The effect of these changes was to insure a satisfactory supply of water to the surge tank without the volume of overflow that was being returned to tails. The mill surge tank will now preferentially (assuming the valves are open) take make-up water from (1) the slimes thickener overflow, (2) the mill water return, and (3) the mine water return, in that order.

The second change made was to add additional piping and valves to allow the CIP sump pump to discharge to the flotation floor sump pump and thus to the slimes thickener for cleanup and re-use. The only options that were originally available were (1) to return this sump's discharge to the pachuca tanks or (2) to pump to the CIP plant tails sump and thus directly to mill tails sump.

The original options were left installed; however the new piping allowed any drainage and spillage (and values) to be recovered immediately rather than be returned from the tails pond. This provided a faster return of any overflows in the CIP plant (and mitigates any value losses as the underflow of the slimes thickener is the feed to the CIP plant), as well as reducing solution volume in the tails line.

The third change that is yet to be implemented is the installation of high and low level alarms and mill control room readouts on the sand surge tank. This tank receives the cycloned sand tails and mine water for repulping these tails prior to their being pumped underground for backfill. There is a 4" (0.10 m) overflow on this surge tank which goes directly to the tails sump. The location of the tank does not allow ready observation of overflows which could allow a large volume of excess solution (and sand) flow to the tails line. It has been easier to allow an excess of mine water to flow to this tank than risk the possibility of bogging the pumps and sand line should the flow drop unexpectedly.

When the water treatment plant is completed, it is intended to install a line to this tank to replace the mine water currently being used for repulping. This again will use a better quality water thus reducing any contaminants being discharged underground. It may be considered prudent at that time to install a float valve system with a back-up water source to ensure a proper steady flow of water to the pumps. The alarms should allow a more closely controlled water flow avoiding overflows and obviating the potential of bogging the pumps and sand lines.

Electrolytic Cell Test Program

In reviewing the literature, it does not appear that any comprehensive fundamental studies of silver-gold electrowinning from CIP desorption liquors has been made in the past. This fact is witnessed by current research and development studies being conducted in South Africa by Anglo-American and the National Institute of Metallurgy (NIM). Both Anglo-American and NIM have developed new gold electrowinning cells for CIP desorption liquors. Pilot studies on these cells are currently underway.

Homestake, in conjunction with St. Joe Minerals has conducted some silver electrowinning tests at Creede. St. Joe's interest in this program was based on their development of the El Indio gold-silver-copper mine in Chile.[1] The El Indio plant includes a mill and a carbon-in-pulp plant that started up in mid-1981. The objectives of these tests were: (1) To determine the optimum operating parameters for "Zadra" type cells. Such parameters will provide for optimum operating economies and produce a product of the maximum possible purity. (2) Determine changes in actual "Zadra" cell design which will improve silver electrowinning operation.

Testwork on this program was conducted using a 4 liter "Zadra" cell which is a scaled-down version of the commercial cells used at Creede. The cathode spool is made from CPVC with numerous holes in the center column for solution flow. Steel wool is wrapped around the center spool and the spool assembly is inserted into a CPVC shell with holes for solution flow. Electrical contact is made by 4 stainless steel rods inserted into the steel wool from the top of the cell. The anode is stainless steel mesh and surrounds the cathode shell. There are six individual cells, each with its own rectifier. The cells were operated in parallel with a common head solution from the Creede plant.

Important results from this testwork are summarized as follows: (1) Current efficiency can be significantly increased (10-15%) by the use of #3 steel wool (coarse) rather than #2 steel wool (fine) which is normally used. (2) In the range of 120-160°F (322°K-344°K) temperature had no effect on silver electrolysis. (3) Silver depositions on steel wool are very loosely adhering. At Creede, approximately 75% of the silver falls to the bottom of the cell as a sludge. The flowrate in our commercial cells is 10 gpm per 20 lbs of steel wool (6.955×10^{-5} m^3/sec/kg). Tests have shown that by increasing the flowrate to 12-14 gpm per 20 lbs of steel wool, (8.345×10^{-5} m^3/sec/kg- 9.736×10^{-5} m^3/sec/kg), virtually all of the silver is "washed" off of the steel wool and falls to the bottom of the cell as a sludge. This sludge can be removed from the cell and refined to dore' bullion without replacing the steel wool cathode.

Figures 5 and 6 show the Zadra cells used. Other types of electrowinning cells are also being evaluated currently such as a fluidized bed cell. Results of this testwork will be reported in a later paper.

Evaluation of Induction Heating for Use in the Fire Refining of Silver (and Gold)

This work was conducted at the Colorado School of Mines by Bruce Brogoitti, a consultant, and Dr. G. P. Martins (1980) of the CSM metallurgical staff, on steel-wool/silver and or silver bearing sludge from the Bulldog Mountain Operation.[2]

The use of induction heating as an alternative to gas or oil fired heating for the fire refining of precious metals is attractive for several reasons: (1) compactness, (2) less waste heat, therefore cooler room temperatures, (3) cleaner operation, (4) no off-gas discharge, (5) energy efficient as less material heated and (6) less time to reach temperature.

The particular application of interest at the Bulldog operation involved the smelting of a steel wool/precious metal and flux charge. Since this particular charge is not an efficient susceptor, the crucible must be of a material that is a susceptor (plumbago or silicon carbide). It is also possible to use a furnace with an independent susceptor. The testwork was conducted to assess the technical potential of induction heating for the Bulldog charges.

The tests were conducted using steel-wool/silver and or silver sludge from the electrolytic cells to which fluxes were added. The objective was to fuse the charge so as to obtain a slag-metal melt which could be evaluated in terms of: dissolution of steel-wool, silver reporting to the slag and fineness of silver (after a suitable contact period). Although silver was used in this evaluation, the results as they pertain to induction heating would be applicable to the refining of gold.

During a recent visit to the Rand Refinery in South Africa, the world's largest precious metal refinery, it was observed that they rely solely on induction heating for the refining, melting and alloying of gold and silver bullion. The area in which several large induction furnaces were operating and containing several hundred kilograms of molten bullion was pleasantly cool and immaculately clean. The furnaces were all mounted with hydraulic rams so that they could be tilted for pouring with minimum difficulty and excellent control. This large scale bullion handling was very impressive and demonstrates the utility of induction heating in this application.

Two induction units were used in the testwork: (1) Westinghouse Corp. 10 kva RF (400 kHz) generation and (2) Inductotherm Corp. 30 kva, 4200 Hz generation.

The Westinghouse unit was used in preliminary testwork with small uninsulated plumbago crucibles (1½" diameter x 3")(0.04 m x 0.08 m). Later work used fire-clay lined plumbago crucibles of the same size but smaller capacity. A single test was made with a cut down fire clay crucible to determine if adequate coupling could be achieved using a cylindrical pellet of steel wool/silver and fluxes as the susceptor.

The Inductotherm unit was used for larger charges with a Model T-30 furnace which contains a graphite susceptor. A plumbago crucible (3" diameter x 4")(0.08 m x 0.10 m) and fire clay crucibles (2½" diameter x 3½")(0.06 m x 0.09 m) were used with this furnace. The unit was also run with an induction coil and a fire clay lined silicon carbide crucible.

The first six tests were conducted to observe the behavior of a charge when subjected to induction heating. Small samples (12 gm (0.12 kg) of steel wool/silver sludge) were heated in a plumbago crucible until molten. The charge including fluxes occupied a volume of about 100 cm^3 (10^{-4} m^3). The charge was kept molten from 15 to 30 minutes after which it was poured into a conical mold. The silver button was then separated from the slag. The silver button weight was typically 43 to 45% of the weight of the silver sludge. The flux composition used was: 1 to 5 parts Borax, 1 to 5 parts Nitre, and 1 part silica by weight to the weight of the steel wool.

The results of these tests indicated that the plumbago crucibles were effective susceptors but drastically reduced the oxygen content of the slag by formation of carbon monoxide. There was insufficient oxidation to completely oxidize the steel wool. In one test an iron button was formed on top of the silver button and the silver exhibited magnetic susceptibility. The temperature of these tests was estimated to be 1400°C (1673.15°K).

The addition of fluorspar to the flux charge (0 to 3 parts by weight to weight of steel wool) and the use of a fire clay lined crucible increased the oxidizing conditions of the slag sufficiently to completely oxidize the steel wool and produce a refined silver button.

In one large scale test a silicon carbide crucible, lined with fire clay, was used to smelt 2 kg of steel wool/silver material together with 1 kg of fluxes and produced a 0.950 kg refined silver ingot. The silicon carbide crucible acted as a susceptor and the volume of the lined crucible was approximately 4ℓ (0.004 m^3). The total dry charge occupied about half of this volume and when molten was again reduced by half to ∿950 cm^3 (9.5×10^{-4} m^3).

A single test on a pressed cylindrical pellet consisting of steel wool/silver plus fluxes indicated that satisfactory agglomeration, and thus contacting, of steel wool and flux components could be achieved. However, the pellet was a poor susceptor and direct heating of the charge was unsatisfactory. This may be a useful technique, nevertheless, to reduce the dry charge volume.

The results of these tests indicate that induction heating can be successfully used to fire refine silver/steel wool and flux composites. Fire clay lined silicon carbide crucibles appear to be the most satisfactory. While specific flux charges will have to be determined for differing steel wool/precious metal charges in order to provide the correct amount of oxidation, a guideline suitable flux charge appears to be: 1-5 parts Borax, 1-5 parts Nitre, 1 part silica, and 0-3 parts Fluorspar by weight to 1 part steel wool. The required temperature and contact time between the slag and precious metal will need to be determined for a specific application. As a guide, temperatures in the range 1100°C to 1250°C (1373°K to 1523°K) and contact times of 1 to 1½ hours (after the charge is molten) appears to be satisfactory.

An estimation of the volumetric requirements of a crucible to treat 1200 Troy ozs (37.32 kg) of precious metals in a single batch would be 20,000 to 25,000 cm^3 (0.020 to 0.025 m^3). The major portion of this volume is required for the fluxes and allows for a 20% freeboard. A crucible with internal dimensions of 11" diameter x 17" (0.28 m x 0.43 m) would be suitable. As an alternative to single large batch treatment, several smaller batches and a small crucible could be used which would reduce the size of furnace required but increase the amount of processing time. This would be an economic decision based on the amount of material to be refined.

Acknowledgements

Particular thanks are due to Bruce Brogoitti, metallurgical consultant and formerly of Greyback Engineering, for his work, both conceptual and design, of the entire pulp handling system modifications; and in particular to his tireless efforts in convincing us that this dredge tank design would really work.

Don Gullickson, former Assistant Mill Superintendent at the Bulldog, and Steve Mitchell, former Homestake metallurgist at the Bulldog, contributed many ideas and suggestions for the developments that have transpired and it is with much appreciation that their work is acknowledged.

Curt Carey, a senior staff metallurgist for Homestake Mining Company did the work on the Electrolytic Cell test program at Creede and contributed the section regarding this work.

Thanks are due to Homestake's drafting section for their help in preparing slides for this report. Thanks to Mary Ellen Polak for her help and assistance in report preparation.

References

1. Anon, 1979, "St. Joe Minerals Will Develop El Indio Au-Ag-Cu Deposit in Chile's High Andes," Engineering and Mining Journal, Vol. 31, No. 4, May, pp. 466.
2. Martins, G. P., 1980, "Evaluation of Induction Heating for Use in the Fire Refining of Silver (and Gold)," Report #GC/GM-180, 1980, unpublished, Colorado School of Mines, Golden, Colorado.

Figure 1. Dredge reclaiming flotation tailings from Upper Tails Dam

Figure 2. Cutter head and hydraulic assembly on dredge

Figure 3. Pulp Storage Tank and Bulldog Mountain Flotation and CIP Plant

Figure 4. DSM Wood Chip Removal Screen

Figure 5. Zadra type 4 liter electrolytic test cell cathode/anode assembly

Figure 6. Zadra type 4 liter electrolytic test cell

Chapter 21

WASTE DUMP LEACHING IN A TROPICAL ENVIRONMENT

Paul Piercy
Bougainville Copper Limited
Bougainville Island, Papua New Guinea

ABSTRACT

Bougainville Copper Limited operates a porphyry copper mine in Papua New Guinea. The waste dumps will contain 400 to 500 million tonnes of rock at the end of mine life.

The paper describes investigations carried out over a period of ten years and highlights the problems associated with waste dump leaching in a tropical and seismically active region. Aspects discussed include mineralogy, chemical reactions, leaching rates, microbiological activity, dump stability, dump structures, effects of tropical conditions, seismic activity and topography.

INTRODUCTION

The mine and processing facilities of Bougainville Copper Limited are located on Bougainville Island, Papua New Guinea. Geographically the island is at the North end of the Solomons chain of islands but politically is part of the Independent State of Papua New Guinea. The mine site is at an elevation of approximately 670 metres and lies on the western slope of the Crown Prince Range, which runs the length of the island.

The ore body is a porphyry type copper deposit with estimated reserves of 650 million tonnes of ore averaging 0.43 percent copper and 0.48 grams per tonne gold. Mining is by the open cast method and metal recovery is by conventional crushing, screening, grinding and flotation.

The mine commenced commercial production in April 1972 and was the culmination of 8 years of exploration, planning and construction carried out under difficult conditions of terrain and tropical environment.

The waste/ore stripping ratio of the mine was designed at 0.85 to 1 which at the end of mine life will develop dumps of 400 to 500 million tonnes of waste rock.

Leaching of the oxidised ores and waste rock has been actively studied for more than ten years.

TOPOGRAPHICAL FEATURES OF BOUGAINVILLE

The island of Bougainville is on the latitude of 6 degrees south of the equator and 980 kilometres to the east of Port Moresby. Bougainville, which is a northwest trending island about 130 miles long and a maximum of 40 miles wide, forms the eastern most part of the Independent State of Papua New Guinea. (Figure 1 Refers)

The island has an equatorial climate with a poorly defined wet season. The terrain is deeply dissected, and side slopes of 30° to 45° are common. The Kawerong Valley is surrounded by hills to an elevation of 1524 metres while elevations within the mine area vary between 610 and 915 metres above sea level. The area is covered by tropical rain forest, with fairly thin undergrowth except in areas of dense secondary growth.

CLIMATE

The rainfall pattern is tropical by nature and is reasonably uniform throughout the year. Although rain can be expected at any time of the year, records point to the heaviest falls occurring between November and May. In general, rain occurs between 23 and 26 days of each month with an average monthly fall of 389 mm. Average rainfall data is presented in Table 1.

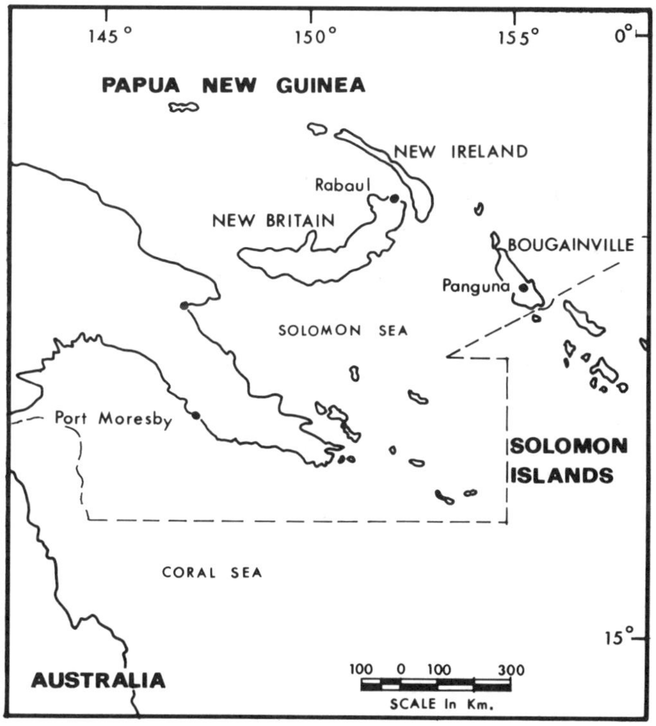

FIGURE 1. LOCATION MAP

TABLE 1

Rainfall Data - Millimetres

Month	Mean Monthly	Mean days of Rain	Max. Monthly	Min. Monthly	Max. in One Day
Jan	514	27	599	357	130
Feb	478	23	645	364	180
Mar	583	26	798	329	172
Apr	449	25	610	216	155
May	334	23	859	132	177
Jun	220	21	433	90	211
Jul	368	29	1218	193	137
Aug	262	23	347	108	88
Sep	328	23	615	180	152
Oct	326	26	596	145	80
Nov	395	24	666	238	136
Dec	416	27	640	228	136
Ave	389	-	-	-	-

The temperature is fairly uniform and varies over a small range. Temperature varies from maximums of 26 to 27.4°C with a mean monthly range of 18.2 to 19.8°C. Extreme temperatures have occurred as high as 32°C with lows of 14°C. Humidity is generally high; the average 9:00 a.m. relative humidity is 82.4 percent and at 3:00 p.m. it is 85.8 percent. Temperature and relative humidity data is shown in Table 2.

TABLE 2

Temperature and Relative Humidity Data

	Mean Temp. °C Mean Max.	Mean Min.	Mean	Extreme Shade Temp °C High	Low	Mean Rel. Humidity % 9 am	3 pm
Jan	26.6	19.8	23.2	31.1	16.1	85.1	85.9
Feb	26.2	18.7	22.4	29.7	16.0	82.6	85.1
Mar	27.2	19.1	23.1	29.8	16.7	82.6	87.8
Apr	26.9	19.3	23.2	29.5	16.2	83.6	86.4
May	26.9	19.6	23.3	29.5	16.7	83.9	87.4
Jun	26.7	18.5	22.6	31.1	15.7	80.7	83.3
Jul	26.3	18.9	22.6	32.2	16.1	84.6	86.9
Aug	26.0	18.2	22.1	29.2	13.9	83.6	83.6
Sep	25.9	18.5	22.2	29.3	13.9	81.7	84.6
Oct	26.4	18.3	22.3	32.6	18.0	79.3	84.7
Nov	27.4	18.3	22.6	32.2	14.4	79.3	85.7
Dec	27.1	19.3	23.2	30.0	15.4	82.2	87.9
Ave	26.7	18.9	22.7	32.6	13.9	82.4	85.8

SEISMIC

Bougainville is a volcanic island with a row of mostly extinct volcanos along its northwest - southeast axis. Mt. Bagana, the remaining active volcano, continues to discharge gases to atmosphere. High seismic activity is experienced because of the island's location within the circum-pacific belt. Records are available for shocks of "felt intensity" since 1967 when two strong-motion accelerographs were installed on the island. Tremors of 7.2 intensity on Richter scale have been experienced. "Felt shocks", of intensity 5, are relatively frequent occurrences which one readily becomes accustomed to.

GEOLOGY

The Panguna deposit is a porphyry copper type, which means its copper content is low and disseminated through rocks which are of igneous origin. The island is essentially an uplifted Cainozoic succession of older volcanic rocks, limestone, and younger volcanic rocks, surrounded by extensive coral reefs. It is part of the volcanic belt which virtually encircles the Pacific Ocean, with several volcanos situated on the island.

The ore minerals at Panguna are chalcopyrite with minor bornite and chalcocite which occur as disseminations, and in fractures and veins.

The mineralisation occurs in andesites, diorites and granodiorites. Pyrite occurs commonly with the copper minerals. Gangue is composed of the various silicate minerals together with their alteration products, which comprise the host rock. Some gypsum/anhydrite occurs in parts of the pit. The magnetite content of the orebody is about 3%. The level of oxidation of the copper minerals is extremely variable but decreases with depth.

MINERALOGY AND PARTICLE SIZE OF PANGUNA DUMP

Superficial material, which is essentially clay and volcanic tuff type material, has been dumped directly into the Kawerong River to avoid stability problems with the waste dumps.

Competent waste rock from the mining operation has been dumped in the Kawerong and Kurua Valleys in the Panguna area. The dumps are essentially in "as blasted" condition with no attempt at size reduction or grading of material.

The waste rock in the dumps has an assay generally greater than 30 percent acid soluble copper or contains sulphide copper below the cut-off level of 0.25%. The cut-off grade for the operation has changed over the years from 0.35% sulphide copper at the commencement of operations to a figure of 0.22% in the current Five-Year Plan. The ultimate cut-off grade is expected to be lowered to 0.2% sulphide copper or less.

Chalcopyrite is the predominant copper mineral with minor bornite and chalcocite occurrences. Digenite and covellite have been recorded in some mineragraphic work, as has arsenic bearing tetrahedrite. Molybdenite is common but is a minor sulphide mineral which has a value of 33 parts per million in ore. Sphalerite and galena occur occasionally but mainly associated with faulting.

Limited mineragraphic studies have shown chalcopyrite weathers to the oxides cuprite and chalcotrichite and ultimately to the copper silicates with traces of carbonates. Quartz crystals have been found with secondary copper silicates and carbonates. The level of soluble silicates is higher than for other known dump leaching operations and may lead to secondary reactions with dissolved copper and iron. Fractured, weathered surfaces also exhibit development of covellite as an intermediate band between chalcopyrite and cuprite - geothite. Adjacent crystal masses of pyrite and chalcopyrite show hydrated areas of intermediate geothite with more iron rich zones nearer the crystal masses.

Preferential embayment of chalcopyrite typifies the greater corrosive attack on chalcopyrite as compared with pyrite.

The rock types found in Panguna are Panguna Andesite (PA), Leucocratic Quartz Diorite (LQD), Biotite Diorite (BD), Biotite Granodiorite (BG), Buiro Granodiorite (BI), Diorites, Quartz Feldspar Porphyry, Feldspar Porphyry (FP), Kawerong Quartz Diorite and Nautango Andesite.

Of these, the PA, BD, BG and BI are the more predominant types in the waste dumps. Apart from the LQD, which is invariably ore and directed to the crushing circuit, the other rock types are minor occurrences.

The dumps are made up of a high proportion of PA and BG, which contains more pyrite than the PA and BG of ore grade. The Panguna orebody is a low sulphide system (less than 1% pyrite) but the level of pyrite in the waste rock appears to be sufficiently high to be economically important to a dump leaching operation.

The make up of rock types and sizes in the dumps are shown in Tables 3, 4, and 5. The information is regarded as only a guide because of the variable nature of the rock types and their mineralisation, and the limited amount of work which has gone into the determination of this data.

TABLE 3

Coarse Rock

Rock Type	% Weight -400 mm	% Weight 400-800 mm	% Weight +800 mm
PA	90	8	2
BD	90	7	3
BG	80	15	5
BI	75	20	5

TABLE 4

Medium Size Fragment

Rock Type	% of Dump Material	% Weight -150 mm	% Weight +150 mm
PA	60 - 70	35 - 45	55 - 65
BG	20 - 25	25 - 35	65 - 75
BD	5 - 10	30 - 40	60 - 70
BI	5 - 10	30 - 40	60 - 70
FP	1 - 5	25 - 35	65 - 75

TABLE 5

Fine Fractions

Mesh Size mm	% Weight BD	% Weight PA
+ 75	22	30
50 - 75	14	15
25 - 50	22	27
16 - 25	11	12
12 - 16	4	4
9 - 12	4	4
- 9	23	8

The ground mass of the rocks is generally made up of variable amounts of quartz, feldspar, micas, and minor calcite. In the case of BI, the ground mass is predominantly feldspar. On the upper benches of the mine, the BD tends to be moderately to highly weathered, with abundant limonite on joints. In the highly weathered material, chrysocolla with minor dioptase, chalcocite and cuprite will be present in varying amounts. The ground mass of the waste rock contains practically no calcite. There is approximately 3-4% calcium in the orebody occurring in calcite and epidote in PA in the eastern section of the pit, and in the feldspar.

The alteration products of the various rock types are shown in Table 6.

TABLE 6

Alteration of the Rock Types

Rock Type	Alteration Type	Silica	Biotite	Chlorite	Feldspar	Epidote	Clay	Sericite	Calcite
		Alteration Minerals							
PA	P	weak	weak	mod.	-	-	-	-	-
	F	-	mod.	mod.	-	weak	weak (faults)	-	weak
BG	P	mod.	mod.	mod.	-	-	weak	weak	-
	F	weak	-	mod.	-	-	mod.	-	-
BD	P	mod.	strong	mod.	weak	-	mod.	mod.	-
	F	weak	-	mod.	weak	trace	mod.	-	-
BI	P	weak	trace	mod.	-	-	-	-	-
	F	-	weak	weak	trace	-	weak	-	-
FP	P	weak	mod.	weak.	-	-	weak	trace	-
	F	weak	-	weak	-	-	mod.	-	-

P - Pervasive F - Fracture

There is little variation in the grain size of the copper minerals and pyrite in the different rock types. The chalcopyrite/pyrite ratio in ore and waste has no relationship to grain size. In waste PA, the ratio of chalcopyrite/pyrite is of the order of 1-4 and in BG the ratio is about 1-10. The amount of bornite has no relationship to grain size of the host rock.

Mineralisation may occur in three ways:

(a) Joints - along fractures or joints
(b) Veins - associated with quartz veins
(c) Disseminated - within the body of the rock

The proportion of the copper minerals in each of these modes of occurrence for waste rock is shown in Table 7.

The copper content of the dumps averages 0.2% to date, however, this is expected to be reduced significantly in the future in the face of changing cut-off grade strategies in the mining operation.

WASTE DUMP DEVELOPMENT

In the early developmental period of the open pit, overburden or superficial material was the main constituent removed. This was done by hydraulicking and trucking to the river system.

Since this initial period, waste rock has been dumped in the Kawerong Valley and side valleys which adjoin it.

Figure 2 shows the development of waste dumps in the Kawerong Valley to early 1980. Although the photo is now outdated, it demonstrates the type of country in which the mine is located and the various areas of importance in the development of the waste dumps.

TABLE 7

Mode of Occurrence for Waste Rock

Rock Type	Mode of Occurrence	Ratio of Sulphide Distribution Minerals Type			Percentage
		Chalcopyrite	Bornite	Pyrite	
PA	Fracture	1	0	4	90-95
	Disseminated	-	-	-	-
	Veins	1	Trace	1	5-10
BG	Fracture	1	0	2	50-55
	Disseminated	1	0	2	25-30
	Veins	1	Trace	2	15-20
BD	Fracture	1	Trace	3	60-70
	Disseminated	1	0	2	5-20
	Veins	1	Trace	2	10-20
BI	Fracture	1	Trace	1	90
	Disseminated	-	-	-	-
	Veins	1	0	1	10
FP	Fracture	1	Trace	2	80-85
	Disseminated	1	0	2	5-10
	Veins	1	Trace	1	5

FIGURE 2. AERIAL VIEW OF KAWERONG VALLEY

The prime factor in the design of the waste dumps, until the possibility of waste dump leaching evolved, was the stability of the dumps. The more stable the dumps, the less likely the dumps will be able to be leached economically.

The initial concept for waste dumps was to utilise the Kurua Valley which was estimated to have a ten year capacity. At the end of this period, it was planned that waste rock would be dumped in the next downstream valley. In 1973, it was determined that the Kurua Valley was below capacity because of several reasons associated with dump stability.

An alternate plan was developed to utilise the Kawerong Valley as the main dump area. To achieve this, it was necessary to relocate the Kawerong River from the valley floor into a concrete lined channel. To achieve this, the dumping of waste rock commenced on the northern rim of the valley and was continued to the Sirakore area with a discharge point into the Sirakore Valley (upstream of valley one). Subsequent to this, the waste dumps have developed to a point in the valley that it is now necessary to enlarge the available dump space. It can be seen in Figure 2 that the dumping has continued along the north rim to a point noted as Dau Valley. The construction of the concrete lined channel to that point is now complete and it is possible to discharge concentrator tailings and the Kawerong River in the Dau Valley. Operating economics, however, will see a continuation of superficial dumping into the Sirakore Valley for the next two years until the waste dumps are physically encroaching the Sirakore Valley.

Construction of the Kurua high level dump road, as noted by the 760 in Figure 2, commenced in 1974 using competent waste rock. The construction of the 760 high level haul road continued until 1978. In the meantime, the construction of lower level haul roads to the Kurua Valley commenced. The construction of these roads was a necessary development for dumping into the valley. Kurua Valley was planned to be dumped in three levels, viz, 520 metres, 640 metres and 760 metres level to maximise stability.

In parallel with the Kurua dump roads, waste dumping at the head of the Kawerong commenced and has progressively extended downstream. The Dapera haul road has now been straightened and the entire area through the level has been filled.

The main factors influencing the design and location of the waste dumps in the latter years were:

(a) the positioning of a public access road
(b) the safe working height of the dumps i.e. bench to bench
(c) stability of the dumps
(d) the effect on the leaching project
(e) the disposal of superficial waste at the most economic cost.

The ultimate dump limits will be adjacent to the ridge between valley 4 and valley 5. An important criterion in the ultimate waste dump structures will be long term stability, either with or without dump leaching. This will be achieved by low angle benching to a safe downstream toe structure.

LEACHING INVESTIGATION

Pre 1976

During the initial investigations into the metallurgical performance of the Panguna orebody, leaching was considered as a possible alternative for the treatment of secondary and oxide ores. Much of the work centred on an evaluation of the leach-precipitation-flotation process.

Other work included vat percolation, waste heap and bacterial leaching. Work was conducted in several laboratories with some pilot heap leaching on site.

Although in some cases copper extractions and solution tenors were quite high, the overall conclusion was that the leaching of secondary and oxide ores was not economic on the scale envisaged for Bougainville Copper Limited.

Copper tenors of 1300 ppm were obtained with large scale evaluation of sulphide waste leaching. Unfortunately, acid consumption was extremely high for the recovery achieved. Observations and measurements of liquors flowing from stockpiles suggested a dump leaching operation was possible. This led to the establishment of a 900 tonnes pilot dump for investigation. The rock used for the dump was 0.17% copper with a high pyrite level of 2-3%. Extraction rates were good, however, iron release from the dump was high, reaching levels of 1200 ppm.

The overall conclusion in the period prior to 1976 was that waste rock leaching was not a viable prospect for Bougainville Copper Limited. This conclusion was based principally on the difficulty of isolating natural drainage, including the Kawerong River, from the leaching areas and the probable loss of liquor by seepage from the system. The capacity for the waste rock to leach autogenously had been demonstrated effectively.

Post 1976

After the North Channel had been commissioned and the Kawerong River diverted into it, it became evident that the waste dumps on the northern rim of the valley had commenced to leach. Some earlier indications had been seen but it was not possible to quantify the phenomena. When the river was diverted the leach streams could be measured in the absence of the massive river flow.

In 1976, a greater interest was sparked in the possibility of waste dump leaching, based on two discrete dumps which had been deposited for the North Channel project. These dumps contain 30 million tonnes of waste rock and were believed adequate as a basis for a small solvent extraction recovery operation.

An investigation programme was commenced, and a small cementation plant installed to recover some of the copper emerging, from the dumps, in solution.

The initial programme involved continuous monitoring of all liquors coming from the dumps to quantify future dump production levels. In parallel to this, studies were undertaken by BC Research of Canada, The University of New South Wales, in Australia, and BCL's Hydrology Department and company microbiologist. Design studies were also undertaken with a consulting engineering group for a SX/EW plant. This was not regarded as an area of doubt with regard to the ongoing project as the technology of solvent extraction/electro-winning was well developed.

The major difficulties associated with studying a potential economic operation were those associated with leaching rates of the waste dumps, the permeability of the dumps, the 'real' volume exposed to leach flows and the ability to collect economic copper tenors with high rainfall and ground water dilution. It must be realised that in the terrain in which the mine is located, many side streams are created from fairly large catchment areas. They are uneconomic as a water resource but sufficient to have a major dilution effect in a leaching situation.

The work undertaken by BC Research, in Vancouver, and the University of New South Wales, in Australia, illustrated it was possible to leach rock at economic levels with good copper tenor solutions. In both cases, the leach trials were in controlled conditions although attempts were made to simulate the problems associated with high rain and dilution effects. Some of the work at the University of New South Wales demonstrated that high irrigation flows can depress copper tenors. It is interesting to note that, although the trials were mostly conducted with acidified leach liquor, if fresh water was used as an irrigation medium the leach rates were not appreciably affected in the long term.

In general, the laboratory evaluations were encouraging for the future of a waste dump leaching operation using Bougainville waste. However, the liquor in the laboratory showed significantly higher iron and silicon in solution when compared to the actual liquors monitored at Bougainville. The work also confirmed the beneficial effect of rest cycles on copper tenors. This was not surprising in the view of the experience of other people in the hydrometallurgical industry.

ON-SITE STREAM MONITORING

Data were accumulated for an extended period on a number of streams in the Panguna dump areas and from it, a typical time/tenor curve was evolved. The difficulty with the curve is that it was not possible to quantify the time element accurately because of some deficiency of information relating to actual time of placement of some dumps. Figure 3 shows what is believed to be the typical development of a Panguna dump based on the best estimate time scale, with marginal dilution effects.

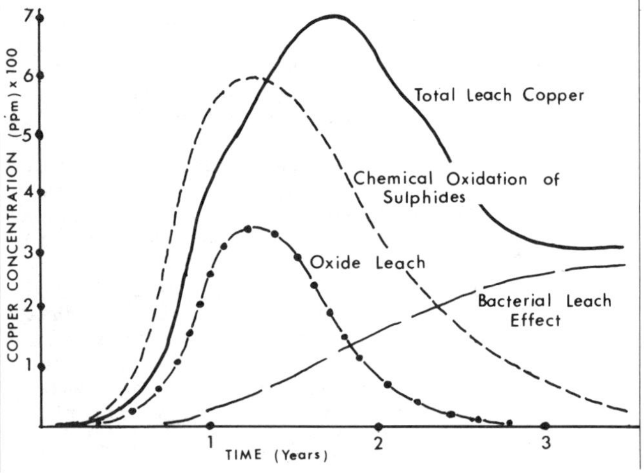

FIGURE 3. TYPICAL DEVELOPMENT OF PANGUNA DUMP

The changes in the dumps are thought to occur in four somewhat imprecise phases; as shown in Figure 3. The curves are presented as an interpretation of available data; a brief description follows.

Phase 1

Initial dump establishment and stabilisation period in which the physical and chemical environment is created for the subsequent dump activity.

Phase 2

Chemical oxidation of pyrite and other sulphide minerals in the dumps follows with a generation of acid in the mass. High dump temperatures are also seen.

Phase 3

The acid generated by the oxidation process attacks the oxide copper minerals in the dumps which are solubilized to join the liquor flows. This leads to a rapid increase in the copper tenor of the solution with peaks around 700 ppm. As the oxide copper is consumed, the copper tenor decreases.

Phase 4

The final phase is the development of bacteria in the dump. The temperature in the mass of the rock during the chemical oxidation of the sulphide minerals is too high for an effective bacterial growth. As the chemical oxidation reduces, the temperature declines, and a more suitable environment for bacteria survival is achieved. Copper tenor continues to drop to about 300 ppm which is held steadily for some years as a result of the bacterial action on these sulphide minerals.

Continued observation of the waste dumps suggest this is a form followed by the later dumps and is not dissimilar to the results of laboratory investigations.

OTHER INVESTIGATIONS

Dapera Leach Trial

A leach trial was conducted on the Dapera dump in the area noted as West DRV in Figure 2. The West DRV trial was to determine the effects of water irrigation and acidified leach liquor on copper production from BCL waste rock. Although the trial was planned for the "dry" months heavy rain affected the overall results. The rainfall was fortunate in that it demonstrated the future potential problems associated with the natural phenomena. It was also demonstrated that copper production from the dumps is more flow dependent than tenor dependent. Unfortunately, the acid trial was subjected to heavy rains and the effect of acid was masked. The acid appeared to offset the dilution effect of the rain as copper tenor was maintained, but in the long term the acid dosage required would not be commercially viable.

Hydrology

Hydrological studies were undertaken to determine flow balances of incoming and outflowing streams. The work left some questions unanswered because it was difficult to achieve a balance as flows were obviously dividing below surface and emerging at different points. A complicating factor in hydrological balance studies was the many sources of water which could not be isolated. As an example, the original exploration adit for the mine located at the head of the Kawerong Valley is some 40 metres below the surface of the waste dump. The adit is known to be discharging water into the dumps at a rate which cannot be determined. The North channel, although installed with waterstop on all joints, is known to be losing water into the valley at an unknown flow rate.

Temperature Readings

One of the features of the chemical activity within the dumps is the high temperatures generated during the early reaction. The most obvious sign of the activity in the dumps can be seen on a cool and wet day. After a rain, the dumps can be seen emitting clouds of water vapour from the dump mass. This is more evident in the newer dumps and on the dump slopes.

Over the past three years, temperatures have been measured in various spots. Temperatures as high as 85°C have been recorded at depth and near the surface of some dumps.

DUMP SETTLEMENT AND STABILITY STUDIES

No attempt is made to grade or compact the rock when dumped. Consequently, the mass will be open and considerable settlement can be expected.

The west DRV area was selected as a test area and survey stations were established for routine settlement measurements. An assessment of the records, by an engineering consultant, suggested settlements of 3-5 metres at the deepest point.

The majority of this settlement occurs early in the life of a dump with high initial rates, which over several months slow down. Seismic activity induces an acceleration in the settlement rates but over a reduced time scale. In 1976, after a major tremor of 7.2 magnitude on the Richter scale the rate returned to that seen in earlier settlement records. The influence of seismic action was also seen in other areas. Longitudinal cracks appeared in the high level haul road and movements in other dumps occurred in areas of weakness.

The duration of increased settlement rates can be broadly related to the tremor magnitude.

High rainfalls also tend to increase the rate of settlement.

Leaching of the waste dumps, either by irrigation or natural phenomena, will cause a breakdown of the rock structures creating a greater percentage of fines in the dumps. The effect is difficult to quantify, however, there is little doubt settlement rates will be greater than with unreactive dumps. This will also lead to differential settlement because of the varying chemical resistance of the rock types.

ROLE OF IRON

Although iron is believed to be an important aid to bacterial leaching of the copper sulphides because of the formation of a porous layer on rock/mineral surfaces, it also has detrimental features.

A typical analysis of the leach liquors monitored during the investigations is shown in Table 8.

TABLE 8

Typical Analysis of Leach Liquor

Element	Concentration - PPM
Cu	300 - 400
Fe	0.6
SiO_2	90
Mn	30
Ca	190
Mg	350
K	20
Na	20
Mo	0.1
PO_4	0.2
Zn	15
pH	3.5

A comparison of early site trials, recent laboratory investigations and the typical analysis suggest that iron is being released in the mass of the dumps in large quantities and is not emerging in the liquors.

The pH range of the emergent liquors is high enough to facilitate the precipitation of iron in the dumps after release by the acid reactions. The basic nature of some of the rocks consumes acid to cause precipitation and conversion to goethite.

The unanswered questions are:

1. "Will the iron continue to transport through the dump mass if acid liquor irrigation is initiated?"

2. "Will the iron remain in the dumps and effectively blind them to solution flow?"

The solution to these questions can only be answered operationally as it is impractical and uneconomic to open a dump to the necessary depth to study the issue adequately. Laboratory test results are difficult to relate to the real life situation where dump heights are 120 metres or greater.

The uncertainty of the effects of iron is a very real area of risk in the assessment of the practicality and economics of initiating a commercial leaching operation.

The other serious aspect of iron is that should it emerge from the dumps in leach liquors the high iron levels will impact in the recovery systems. The recycle of spent liquors (raffinate), containing high iron levels to the dumps will also facilitate dump blinding with iron oxide deposits.

MICROBIOLOGICAL ASPECTS

The laboratory studies in test columns gave results not dissimilar to the experience of others. Bacterial growth was negligible above pH 3. High pH sulphur oxidising bacteria appeared initially in column effluents before Thio bacillus ferroxidans emerged and before copper release was significant. Innoculation with leaching bacteria increased the level of viable bacteria in effluent but did not, in the short term, affect copper release.

Added nutrients had no effect on copper release or bacterial activity.

The major bacterium was shown to be Thiobacillus ferroxidans species. The growth of leaching bacteria was not as high as expected.

Low concentrations (3 ppm) of molybdenum were toxic to Bougainville bacteria.

Site studies revealed the presence of Thiobacillus ferroxidans and Thiobacillus thiooxidans species but it has not been possible to correlate viable counts with copper and iron release in the dumps. This is not totally unexpected as the progress of new dumps must effect the pH of liquor passing through the new rock. Bacteria would be destroyed and some precipitation of copper can be expected.

Limited studies were made of algal growth on waste dump surfaces and in the effluent streams. An algal sample taken from the West DRV dump top contained 266 ppm copper, on a dry basis, whereas Hormidium fluitans from the effluent contained 4030 ppm copper.

Although site results are not clear the observations combined with laboratory data confirm bacterial activity is established in the older waste dumps.

It is unfortunate that "dead" bacteria cannot be readily identified in streams. This information could be equally as valuable as knowing the level of viable bacteria.

RELATED ISSUES

Because of the narrow valley and waste dump heights haul roads are operational for long periods. The compaction of the surface rock, together with paving material is quite significant, in places several metres deep. All disused haul roads and dump tops are routinely ripped before overdumping. Unfortunately, there is an economic limit to the depth which can be ripped. This leaves a question which is difficult to answer, "What effect will this have on a leaching operation?"

Dump movement due to normal settlement and rock breakdown is expected to overcome the effect of these potential impervious layers.

An issue continuously under review at Bougainville is that of mine cut-off grade. A reduced cut-off grade will potentially increase mine life at the expense of copper directed to the dumps in waste. The effect on a leaching operation is obvious.

SUMMARY

The geographical location of Bougainville in the tropics presents a number of problems not normally encountered in dump leaching operations in other parts of the world.

All of these have a direct impact, and are very real risk areas, in the decision to proceed with a commercial dump leaching operation. That decision has not been made and it is difficult to further quantify the risk areas because of the issues discussed above.

It is beyond argument that the Bougainville waste rock leaches autogenously but the commercial decision is not so clear.

ACKNOWLEDGEMENTS

Many people and organisations have contributed to the leaching studies for Bougainville Copper Limited over the years. The culmination of the studies was a technical feasibility report which lists 117 references in its bibliography.

The author wishes to record his appreciation to all of those who have contributed to the project in the past and to the management of Bougainville Copper Limited for permission to publish this paper.

BIBLIOGRAPHY

1. Hopkins, W. R., et al, Unpublished internal consolidation report for Bougainvile Copper Limited "Technical Feasibility Study - Waste Dump Leaching at Panguna Mine Site" Volumes 1 and 2., 1980

Chapter 22

DEVELOPMENT OF IN SITU LEACHING BY GUNPOWDER COPPER LIMITED, QUEENSLAND, AUSTRAILIA

John E. Butler
Renison Goldfields Consolidated Limited
Sydney, New South Wales, Australia

Martin C. Ackland
Gunpowder Copper Limited
Gunpowder, Queensland, Australia

Peter C. Robinson
Renison Goldfields Consolidated Limited
Sydney, New South Wales, Australia

ABSTRACT

The development of in situ copper leaching at the Mammoth Mine of Gunpowder Copper Limited, following the suspension in 1977 of conventional mining/milling operations, is described.

Viability of water leaching of the chalcocite/pyrite orebody is currently being assessed by means of laboratory testwork, 'pilot scale' heap leaching, large scale leaching of remnant mineralized material, and the processing of a recently developed large scale test stope. Additional planned developments will expand the scope and nature of the remnant ore and trial stope programmes.

Current production is at the rate of about 3,000 tonnes per annum, the saleable product being cement copper which is recovered in a traditional launder system. The spent solution from cementation is recycled for further leaching.

Results of the testwork and their influence on the future scale of leaching operations and method of copper recovery are discussed.

INTRODUCTION

Due to the present economic climate, particularly the depressed copper market, conventional mining, milling and flotation operations have been suspended. However, due to the apparent amenability of the ore to 'natural' leaching, investigations have been initiated into establishing an in-situ leaching operation as a preferred alternative to conventional operations.

Cement copper production has been developed in parallel with the leaching trials and is presently at the rate of approximately 3,000 tonnes of contained copper per annum. Testwork includes controlled leaching of ore both in the laboratory and in-situ following blasting.

MINE DESCRIPTION

Gunpowder is situated approximately 115 km north of Mount Isa in north west Queensland. Current operations are centred around the Mammoth orebody some 3 km north west of Gunpowder township. Leases extend west 2 km to include the Esperanza orebody, which is somewhat lower grade, and cover an area 26 km north which includes Mount Oxide, a considerably lower grade orebody.

The region is characterised by high daytime temperatures during the summer months, and irregular rainfall. The area occasionally undergoes drought conditions for long periods, but adequate water supplies, capable of sustaining a conventional milling operation or an in-situ aqueous leaching operation, are available from a dam on the Gunpowder Creek.

Three major orebodies have been outlined at the Mammoth Mine.[1,2] No. 1 Orebody, which contains the current areas of leaching testwork, outcrops on surface and dips at about 70^0 to a depth of approximately 220m. No. 2 Orebody is down plunge and south of No. 1 Orebody, and extends from approximately 200m to at least 500m below surface. No. 3 Orebody is vertically below No. 1 Orebody and is thought to coalesce with No. 2 Orebody at depth.

Proven and probable ore reserves (>0.5% Cu) at Mammoth Mine total almost 33 million tonnes averaging 1.74% Cu.

MINERALOGY

The host rocks at Mammoth are impure sandstones, arkoses and quartzites.

The major copper bearing minerals are chalcocite and chalcopyrite, with minor amounts of bornite, digenite and covellite. In the upper sections of the mine, chalcocite is generally the dominant mineral, but in some areas, particularly towards the north, chalcopyrite predominates. At depth, chalcopyrite becomes more abundant, and in the lowest parts of the orebody, becomes the dominant copper mineral.

Pyrite is abundant and content shows marked variations up to 30%. In the areas mined to date the pyrite content has averaged approximately 10%; however for the remaining areas it will probably reduce to about 5%. As the pyrite content of the ore increases, complex associations of chalcocite with pyrite become highly significiant.[3]

Minor quantities of cobalt are present, generally as a nickel cobalt arsenide which tends to 'rim' pyrite grains.

BACKGROUND AND DEVELOPMENT OF IN-SITU LEACHING

The first recorded production from Mammoth Mine was in 1927 from an open cut 12m deep, yielding ore of approximately 15% Cu. Between 1948 and 1969 two shafts were sunk to about 33m and a limited amount of underground development took place. Total ore production to 1969 was only 6,719 tonnes averaging slightly more than 10% Cu.

In 1969 the leases were acquired by Surveys and Mining Limited and later that year an open cut mine was commenced in the vicinity of one of the original shafts, whilst an exploration decline was advanced 330 metres. In November, 1971 the mine was purchased by Gunpowder Copper Limited, and a mechanised, sub-level, overhead benching, mining method with access via a footwall decline was employed.

The Gunpowder concentrator was commissioned in April, 1970 and initially a large amount of oxidised ore from Mount Oxide was treated. Operation of Mount Oxide ceased in December, 1971.

Operation of the Mammoth open cut ceased in January, 1972, all subsequent production coming from underground.

Initially, run-of-mine ore was prepared for conventional flotation by three stages of crushing and two stages of ball milling. Nominal treatment rate was 350,000 tonnes of ore per annum, and in October, 1976 a rod mill was installed to increase capacity to 500,000 tonnes per annum. Concentrate was toll smelted in Japan.

Between 1974 and 1977 ore grades averaged approximately 2.7% Cu and between 5 and 15% pyrite. Copper recovery varied from 80 to 87% depending on acid soluble copper levels, which at times exceeded 10% of total copper, and pyrite content of the ore.

In addition to the effect on recovery, the rapid post-mining oxidation and decrepitation of the ore led to degradation of surface stockpiles, resulting in 'sticky' ore. Additionally erratic pH control in the flotation section resulted from acid generation within the ore.

High transportation costs and low copper prices resulted in the decision to place the mine on a care and maintenance basis in September, 1977. At that time no serious consideration was given to an in-situ leaching operation, although it was considered that copper could probably be effectively leached from the ore, since it was felt that there was a strong possibility of resuming conventional operations.

However, by mid 1978 it became evident that conventional reopening was becoming less likely and some preliminary testwork was undertaken by the University of New South Wales to determine the feasibility of aqueous leaching of Mammoth ore. This work, utilising finely ground ore, was undertaken between September and November, 1978, and indicated that the reactivity of the ore was extremely high. Under ideal laboratory conditions copper extractions approaching 100% were obtained.

Encouraged by these results a test heap of approximately 770 tonnes of ore from No. 1 Orebody was established. Irrigation of the heap continued for some 38 weeks and it was found that rapid copper dissolution occurred.

Consequently it was decided to extend the trial to include irrigiation of some 660,000 tonnes of broken material in No. 1 Orebody. This material was readily accessible to liquor sprayed directly into the old open cut. Liquor circulation commenced in July, 1979 and copper recovery, utilising 4 cementation launders, started in November, 1979. Production was at a nominal rate of 2 tonnes of contained copper per day.

In April, 1980 the decision was taken to expand operations to a production rate of approximately 2,000 tonnes of contained copper per annum, by installation of an improved liquor distribution system, 8 additional cementation launders and a belt filter for cement copper dewatering. This objective had been realised by August, 1980.

In order to determine the response of freshly broken ore to aqueous leaching an area of No. 1 Orebody was selected for a test leach stope, the area to be completely isolated from existing leaching operations. In December, 1980 approximately 55,000 tonnes of ore was broken into an existing void, and water irrigation of this material commenced in January, 1981.

Coincident with this test further laboratory testwork was initiated at the University of New South Wales, in an attempt to correlate laboratory leaching results and actual results from the stope leaching. As a natural progression, testwork on samples from Nos. 2 and 3 Orebodies would be used to predict the feasibility of aqueous leaching in the lower levels of the deposit.

DESCRIPTION OF TESTWORK AND FACILITIES

Laboratory Testwork

Initial testwork, in late 1978, was undertaken by the School of Biological Technology, at the University of New South Wales.[4]

Results from preliminary 'shake-flask' experiments confirmed by limited work on stirred reactors, indicated greater than 90% copper dissolution following grinding to minus 200 mesh. Additionally, the high acid generating capacity of the ore was demonstrated to be not only sufficient to sustain copper dissolution, but also to markedly affect iron dissolution.

Current testwork also at the University of New South Wales is centred on material emanating from the test stope and from Nos. 2 and 3 Orebodies. In addition to reactor runs, tests are being conducted in 50mm and 150mm diameter x 1.5m columns containing approximately 10mm material, and 'weathering' tests on sectioned large single particles of approximately 100mm diameter. It is hoped to correlate the results on No. 1 Orebody material from these tests, with actual results from the test stope, and then predict the amenability to leaching of material from Nos. 2 and 3 Orebodies.

Initial indications from reactor runs are that ore from the upper levels of Nos. 2 and 3 Orebodies behaves in a similar manner to that from No. 1 Orebody.

Optimisation of leaching conditions, by control of variables such as pH, temperature, oxygen availability and bacteria levels and types, will also be investigated in the present testwork programme.

Heap Leach Test

Following the success of the initial laboratory testwork, a test heap containing some 770 dry tonnes of material from No. 1 Orebody, was established in January, 1979.

Although the ore had been broken prior to the mine closure in October, 1977, efforts were made to obtain unoxidised material from deep within the relevant stopes. The material assayed 2.01% Cu; 6.5% Fe and 5.3% S.

The test heap was constructed in a redundant concrete settling vat to a height of approximately 2m and was irrigated by means of nylon soaker' hoses. Lixiviant was recirculated in a closed loop and evaporation, seepage and wind borne losses were compensated for by make up with fresh water by automatic level control in the liquor collection sump. Liquor tenors were monitored and from measured make-up and occasional bleed-off quantities, copper recovery was calculated. The initial test duration was 264 days.

In November, 1980 the heap was relocated on an impervious clay base, the objective being to determine the longer term leaching characteristics of the ore in order to develop a mining plan for the orebody. The relocation was due to the desire to have the operation close to the newly constructed cementation launders, thus rationalising the use of available personnel. The new construction is approximately 16m diameter and 2.5m high. The heap was resampled and was found to contain 1.35% Cu, indicating a leaching recovery of approximately 33% during the first 'stage' of the trial.

This second 'stage' of leaching continued for 154 days, during which time a further 11% of the copper was recovered. However, the bulk of this copper was recovered in the first 6 weeks of the trial, following which copper recovery reduced significantly. This was thought to be due to blinding of the heap by the accumulation of fines in the surface layer, a condition which was only marginally improved by ripping of the surface.

Rate of copper recovery is indicated in Figure 1.

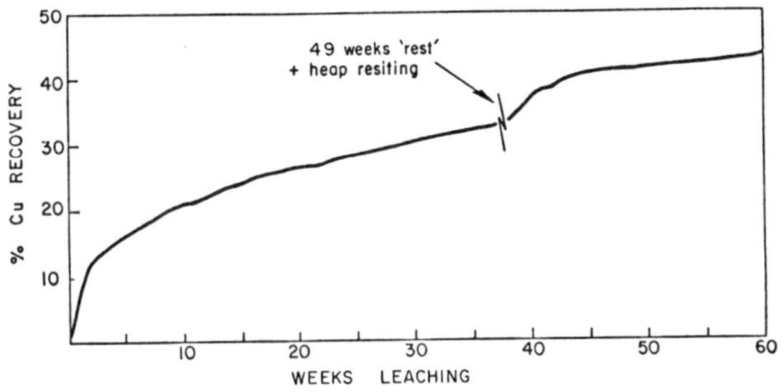

FIGURE I. CUMULATIVE COPPER RECOVERY FROM TEST HEAP.

Broken Ore Leaching

Approximately 660,000 tonnes of nominally 1% Cu ore had previously been broken but not removed from No. 1 Orebody; the top surface of this material is exposed in the floor of the old open pit, to which access has been cut off by the collapse of the floor into the underground workings.

In view of the encouraging results from the heap leach test, it was decided to conduct a large scale leaching trial on this material. The trial commenced in July, 1979, and is still underway.

Initially water was introduced through 100mm polypropylene pipes perforated in a spiral pattern with 6.35mm holes at 0.35m intervals, the pipes being laid on top of the broken ore. This system provided reasonable cover of the accessible surface, but was susceptible to damage from rockfalls from the faces of the open pit. The pipes have since been replaced by monitors and agricultural irrigators located on the upper levels of the pit, about 50m above the surface of the broken ore. These units are positioned in an attempt to maintain, as far as possible, even wetting of the broken ore surface. However, it is estimated that, due to the configuration of the open cut, less than 50% of the available ore is currently being wetted.

Pregnant liquor storage of approximately 10 megalitres was facilitated by the construction of bulkheads in the lower levels of the mine and liquor is pumped to surface using one of the two existing mine dewatering pumps, at the rate of approximately 90,000 litres per hour.

Leaching of the broken ore has now been extended to cover underground stopes which have recently been blasted to ensure accessible pockets of fractured ore. This is the first stage of a progression towards a possible full scale in-situ leaching operation and is intended to serve as a proving period for the mining and leaching operations whilst confirming copper recovery rates under actual conditions.

Figure 2 shows in simplified form the areas currently under irrigation.

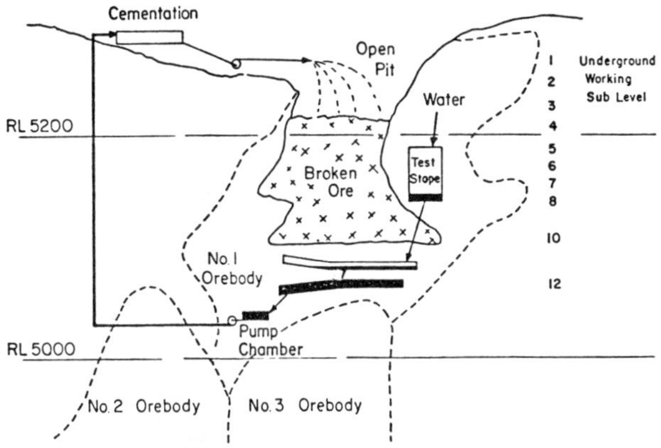

FIGURE 2. DIAGRAMMATIC REPRESENTATION OF GUNPOWDER LEACHING OPERATION.

Test Stope Leaching

By the middle of 1980, the practical feasibility of in-situ leaching had been demonstrated from the broken ore leaching trial. However this trial had been confined to the mass of broken ore in the open-cut and connected underlying stopes and, in view of the likelihood of considerable oxidation with standing time, it was thought imprudent to use these results to predict the behaviour of freshly broken ore in underground stopes.

As a consequence arrangements were made to fracture a block of approximately 55,000 tonnes of No. 1 Orebody material into a pre-existing underground void discrete from previous mining operations. Leachable material has since been calculated at 53,500 tonnes with assays averaging 1.0% Cu, 3.1% Fe, 2.7% S and 0.02% Co. A separate collection sump and liquor reticulation system were installed such that the trial is independent of other leaching operations and as a result little possibility of contamination of liquor with mine water or leaching solutions from other areas exists.

Fragmentation of ore in the stope was excellent, with the majority of visible material being in the 50-100mm size range.

Water, recirculated liquor or cementation launder tailings, can be introduced onto the top surface of the broken ore through a system of sprays which ensure complete coverage of the surface. Liquor collected in the sump beneath the stope, overflows a vee-notch weir, where it is sampled and flow is recorded. Pregnant liquor which leaves this sump finds its way into the sump receiving liquors from the main broken ore leaching operation.

The trial can be conducted either in a once-through 'flushing' mode or a 'recirculating' mode, dependent on liquor tenors etc. Flow rate of liquor to the top of the stope can be controlled as required.

Since it is envisaged that a solvent extraction/ electrowinning copper recovery system would be employed for larger scale operations testwork to date has been mainly in a 'flushing' mode utilising fresh water. Cementation launder tailings were utilised for a 9 day period from day 141 and further trials with this liquor are planned.

Although the feed water pH is approximately 6.7 it has been found unnecessary to acidify it as it has been determined that acid generation within the stope is extremely rapid.

Testwork to date has centred on determining the effect of flow rate on various leaching parameters. It has been found necessary to operate at a particular flow rate continuously for approximately 3 weeks before meaningful results can be obtained; the first week being required to achieve steady state conditions. Consequently for a large proportion of the available time the stope has been operated under sub-optimum conditions. Nevertheless during the first six months of operation copper recovery has been approximately 33%.

Figures 3 through 6 indicate the variation in various leaching parameters with the time of leaching and irrigation rate. In order to establish trends the results have been averaged over steady state periods. These relationships are discussed later in detail.

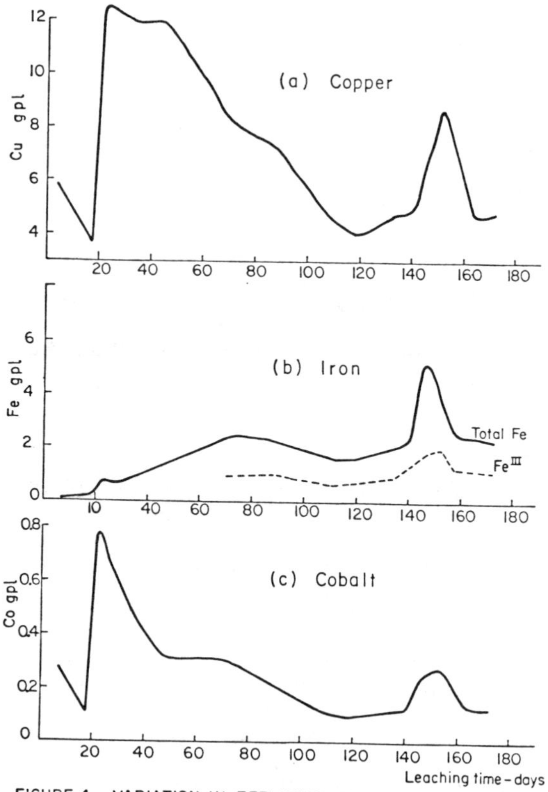

FIGURE 4. VARIATION IN EFFLUENT LIQUOR TENORS WITH TIME — TEST STOPE.

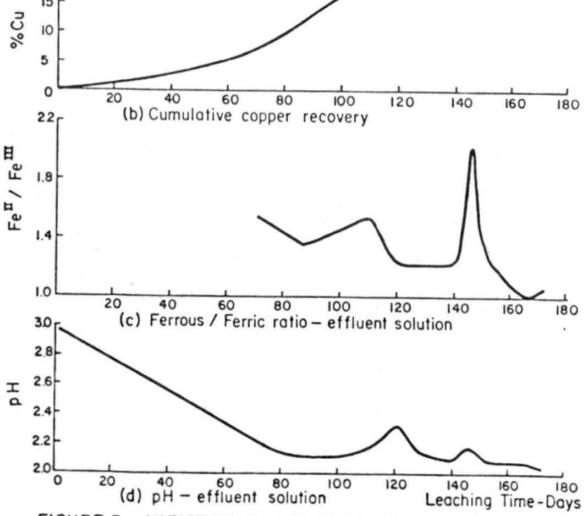

FIGURE 3. VARIATION IN LEACHING PERAMETERS WITH TIME– TEST STOPE.

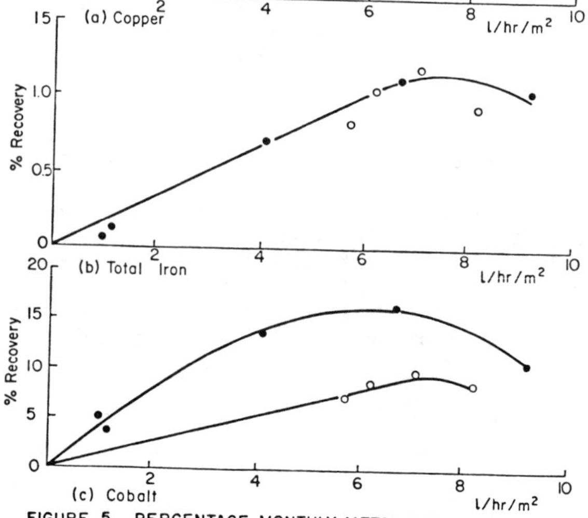

FIGURE 5. PERCENTAGE MONTHLY METAL RECOVERY RATE vs. IRRIGATION RATE – TEST STOPE.

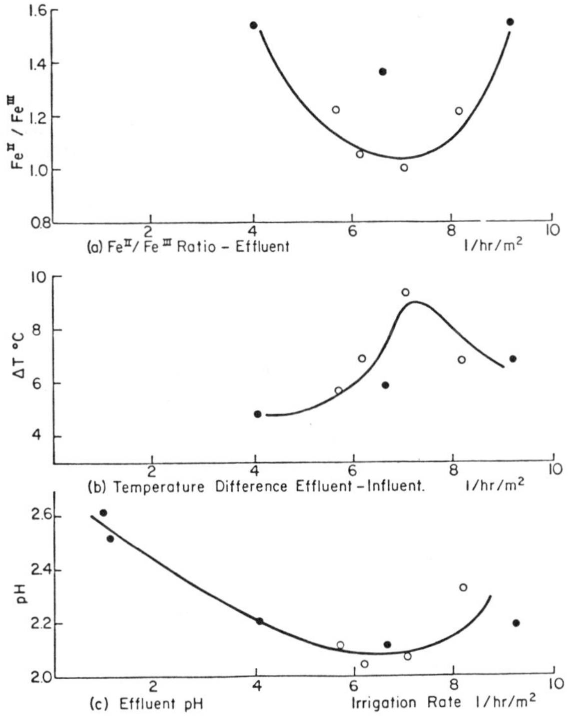

FIGURE 6. VARIATION IN LEACHING PARAMETERS WITH IRRIGATION RATE — TEST STOPE.

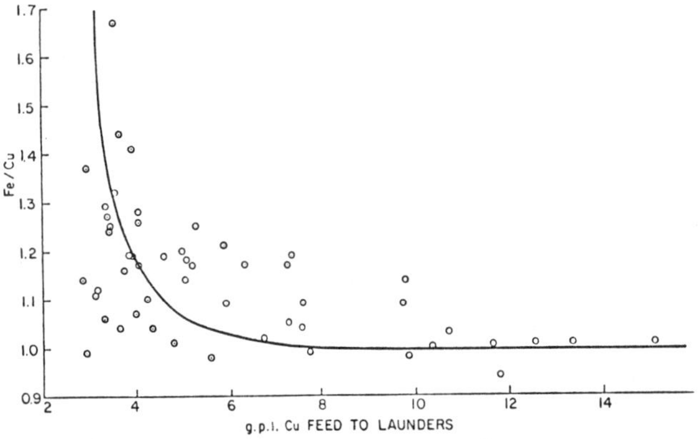

Figure 7 - Relationship between liquor feed grade to cementation launders and ratio of Fe dissolved to Cu precipitated from solution analyses.

Copper Recovery

Liquor from the various leaching areas, after collection in the main pregnant solution storage sump, was initially fed through four cementation launders, 10.40m long and 1.88m wide with a liquor depth of 0.7m above the scrap supporting screen. This facility was increased by an additional eight launders, of similar dimensions, in August, 1980, in order to increase copper production to a nominal 2,000 tonnes per annum.

Table 1 illustrates average launder performance over the past 12 months operation.

Table 1 - Average Cementation Performance 1980/81

	pH	gpl Cu	gpl Fe^{++}	gpl Fe^{+++}	gpl Fe_T	Fe^{++}/Fe^{+++}	gpl Co
Influent	2.48	5.99	13.25	2.23	15.48	5.9	0.32
Effluent	2.91	3.04	17.92	0.85	18.77	21.1	

Due to the relatively high copper tenor of the feed liquor to the launders, the "scrap-factor" for the operation has been comparatively low, averaging 1.5 during 1980/81. This ratio is extremely sensitive to liquor grade as indicated in Figure 7.

To date little effort has been made towards optimising launder production capacity, however as copper recovery from underground increases to above 3,500 tonnes per annum, this section will become a bottleneck in production.

For this reason, and also due to the high cost of supplies of detinned steel scrap to a location such as Gunpowder, a study has recently been initiated into the feasibility of replacing the cementation launders with a solvent extraction/electrowinning operation.

OBSERVATIONS AND CONCLUSIONS

The following observations and conclusions embrace all aspects of the testwork programme although most of the data referred to emanates from trial leaching in the test stope (refer Figures 3 to 6).

Leach liquor application rate

(i) Due to the high reactivity of the ore no limitation has been imposed on irrigation rate by decreasing copper tenor in the pregnant solution. During the first 6 months tenors were maintained to above 4 gpl Cu even at an application rate of 9 $l/m^2/hr$, although a fall-off in grade to below this level has been noted recently.

(ii) In agreement with previous observations,[5] copper production appears to increase linearly, with irrigation rate. However, copper solubilisation rate tends to tail off above an application rate of approximately 7.5 $l/m^2/hr$. Similar trends are also observed with cobalt and iron solubilisation rates.

(iii) As would be expected, the relationship between application rate and copper recovery rate is dependent on the absolute level of copper recovery. However, it can be seen from Figure 5(a) that, although there is a marked difference in copper recovery rate above and below 20% overall recovery, the maximum recovery rates occur at similar application rates.

Cobalt appears to behave in a manner very similar to copper but it is evident that iron solubilisation rate does not exhibit the same decline over the leaching period to date, presumably due to the relatively low level of overall iron solubilisation.

(iv) Corresponding with the maxima recognised for metal solubilisation rates are minima in effluent ferrous: ferric ratio and pH together with a maximum in temperature increase for the leaching solution passing through the stope.

The reasons for this combination of optimum leaching conditions at this particular solution application rate are unclear, but it is evident that acid generation is one of the primary rate determining factors at high irrigation rates. (Refer Figure 3(d)).

Intermittent solution application

Since there has been no need, to date, to improve liquor copper tenors, the effect of 'rest periods' has not yet been tested on the test stope. There are, however, some indirect indications that intermittent irrigation could be beneficial:-

(i) Between day 142 and 152 for test stope leaching, attempts were made to test the effects of irrigation with launder tailings. For a variety of reasons maintenance of continuous solution application was impossible and in consequence the stope underwent several periods of relative rest. On reinstatement of fresh water irrigation there was an immediate and sustained increase in copper recovery rate over that obtained immediately prior to day 142. Although this implies some benefit from the periods of rest, obviously the effect of iron injection into the stope could itself have a significant impact on reactivity.

(ii) Single particle 'weathering' tests in the laboratory indicate only minimal copper solubilisation and sample decrepitation over periods in excess of 100 days. These tests were conducted with a continuous drip feed to the 100mm sample such that a 'wet' surface is maintained. As a comparison, mineralised rocks at Gunpowder decrepitate extremely rapidly when left exposed to the atmosphere.

Orebody Permeability

Perhaps the most significant factor which will ultimately determine absolute metal recovery at Gunpowder will be the rate of compaction of the orebody under leaching conditions. Despite the absence of clays in the deposit decrepitation of material on the test heap virtually plugged the heap pores towards the end of the second period of leaching.

Although there are indications on the surface of the test stope that fines generation is significant, this does not yet appear to have had a major impact on copper recovery rate. However in line with findings on heap design it may be that stope height at 45m is excessive for optimum recovery, although the traditional difficulties associated with segregation of particle sizes during heap construction is avoided with in-situ leaching after blasting.

Oxygen Availability

No attempts have been made at air injection into the orebody over and above normal ventilation requirements.

However, it is likely that oxygen diffusion is a limiting factor during leaching at high irrigation rates when the ore surface is 'saturated' with leach liquor. This situation is similar to conditions applied in the single particle weathering tests.

Mineralisation

Whilst the pyrite content of the orebody is sufficiently high to eliminate the need for any acid addition, it would appear that this is not detrimental to leaching, since temperatures within the broken mass of ore do not seem to have exceeded the level critical to bacterial activity. The maximum temperature recorded for the effluent solution has been $40^{0}C$.

Highly significant are thought to be the intimate associations of both copper and cobalt minerals with pyrite. The creation of galvanic couples at these associations would tend to accelerate copper and cobalt dissolution whilst at the same time passivating pyrite[6] This factor may explain the strong correlation between copper and cobalt solubilisation for both the test stope and the test heap.

Bacteria

Gunpowder bacteria have been shown to be particularly vigorous and, although not yet categorised, are thought to comprise a mixture of T. Ferro-oxidans and T. Thio-oxidans species.

Although there is an absence of elements toxic to bacteria, it is noteworthy that the Gunpowder strains maintain a high degree of activity even in solutions with extremely high copper tenors.

Iron Levels

Figure 4 (b) confirms previous findings[6] that a ferrous iron concentration of approximately 0.4 gpl is required to initiate bacterial catalysis.

Although there has been an overall gradual decline in the ratio of ferrous/ferric iron in the test stope effluent, this ratio is generally greater than unity, which is somewhat unusual for this type of operation. As would be expected, pH follows a trend similar to the ferrous/ferric ratio.

Cementation Launder Operation

Efforts to date to optimise launder operations have been confined to increasing internal agitation by the installation of high pressure under-screen liquor injection. This resulted in a significant increase in launder capacity although, predictably, a finer cement was produced. Further avenues to be investigated include methods and cycles of scrap charging, flow patterns and scrap classification.

Low copper recovery (<50%) in the launders can be tolerated since upgrading of the recycled launder tailings during passage through the mass of broken ore presents no difficulties. This is in contradiction to previous observations.[7] Additionally, since there is a direct correlation between launder feed grade and scrap consumption (Figure 7) there is considerable incentive to maintain pregnant solution tenor above approximately 6 gpl Cu.

FUTURE DEVELOPMENTS

As a first step to determining the viability of an in-situ leaching operation at Gunpowder, a tentative production plan has been developed in the light of experience gained and from extrapolation of test results. Obviously, this plan will be modified dependant on future developments, but at present, the following basic assumptions have been made for the 1981/82 financial year:-

(i) An annual ore fracturing rate of approximately 700,000 tonnes utilising a long hole open stoping method, with only sufficient ore being removed from undercuts and slot raises to provide for an expansion of 20-25% on fragmentation. Various drilling patterns and powder factors will be evaluated to monitor the effect of fragmentation on leaching performance. The ore removed will either be tipped into old stopes or stacked in dumps on the surface, and in each case leached.

(ii) A copper recovery rate of 45% for the first year after breakage, thereafter recovery falling off rapidly to 10% in year 2 and 5% in year 3. Additionally a 'wetting efficiency' factor of 90% resulting from anticipated difficulties in solution application in some areas. This effectively means that copper recoveries assumed are 40.5, 9.0 and 4.5% respectively for years 1, 2 and 3 after breakage of individual stopes.

(iii) Copper production limited by existing cementation launder capacity to 4,000 tonnes per annum.

Investigations which have been completed so far have indicated that production of copper from the Mammoth orebodies by in-situ leaching is likely to be more successful financially than that from the resumption of conventional milling to produce sulphide concentrates. In order to establish greater confidence in the prediction of necessary investment, operating costs and returns, further studies are underway utilising the data generated from the testwork and operations described above.

Other areas requiring investigation are mine design, irrigation and liquor collection systems for large in-situ fragmented ore blocks, expanded power and service requirements, and market analysis. Studies on these aspects are proceeding in parallel with the metallurgical investigations in order to provide the inputs necessary to complete a full feasibility study on the future of Gunpowder as an in-situ leaching operation.

REFERENCES

1. Moore, G.P., "Geology and Mineralisation of the Mammoth Copper Deposit". The Australian Institute of Mining and Metallurgy, N.W. Queensland Branch - Regional Meeting, August, 1974.

2. Alexander, E.G. and Mitchell, M.A., "Mining Practice at Mammoth Mine". The Australian Institute of Mining and Metallurgy, N.W. Queensland Branch - Regional Meeting, August, 1974.

3. Jillett, I.M., "Metallurgical Operations at Gunpowder". The Australian Institute of Mining and Metallurgy, N.W. Queensland Branch - Regional Meeting August, 1974.

4. Harris, B. and Ralph, B.J., "Report on the Suitability of Gunpowder Mine Ores for in-situ Bacterial Leaching", March 1979 (Internal).

5. Catanach, C.B., "Development and In Place Leaching of Mountain City Chalcocite Orebody". Extractive Metallurgy of Copper Volume 2.

6. Murr, L.E., "Theory of Practice of Copper Sulphide Leaching in Dumps and In-Situ". Minerals Science and Engineering, Volume 12 No. 3, July, 1980.

7. Fletcher, J.B., "In Place Leaching at Miami Mine, Miami, Arizona". Society of Mining Engineers AIME Transactions, Volume 250, December, 1971.

Chapter 23

PILOT–PLANT SOLUTION MINING AND ENVIRONMENTAL MONITORING AT THE MUTOOROO COPPER MINE, SOUTH AUSTRALIA

J. H. Canterford
A. N. Smith
CSIRO Division of Mineral Chemistry
Port Melbourne, Victoria, Australia

K. F. Bampton
F. Bologiannis
Adelaide and Wallaroo Fertilizers Ltd.
Adelaide, South Australia, Australia

ABSTRACT

The Mutooroo copper deposit contains an oxidized zone, a supergene sulfide zone, and a primary sulfide zone. The deposit was worked between 1887 and 1917, and the workings have since filled with acidic saline water containing copper. The mine is in an arid, inhospitable area with no infrastructure. A pilot in situ leaching circuit for copper production and an environmental monitoring program have been in operation for about a year to provide experience for possible application to other copper deposits.

For copper production, acidic liquor (pH 2.5 – 3.0) is fed to seepage ponds along the line of the orebody. Percolation of the liquor through the bottom of the ponds to the workings was initially neglegible, so boreholes were drilled to direct the liquor into the orebody. Blasting at the bottom of some of the boreholes has increased the penetration rate. Pregnant liquor is airlifted from a production borehole, settled to remove solids, acidified to pH 2.5, and cemented. Scrap tinplate (~2 kg per kg of precipitated copper) is used for the cementation, and the current production of cement copper is 1 tonne/month. The liquor leaving the cementation circuit is sent to a holding/aeration pond before being reacidified and returned to the seepage ponds. Pumping rates and copper concentrations have varied from 50 to 300 1/min and from 80 to 200 ppm respectively.

Environmental monitoring is done by recovering solution samples from known depths in boreholes drilled adjacent to the line of the orebody, then analyzing the samples for heavy metals. The results to date show no unacceptable changes in heavy-metal concentrations. Variations in water–table levels are ascribed to climatic conditions rather than to lateral seepage of process liquors.

INTRODUCTION

To date, large-scale solution mining has received little attention in Australia, although the potential of this technique has not gone unnoticed. The possibility of long-term damage to the local environment has also been recognized. Because no two orebodies are the same, it is clear that the application of solution mining techniques and the assessment of environmental damage will be somewhat different for different orebodies, even though the general principles will be the same.

At present the most significant copper solution mining operation in Australia is at the Mammoth (Gunpowder) Mine in northern Queensland. This mine currently produces more than 3 000 t of cement copper a year.[1]

Adelaide and Wallaroo Fertilizers Ltd (AWF) operates a reduction roast/ammoniacal leach plant at Burra, South Australia, using locally mined oxidized copper ore. The plant produces high-purity copper oxide both for direct sale and for manufacturing a range of copper salts. As one facet of its program to ensure a continuing supply of raw materials for the Burra plant, AWF decided to investigate the feasibility of producing cement copper by in situ leaching. Whereas the relatively low production rates and the necessary smelting and refining costs tend to make this route unattractive for producers of copper metal (except in an auxiliary role), both the product and the scale of operation seem well matched to the requirements of producers of copper chemicals.

South Australia has hundreds of abandoned copper mines, many of which still contain appreciable quantities of a wide range of oxide and sulfide copper minerals. Few of these mines could be economically reopened and worked by conventional mining techniques, but there is every chance that a number of them could be subjected to solution mining techniques to produce, in aggregate, the several thousand tonnes a year of cement copper that AWF could use at the Burra plant.

Because of the lack of technical information relating to Australian conditions, and because of the financial risk involved in setting up a scientifically based solution mining operation, AWF applied to the Australian Industries Research and Development Incentives Board for financial support to carry out solution mining and recovery of cement copper at the Mutooroo copper mine, one of a number of apparently suitable copper mines surveyed by AWF. The application was referred to the CSIRO for comment on the technical feasibility of the proposed project.

Eventually, a collaborative research project was arranged between AWF and the CSIRO Division of Mineral Chemistry. This project has two major aims:

- To maximize copper production by optimizing the processing parameters throughout the overall circuit.
- To monitor the depth and composition of the groundwater in the surrounding country rock in order to ascertain if the leaching and metal-recovery operations are likely to have deleterious effects on the local aquifers, and, if so, how these might be minimized.

This paper discusses operations to date and outlines the development of the monitoring program. The Mutooroo project is regarded as experimental, and it is anticipated that both the mining technique and the monitoring program will be transferable to other, larger operations. The pilot plant is intended to become self-supporting after the experimental period.

MUTOOROO COPPER DEPOSIT

Location

The Mutooroo copper deposit (32°15'S, 140°56'E) is about 20 km SSW of Cockburn, a small town on the South Australia-New South Wales border on the Adelaide-Broken Hill road/rail link. The top of the deposit is about 210 m above sea level. The deposit occurs in an area that is sparsely vegetated with grasses, saltbush, bluebush, and a few scattered black oaks. The topography of the surrounding area is gently undulating with a maximum local relief of 10 m. The topography and sparse vegetation are such that the area is subject to significant air movement. The climate is typically continental, with overnight winter minima of less than 0°C and daytime summer maxima of more than 40°C. The average annual rainfall recorded over 30 years at Mingary (about 20 km NW of the mine) is 200 mm, while the average annual evaporation is of the order of 2 500 mm. Rainfall in individual years varies widely from the average, and a whole year's rain may fall in a matter of hours, with consequent flooding. The region is also prone to prolonged drought.

AWF holds Mineral Lease 4211 (14 hectares, or approximately 500 × 270 m elongated N-S), which covers the northern section of the mineralized zone and incorporates most of the previous workings. ML 4211 is totally enclosed by Exploration Licence 650 (24 km^2), which is held by Mines Exploration Pty Ltd (MEPL). EL 650 was extensively drilled during the mid-1960s, and a primary sulfide deposit of 8.7×10^6 t of ore grading 1.8% copper has been delineated south of ML 4211.

Geological and Mineralogical Description

Little is known about the detailed geology of the Mutooroo copper deposit and the surrounding country rock. Exploration and surface mapping in the 1950s was followed by an aeromagnetic survey and reconnaissance mapping in 1962. A substantial anomaly was found over the southern workings, and this discovery was followed up by the diamond drilling of 28 holes totalling about 14 000 m. The most detailed description of the geology of the area is given by Brooke,[2] who mapped the area on a scale of 1 in. = 400 ft and studied diamond-drill cores made available by MEPL.

The country rock belongs to the Middle Proterozoic (Carpentarian) Willyama Complex and consists of gneisses, schists, amphibolite, and pegmatite. The country rock, as well as the orebody, dips 45° to 70° with a NNE to NE strike. The mineralized zone extends for more than 2 km, with the ore width varying from less than 2 m to about 60 m. The average ore width seems to be less than 4 m. Brooke[2] described the orebody as a vein system emplaced in a set of overlapping shears that conformed with the bulk foliation of the country rock, although they were sometimes discordant in detail. However, on the basis of similarity to many other massive sulfide orebodies, AWF considers it to be a metamorphosed volcanogenic stratiform deposit.

The orebody displays well developed oxidized, supergene sulfide, and primary sulfide zones. The depths of the oxidized/supergene and supergene/primary transitions are of the order of 20 m and 45 m respectively. The water-table level is about 18 m. The principal minerals are reported[2] to be as follows:

Oxidized zone - atacamite, azurite, malachite, chrysocolla, cuprite, quartz, hematite.
Supergene zone - azurite, malachite, chalcopyrite, bornite, quartz, hematite, pyrite.
Primary zone - chalcopyrite, quartz, pyrite, pyrrhotite.

The primary sulfide zone is characterized by sharp contacts with the country rock and by the absence of significant disseminated mineralization.

Previous Mining Operations

The deposit was discovered in 1887, and 200 t of ore of unknown grade were produced in that year. The Mutooroo Copper and Silver Mining Company was formed and began mining in March 1888. Complete production data are not available, but the grade of ore mined was apparently quite high and the initial operations were regarded as highly successful. In 1894 the property was purchased by the Dry Creek Smelting Company, which almost immediately sold it to a small syndicate that in turn leased it to tributers. In 1906 the Mutooroo Copper Corporation acquired and rehabilitated the rundown mine. This company ceased active mining at the end of 1908 but continued to produce cement copper from acidic mine water (with sporadic ore production by tributers) until 1917. There was little activity at the mine from then until 1937, when the Copper Recovery Company undertook the recovery of cement copper from mine water. Apparently, little production resulted. From 1945 onwards several small groups worked the lease, with the last lessee attempting to heap leach dump materials in the early 1970s.

Most of the underground development at Mutooroo was carried out at the northern end of the mineralized area and is contained in ML 4211. Figure 1 shows a longitudinal section and a cross section of the workings. Within ML 4211, which has its southernmost boundary at Green's No. 2 shaft, a total of 7 shafts, 3 winzes, 9 drives, and 12 crosscuts were developed, resulting in a total interconnected volume of about 2 800 m^3. It is estimated that a further 4 100 m^3 were stoped, mainly in the oxidized and supergene sulfide zones, giving a total volume of 6 900 m^3 within the lease. The 18.3 m and 28.4 m levels were extended past the southern boundary of ML 4211, and it is estimated that these interconnected shafts, drives, and stopes are equivalent to a further 890 m^3.

Visual examination of ML 4211 reveals that all of the original shafts have either collapsed or been substantially filled in. Some shafts now exist only as slight surface depressions, while others are open to less than 10 m. It is likely that most of the collapsed areas are restricted to the oxidized zone and that most of the stoped areas have either been backfilled or filled in by the collapse of overlying rock. It is concluded that about 2 700 m^3 of development remains open below the water table.

Many of the waste dumps around the more extensively worked shafts now show signs of autoleaching, probably because of the generation of sulfuric acid from pyrite and pyrrhotite.

Table I. Depth, pH, and composition of typical groundwater and mine-water samples.

	Groundwater	Mine water
Depth (m)[a]	48.66	45.69
pH	6.56	4.71
Composition (ppm)		
Cu	0.60	158
Fe	5.15	288
Ni	0.20	2.35
Co	0.18	12.2
Zn	0.00	0.90
Na	6 790	8 100
Mg	1 210	1 820
Ca	850	440
Cl	11 380	12 780
SO_4	6 360	8 030

[a]Water table ≈ 16.5 m.

CURRENT OPERATIONS

The production optimization and environmental monitoring programs described below have been dictated by a number of factors, including the "tight" geological nature of the orebody and the surrounding country rock (in terms of their permeability and porosity), the ore reserves, the climatic conditions, the lack of infrastructure, the corrosive nature of the groundwater and the mine water, and the lack of access to the underground workings.

There is no power or water at Mutooroo. Thus, it is necessary to generate power on-site and a water-conservation program has had to be developed. The latter includes the establishment of a meteorological station and the construction of a 50 Ml dam about 400 m away on the headwaters of Pine Creek. The dam was built in May 1980 but is still empty as a result of drought. Only limited workshop facilities have been established on-site.

Both the groundwater outside the mineralized zone and, more particularly, the mine water are extremely corrosive, mainly as a result of the very high salt content of the water (Table I). As expected, the composition and acidity of the mine water vary as a function of depth; Fig. 2 gives typical data obtained before production pumping was begun. The sharp increase in heavy-metal concentrations in the 45 m region corresponds to a transition from the supergene sulfide zone to the primary sulfide zone. The mine water is characterized by relatively low acidity (pH 4-6), which indicates that there is a low level of *in situ* acid generation by sulfide oxidation. This may result from insufficient oxygen in the sulfide zone, but it can also result from a lack of bacterial activity. Most strains of *Thiobacillus ferrooxidans* have a low chloride-ion tolerance level.

The corrosive nature of the liquors to be handled at Mutooroo led AWF to conclude that metal centrifugal pumps and pipes would not be appropriate and that airlift pumping and acid-resistant polymers should be used. It is appreciated that airlift pumping is relatively inefficient and that it leads to inefficiency in the cementation circuit since it induces oxidation of ferrous iron to ferric iron in the pregnant liquor and results in increased iron consumption.

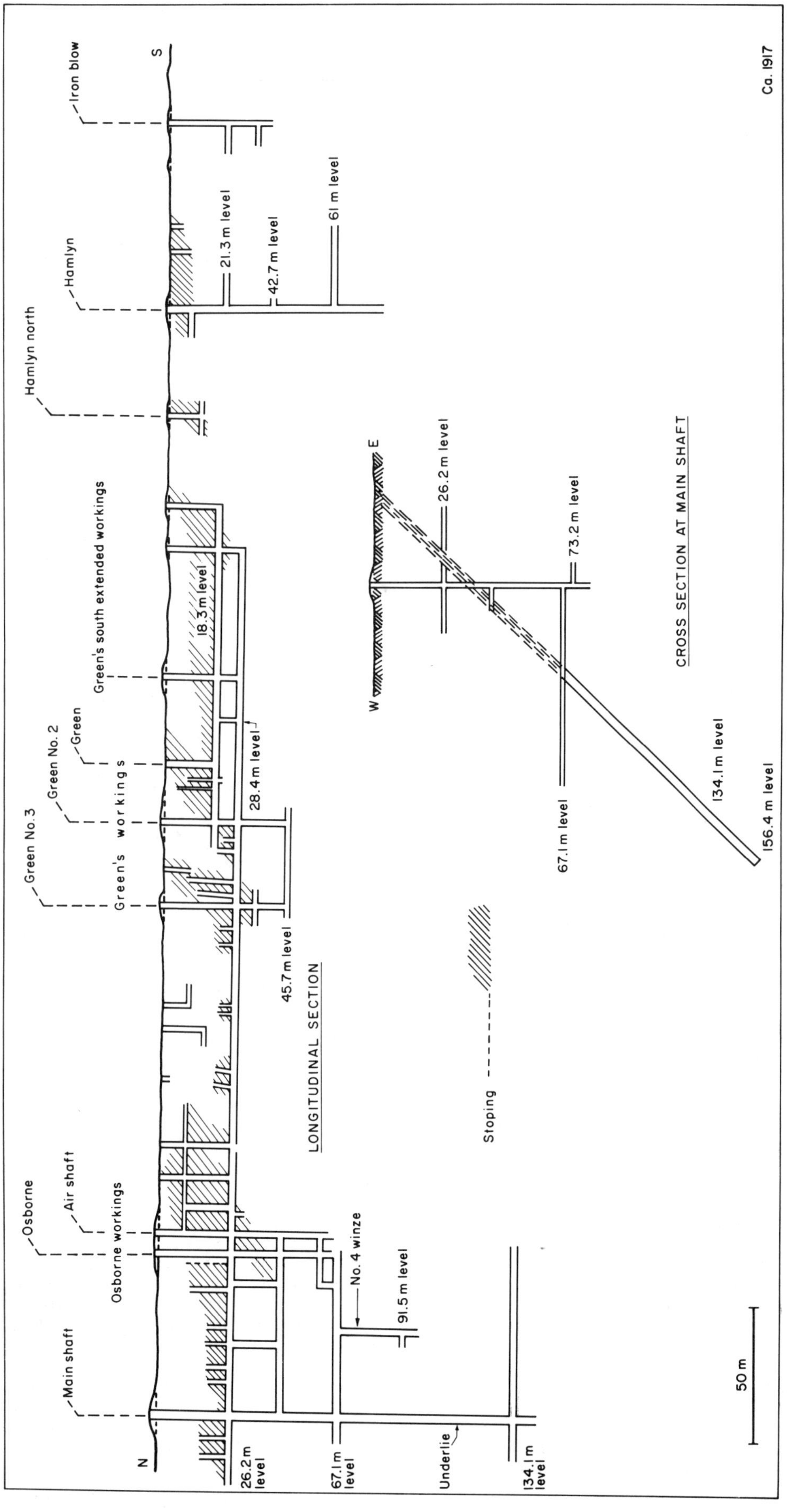

Fig. 1. Longitudinal and cross sections of the original Mutooroo workings.

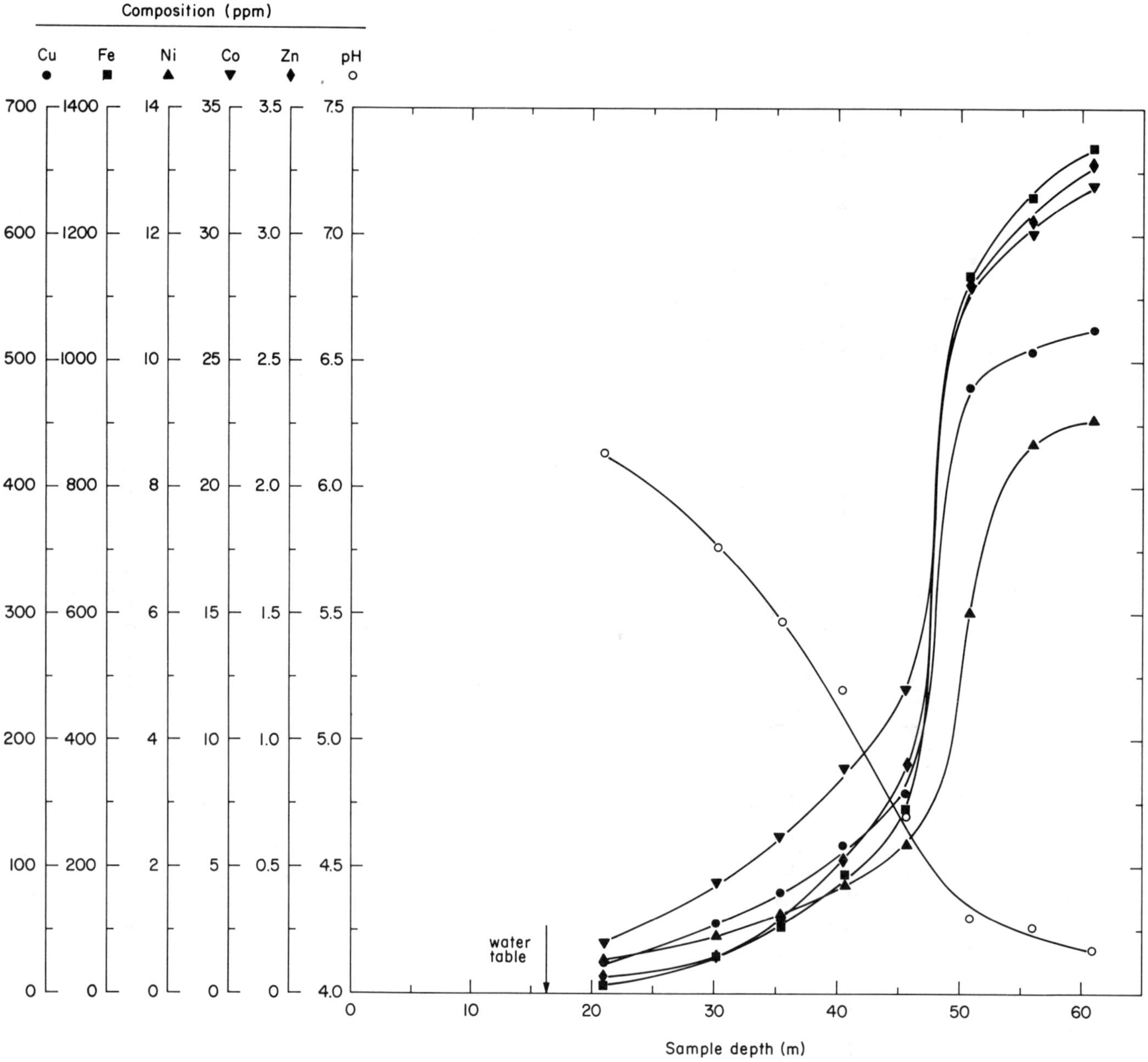

Fig. 2. Composition profile for borehole OH1 prior to the beginning of production pumping.

The shape of the orebody and the presence of the original underground workings have been used to advantage in siting the production borehole and in choosing the method of feeding the return liquor to the mineralized zone. There was little point in using, for example, a five-spot injection/production pattern. The lack of access to the underground workings and the almost total obliteration of the original shafts have created some problems in locating, for example, the inclined shaft sunk off the main shaft (also known as the engine shaft) (Fig. 1). These factors also mean that determinations of the liquor flow within the workings have been only partially successful since they are based on determinations of the water table and on composition profiles of several boreholes that intersect the workings.

It must be kept in mind that the Mutooroo operation is small and experimental. The size of the project is restricted not only by design but also by the ore reserves available within ML 4211. Allowing for 50% recovery from the oxidized zone and 20% recovery from the sulfide (supergene plus primary) zone, it is estimated that 1 000 t of copper should be recoverable from the lease without any additional underground work.

Wherever possible, equipment (e.g., tanks) from previous Mutooroo operations has been refurbished instead of new units being designed and built. Such equipment is not always suitable for long-term operation, but its use has resulted in minimal delay and expenditure. Establishment of a production bore, preliminary pumping tests, construction of seepage ponds, and other site establishment work was carried out in November and December 1979. Site activity resumed in May 1980 with the drilling of environmental monitoring bores and further construction work. The CSIRO began collecting data in July, and operations began in August.

The Production Circuit

The original concept at Mutooroo was to incorporate production pumping from one or more bores drilled so that they intersected the original interconnected underground workings. The latter were to be used as collection channels for the recovery of pregnant liquor. Barren liquor was to be returned to the mineralized zone and the workings via a number of seepage ponds constructed along the strike line of the orebody. Once pumping began, however, it became apparent that modifications were required, particularly to improve the percolation rates of the return liquor. Figure 3 shows the present production circuit; each major unit and any modifications made are discussed below.

The water-table level and heavy-metal concentration as a function of depth within the workings are determined daily by monitoring borehole OH2. Periodic monitoring of the air-shaft bore has also been carried out.

Production Pumping. Preliminary pumping tests were carried out at the air-shaft bore, which was drilled to 70 m and cased with 150-mm-diam PVC pipe to 67 m. The bottom 12 m of the casing were perforated to allow the ingress of copper-bearing liquor. The initial water-table depth and pumping rate were of the order of 18 m and 400 l/min respectively. After several weeks of pumping, the corresponding figures were 42.5 m and 100 l/min. It was decided that pumping should be carried out from further down in the workings in order to give greater submergence to the airlift pump. The best location for the production bore is obviously at the bottom of the inclined shaft sunk off the main shaft (Fig. 1). Attempts to locate this area were unsuccessful, although borehole OH1 intersected the mineralized zone at 82 m. Blasting at this depth resulted in "holing through" to the workings.

OH1 is currently used as the production borehole and has been cased to 77 m with 100-mm-diam PVC pipe, of which the bottom 3-4 m are perforated. The airlift pump consists of a 32-mm-diam polyethylene air line fitted with a stainless-steel "U" tube, the outlet of which is fixed at a depth of 76 m. The pumping rate has remained unsatisfactory (after initial improvement) at about 100 l/min because of further lowering of the water table to about 53 m. This has been caused by the slow return rate of the barren liquor to the workings. Methods of increasing the return rate are being investigated. The possibility of drilling an alternative production borehole is also being investigated.

Cementation. From the production bore, pregnant liquor is transferred to a series of three settling tanks in which suspended solids are allowed to settle and some of the ferric iron is precipitated as goethite and jarosite. The pH of the liquor drops to about 3.8 in the first tank as a result of hydrolysis reactions. Acid is added to the second tank to lower the pH of the liquor in the final tank to 2.5. The liquor in the final tank contains 80-200 ppm of copper. The settled liquor then goes to a series of four cementation tanks. Each of these tanks is constructed from an open concrete box culvert (2.4 × 1.2 × 0.8 m) to which rubber-lined steel endplates are bolted. To reduce the short circuiting of liquor flowing through the tanks, each tank is fitted with a tightly fitting epoxy-coated wooden baffle. The gravity flow from tank to tank is via cement-lined channels and is such that liquor enters one end of a tank and exits from the other.

The cementation tanks are filled with tinplate scraps from can manufacturers. The tinplate is not detinned and is supported on fiberglass mesh screens. Initially, air agitation was used in all of the cementation tanks to ensure that the product did not adhere to the tinplate surfaces and fell to the collection zone beneath the screens. The air agitation was used, along with air-operated diaphragm pumps elsewhere in the circuit, to keep the operation simple with a single compressed-air source. However, air agitation leads to excessive precipitation of iron oxide in the cementation tanks, so agitation by liquor recirculation has been substituted in tanks 1 and 2. Higher pressure pumps than those presently in use will be required to provide a satisfactory degree of agitation when/if higher copper concentrations in the pregnant liquor are achieved. At the present time, about 2 kg of tinplate are consumed per kilogram of precipitated copper. The overall cementation efficiency is of the order of 90%, and the bulk of the copper is precipitated in the first cementation tank.

The addition of tinplate and the removal of copper slurry from the cementation tanks are carried out regularly and are dependent upon the liquor flowrate and the copper concentration. The copper slurry is collected in a recovery trough and is spread out to dry on a concrete pad. The air-dried product is then packed in drums and trucked to the Burra works. Product quality has improved markedly since the beginning of production, but chloride and sulfate ions remain as significant impurities because there is no fresh water for washing the product.

Return Liquor. Barren liquor from the cementation circuit gravitates to an aeration pond. At present, no action is taken to induce oxidation and precipitation of the iron in the liquor, although some precipitation takes place spontaneously. From the aeration pond, the liquor is either siphoned or pumped to the five seepage ponds along the orebody. Acid is added to this liquor to give it a pH of less than 3. The amount of acid added to the return liquor (and to the pregnant liquor entering the cementation circuit) is about 10-15% higher than the theoretical amount because of the buffering caused by the high ionic concentrations of the liquors.

The return liquor is sequentially transferred to each seepage pond in order to maximize the contact area with the mineralized zone. A rest period is included in the cycle in order to induce aeration of the mineralized zone.

Initially, it was anticipated that the return liquor would percolate fairly rapidly from the seepage ponds into the workings, particularly since gossan was visible in the floors of several of the seepage ponds. A certain amount of liquor was expected to be lost by lateral migration. However, it soon became apparent that, except for ponds 3 and 5, little percolation was occurring. The water table within the workings remained fairly constant at 50-55 m, and pond-depth measurements (corrected for evaporation) indicated that only a small proportion of the liquor (<10%) was percolating through the pond floors. The rate of seepage from pond 5 has remained quite high (>100 l/min), and on two occasions caved-in areas have appeared (presumably close to stopes). However, little of this liquor seems to have returned to the pump at borehole OH1.

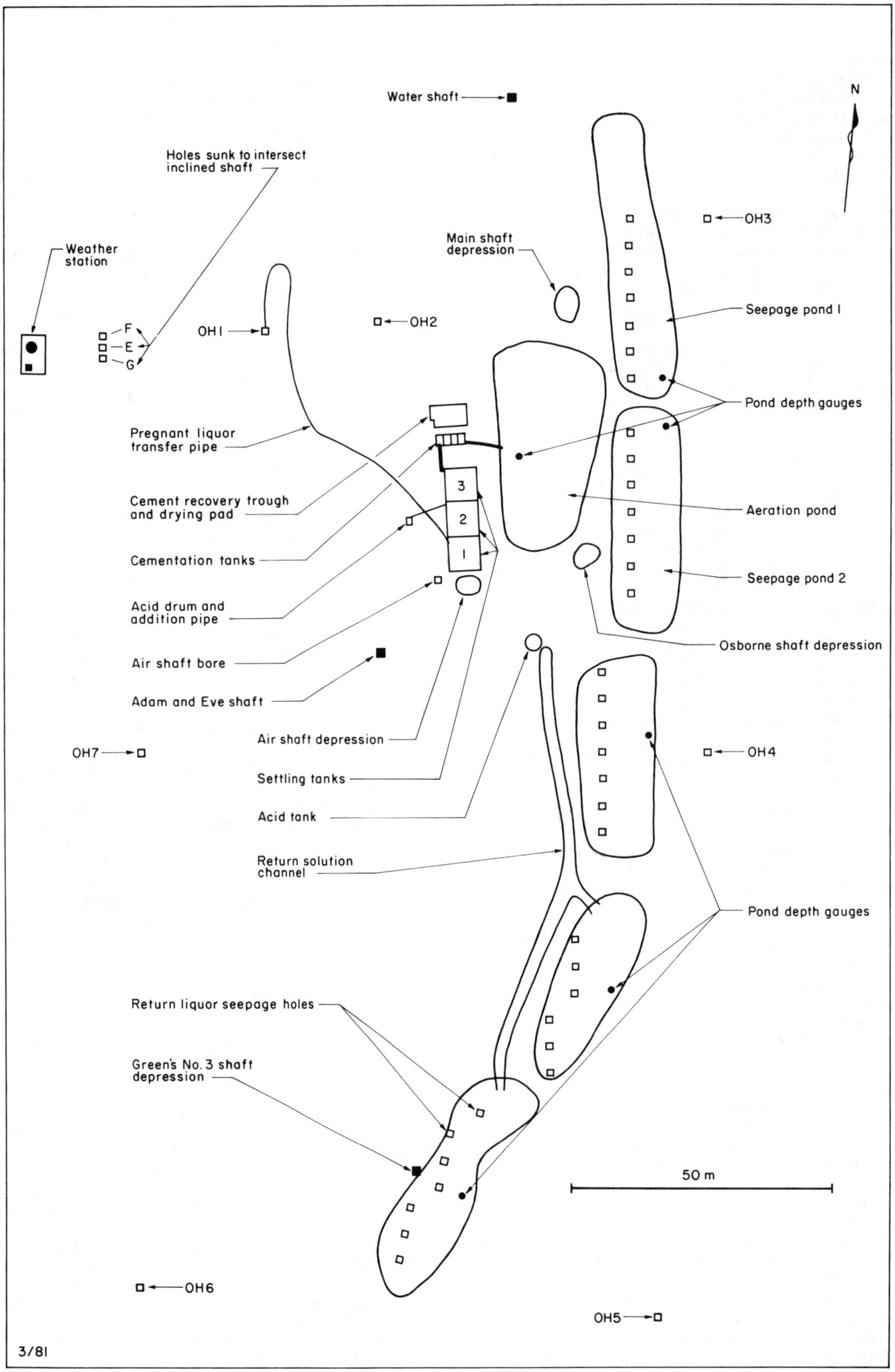

Fig. 3. Plan of current operations.

Some liquor was returning to the workings, particularly from ponds 1 and 2, for through borehole OH2 water could be heard running back into the workings. This borehole intersects the orebody and the interconnected workings and is used to monitor the water table and the composition profile in the workings. The borehole is cased with perforated 100-mm-diam PVC pipe. It was established that the water was entering at a depth of about 26 m, the same depth as the first level in the workings (Fig. 1), and that its composition was consistent with that of the liquors in the seepage ponds.

To increase the liquor return rate, seven boreholes cased with 50- or 80-mm-diam PVC pipe were sunk 6 m through the floor of each seepage pond. Blasting was carried out in the boreholes in ponds 1 and 2 to increase the permeability. The tops of the boreholes are all at the same level and are just above the pond floors, and the return liquor is fed in either directly or by overflow from the flooded ponds. These modifications reduce water losses from evaporation while ensuring that the return liquor has a high dissolved-oxygen content and makes direct contact with the mineralized zone.

The liquor return rate from ponds 1 and 2 has improved, but when ponds 3, 4, and 5 have been used, there has been no significant rise in the water table at the production borehole. It is concluded that the natural or artificial filling of the shafts (and possibly the drives) with waste material, which has a clay-like consistency, forms a physical barrier to the movement of liquor northward within the workings toward the production borehole. This barrier is associated with the Osborne shaft, which divides ponds 1 and 2 from ponds 3, 4, and 5. Evidence for this contention is provided by the fact that there has sometimes been a significant rise in the water table in the workings south of the Osborne shaft. At one stage, liquor was observed to well up through several of the boreholes sunk in pond 4. Since there has not been any significant change in the water table outside the mineralized zone, it is probable that a certain amount of water has flowed southward along the interface between the country rock and the orebody.

At the time of writing, attempts to blast the 26-28 m levels free of waste material in the region of the Osborne shaft are about to be made. At the same time, it is proposed to sink a new production borehole in the region of Green's No. 3 shaft. If it is not possible to open a channel to allow the free movement of liquor within the workings, production pumping will be done at both the old and the new production boreholes.

Environmental Monitoring

One of the potential dangers of any solution mining operation is contamination of the surrounding environment with acidic metal-bearing liquors. Because of the location and geology of the Mutooroo deposit, it might have been justifiable to view the development and implementation of a continuous environmental monitoring program as a wasted effort. However, it was decided right from the start that environmental monitoring should have a high priority. It was anticipated that no significant environmental problems would be encountered and that the experience gained would be invaluable when the solution mining techniques were transferred to other locations where the potential for damaging local aquifers was much greater.

The environmental monitoring program at Mutooroo centers around the frequent measurement of water-table levels and heavy-metal concentrations in a number of monitoring boreholes sunk in the country rock parallel to the strike line of the orebody. The location of the monitoring boreholes (OH3 to OH7) is shown in Fig. 3. Each borehole has been cased with 50- or 100-mm-diam perforated PVC pipe to a depth of more than 55 m (i.e., to at least 40 m below the natural water table).

Prior to the beginning of pumping, baseline data were collected over several weeks, with several sets of data being collected for each borehole. During the initial stages of production pumping, water-table levels and the heavy-metal concentrations of borewater samples recovered from 5 m below the water table were determined every 2-3 days, with complete profiling being done at less frequent intervals. At present, complete profiling is carried out every 2-3 weeks, with more frequent determinations of water-table levels.

Prior to the beginning of on-site operations, some time was spent in considering the most appropriate form of borehole monitoring for Mutooroo. It was immediately realized that because both groundwater and mine water were to be monitored, the use of direct-reading downhole devices was not possible since the heavy-metal concentrations (particularly in the mine water) would be beyond the operating ranges of such devices. Thus, samples of borewater would have to be recovered, diluted, and analyzed.

Consideration of the commercially available borewater samplers[3] indicated that none of these were suitable for use at Mutooroo, so it was decided to design and construct our own borewater sampler. This sampler[4] has been in continuous use at Mutooroo and has proved to be extremely reliable.

Since production pumping began, there has been no unacceptable change in pH or heavy-metal concentration as a function of depth in any of the monitoring bores sunk in the country rock. The greatest change has been in the iron concentration, which has increased by a factor of 2-3 to about 5 ppm. There have been changes of up to 0.75 m in the water-table levels, but this appears to be related to climatic conditions rather than to the lateral seepage of return liquor from the seepage ponds. Table II gives typical monitoring data. The lack of contamination of the country rock surrounding the mineralized zone is undoubtedly a result of its tight geological nature. Should there ever be a significant increase in heavy-metal concentrations in the monitoring boreholes, a suitable means of preventing it from occurring again or from spreading further would be implemented. In the latter case, for example, continuous pumping from a suitably placed borehole would be instituted.

The large differential between the water table in the workings and that in the surrounding country rock (>30 m) indicates that little, if any, mine water is moving into the surrounding country rock. It is not possible to totally exclude the presence of an aquifer carrying metal-bearing liquors away from deep in or below the workings, but if such an aquifer exists the present operations would not increase its flowrate or heavy-metal content.

Table II. Typical environmental monitoring data (from borehole OH3).

Date[a]	Water table (m)	Sample depth (m)	pH	Composition (ppm)				
				Cu	Fe	Ni	Co	Zn
21 July 1980	13.85	18.12	6.42	0.37	0.56	0.16	0.37	0.06
		23.22	6.49	0.33	1.83	0.16	0.32	0.05
		28.33	6.57	0.15	2.29	0.16	0.25	0.03
		33.45	6.57	0.17	2.24	0.16	0.25	0.03
		38.45	6.60	0.07	2.42	0.13	0.19	0.01
		43.63	6.64	0.09	2.22	0.13	0.18	0.00
		48.76	6.64	0.12	2.18	0.15	0.25	0.00
		53.87	6.53	0.35	2.11	0.16	0.26	0.01
14 December 1980	14.16	18.94	6.44	0.67	3.68	0.16	0.27	0.00
		24.04	6.49	0.49	4.22	0.15	0.27	0.00
		29.15	6.50	0.35	4.64	0.14	0.25	0.00
		34.26	6.48	0.35	4.79	0.15	0.19	0.00
		39.37	6.56	0.30	5.00	0.12	0.19	0.00
		44.91	6.55	0.21	4.97	0.14	0.22	0.00
		49.58	6.56	0.24	5.06	0.16	0.25	0.00
		51.82	6.54	0.26	4.94	0.15	0.24	0.00
20 March 1981	14.35	18.02	6.37	0.60	3.84	0.20	0.22	0.00
		23.12	6.46	0.34	4.58	0.22	0.20	0.00
		28.23	6.48	0.15	4.74	0.20	0.20	0.00
		33.35	6.52	0.00	4.77	0.19	0.20	0.00
		38.45	6.53	0.07	4.85	0.19	0.20	0.00
		43.55	6.55	0.00	4.86	0.19	0.23	0.00
		48.66	6.56	0.01	5.15	0.20	0.18	0.00

[a]Production pumping began on 19 August 1980.

CONCLUDING COMMENTS

In terms of the current production rate -- 1 t of cement copper per month -- the Mutooroo project cannot be regarded as a economic success. On the other hand, a considerable amount of knowledge has been gained in terms of production optimization, environmental monitoring, and operating costs. Since the accumulation of this knowledge was a major aim, it is appropriate to consider the project at least a partial success.

Despite the problems encountered at Mutooroo, most of which relate to the lack of access to the underground workings and to the previous mining history, the belief that solution mining has enormous potential has not waned. AWF is likely to undertake other solution mining projects and in so doing will undoubtedly put the experience gained at Mutooroo to good use. The production and monitoring programs will, of course, have to be tailored to the local conditions. This is not an original conclusion, but there is no doubt that practical experience reinforces what can be learned from the technical literature.

REFERENCES

1. Butler, J. E., Ackland, M. C., and Robinson, P. C., "Development of In-Situ Leaching by Gunpowder Copper Limited," this volume, pp. 253-261.

2. Brooke, W. J. L., "An Investigation of the Mutooroo Copper Mine, S.A.," B.Sc.(Hons.) thesis, University of Adelaide, 1966, 89 pp.

3. Riha, M., "Groundwater Sampling and Testing," Drill-80: Drilling, Sampling and Communication Problems, 10th Anniversary Conference of the National Water Well Association of Australia, Cowes, Victoria, 11-16 October 1980, 8 pp.

4. Canterford, J. H., Kuester, A., and Miles, J. G., "Borewater Sampling Device," Australian Patent Application PE 9181, 4 June 1981.

Chapter 24

EXPERIMENTAL OBSERVATIONS OF SOLUTION FLOW IN THE LEACHING OF COPPER–BEARING WASTE

L. E. Murr
Oregon Graduate Center
Beaverton, Oregon 97006

W. J. Schlitt
Process Technology
Kennecott Minerals Company
Salt Lake City, Utah 84147

L. M. Cathles
Department of Geochemistry
The Pennsylvania State University
University Park, Pennsylvania 16802

ABSTRACT

Large-scale (3.1 m dia. x 10.8 m high) column leach studies were conducted on two types of copper-bearing mine waste. One type was a quartzitic material containing chalcocite and pyrite; the other was a high carbonate intrusive rock containing chalcopyrite and pyrite. In addition to studying their copper leaching behavior, considerable effort was devoted to establishing the flow characteristics associated with each waste type. Factors studied in relation to flow behavior included the rock size distribution, rock decrepitation during leaching, flow channel size and location, column permeability, and formation of various alteration products resulting from solution-rock reactions. Both the mineralogical aspects and flow characteristics must be considered in order to explain the difference observed in the leach behavior of the two waste types.

INTRODUCTION

Solution flow in leach dumps and heaps or in columns of representative rock has been the subject of only limited studies. This is a regime involving unsaturated flow in semiconsolidated rock with a wide size distribution. Even the concepts of continuous flow systems are fairly recent, and have evolved from the search for better methods of operating industrial chemical reactors. (1-3) Waste rock consolidation and permeability play a key role in solution flow characteristics. Harris (4) suggested that bulk density variations in a leach dump could alter the particle-to-particle contact, and that large particle aggregates could act like single, large particles. In the limiting case where the bulk density approaches that of the undisturbed, consolidated rock, the entire waste unit would behave as a single particle. Roman (5) treated the phenomenon of bulk density and its effect on consolidation in terms of lixiviant flow paths or solution channeling in leach dumps. Like Harris, he assumed that no solution moved through particles or particle aggregates associated with zones of high bulk density. Solution flow paths are thus effectively separated by particle regimes, creating solution channels in the waste body.

While some attempts have been made to describe solution flow in leach dumps and related systems by utilizing computer model approaches and laboratory-scale simulations (5,6), there have been no direct, large-scale observations of solution movement in channels or of the actual channel sizes and distributions within a waste body. This paper describes a series of experiments involving large-scale, controlled waste leaching tests, conducted over a period of several years, and aimed at providing some direct observation of the changes in waste consolidation and permeability, the size and distribution of solution channels, the movement of such channels with continued solution application, and other features relating to solution flow during the leaching of low-grade waste. These results are then combined with mineralogical information to explain observed leach behavior.

A HISTORICAL PERSPECTIVE OF THE EXPERIMENTAL PROGRAM

Kennecott Waste Leaching

The experimental program was initiated in 1975 specifically to provide data to verify a waste leaching model devised by Cathles and Apps. (7) The initial phase of the experiments involved loading 1.6 x 10^5 kg of waste rock from Kennecott's Chino Mines Division into a 10.2 x 10^4 liter capacity, insulated stainless steel tank to provide a column of material 3.1 m in diameter and 10.8 m in height. This tank was part of a four-tank complex forming the experimental facility at the John D. Sullivan Center for In-Situ Mining Research at New Mexico Institute of Mining and Technology in Socorro (Fig. 1). This waste rock was sized and analyzed at the time it was loaded into the column; the loading was done carefully to ensure a fairly homogeneous distribution of rock sizes and to avoid load coning or size classification as the rock dropped into the column from the loading elevator.

After the loading column permeability experiments were conducted by blowing air into the bottom of the column and measuring the pressure difference at vari-

81-306

ous levels (heights) using different, equilibrated flow rates (flow rate in equal to flow rate out of the column at specific locations). Following these experiments, the column was flooded with water and allowed to drain down, while monitoring the drain-down rate. This provided an indication of the water retention and void space in the waste body. The permeability measurements described above were then repeated for the "wet" waste body. After this, leaching studies on this column were conducted for a period of about two years; these experiments have been described in detail in several publications (8-12) and summarized in a recent review. (13) Data were collected on copper recovery, oxygen consumption, temperature, solution composition, pH, and Eh, and bacterial activity throughout the program.

At the conclusion of the Kennecott waste-rock experiments, the permeabilities throughout the waste column were again measured, and a tracer, consisting of a pulse of concentrated NaCl solution, was added to the column influent in the final solution application cycle. The Cl^- ion concentration in the effluent solution was monitored as a function of time. Following a period of fresh water application when the final Cl^- ion concentration was reduced to the initial background level, a rhodamine B dye having ultraviolet sensitivities, was added to the influent and applied until recognizable concentrations appeared in the effluent solution. The dye was added to stain the waste in the column in order to mark the final solution flow path or paths. Following dye staining, the column was completely drained and unloaded by hand. As unloading progressed, photographs of the exposed surface were taken at successive levels in the column using ultraviolet illumination. In addition, a grid system was constructed at these locations to facilitate recording of channel size and location. Waste rock samples were also taken from specific grid locations for analysis. Larger rock samples were taken at the instrumented access ports through which samples had been extracted during the course of the leaching experiments. These samples, which weighed roughly 10^3 kg, were coned, quartered, and screened to determine the post leach rock size distribution.

These "post-mortem" analyses were designed to provide some information on column performance during leaching, particularly how the chemical and physical changes in the waste may have altered its leaching behavior. More importantly, these results were associated with changes in solution flow during leaching. Although little information could be gleaned which relates to the years of leaching, end points could be established which provide direct evidence of the changes in flow parameters that occurred, and the trends that developed during extended leaching.

Duval Waste Leaching

Before the end of the first year of leaching in the Kennecott waste column, a second column was readied for additional testing. This involved 1.7 x 10^5 kg of Duval-Sierrita, high-acid-consuming waste (2% carbonate material) which was loaded into another of the insulated stainless steel vessels shown in Fig. 1. This column differed from the Kennecott column in a number of ways. First, since solution flow was considered to be one of the most important aspects of the leaching process, an attempt was made to more directly monitor this by placing moisture probes (consisting of stainless steel electrodes spaced in a gypsum block) in a specific array at particular locations within the waste body as shown in Fig. 2. These probes were hand placed and waste rock was hand placed over them for some depth to ensure their integrity. They were connected to a central data collection system and the moisture levels at specific locations were measured every other day. The probes were calibrated in pH environments which were considered typical of the leaching regime (pH 1.8 to 2.6), and the resistance calibration was compared with actual moisture levels. Readings were taken in resistance values and automatically converted to calibrated percent moisture.

Table 1 compares the chemical analyses and other experimental parameters associated with the Kennecott and Duval waste used in these experiments. The analysis for a Phelps-Dodge waste which was included in the initial leaching experiments is also shown.

The Duval column leaching experiment was also scaled with smaller columns, and these experiments have been described in detail. (10-13) However, since drain-down characteristics, water retention, and post-mortem features were included in the present study, the Duval scale-test data is compared with that from the large column Kennecott and Duval experiments shown in Table 2.

TABLE 1

EXPERIMENTAL WASTE-BODY ANALYSES

Analysis	Kennecott-Santa Rita Waste	Duval-Sierrita Waste	Phelps-Dodge-Ajo* Waste
Quartz	68	29	32
K-feldspar (wt%)	2	38	40
Biotite (wt%)	2	5	-
Dickite (wt%)	-	5	-
Total Cu (wt%)	0.36	0.34	0.32
Non-sulfide Cu (Acid-soluble % of Total Cu)	0.22	0.03	-
Chalcocite (wt%)	<0.3	-	-
Chalcopyrite (wt%)	<0.1	0.8	0.7
Pyrite (wt%)	4	3.3	4
Carbonate (wt%)	<0.1	2	<1
Total SiO_2 (wt%)	75	71	-
Total Fe (wt%)	5.6	2.7	-

*Medium column (0.38 m diameter) and small column (0.10 m diameter) leaching.

TABLE 2

WASTE LEACHING TEST DATA

Parameters	Duval Large Column	Duval Medium Column	Duval Small Column	Kennecott Large Column
Diameter, d (meters)	3.05	0.38	0.10	3.05
Waste-body Height, h (meters)	9.75	2.64	0.69	10.5
d/h	0.290	0.145	0.145	0.290
Maximum Particle (Rock) Size (inches)	6	0.75	0.20	<4
Initial Drain-down Rate (ℓ/min)	4	0.20	0.10	40
Initial Onfluent Solution (Leach) Rate (ℓ/min)	2	0.100	0.005	2.6
Initial Air-flow Rate (ℓ/min)	17.5	0.875	0.044	35
Recovered/saturated Solution*	0.77	0.72	0.68	0.83
Waste-body Weight (kg)	1.7×10^5	4.1×10^2	7.0	1.6×10^5

*Recovered solution indicates water recovered; saturated solution indicates total solution (water) to fully saturate the waste body.

Like the Kennecott column described above, the large and medium Duval columns (Table 2) were subjected to chloride tracer tests following approximately 735 days of leaching. The tests were conducted in the manner described above. Following the chloride tracer tests, both the Duval columns were flushed with concentrated solutions of rhodamine B dye until the dye could be detected by ultraviolet illumination in the effluent. The large Duval column was then unloaded by hand as in the case of the Kennecott column and visible and ultraviolet photographs taken at various levels. Grids and references were installed at each level selected and hand samples as well as large samples for size analysis were collected at specific locations. Prior to unloading, permeabilities were also measured in the Duval large column as described previously.

The Duval medium column was systematically dismantled by sawing at 1-ft (0.31-m) intervals and photographed in ultraviolet light. A composite sample of the leached waste was also coned and quartered for size analysis and results were compared with the initial and final size analyses for the large Duval column.

RESULTS AND DISCUSSION

Fig. 3 illustrates the solution circuits characterizing the large experimental columns (Kennecott and Duval waste bodies). Fig. 4 compares the initial drain-down characteristics for the waste columns and smaller-scale waste units used in leaching of the Duval waste. Fig. 5 compares the complete leaching histories for the large Kennecott and Duval waste columns, along with smaller-scale tests involving other waste. Fig. 5a also shows the solution application cycles, referred to as solution flushes. The specific features of several representative solution flushes for the Kennecott and Duval large waste columns are illustrated in Fig. 6. These flush histories are indicative of the chemical changes taking place in the system as leaching continues. Fig. 6 shows that the Eh and iron levels in solution increase with time, while the pH decreases.

The iron and pH effects are the result of column aeration which gradually oxidizes the large amount of pyrite contained in the waste. As detailed by Schlitt and Jackson (14), pyrite oxidation will generate acid and add iron to the leach solution. Since the solution is recycled in a closed loop, iron and acid will continue to build up as leaching continues. Unlike large dumps which eventually reach a steady state that balances acid generation and iron solubilization with neutralization and hydrolysis, the intense column aeration and absence of any solution bleed* did not allow this to occur during the two-year column test.

Another feature of the large column leach tests is the observation that copper recovery efficiency (or rate of copper recovery) declines with the number of flushes for the Kennecott column (Fig. 6a-6c) but increases at first for the Duval column (Fig. 6d-6f). This and other differences observed when comparing the Kennecott and Duval large column leaching responses are the result of the fundamentally different chemistries and mineral compositions. In particular, the large carbonate fraction in the Duval waste must

*Solution was intentionally bled from the Kennecott column after flush 16 to reduce the high circulating load of iron and acid.

81-306

be neutralized before copper and iron can be solubilized without reprecipitating due to the low acidity (high pH) of the leach solution. On the other hand, the Kennecott material exhibits typical diffusion controlled behavior in which copper extraction declines as the readily leachable and most accessible copper is removed preferentially.

There are also differences in the effluent solution flow rates observed for the Kennecott and Duval waste flushes as shown in Fig. 6. For the Kennecott column, flushing starts faster and ends sooner in the early stages of leaching. In the Duval column, early flushes show solution is held up--in flush 3 the flush rate exceeds the application rate, flush solutions are delayed approximately 20 hours. With continued leaching this delay is progressively diminished. Fig. 7 suggests that the permeabilities actually change upon wetting and leaching in opposite directions for the two large columns, increasing for the Kennecott column while decreasing for the Duval column. The changes in permeability and flush behavior suggest that in the Kennecott column mineral dissolution increased column permeability with time but nevertheless made the flow through the column somewhat more homogeneous if it was changed at all. In the Duval column, on the other hand, limey material that initially blocked flow was evidently dissolved or broken, but gypsum precipitation progressively reduced the overall column permeability.

The decrease in permeability of the Duval column is consistent with the waste rock alteration impled by the size analyses shown in Fig. 8. Considerable decrepitation occurred throughout the column in the Duval waste. The Kennecott rock did not decrepitate to anywhere near the same degree. This is expected since the Kennecott material is high in quartzite which is relatively stable in acid. By comparison, the Duval waste has a high intrusive rock and limestone content which is subject to decomposition and alteration in acid solution. The fact that the Duval waste size distribution was altered everywhere, not just in flow channels, suggests much of the waste breakdown occurred as a result of the initial inundation of the column.

Chloride (Cl^-) tracer data for the Duval and Kennecott columns are shown in Fig. 9. The peak chloride effluent concentrations are greater for the Kennecott column, and the effluent chloride "pulses" are slightly narrower. This indicates that after leaching, the major flow channels of the Kennecott column are more permeable than those of the Duval column and flow through these channels occurs with less water saturation even though the channels occupy a much larger volume fraction of the waste than they do in the Duval column. Since the solution influent flow rates and the tracer pulse concentrations were identical for the two large column tests, the differences in pulses noted are indeed real.

The solution channel sizes and distributions observed in the reconstructed dye-stained composites for the two large columns are shown in Fig. 10. The Kennecott column shows much larger channel area in the composite-model views in Fig. 10a. Overall about 40 volume percent of the Kennecott waste was stained whereas only approximately 14% of the Duval waste was stained.

An examination of the moisture data associated with the Duval column indicates that the channel areas were small and that they shifted their position within the waste body. These features are illustrated by computer-generated moisture profiles, examples of which are reproduced in Fig. 11. Fig. 11a shows the calibration scheme utilized in generating moisture profiles from alternate daily moisture data while Fig. 11b and 11c illustrate some examples of moisture variations in the Duval column. The calibration shown in Fig. 11a was performed by creating ellipsoids having various eccentricities at each sampling location within the column so that at 100% moisture (fully saturated condition) the ellipsoids throughout the column would touch, establishing a continuous profile of moisture. In all cases, the minor axis of the ellipsoids (the ellipsoid diameter) corresponds to a calibrated percent moisture at each sample location. This is readily apparent in Fig. 11a by comparing the corresponding moisture levels shown. Fig. 11b illustrates the draining of water from the column, together with a gradual establishment of solution channels. Fig. 11c shows the movement of a solution channel from one side of the waste column to the other, corresponding to a solution flush and the subsequent movement of water through the column. An examination of the moisture profiles throughout the Duval waste column during the entire leaching period shown in Fig. 5b revealed that although the solution channels moved, this movement was erratic and infrequent. Only one other unambiguous channel shift, similar to that in Fig. 11c, was observed during the course of the Duval column leaching.

There is some evidence that the major flow channels in the Kennecott column may be migrated also. Fig. 5a indicates that after nearly two years of leach cycles in the Kennecott column, roughly 60% of the total copper was extracted. On examining Fig. 10, the average channel volume is observed to be roughly 40% of the waste volume (the average for the dyed zones at all levels throughout the waste column). This suggests that the channels moved as they did in the case of the Duval column shown in Fig. 11. However, at least part of this contact may have been associated with the initial inundation of the column. This flooding of the entire rock mass is believed to have removed much of the readily soluble copper which was low in both the stained and unstained regions. The flow channels need not have migrated very much in the Kennecott column. Because of the strong decrease in leach rate (Fig. 5) the total waste volume actually contacted by the lixiviant probably did not greatly exceed 60%.

There is additional evidence for the relative permanence in the solution channel geometries in the Kennecott waste column. The particle size distributions obtained for regimes clearly identified with the channel areas and with the areas outside the channels are shown in Fig. 12. The implications of Fig. 12 are that the wet areas within the solution channels leached in obvious pereference to the relatively unleached areas outside the channel perimeters. This gave rise to a more altered rock mass producing smaller particle sizes. Consequently, the whole rock analyses, especially the copper concentration within the channel zones, should be lower than that in the relatively unleached areas outside the channels. This is indeed observed experimentally as shown in Table 3. When data in Table 3 and Fig. 12 are combined, they suggest the relatively permanent, long-term nature of the solution channels reconstructed from dye-staining observations as shown in Fig. 10a.

TABLE 3

KENNECOTT LARGE COLUMN POST-MORTEM WASTE ANALYSIS

		Assays, Wt %									
		As Received Material*					Water Leached Material*				
		Copper		Iron	Sulfur		Copper		Iron	Sulfur	
Sample Designation†		Total	Nonsulfide	Total	Sulfide	Sulfate	Total	Nonsulfide	Total	Sulfide	Sulfate
Port #1	Dry	0.071	0.048	2.45	2.24	1.17	0.056	0.033	2.25	2.02	0.50
	Wetted	0.063	0.054	3.05	2.61	2.55	0.035	0.025	2.01	1.97	0.64
Port #2	Dry	0.023	0.012	2.35	1.60	0.57	0.016	0.005	2.25	1.46	0.14
	Wetted	0.019	0.010	4.20	4.24	2.58	~0	~0	3.59	0.68	~0
Port #3	Dry	0.022	0.014	3.00	2.66	1.02	0.006	~0	2.71	2.37	0.14
	Wetted	0.018	0.010	3.65	3.96	2.37	~0	~0	3.06	3.12	~0
Port #4	Dry	0.053	0.028	3.20	2.57	1.02	0.038	0.013	3.05	2.40	0.52
	Wetted	0.020	0.014	3.35	5.84	1.59	~0	~0	2.98	5.45	0.42
25 Feet	Dry	0.051	0.038	2.71	2.18	1.17	0.034	0.021	2.57	2.01	0.66
	Wetted	0.087	0.062	5.20	6.11	2.82	0.059	0.034	4.94	5.62	1.35

*As-received material refers to material collected during the post-mortem analysis while the water leached material refers to the same samples after water flushing to remove water soluble salts, etc.

†Dry designations refer to material outside the dyed channel zones as shown in Fig. 10a; wetted designations refer to material within the dyed zones and presumed to have been in a solution channel. The port designations refer to the following distances from the upper surface of the waste column: Port #1, 6.2 m; Port #2, 3.9 m; Port #3, 2.4 m; Port #4, 1.5 m; 25 feet, 7.7 m.

Fundamentally different behavior in the Duval medium sized column is suggested by the tracer data (Fig. 9c) and the dye-staining observations shown in Fig. 10c. These indicate that uncharacteristic wall channeling dominates the solution flow in the medium column. The fact that small columns (the Duval medium column in Fig. 10) exhibit significant wall channeling indicates that laboratory scale column leaching cannot provide dependable scale-up data for copper extraction.

Another interesting aspect of solution flow in both large columns, particularly the Kennecott one, is the variation in the copper content of the column effluent with time. The copper level is high initially, goes through a minimum, then increases at the end of the drain-down when the outflow rate is dropping (Fig. 6). One explanation for this is that the large channels with very high void volume will take the bulk of the flow. The first solution to pass through these zones will remove most of the solubilized copper and account for the high initial copper content. As water application continues, solution will short-circuit through the major channels without picking up much additional copper. However, these zones will drain quickly once solution application ceases. Then the major contribution to the column effluent will be the smaller channels where capillary forces retard flow. Since these paths have a smaller ratio of cross sectional area to rock surface contacted, there will be less dilution and higher copper concentration. The smallest interstices will drain the slowest and will have the least dilution, causing effluent copper concentration to increase continually until drain-down ceases. This indicates leaching and flushing will occur from minor channels, even though major channels are present.

SUMMARY AND CONCLUSIONS

Observations made regarding the flow and leaching behavior of the two large columns are summarized in Table 4. The fluid flow observations clearly help explain observed differences in leaching. The low (4%) copper extraction from the Duval column is consistent with the low percent of waste contacted by fluid (14%). In the Kennecott column, 60% of the copper was extracted and 40% of the waste was contacted by leach solution at the end of leaching. The two wastes also behaved quite differently during leaching and this helps explain the poor fluid leaching reputation of limey waste. The Kennecott waste tended to become more permeable and the solution distribution somewhat more uniform with leaching. The Duval waste becomes less permeable due to the precipitation of large amounts of gypsum. Thus the lime content of the Duval column delayed leaching by initially keeping the pH too high to allow appreciable copper solubility. However, the lime content of the waste made later leaching more difficult as well by dissolving carbonate and precipitating gypsum in such a fashion that the overall permeability of the column was reduced. The tendency for the Duval column to plug with gypsum is probably reflected in the Cl^- tracer tests which showed a slower through-flow for the Duval column despite a higher initial permeability, and despite the fact a much smaller percentage of the waste was wetted.

TABLE 4

SUMMARY OF OBSERVATIONS ON BEHAVIOR OF LARGE COLUMNS

		Kennecott	Duval	Reference
Total Cu Extraction, %		60%	4%	Fig. 5
Initial Time to Reduce Maximum Drainage Rate by 90%		-15 hrs	-140 hrs	Fig. 4
Drainage Rate After Leaching		A bit slower	Faster (-20 hrs)	Fig. 6
Permeability (Darcies)	Dry	.2 (bottom)-.4 (top)	.8 (bottom)-2.7 (top)	
	Wet	.2 (bottom)-.4 (top)	.6 (bottom)-1.0 (top)	Fig. 7
	After Leaching	.4 (bottom)-1.2 (middle)-.5 (top)	.3 (bottom)-.5 (top)	
Waste Size Distribution After Leaching		Slightly broader (more fines) in wetted zones after leaching	Much broader everywhere after leaching	Fig. 8 Fig. 12
Slug Chloride Tracer Test Conducted After Leaching		Higher peak concentration, slightly narrower shape	Lower peak concentration, slightly broader shape	Fig. 9
Dye Staining After Leaching		-40% total waste contacted	-14% total waste contacted	Fig. 10
Evidence for Flow Channel Migration		60% copper extracted but only 40% waste contacted at end of leaching	Migration shown by shifts of high moisture areas	Discussion and Fig. 11

Gypsum plugging may have promoted channel migration in the Duval column, although the flow channels may have also migrated in the Kennecott column.

This paper is mainly intended to report flow observations made on two large columns containing waste material of quite different mineralogy. The results are admittedly preliminary. Not all the important alteration reactions or the precise means by which they effect changes in column permeability have been elucidated. The observations do however indicate that the flow distribution and the changes in flow distribution during leaching can be very different for different waste types. The physical aspects of fluid flow help explain differences in leaching in the columns and in the field, and are clearly important if the leaching process is to be understood and optimized.

REFERENCES

1. Levenspiel, O., Chemical Reaction Engineering, Wiley, New York (1962).
2. Denbigh, K. G., Chemical Reactor Theory, Cambridge Univ. Press, England (1965).
3. Szekely, J. and Themillis, N. J., Rate Phenomena in Process Metallurgy, Wiley-Interscience, New York (1971).
4. Harris, J. A., "Development of a Theoretical Approach to the Heap Leaching of Copper Sulfides,: Proc. Austrailian Inst. Min. Metall., 230, pp. 91-92 (1969).
5. Roman, R. J., "Solution Channeling in Leach Dumps," Trans. Soc. Mining Engrs. AIME, 262, pp. 73-74 (1977)
6. Jacobsen, R. H., et al., "Optimization of Dump and Heap Leaching--Theoretical and Practical Studies," Print No. T IV a3, Joint Meeting of MMIJ-AIME, Tokyo, Japan, pp. 1-14 (May 24-27, 1972).
7. Cathles, L. M. and Apps, J. A., "A Model of the Dump Leaching Process that Incorporates Oxygen Balance, Heat Balance, and Air Convection," Met. Trans. B, 6B, pp. 617-625 (1975).
8. Murr, L. E., et al., "Chemical, Biological and Metallurgical Aspects of Large-Scale Column Leaching Experiments for Solution Mining and In-Situ Leaching," In-Situ, 1, pp. 209-233 (1977).
9. Cathles, L. M., Reese, D. A., and Murr, L. E., "Dump Leaching--Theory, Experiment, and Practice," American Nuclear Society Topical Meeting on Energy and Mineral Resource Recovery (Proc.), April 12-14, 1977, CONF-770440, pp. 584-595.
10. Murr, L. E., "Observations of Solution Transport, Permeability, and Leaching Reactions in Large, Controlled, Copper-Bearing Waste Bodies," Hydromet., 5, pp. 67-93 (1979).
11. Murr, L. E. and Berry, V. K., "Observations of a Natural Thermophilic Microorganism in the Leaching of a Large, Experimental Copper-Bearing Waste Body," Matall. Trans. B,10(B), pp. 523-531 (1979).
12. Cathles, L. M. and Murr, L. E., "Evaluation of an Experiment Involving Large Column Leaching of Low-Grade Copper Sulfide Waste: A Critical Test of a Model of the Waste Leaching Process," Leaching and Recovery of Copper From As-Mined Materials, Schlitt, W. J. (ed.), Soc. Mining Engineers of AIME, pp. 29-48 (1980).
13. Murr, L. E., "Theory and Practice of Copper Sulfide Leaching in Dumps and In-Situ," Minerals Sci. Engr., 12 (No. 3), pp. 121-189 (1980).
14. Schlitt, W. J. and Jackson, J. S., "In Situ Generation of Acid During Dump Leach Production of Copper," In-Situ, 5 (No. 2), pp. 103-131 (1981).

ACKNOWLEDGEMENTS

This research was supported in part by Kennecott Minerals Company and by the National Science Foundation (under RANN Grants AER-27-03758 and AER-76-03758-A01, and DAR Grant 7826167). It is also a pleasure to acknowledge members of the NSF Project Oversight Committee who influenced portions of this research endeavor, namely Dr. M. E. Wadsworth, Dr. N. Arbiter, Dr. James Grunig, Dr. K. L. Temple, Dr. J. A. Apps, D. A. Reese, and R. D. Groves. The help of Dr. J. D. Stephens is also greatly appreciated.

Fig. 1. View of the experimental column leaching facility at the John D. Sullivan Center for In-Situ Mining Research. The facility consists of four 10.2 x 10^4 ℓ (27,000 gal.) capacity insulated stainless steel towers.

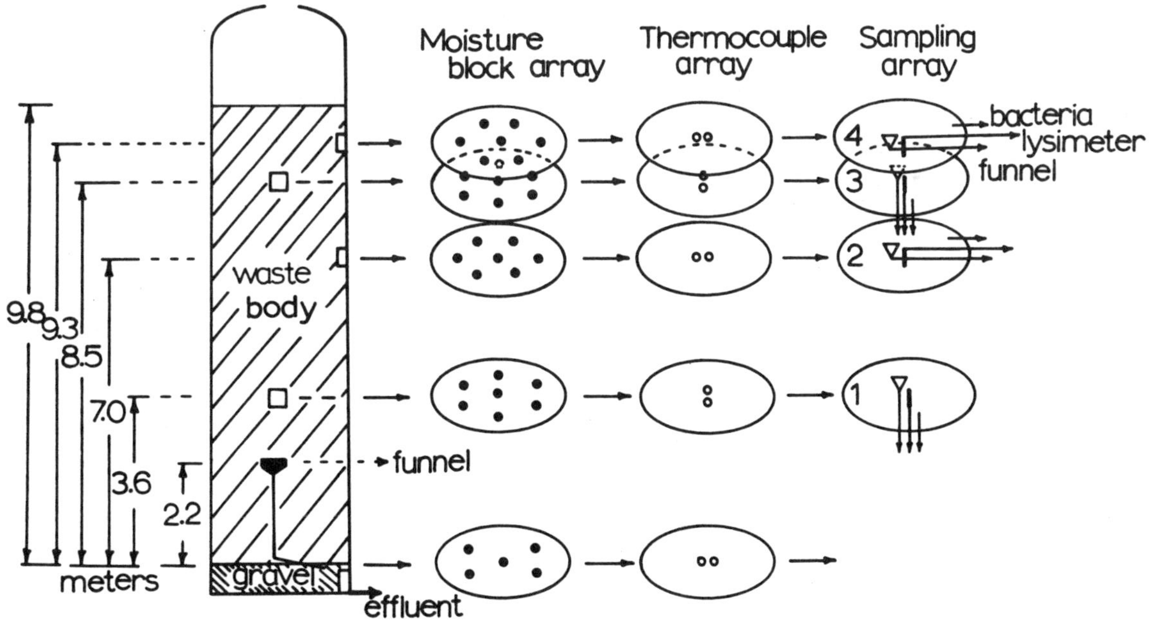

Fig. 2. Schematic view of the Duval waste column sampling and instrumentation arrays.

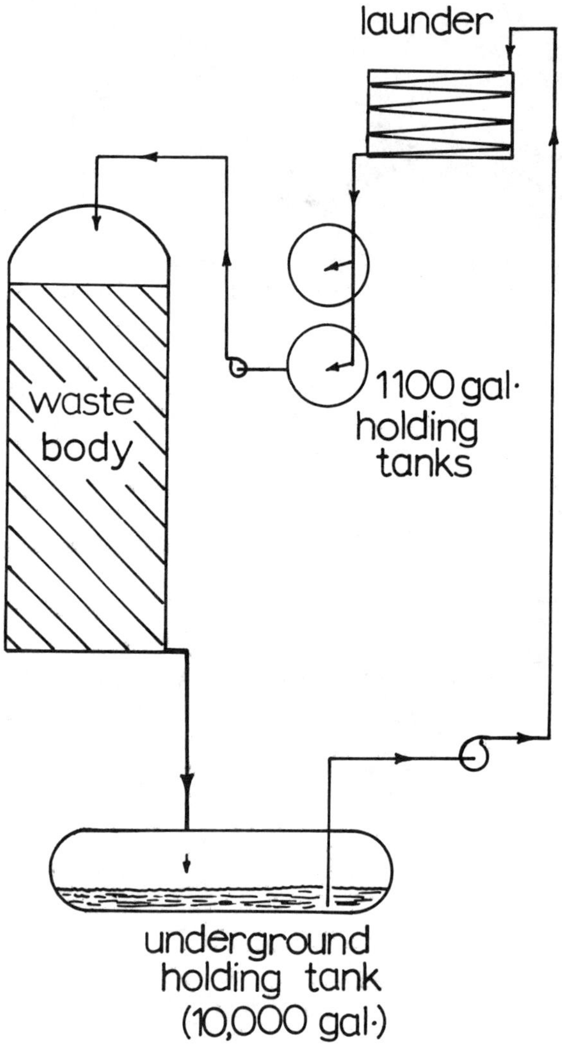

Fig. 3. Solution circuit schematic for the large waste column leaching experiments.

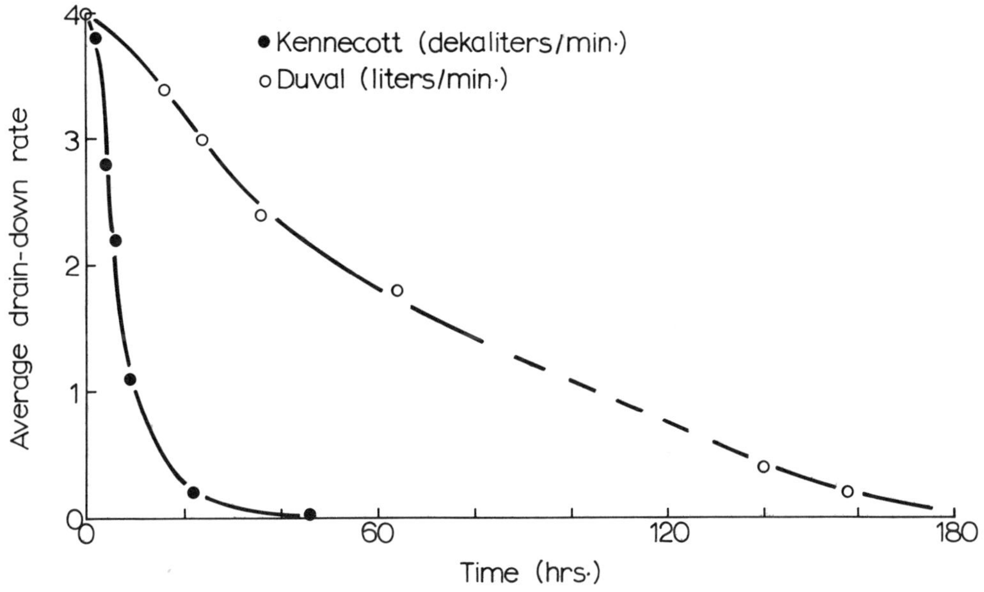

Fig. 4a. Kennecott and Duval large waste columns.

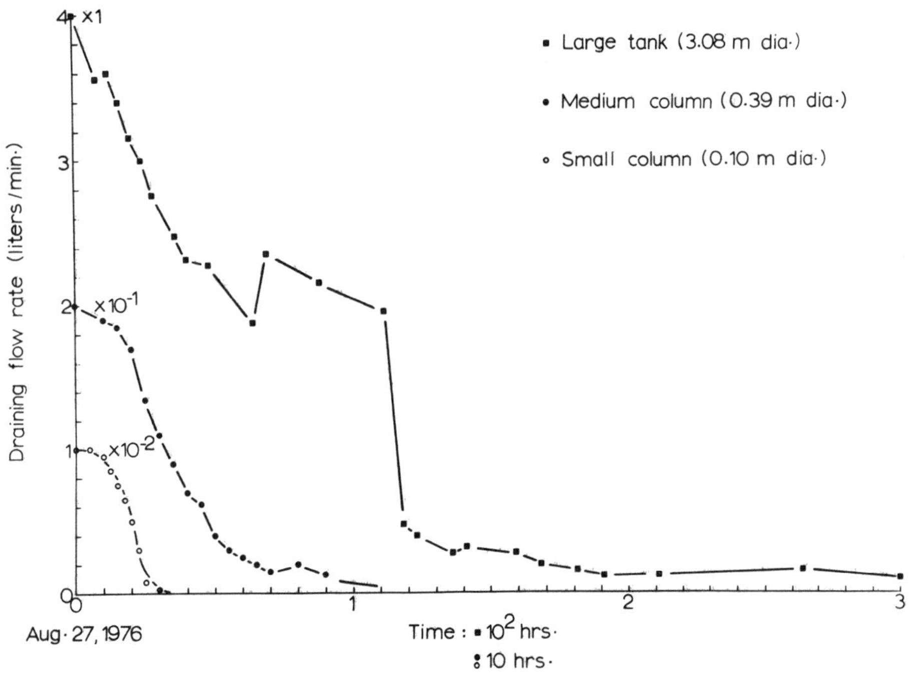

Fig. 4b. Comparison of the Duval large column draining characteristics with smaller scale columns.

Fig. 4. Initial waste column water drain-down characteristics.

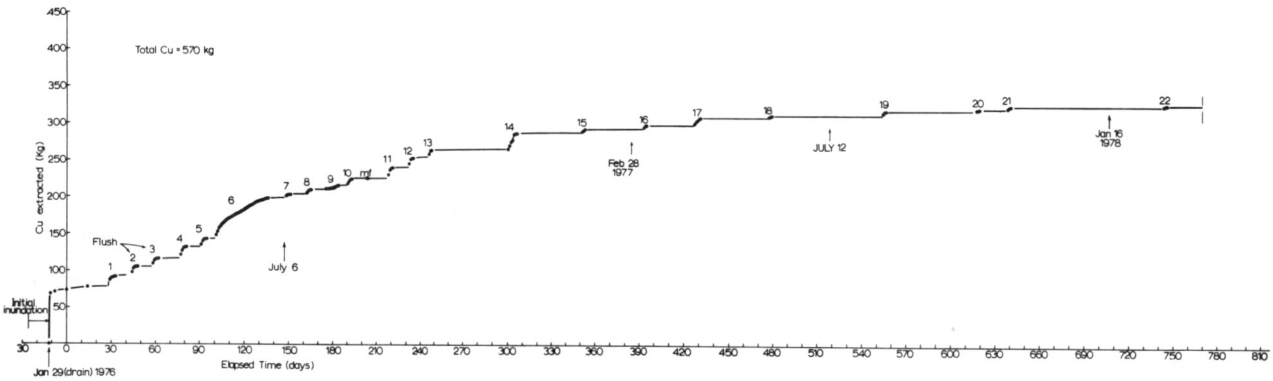

Fig. 5a. Kennecott large column copper recovery data as a function of solution cycles or flushes. The percent recovered can be determined from the total copper shown.

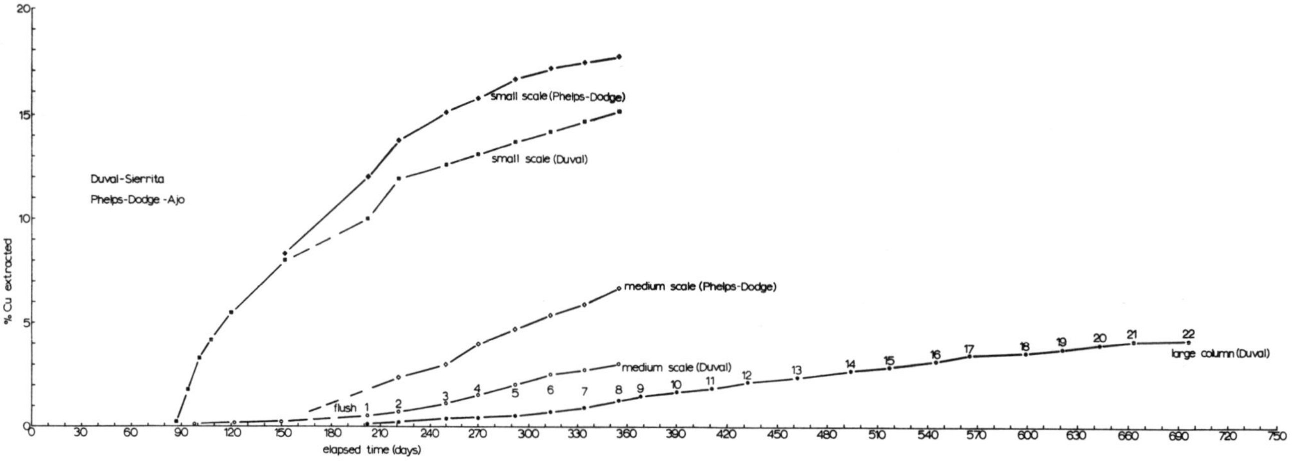

Fig. 5b. Comparison of copper recovery data for Duval and Phelps-Dodge waste columns.

Fig. 5. Experimentally observed copper leach extractions.

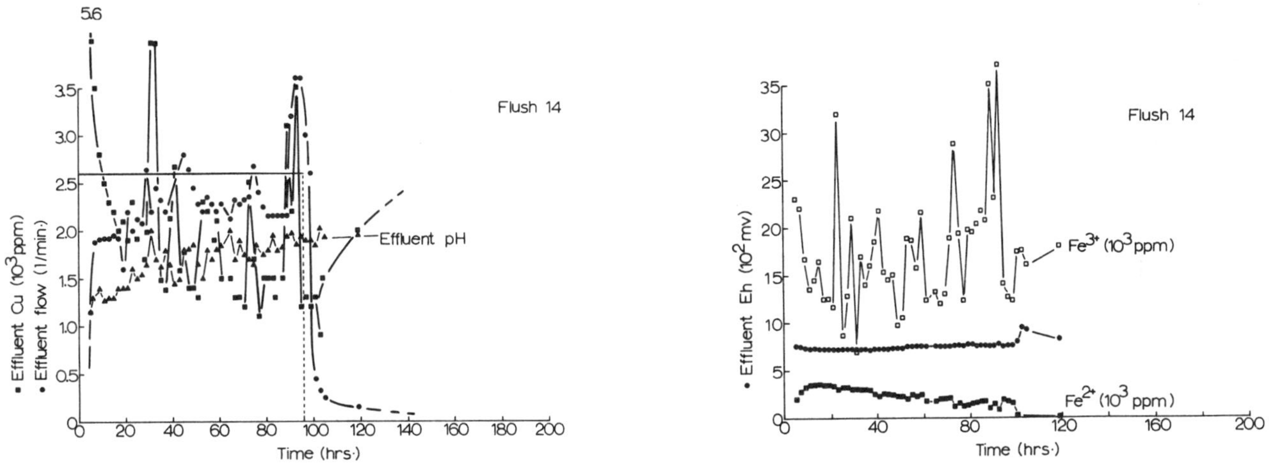

Fig. 6a. Kennecott large column - flush 14.

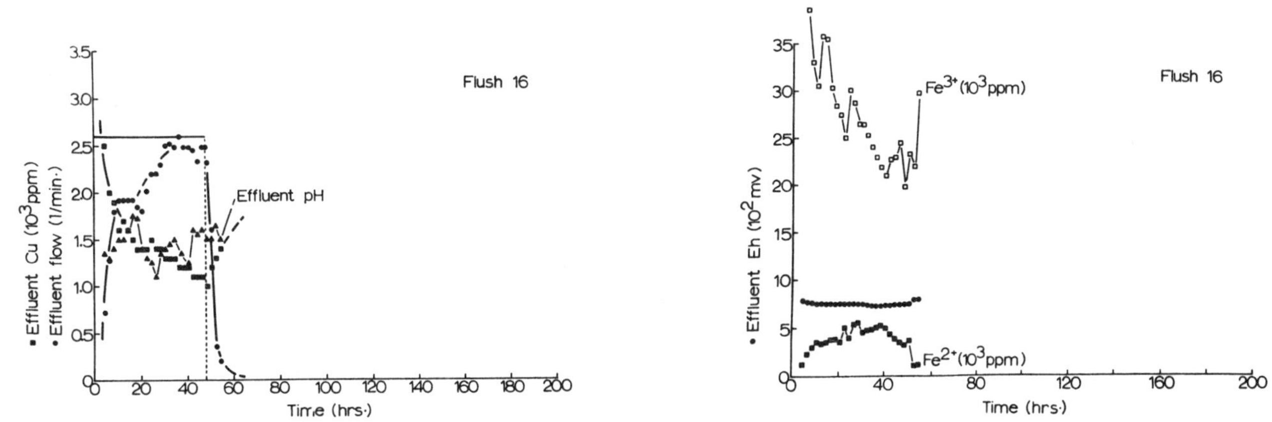

Fig. 6b. Kennecott large column - flush 16.

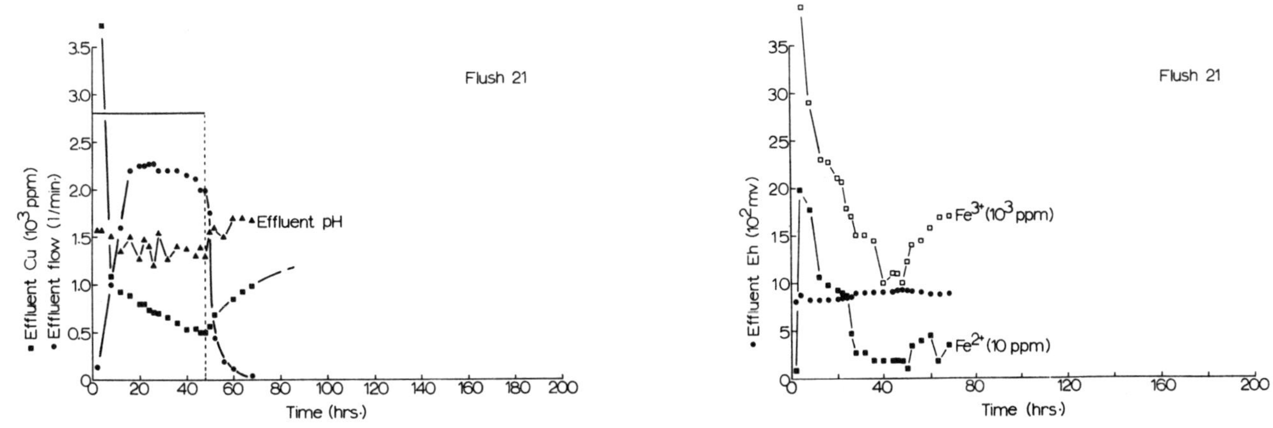

Fig. 6c. Kennecott large column - flush 21.

Fig. 6. Comparison of leaching characteristics associated with specific leach cycles or flushes shown in Fig. 5. The onfluent solution flow rates are shown by the heavy, solid line which ends at the dashed-line step function.

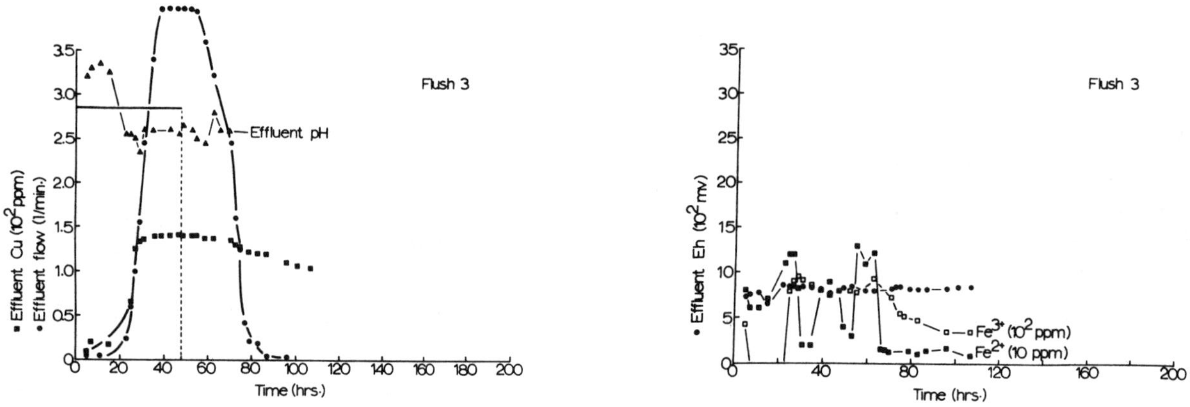

Fig. 6d. Duval large column - flush 3.

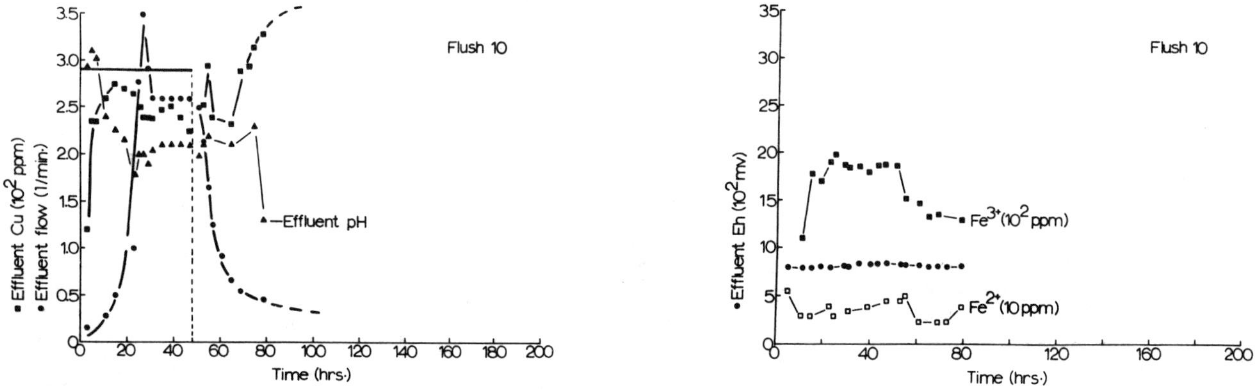

Fig. 6e. Duval large column - flush 10.

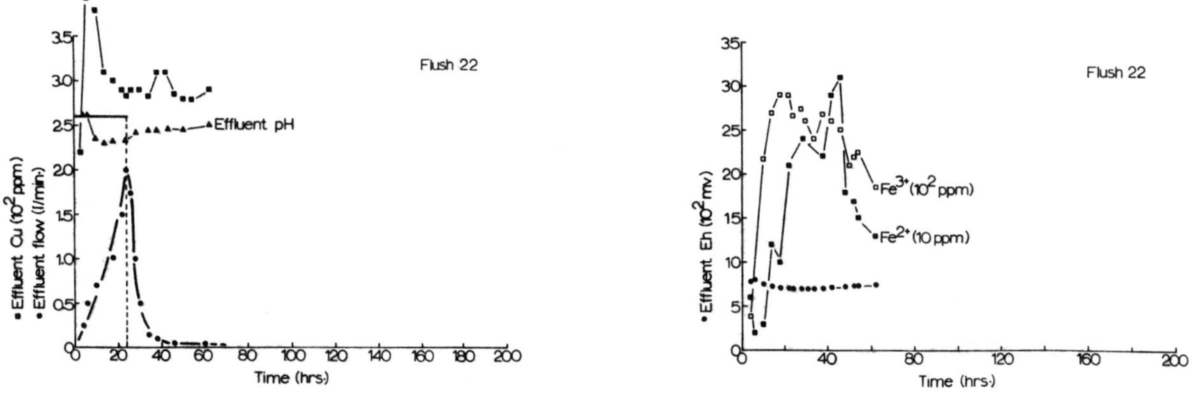

Fig. 6f. Duval large column - flush 22.

Fig. 6. Comparison of leaching characteristics associated with specific leach cycles or flushes shown in Fig. 5. The onfluent solution flow rates are shown by the heavy, solid line which ends at the dashed-line step function.

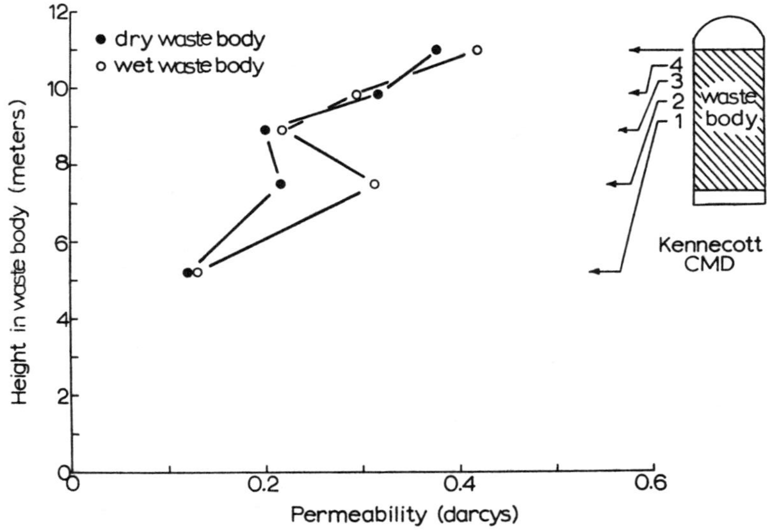

Fig. 7a. Kennecott waste column permeability for the dry waste and wet waste following column saturation and drain-down.

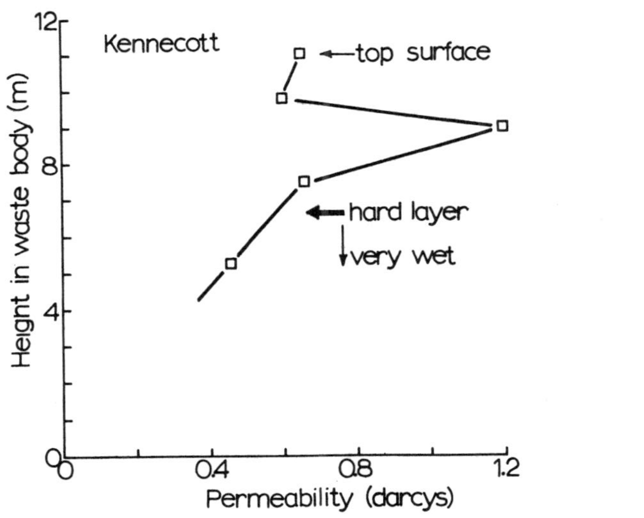

Fig. 7b. Kennecott waste column permeability following flush 21.

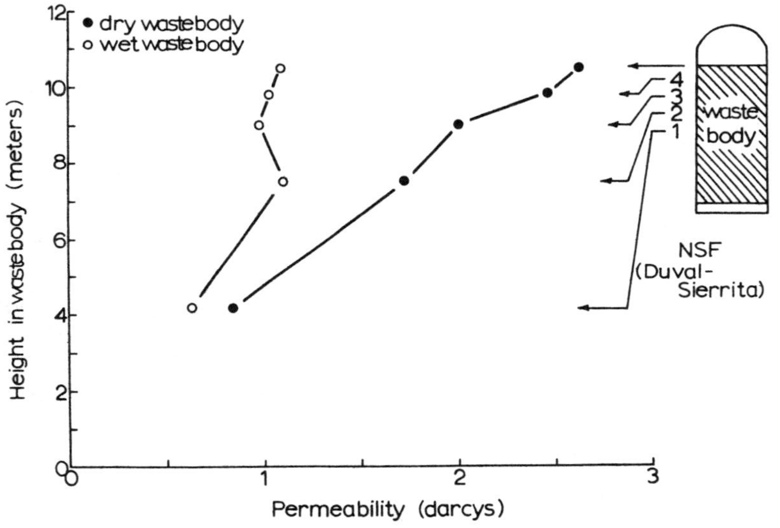

Fig. 7c. Duval waste column permeability for the dry waste and wet waste following column saturation and drain-down.

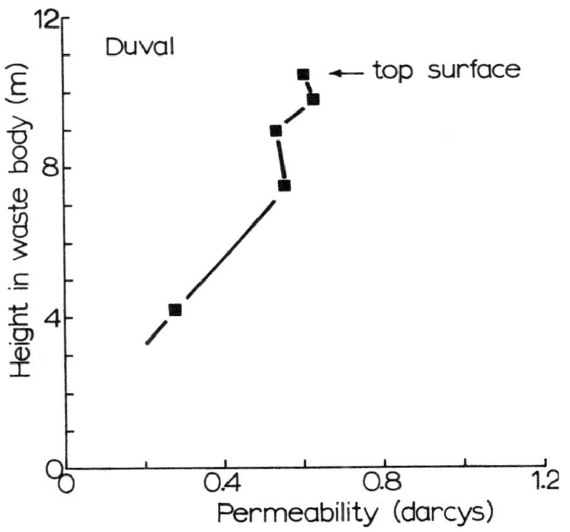

Fig. 7d. Duval waste column permeability following flush 22.

Fig. 7. Large column permeability data.

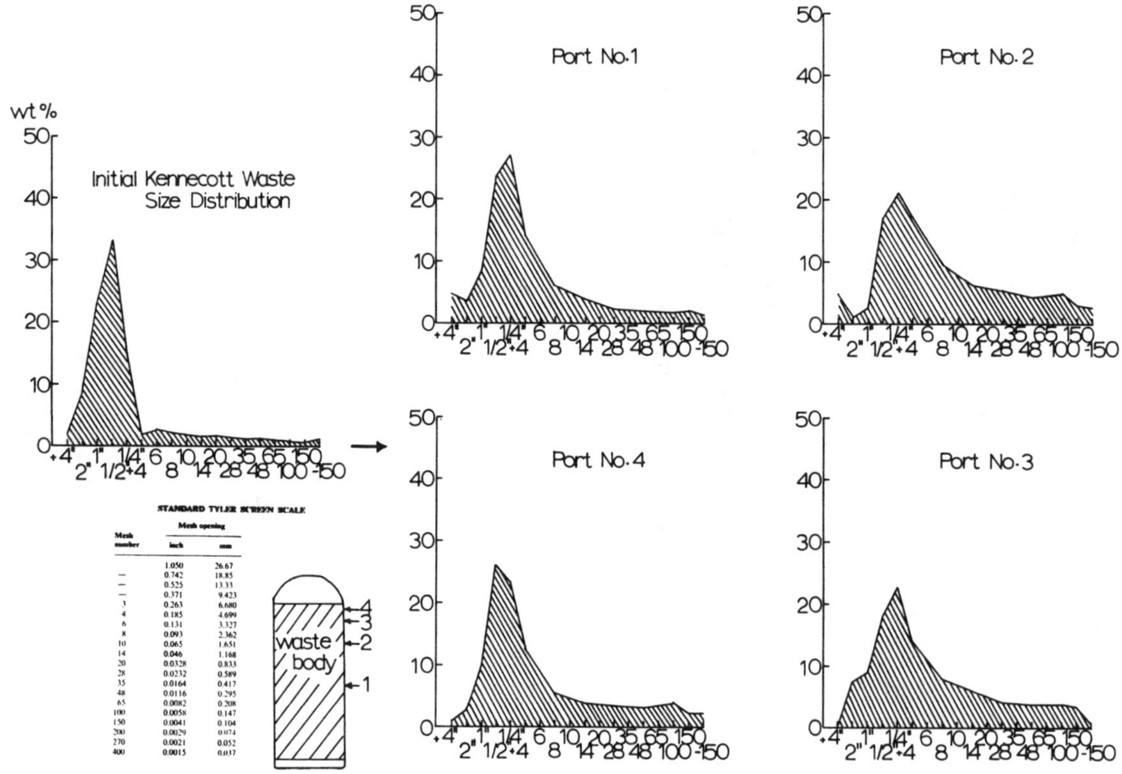

STANDARD TYLER SCREEN SCALE

Mesh number	Mesh opening inch	Mesh opening mm
	1.050	26.67
—	0.742	18.85
—	0.525	13.33
—	0.371	9.423
3	0.263	6.680
4	0.185	4.699
6	0.131	3.327
8	0.093	2.362
10	0.065	1.651
14	0.046	1.168
20	0.0328	0.833
28	0.0232	0.589
35	0.0164	0.417
48	0.0116	0.295
65	0.0082	0.208
100	0.0058	0.147
150	0.0041	0.104
200	0.0029	0.074
270	0.0021	0.052
400	0.0015	0.037

Fig. 8a. Kennecott waste column data.

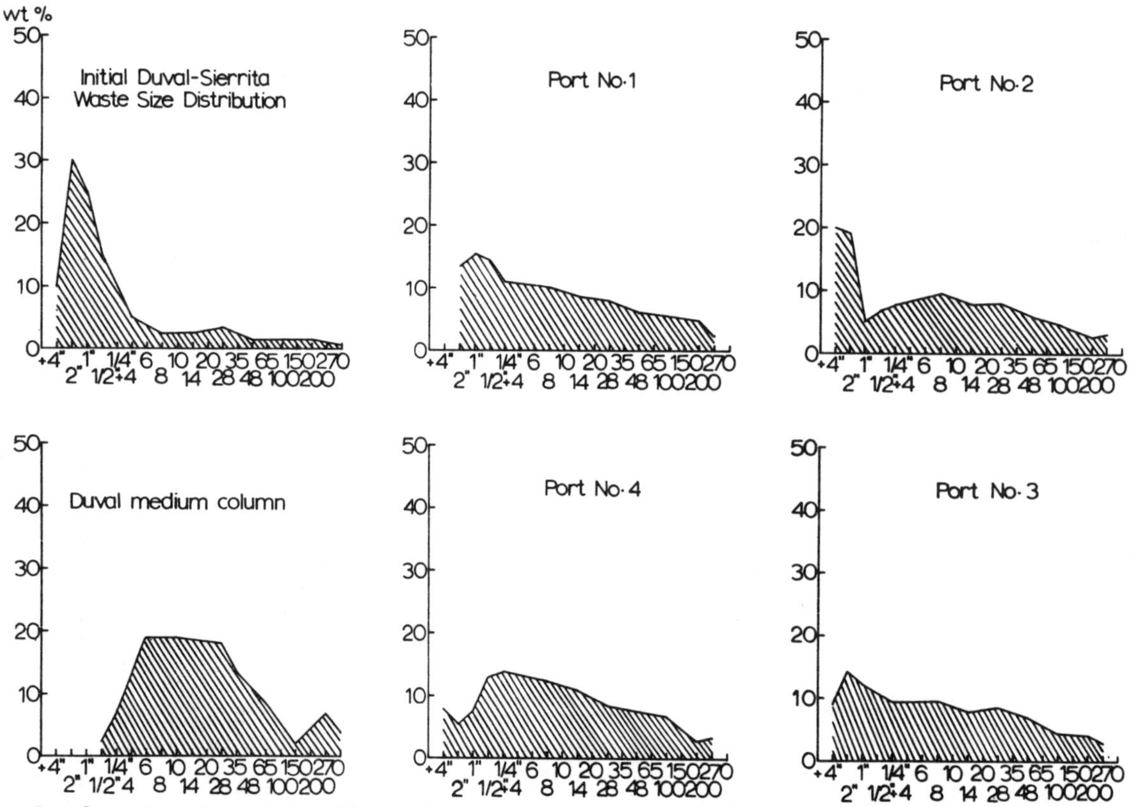

Fig. 8b. Duval waste column data. The medium column data shown corresponds to the condition following flush 22.

Fig. 8. Particle (rock) size distribution within the large columns before and after leaching. The waste size convention is shown in (a).

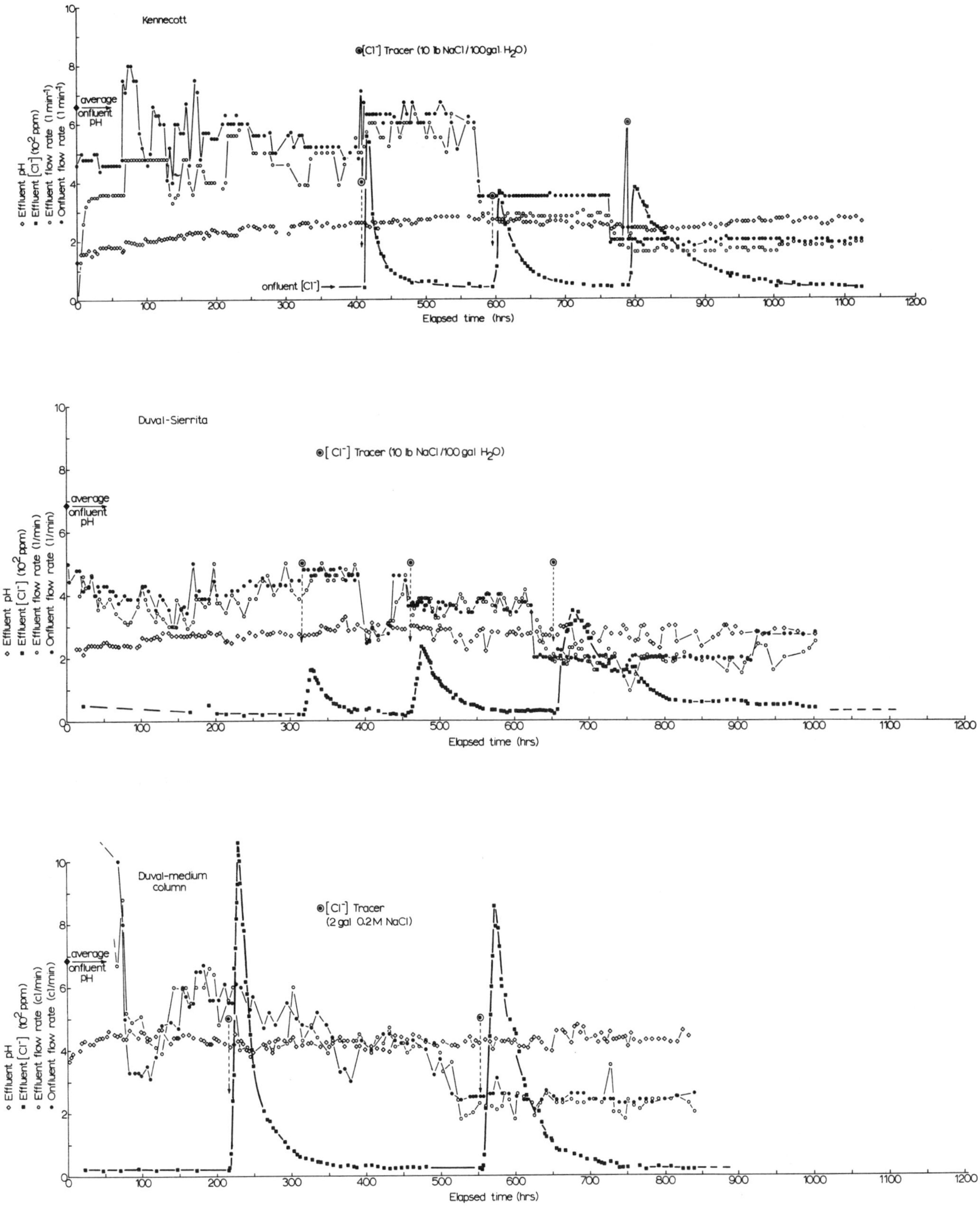

Fig. 9. [Cl⁻] tracer data for the Kennecott and Duval large column waste regimes and the Duval medium column waste regime. The effluent pH is indicative of the rate of the acid dilution within the respective columns. The tracer pulse application times are noted at the concentrations specified. These markers do not indicate the instantaneous onfluent or effluent Cl^- concentration.

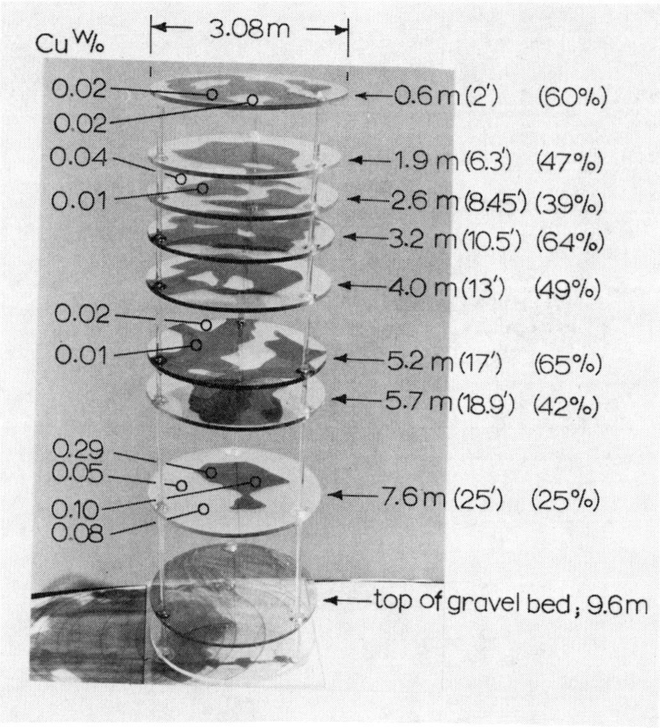

Fig. 10a. Kennecott large waste column.

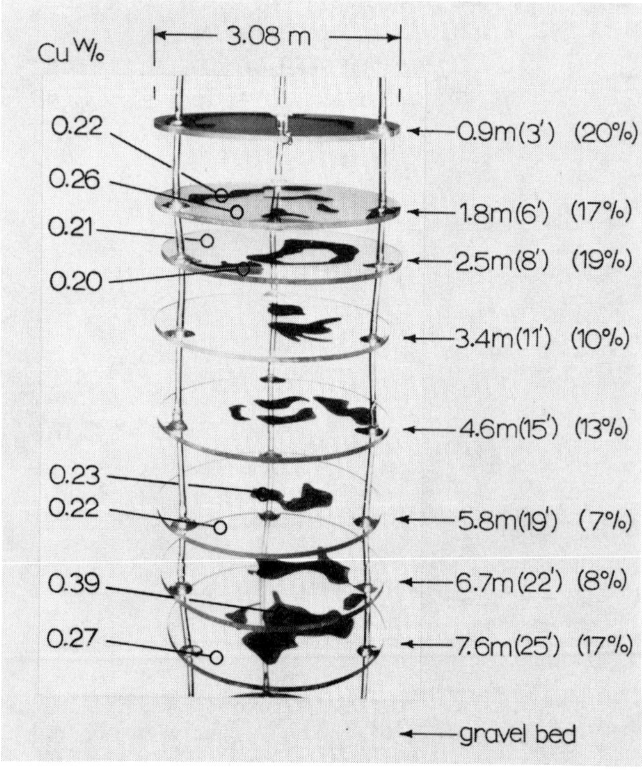

Fig. 10b. Duval large waste column.

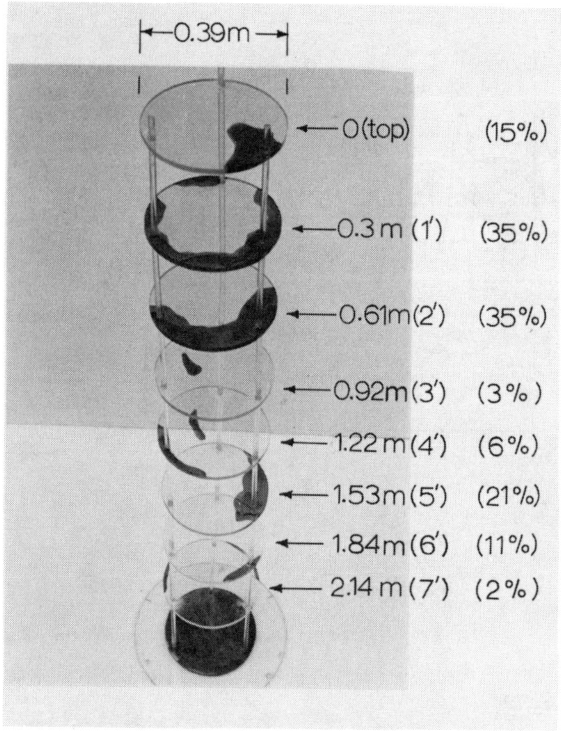

Fig. 10c. Duval medium waste column.

Fig. 10. Models of rhodamine-B ultraviolet dye staining sections reconstructed from ultraviolet photographic composites. The darker areas are considered to represent solution channels. The weight percent copper inside and outside these zones is indicated in (a) and (b) while the actual fraction of the specific cross-section characterized by the solution channel is shown in percent (%).

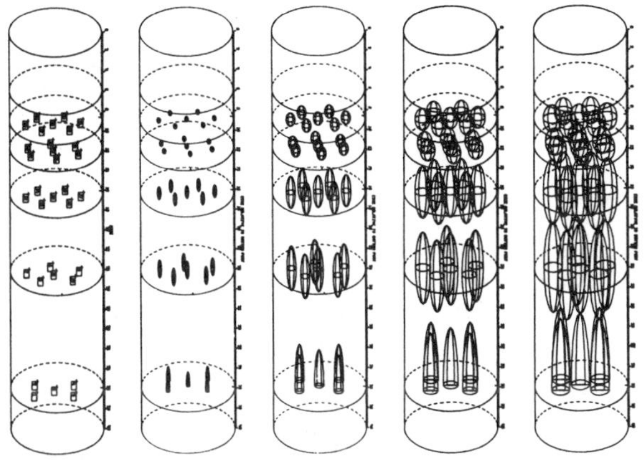

Fig. 11a. Calibration profiles. The plot at the extreme left shows the position of specific moisture probes located within the actual waste body. The other four views show characteristic constant moisture values at each such location corresponding to specific moisture contents of 20, 50, 80, and 100%, respectively.

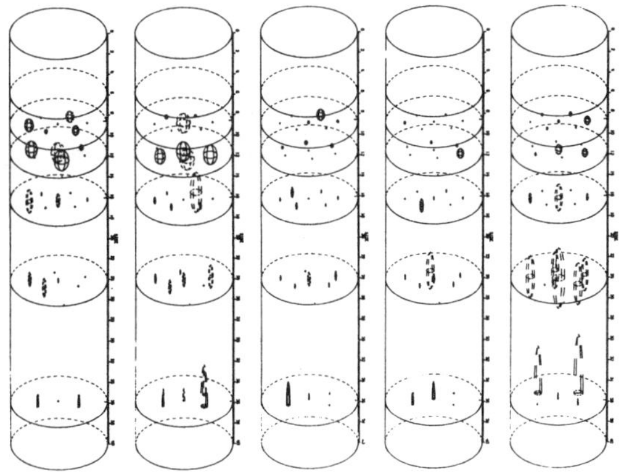

Fig. 11b. An example of actual moisture profiles each representing a two-day comparison covering a period from June 25, 1977 to December 6, 1977 as indicated. This period corresponds to flushes 4 to 12 shown in Fig. 5(b). Variations in the positions of high mositure are indicative of solution flow and changes in the flow patterns or solution-channel positions.

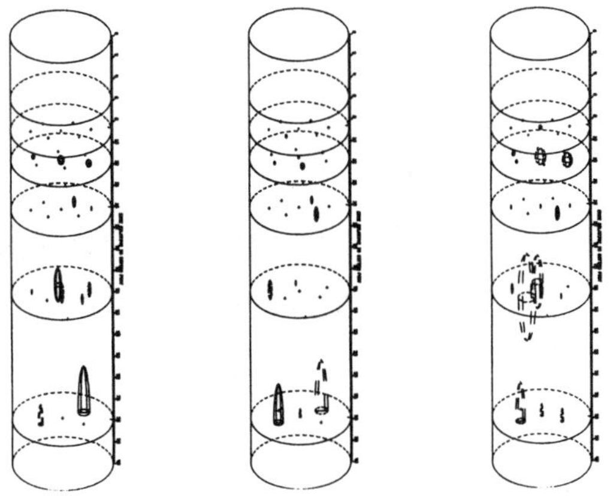

Fig. 11c. Additional examples of solution channel movement during the period May 13, 1978 to May 30, 1978.

Fig. 11. Computer-generated moisture profiles in the Duval large waste column.

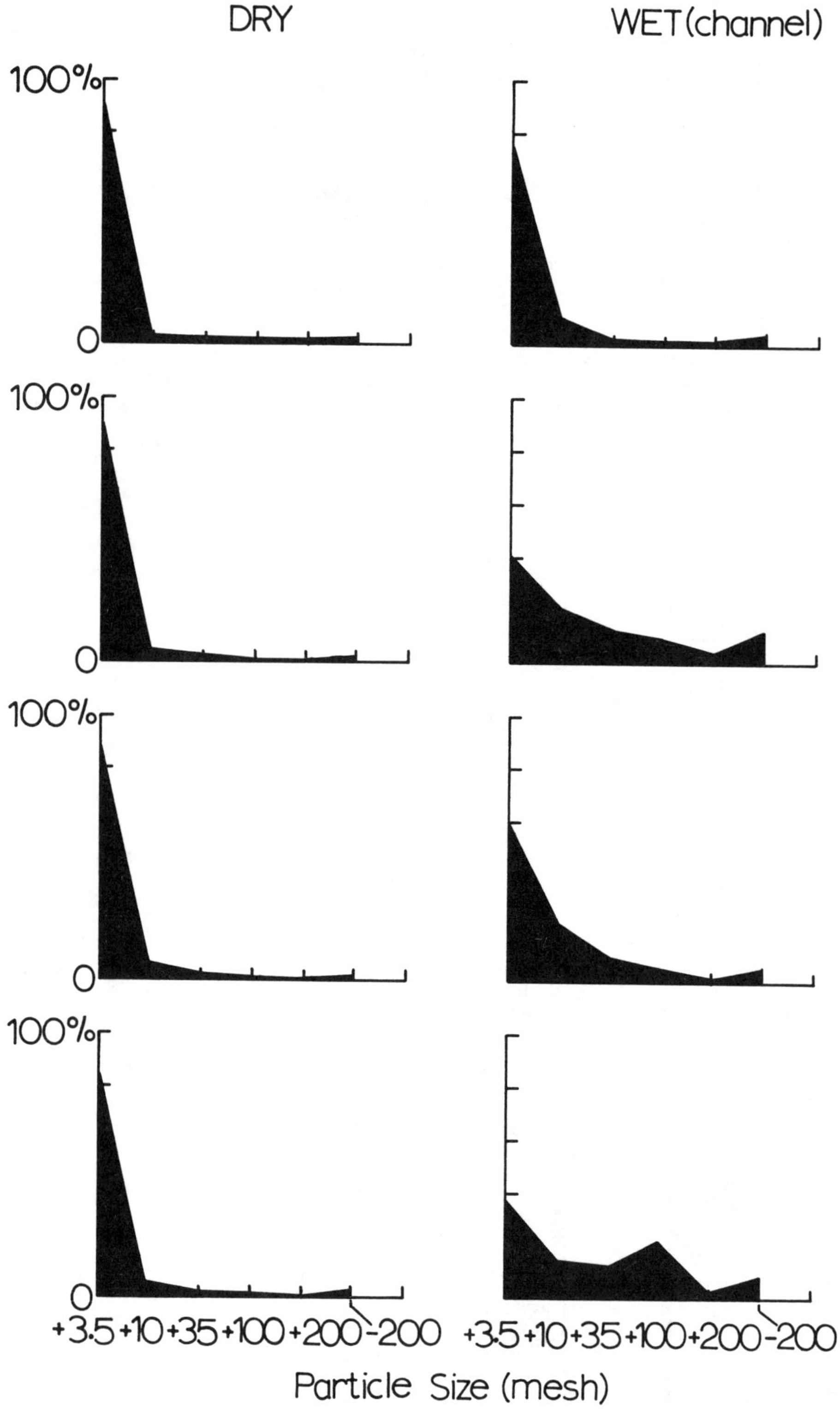

Fig. 12. Comparison of particle size distributions in the Kennecott large column corresponding to the solution channel areas (wet zones) and areas outside the channel areas (dry zones) during the waste column leaching corresponding to Fig. 10. The comparisons are made at specific port locations or distances from the top of the waste column shown.

Chapter 25

COST AND SENSITIVITIES ANALYSIS OF URANIUM IN SITU MINING

G. W. Toth
NUS Corporation
Rockville, Maryland 20850

W. C. Larson
U. S. Department of the Interior
Bureau of Mines
Minneapolis, Minnesota 55417

ABSTRACT

Uranium in situ leach mining has recently emerged as a viable alternative to conventional mining techniques. This paper presents the results of an assessment of uranium in situ leach mining costs through the application of process engineering and discounted cash flow analysis procedures. A computerized costing technique was developed to facilitate rapid cost analyses. Application of this cost model will generate mine life capital and operating costs as well as solve for economic production cost/lb U_3O_8. Conversely the rate of return may be determined subject to a known selling price. The data bases of the cost model have been designed to reflect variations in Texas vs Wyoming site applications. The results of applying the model under numerous ore deposit, operating, wellfield, and extraction plant conditions for Texas and Wyoming are summarized. Sensitivity analysis of changes in key project parameters is also tested and reported.

INTRODUCTION

In a recently completed research project for the U.S. Bureau of Mines - Twin Cities Research Center, a computerized costing procedure for uranium in situ leach mining was developed by the NUS Corporation. This costing procedure, termed a cost model, employs a process engineering approach for estimating total project costs as well as equipment and manpower requirements for uranium in situ leaching operations in either Texas or Wyoming. Both capital and operating costs are generated by the model along with the overall production cost/pound U_3O_8 subject to a known rate of return. Conversely, rate of return on equity can be solved for, subject to a given price of yellowcake.

During the course of the research effort, this cost model was applied to typical site conditions in Texas and Wyoming. Sensitivity tests were conducted for numerous key parameters to determine the cost influence attributable to incremental changes in parameter values. The results of these sensitivity tests along with a description of the in situ leaching cost model have been documented as Bureau of Mines, Mineral Research Contract Report No. J019912, "Cost and Sensitivities Analysis for Uranium In Situ Leach Mining," dated March 1981.

In the remainder of this paper, the various features and capabilities of the cost model will be highlighted, along with a review of the sensitivity analysis findings.

COST MODEL FEATURES

The design of the uranium in situ cost model required the development and integration of a process engineering approach, a discounted cash flow procedure and a regionalized cost data base. The resulting model has also been designed to offer the user maximum flexibility in specifying site conditions. The model accepts this site information, sizes both well-field and extraction plant, determines equipment and manpower requirements, and assigns appropriate costs from the model data base.

The following points provide further detail regarding features of this cost model.

- o Develops detailed costs (both capital and operating) and requirements for any user-specified project condition for the life of the project.
- o Solves for minimum required sales revenue per pound U_3O_8 (production cost) or rate of return on equity.
- o Contains regionalized data base for both Texas and Wyoming site applications.
- o Allows for cost analysis applications when only minimal information is known, as well as for cases in which detailed project data are available.
- o Accepts and accounts for either static or dynamic site conditions throughout project life.
- o Accepts user-specified capitalization structure options.

Each of these points is further explained in the following discussion.

<u>Develops detailed costs</u>. Beginning with the point in time when a decision is made to start a pilot plant facility, all capital and operating costs are estimated by this model. The categories of costs generated are listed below:

<u>Capital Costs</u>	<u>Operating Costs</u>
Process equipment	Well field replacement
Equipment installation and/or site improvement	Manpower
Building	Chemical (reagent)
Initial well field	Utility
Permitting	Operating and maintenance supplies
Pilot plant	Makeup water
Restoration system	General and administrative
Engineering and/or project management	
Fixed capital	
Contingency	
Deferred capital	

Operating costs listed above are considered direct operating costs. Other noncash costs, such as depreciation and depletion, are also calculated in the discounted cash flow analysis (DCF). Royalties and local taxes are likewise estimated in the DCF analysis. The costs presented above represent those categories computed by submodels dealing with capital and operating cost estimates which all feed into the DCF analysis.

The process analysis component of the model essentially sizes the project. The basis user-specified site conditions are translated into requirements for extraction plant size (gallons per minute feed rate) and well field size (number of patterns and well fields) to meet the user-specified production level. Manpower, horsepower, and well field replacements are also computed.

These requirement calculations serve as the basis for appropriate cost assignments from the model data base. A simplified overview of the cost model process is presented in Figure 1.

Figure 1

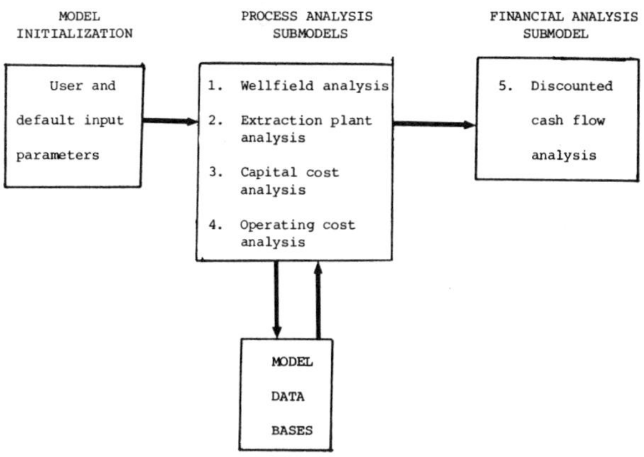

In Figure 1, submodels 1 and 2 develop requirements and costs for the respective categories listed. Submodels 3 and 4 take both capital and operating cost components of the well field and extraction plant and generate total costs for each of these categories. The DCF analysis of submodel 5 solves for the sales revenue per pound U_3O_8 or the rate of return on equity.

Solves for minimum required sales revenue per pound of U_3O_8 or rate of return on equity. The discounted cash flow analysis submodel provide the mechanism for either of the above solution options. Solving for one of these options requires knowledge of the other as a model input. The DCF analysis uses a profit and loss statement structure for financial analysis of the project.

A user solving for sales revenue per pound of U_3O_8 is interested in determining the minimum sales price that is adequate to cover all operating costs and capital recovery expenses and to provide a specified rate of return on equity invested in the project. In this situation, sales revenue per pound assumes an interpretation of economic cost of production.

A user solving for rate of return on equity will be employing a known market price of U_3O_8 as input to the model. This situation will be testing the viability of a project in terms of its rate of return yield at the anticipated market value.

Either of the solution options can provide valuable planning information for property screening or for testing alternative production levels or other project design factors for a given ore body configuration.

Contains regionalized data bases. The distinction between costs incurred in Texas and Wyoming project sites has been incorporated into the data base of the cost model. The primary variations recognized by the data base account for equipment price differences for similar equipment as well as for equipment and/or process system preferences typically associated with each region.

A summary of the key data base distinctions for Texas and Wyoming sites follows:

Extraction Process Equipment

- Defines ion-exchange system typically employed.
- Allows user four options (Upflow fixed bed, Downflow, Upflow Porter, Upflow USBM)
- Defines cost differences for each ion-exchange system in Texas versus Wyoming for each equipment item for three plant sizes (400, 1,000, and 2,000-gpm plant feed rates).

Well Field Equipment

- Defines drilling and casing cost differences for each region for three depth categories.
- Defines surface piping cost differentials for insulated piping (Wyoming) versus standard PVC piping.

Capital Costs

- Defines schedule of expenditures for each capital item according to Texas versus Wyoming site location.
- Defines permitting cost differentials and time involved according to region.
- Restoration system selection and therefore cost is based on region (deep well for Texas versus evaporation pond for Wyoming).

Operating Costs

- Incorporates differentials in chemical reagent costs between regions.
- Incorporates preferences for leaching solution between regions (ammonium carbonate-bicarbonate for Texas, sodium carbonate-bicarbonate for Wyoming).
- Includes differences in power and labor costs between regions.

Allows for cost analysis applications under a wide range of information availability conditions. To accommodate the broadest possible applications, the input structure of the cost model has been organized into three categories:

1. Required input.
2. Optional input (default values).
3. Calculation override input.

These three categories represent the range of information availability regarding a uranium in situ leach mining project. Category 1 contains the basic input parametes that must be known about a project in order to initiate a model run. There are 13 input parameters in category 1, organized according to physical, operating, and financial characteristics. Examples of these parameters are depth, ore grade, ore thickness, annual production, productive life, and capitalization parameters. There is also a series of cost update factors which may be used for cost base years other than 1980.

Category 2 input included more detailed characteristics of the project which may not always be known. These parameters are assigned default values in most cases according to the site location. These default values will be used by the model calculation procedures unless the user specifies another value in the input sequence. Examples of parameters in this category include leach solution, pattern type, solution grade, oxidizer, IX process, injection - production well ratio, pattern spacing, production well flow rate and operating schedule.

Category 3 includes those cost and requirement parameters that are calculated by the model. If information is available, however, on the specific costs of one or more of the parameters included in this listing, the model user may input the value when initiating the model run. This procedure will then negate any model calculation required for the subject parameter and instead use the value established by the user input.

Accepts and accounts for either static or dynamic site conditions throughout the project life. This feature relates to the ore deposit geometry, chemistry, well field design, and anticipated flow rates and solution grades. The model user is given the option of specifying constant conditions throughout the project life for 10 parameters or of varying the conditions for each succeeding set of well field patterns.

When changing site conditions is more appropriate than using average values for the selected input parameters, the model user may specify changing values in terms of absolute or percentage values. To further demonstrate this option, the following example of ore body depth changes over the life of the project is presented:

Depth = 400, 420, 480, 410

In this case, the input indicated that well field 1 is 400 feet deep, well field 2 is 420 feet deep, and wellfields 3 and 4 are 480 feet and 410 feet respectively. A constant percentage increase (or decrease) from the starting value also could have been input.

Values for up to 30 well fields may be input in this manner. Other parameter inputs that may be varied in a similar fashion include:

Depth of Deposit	Injection to production well ratio
Thickness of Deposit	Injection to production well spacing
Ore Grade	Production well flow rate
Solution Grade	Monitor well fraction of total wells
Well field pattern type	Recovery or percent of contained reserves

Accepts user-specified capitalization structure options. The model user may indicate any debt-equity capitalization structure for the project being analyzed. Further, the length of loan payback as well as the debt servicing rate may also be established by user input.

This flexibility offers obvious advantages for testing the effects of alternative project financing arrangements and the sensitivity of rate of return or cost per pound of U_3O_8 to variations in any of the capitalization parameters.

All of the above-mentioned features are indicative of a costing tool which has been designed for maximum user flexibility.

The model will facilitate rapid sensitivity testing of the effect of the project parameter changes on cost results for a specific site. It will serve a useful function in preliminary screening of properties for economic viability. Alternative well field designs or extraction plant systems may likewise be quickly examined.

The model is not designed to predict key project parameters such as production well flow rate or solution grade based on permeability, depth, or other influencing factors. Many parameters, however, have been assigned default values based on regional location which may be used or overriden by model users.

The primary value of this model is its ability to quickly translate any user-dictated values for such parameters into overall project cost and design implications.

MODEL APPLICATION

When applying the model to hypothetical but representative site conditions found in Texas and Wyoming, the following project parameters were employed:

Category 1 Input	Representative Texas Parameter Values	Representative Wyoming Parameter Values
Depth	400 ft	400 ft
Ore Thickness	10 ft	10 ft
Ore Grade	.1%	.1%
Annual Production	625,000 lbs/yr	625,000 lbs/yr
Productive Life	8 yrs	8 yrs
Rate of Return	20%	20%
Project Start Year	1980	1980
Cost Base Year	1980	1980
Debt Financed Portion	0	0
Category 2 Input (Default)		
Recovery	70%	70%
Leachant	Ammonia	Sodium
Oxidizer	O_2	O_2
Extraction Plant Process	Downflow	Upflow - USBM
Solution Grade	50 ppm	80 ppm
Pattern Type	5 spot	7 spot
Inj/prod. Ratio	2:1	3:1
Inj/prod. Dist.	50	40
Prod. Well Flow	20 GPM	10 GPM
Operating Schedule	350 days/yr	340 days/yr
State & Local Tax Rate	7½% of Sales Revenue	Per AEC Circular 5
Royalty Payments	10% of Sales Revenue	Per AEC Circular 5

These conditions reflected the considerations potentially encountered in recovering 5,000,000 lbs. U_3O_8 in 8 years. The results of these cases are presented in Figure 2.

In examining these results, the Texas costs are predictably lower than Wyoming costs ($25.59/lb. vs. $28.32/lb.).

The primary factors contributing to the cost differentials for the Texas vs. Wyoming cases must be concluded to be the result of the wellfield design requirements and the higher unit cost for wellfields in Wyoming. These wellfield considerations overcompensate for the solution grade advantage assumed for Wyoming (80 ppm vs. 50 ppm for Texas).

The greater number of wells required for the Wyoming project (due to the higher injection to production well ratio and the 7 spot pattern) accounts for the largest single cost variance between the cases. However, the significantly lower royalty and state and local taxes in Wyoming (per AEC Circular 5 assumptions serve to offset some of the higher wellfield

Figure 2

COST BREAKDOWN COMPARISON OF TEXAS VS. WYOMING PROJECT CONDITIONS FOR RECOVERING 5,000,000 Lbs. U_3O_8 IN 8 YEARS

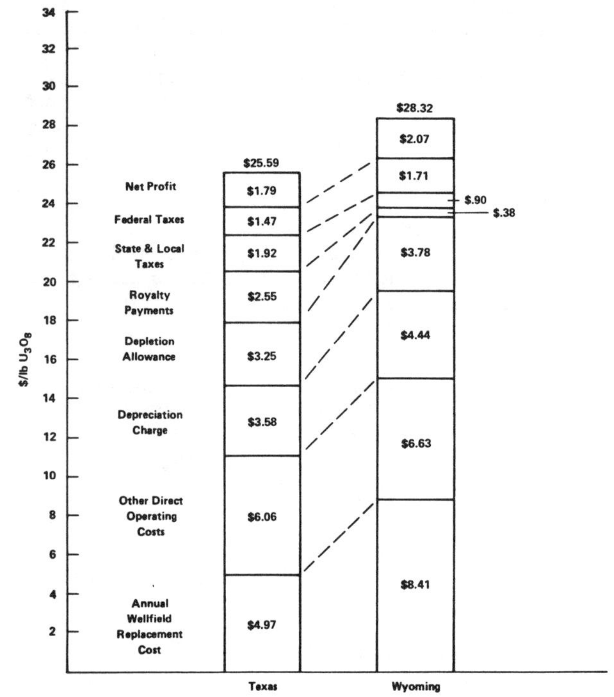

costs.

Other operating costs for Texas and Wyoming sites reveal nearly identical percentages of total cost, while in actual values Wyoming costs are approximately 10% higher. Similar findings are apparent for depreciation, depletion, federal taxes and net profit.

When identical site conditions for Texas and Wyoming are simulated and compared, however, the overall economic advantages appear to shift in favor of Wyoming. This result was demonstrated by employing the cost model for a Texas and Wyoming site having the following characteristics.

<u>Category 1 Input</u>

Depth	= 400'
Thickness	= 10'
Ore Grade	= .1%
Annual Prod.	= 500,000 lbs/yr
Production Life	= 10 years
Rate of Return	= 20%
Project Start year	= 1980
Cost Base Year	= 1980

<u>Category 2 Input</u>

Leachant	= Ammonia (Texas) Sodium (Wyoming)
Solution Grade	= 80 ppm
Oxidizer	= H_2O_2 (O_2 also an option)
IX Plant	= Downflow Fixed Bed
Pattern Type	= 5 spot (line drive or 7 spot are also options)
Prod. Well Flow	= 15 gpm
Inj/Prod. well spacing	= 50 ft.
Inj/prod. well ratio	= 2:1
Prod. Well Flow Rate	= 15 gpm
Operating Sched.	= 350 Days/year
State & Local Tax Rate	= 7½% of Sales Rev. (Texas)
	= Per AEC Circular 5 (Wyoming)
Royalties	= 10% of Sales Revenue (Texas)
	= Per AEC Circular 5 (Wyoming)

A review of this listing indicates the only variation between Texas and Wyoming cases is the use of leachant and the tax and royalty assumptions (state owned land in Wyoming and private land in Texas).

In addition to illustrating the results of nearly identical site conditions for Texas vs. Wyoming, the results will also establish a reference point for conducting sensitivity analysis.

The results of the two base cases are presented in Figure 3.

Figure 3

COST BREAKDOWN COMPARISON OF TEXAS VS. WYOMING BASE CASES

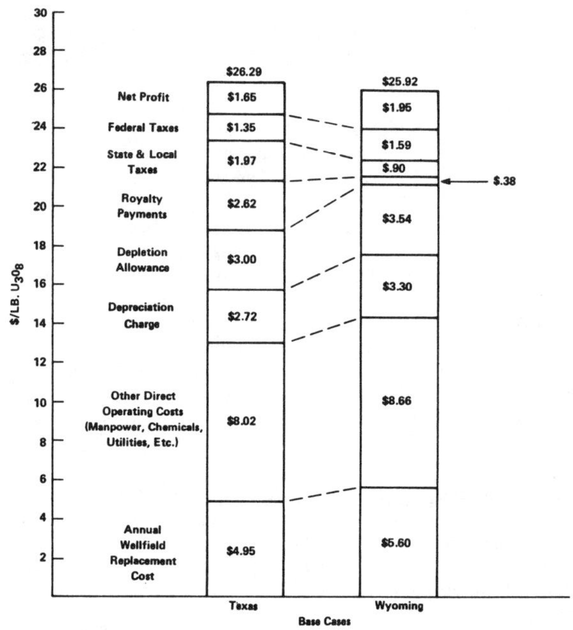

The results presented in Figure 3 are the following:

	Production Cost/lb. U_3O_8
Texas Base Case	$26.29
Wyoming Base Case	$25.92

While the total cost/lb. comparison may indicate results contrary to what might normally be expected from Texas and Wyoming sites, (i.e., Wyoming would normally be expected to be higher), the individual components of the costs do follow the traditional patterns.

For example, wellfield costs, other operating costs and capital recovery charges are all higher for Wyoming than Texas. Even though the extraction plants and wellfield designs are calculated to be identical for the two base cases, Wyoming costs are 13% higher for all other direct operating costs, and 21% higher for capital recovery (depreciation). Depletion allowances are also 18% higher in Wyoming - primarily owing to the statutory depletion allowance limit restriction guidelines. The 22% of sales revenue principle would seemingly mean a higher depletion allowance for Texas than Wyoming since the cost/lb. is higher. The upper limit on the depletion allowance however is 50% of Net Income before taxes minus state and local taxes. In Texas , the local ad valorem tax assumption of 7½% of sales revenue in this model, provides for a greater deduction ($1.97 vs. $.90 in Figure 3) from Net Income before taxes than the Wyoming state and local tax structure (calculated according to the AEC Circular 5 guidelines). This phenomenon lowers the depletion allowance limit for Texas while raising it for Wyoming projects.

Royalty payments are likewise greater for Texas sites where it is assumed that a royalty payment of 10% of sales revenue is required to private property owners. In Wyoming, it is assumed that project development will take place on state owned lands. Royalty payments in this case are again based on the AEC Circlur 5 guidelines (keyed to ore grade).

The combination of royalty and state and local tax structure assumptions then appear to offset the cost advantages Texas sites normally enjoy over Wyoming sites. When direct operating costs and capital recovery charges are isolated from other project costs the regional comparisons are:

	Direct Operating & Cap. Recovery Costs
Texas	$15.69/lb.
Wyoming	$17.56/lb.

These results indicate that direct project costs are approximately 12% higher for the Wyoming base case.

The Texas case also indicates slightly lower federal taxes paid and net profit per pound. The lower federal tax liability is due, once again, to the higher deduction for royalty and state and local tax payments. The slightly lower net profit/lb. has its basis in the fact that although the rate or return on equity was identical for both base cases (20%), there was a lower capital requirement against which ROR could be computed for the Texas base case.

The cost results of the two base case analyses, serve as the reference points against which the subsequent sensitivity trials are compared. The parameters examined for sensitivity influence on the resulting production cost/lb. U_3O_8 include the following:

- Annual Production Level
- Solution Grade
- Ore Depth
- Ore Thickness
- Ore Grade
- Pattern Spacing
- Production Well Flow Rate
- Rate of Return On Equity

This listing does not suggest that the operator has control over the variability of all these parameters. Rather, the effects of changes in these parameters is being investigated - regardless of the degree of control available over such changes.

Each of these parameters are tested by incrementally increasing and decreasing their values around the base case condition. In order to be valid, all other parameters must be held constant while the parameter being tested is alone varied. It is acknowledged that a change in one parameter may normally induce a cause and effect relationship with another. This illustration does not, however, attempt to keep the system in balance. Rather it attempts to investigate the "what if" situations regardless of probability of occurance. Of course, if joint sensitivities are being investigated, then the subject joint parameters may alone be varied in equal percentage increments.

It is important to stress that the results of conducting a sensitivity analysis of this nature are valid only for the base case used. A different base case will yield different cost influencing effects.

In an attempt to evaluate the degree of cost influence attributable to each parameter, a common basis for comparison was established. This was accomplished by converting both parameter value changes and resulting production cost changes to percentage variations from the base case.

A summary of these results is presented in Table 1. A graphic illustration of these same findings is presented in Figure 4. These sensitivity analysis results identify pattern spacing as having the most significant effect on cost results. In other words, each percentage variation of pattern spacing from the base case value, creates the largest percentage change in production cost among those parameters being tested.

Table 1

BASE CASE SENSITIVITY ANALYSIS RESULTS

	Production Cost Change % Subject To Parameter Change of:	
	+30 %	-30 %
Production Level	- 5.5	+11.1
Solution Grade	- 9.0	+14.1
Ore Depth	+ 7.5	- 8.3
Ore Thickness	- 6.0	+11.0
Ore Grade	- 3.5	+ 9.5
Pattern Spacing	-10.0	+17.8
Prod. Well Flow	- 2.5	+ 4.5
Rate of Return	+11.0	-11.0

Figure 4

SENSITIVITY ANALYSIS SUMMARY OF PARAMETER CHANGES ON BASE CASE COSTS

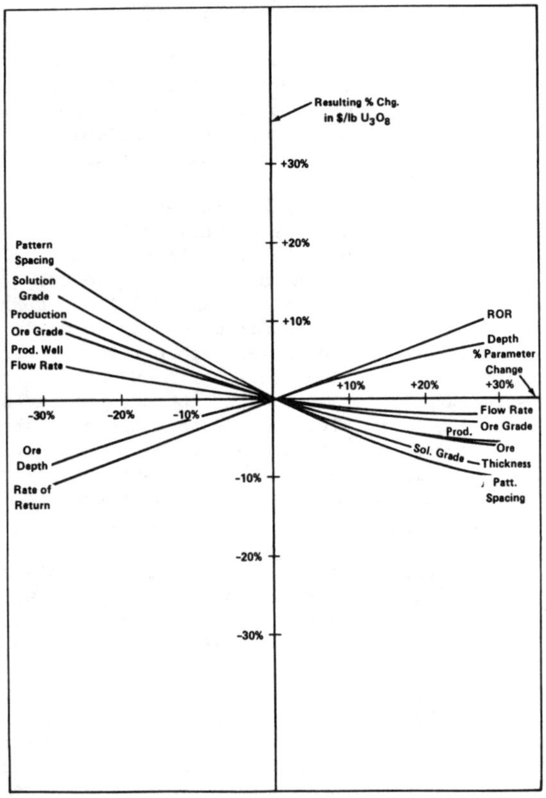

The cost influencing ability of all parameters tested are listed below in descending order of the percentage cost change each created (either positive or negative) for both positive and negative parameter changes.

Negative Parameter Variations

- Pattern Spacing (positive)
- Solution Grade (positive)
- Production Level (positive)
- Ore Thickness (positive)
- Rate of Return (negative)
- Ore Grade (positive)
- Ore Depth (negative)
- Production Well Flow Rate (positive)

Positive Parameter Variations

- Rate of Return (positive)
- Pattern Spacing (negative)
- Solution Grade (negative)
- Ore Depth (positive)
- Ore Thickness (negative)
- Production Level (negative)
- Ore Grade (negative)
- Production Well Flow Rate (negative)

While the results of this sensitivity analysis is indicative of definite patterns of cost behavior, it must be emphasized that the results cannot necessarily be universally applied to any in situ project. A variation in the base case conditions could cause a reodering of these same paramenters.

Perhaps the most significant aspect of these findings is the demonstration of the cost model capability to quickly provide cost results for any user specified site condition and/or sensitivity considerations. It is anticipated that wider industry application of this model in the near future will attest to the value of this cost estimating tool.

Chapter 26

GROUNDWATER PROBLEMS AT URANIUM IN SITU LEACH MINING OPERATIONS IN THE POWDER RIVER BASIN, WYOMING

Terry D. Vandell
Kennecott Minerals Company
Salt Lake City, Utah 84126

ABSTRACT

In situ leaching or solution mining is a relatively new addition to the list of mining methods currently used to extract uranium in Wyoming's Powder River Basin.

The environmental advantages of in situ leach mining of uranium can be significant. While the conventional extraction of minerals produces significant impact on the environment, solution mining may result in considerably less impact. In situ leach mining permits economical recovery of low-grade uranium deposits currently unrecoverable by conventional mining techniques.

In situ leach mining, however, can be conducted only on those uranium ore deposits that meet certain criteria, which generally include: (1) the ore deposit must be located in a water saturated zone, (2) the ore deposit must be confined both above and below by impervious layers, (3) the ore deposit must have adequate permeability, and (4) the ore deposit must be amenable to chemical leaching.

The purposes of this presentation are: (1) to discuss actual as well as potential incidents that have caused and that could result in unnecessary ground water contamination at uranium in situ leach mining sites, particularly to aquifers that surround the ore zone, and (2) to discuss remedial actions and solutions that have been or that could be implemented in the future to stop or prevent serious groundwater contamination at uranium in situ leach mining sites in Wyoming's Powder River Basin.

INTRODUCTION

The purposes of this paper are: (1) to discuss the actual as well as potential incidents that have caused and that could cause groundwater contamination at uranium in situ leach mining sites and (2) to discuss remedial actions and solutions that have been or that could be implemented in the future to stop or prevent serious groundwater contamination from in situ leach mining in Wyoming's Powder River Basin. Open pit, underground, and existing and proposed in situ leach uranium mine sites in the Powder River Basin are shown in Figure 1.

In situ leach mining or solution mining for extracting minerals in liquid form is a relatively new addition to the list of conventional mining methods currently used to extract uranium in Wyoming's Powder River Basin. Basically, the in situ leach mining method involves: (1) the injection of a leach solution lixiviant (an acidic or basic oxidizing solution) into a uranium-bearing ore body, via injection wells, to complex the contained uranium, (2) mobilization of the uranium complex from the host material via creation of a soluble complex salt, and (3) surface recovery of the solution bearing the uranium complex via production wells. Uranium is then separated from the leach solution by conventional milling unit operations (ion exchange).

The environmental advantages of in situ leach mining of uranium can be significant. While the conventional extraction of minerals produces significant impact on the environment, the solution mining method generally results in considerably less impact. In situ leach mining also permits economical recovery of low-grade uranium deposits currently unrecoverable by conventional mining techniques.

In situ leach mining, however, can be conducted only on those ore deposits that meet certain criteria. These generally include: (1) the ore deposit must be located in a saturated zone. (2) the ore deposit must be confined both above and below by impervious layers for the deposit to be leached and to prevent lateral and vertical excursions (the movement of lixiviant out of the production zone due to in situ mining activities), (3) the ore deposit must have adequate permeability, and (4) the ore deposit must be amenable to chemical leaching.

GEOLOGY

The Powder River Basin covers approximately 12,000 sq miles, and, as shown on Figure 1, it is bounded on the south by the Laramie Range Hartville Uplift, on the east by the Black Hills Uplift, on the west by the Big Horn Mountains and Casper Arch, and it is open to the north where it continues into the State of Montana. The basin incorporates a sedimentary rock sequence ranging in age from Recent (Holocene) to early Paleozoic (Cambrian) which has a maximum thickness of about 15,000 ft along the northwest-southwest synclinal axis. A geologic cross section through the Big Horn and Powder River Basins is shown in Figure 2.

Three stratigraphic units of Tertiary age are recognized in the Powder River Basin. The Fort Union Formation, which is the host unit for the uranium deposits in the southern Powder River Basin, thickens from the southern margin of the basin to 3,300 ft (1,000 m) in the central portion of the basin. The unit overlies the Cretaceous Lance Formation and is, in turn, overlain unconformably by the Eocene Wasatch Formation which is the main uranium host unit in the central part of the Powder River Basin.

The uranium ore deposits currently being leached in the Powder River basin are generally at depths ranging from 100 to 500 ft. A plan view and cross sections of uranium roll front and uranium ore deposits are shown in Figure 3.

The Fort Union Formation is composed of interbedded fine-to coarse-grained sandstone, siltstone, claystone, lignite, and coal. The materials which make up the formation in the southern part of the basin were derived from Mesozoic and Paleozic sedimentary rocks, from Precambrian crystalline rocks of the Laramie Range and the Granite Mountains, and from volcanic debris from the Absaroka volcanic center near Yellowstone. The upper portion of the Fort Union Formation, containing most of the uranium ore is locally referred to as the Highland Sandstone. The Highland Sandstone is relatively consistent laterally and underlies a large area in the southern portion of the basin. The sandstones which comprise the Highland Sandstone are arkosic, loosely consolidated, and contain substantial organic debris and tuffaceous material. Individual sandstones that contain the uranium deposits are tonguelike extensions off the edge of the central Highland Sandstone Unit. The massive Highland Sandstone Unit consists of discontinuous sands separated vertically by siltstones, claystones, and lignites; laterally, the sandstones exhibit facies changes into carbonaceous claystones and lignites.

The Wasatch Formation, of early Eocene Age, attains a thickness of approximately 1,000 ft (305 m) within the basin and unconformably overlies the Fort Union Formation. The unit consists of clays and siltstones containing abundant, thick lenses of coarse, crossbedded, arkosic, poorly sorted sandstone with calcitic cements. Thin beds of coal or carbonaceous shale are common in some areas.

HYDROGEOLOGY

Zones of uranium mineralization follow the general trend of old drainage channels. The geologic environment favoring the deposition of uranium ore is deficient in oxygen, has zones with less permeable siltstones and shales, and contains reducing agents such as carbonaceous material, hydrogen sulfide, or pyrite. It is believed that oxygenated groundwater percolating through the Powder River Basin sediments, (which were derived from granite), oxidized and mobilized the uranium. The uranium enriched groundwater then moved through the aquifer towards the basin center, coming in contact with carbonaceous material and pyrite. During active anaerobic decomposition of the organics, the groundwater may have encountered hydrogen sulfide and methane, which reacted to reduce the uranium to insoluble uraninite (UO_2). The pH and Eh of the solution decreased at the Oxidation-Reduction front and freshly precipitated uranium was dissolved, by oxygenated groundwater, and displaced further downgradient. Ultimately, uranium of economically recoverable grade may be deposited at the oxidation-reduction interface as shown in Figure 3, Plan View and Cross Section A. The physical shape of an ore roll is dependent on the local permeability of the matrix material and its continuity and distribution in the geologic unit. Such ore bodies are prevalent in most of the established uranium mining districts in the western United States. Figure 3, Cross Section B, illustrates a typical injection-recovery well system and

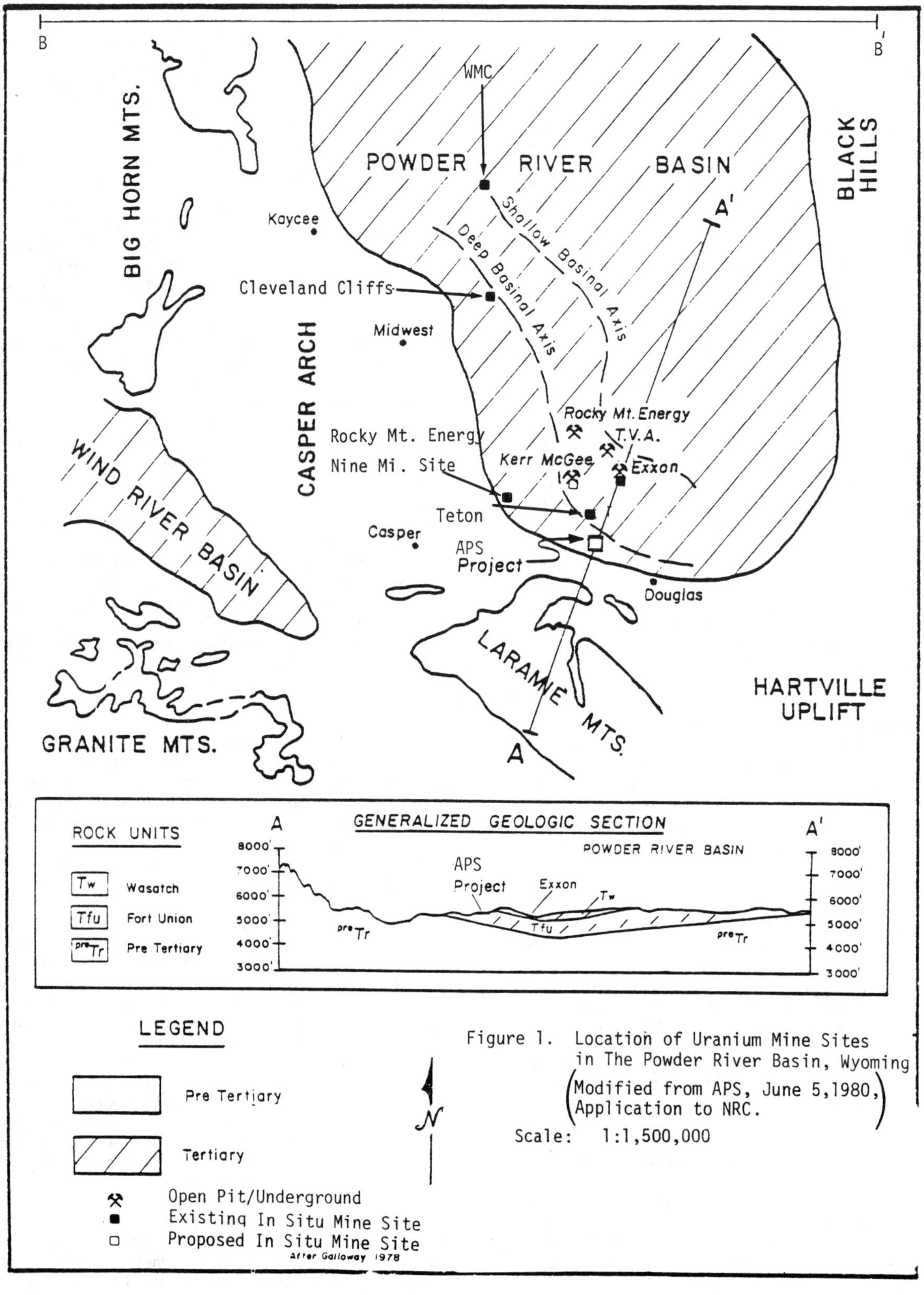

Figure 1. Location of Uranium Mine Sites in The Powder River Basin, Wyoming (Modified from APS, June 5,1980, Application to NRC.

Scale: 1:1,500,000

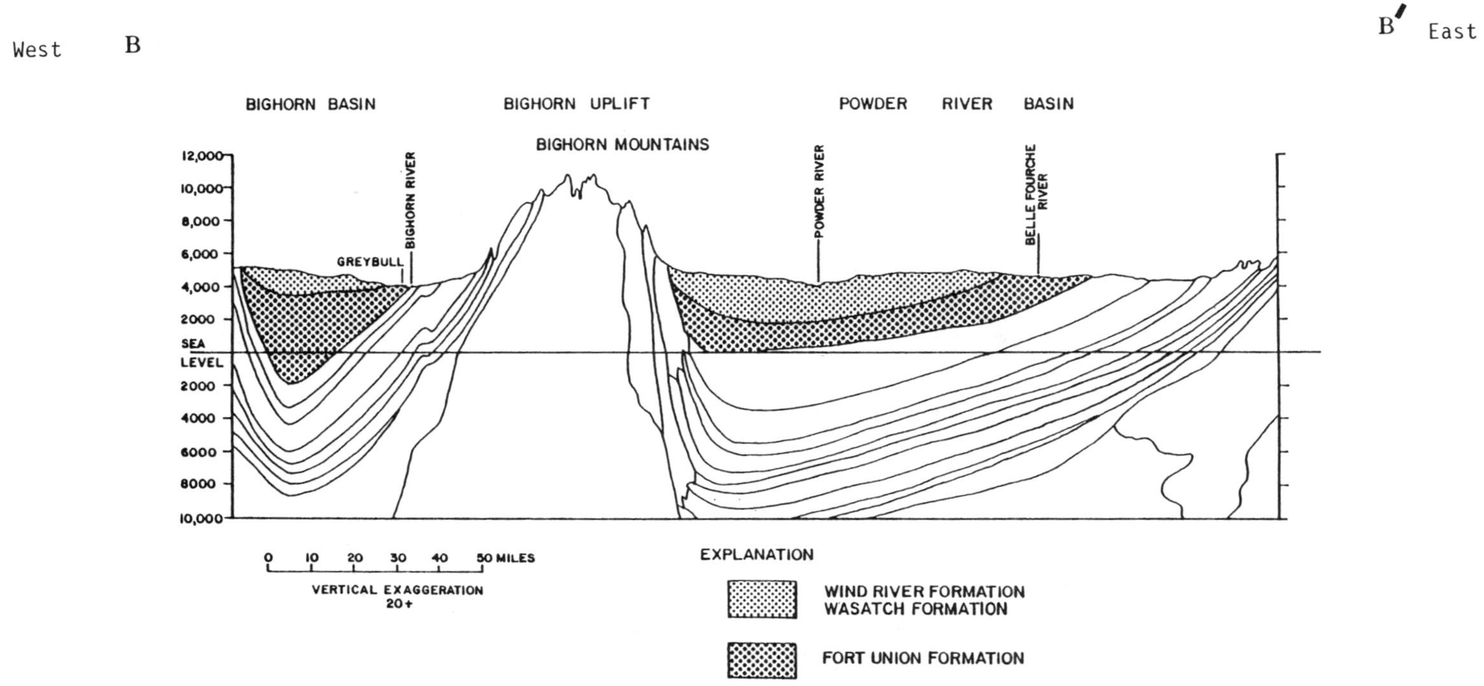

Figure 2. Geologic cross section (B-B') through the Big Horn and Powder River Basins, Wyoming. Ref: NUREG/CR-0311

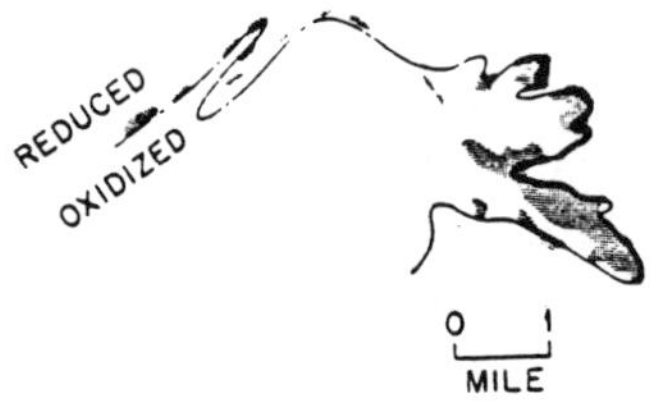

PLAN VIEW OF DEPOSIT

Ref: FES, NUREG-0481

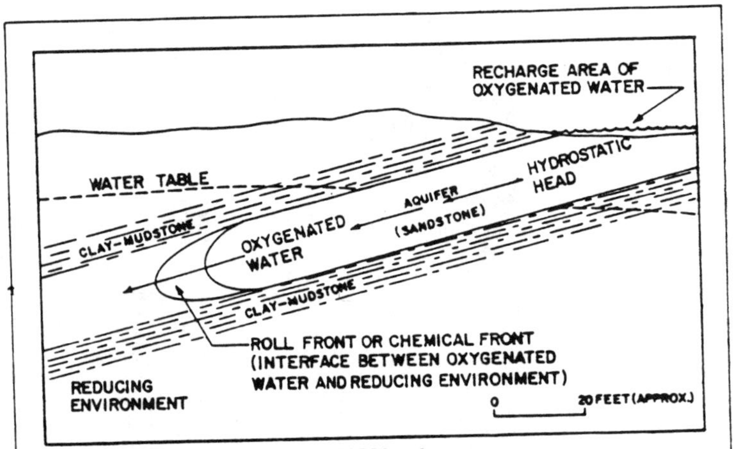

CROSS SECTION A

Plan View of uranium ore deposit and cross section of geochemical cell or roll front. Ref: NUREG/CR-0311.

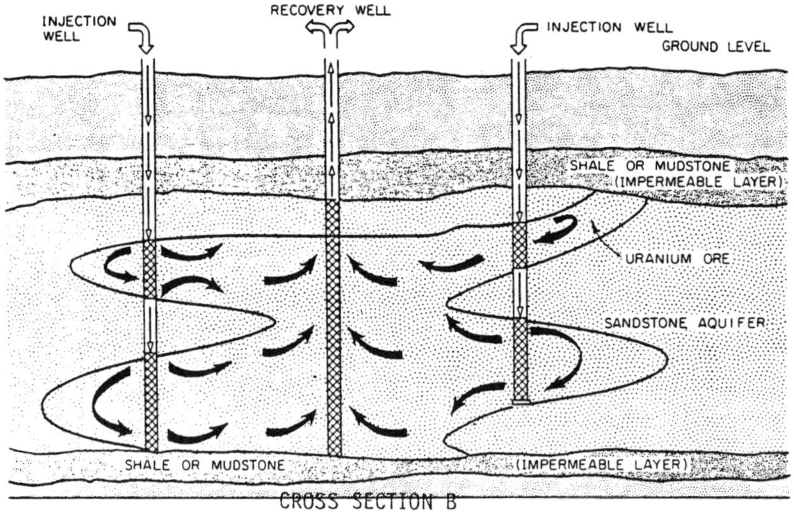

CROSS SECTION B

Figure 3. Cross section of a typical uranium roll front deposit and the solution mining unit. Ref: FES, NUREG-0481 the generalized groundwater flow lines which generally occur during mining.

WATER USE

Groundwater in the Powder River Basin is primarily used for livestock watering and private water supply. Some shallow alluvial wells, yielding between a few gpm to 1,000 gpm, are located on the floodplain of the Powder River. However, groundwater from the alluvial aquifers is generally of poorer quality than groundwater from deeper aquifers in the Wasatch and Fort Union Formations. Groundwaters from these two formations are generally good, although they may be high in TDS and are frequently defined as calcium, sodium-bicarbonate or sodium-sulfate types. Groundwater in the ore zone of the Wasatch Formation may be as high as 8,000 ppm TDS and groundwater in the ore zone of the Fort Union Formation may be as high as 2,000 ppm TDS. Wells in the Wasatch Formation yield from 10 to 500 gpm and wells in the Fort Union Formation yield a maximum of around 150 gpm.

In summary, because groundwaters in the aquifers surrounding and in some cases within uranium ore zones in the Powder River Basin are generally good (generally meet Wyoming DEQ Standards for at least livestock use), it is important that mining operations be conducted in such a manner that lixiviant be restricted, as much as possible, to the ore zone being leached. Lateral movement of lixiviant in the ore zone aquifer away from the well field, and vertical movement of lixiviant into aquifers above or below the ore zone, must therefore be avoided if at all possible. If accidental lateral or vertical excursions do occur, the affected aquifer(s) must be restored at least to their potential pre-mining water use condition or to a condition acceptable to the regulatory agencies involved.

HYDROGEOLOGIC PROBLEMS

Excursions have occurred, and probably will continue to occur, at uranium in situ leach mine sites in Wyoming's Powder River Basin. Lateral and vertical excursions have already occurred as a result of: (1) the lateral movement of lixiviant out of cracks in well casings and (2) the vertical movement of lixiviant through improperly abandoned exploration drill holes.

Cracked well casings are generally the result of faulty well construction and development. Faulty well construction practices, including the use of improper undereaming tools, improper sealing above and below open intervals, flaws in casing joints or the use of improper casing (i.e., using PVC when fiberglass or steel casing should be used) can lead to casing failures which can and have resulted in the movement of lixiviant to zones other than the ore zone.

Therefore, before injection and recovery wells are placed into operation, high pressure (at least the maximum well head pressure to be used during production) well casing integrity testing should be conducted. A typical arrangement for a high pressure well integrity test using packers is shown in Figure 4. Well integrity is assumed to be good if the pressure in the well does not decline by 2 to 3 psi or generally by 10% or more of the maximum pressure that was injected, over a 10 to 30 minute period. If a significant pressure drop does occur, the casing should then be checked using a downhole TV camera or other means. Most pressure drops occur in the first 10 minutes after injection. If a significant pressure drop occurs, the casing must be repaired, and then re-tested. If the flaw cannot be located or corrected, then the well must be properly abandoned and a new properly constructed well installed in its place.

Improperly abandoned drill holes have and probably will continue to be the major factor contributing to the occurrence of excursions at Wyoming's uranium in situ leach mine sites. As illustrated in Figure 5, improperly abandoned drill holes can serve as major channels for lixiviant migration into surrounding aquifers. If vertical excursions result in the contamination of aquifers of good water quality, costly clean-up programs will have to be implemented to return the aquifers to their potential or existing pre-mining water uses(s).

As is typically the case in the Powder River Basin, exploration holes have been drilled by mining companies other than those currently conducting the in situ leach mining. Consequently, the locations and drill hole abandonment techniques are frequently unknown or only very sketchy information is available. Consequently, locating and plugging abandoned drill holes is costly and time consuming. Where the number of improperly abandoned exploration holes is extensive and their locations not well defined, it may be impossible to locate and properly plug all of the holes.

Table 1 lists some of the causes, methods of defining the cause(s), and possible solutions for correcting excursions, should they occur, at a uranium in situ leach mine site. As indicated in Table 1, if cracked well casings or faulty well construction is believed to be the cause of an excursion, high pressure packer tests, down hole TV cameras, or dye tracer injection tests can be used to evaluate the casing integrity, and the casing can either be fixed or the well properly abandoned.

If an excursion occurs and is thought to be due to improperly abandoned exploration drill holes, the methods for verifying the cause and possible solutions to the problem are as listed in Table 1. Maps and drill hole plugging records should be reviewed and a visual inspection in the field for the holes should be conducted. The solution to this problem is to plug, seal, and cap the drill hole(s) properly.

If the confining units which surround an ore zone are vertically or laterally discontinuous and are not thick enough, or the permeability, especially vertical permeability, is not low enough to prevent vertical excursions, then in situ leach mining at a site may not be feasible. This is also true of sites where the ore zone being mined is located near a geologic fault or other geologic feature along which lixiviant could migrate and ultimately contaminate surrounding aquifers.

Where the geohydrology is not suitable for a particular mining operation plan, mining still might be feasible if: (1) the injection/recovery rates are reduced, (2) if mining is avoided in those areas that are least suitable hydrogeologically (i.e., in those areas with insufficient confinement, or where there are high permeability geologic pathways for lixiviant movement out of the ore zone), or (3) the occurrence of excursions are unavoidable but are acceptable environmentally. In such a case as (3), a groundwater pump-back "clean-up" system to recover lixiviant so that the environmental impacts would be minimized, would generally have to be designed and implemented.

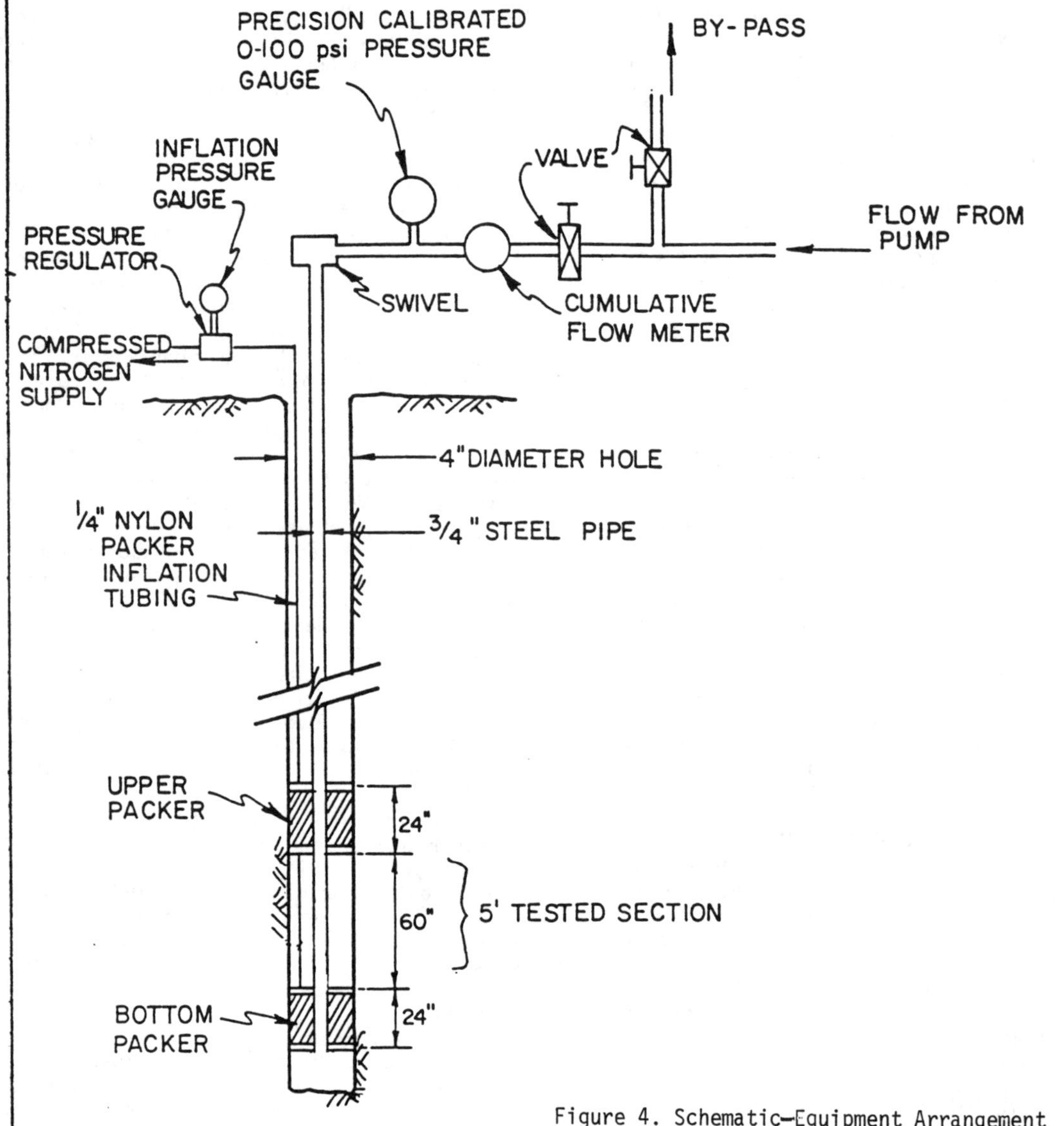

Figure 4. Schematic—Equipment Arrangement For High Pressure Packer Testing
(Modified from WMC application, Aug. 21, 1980, to NRC)

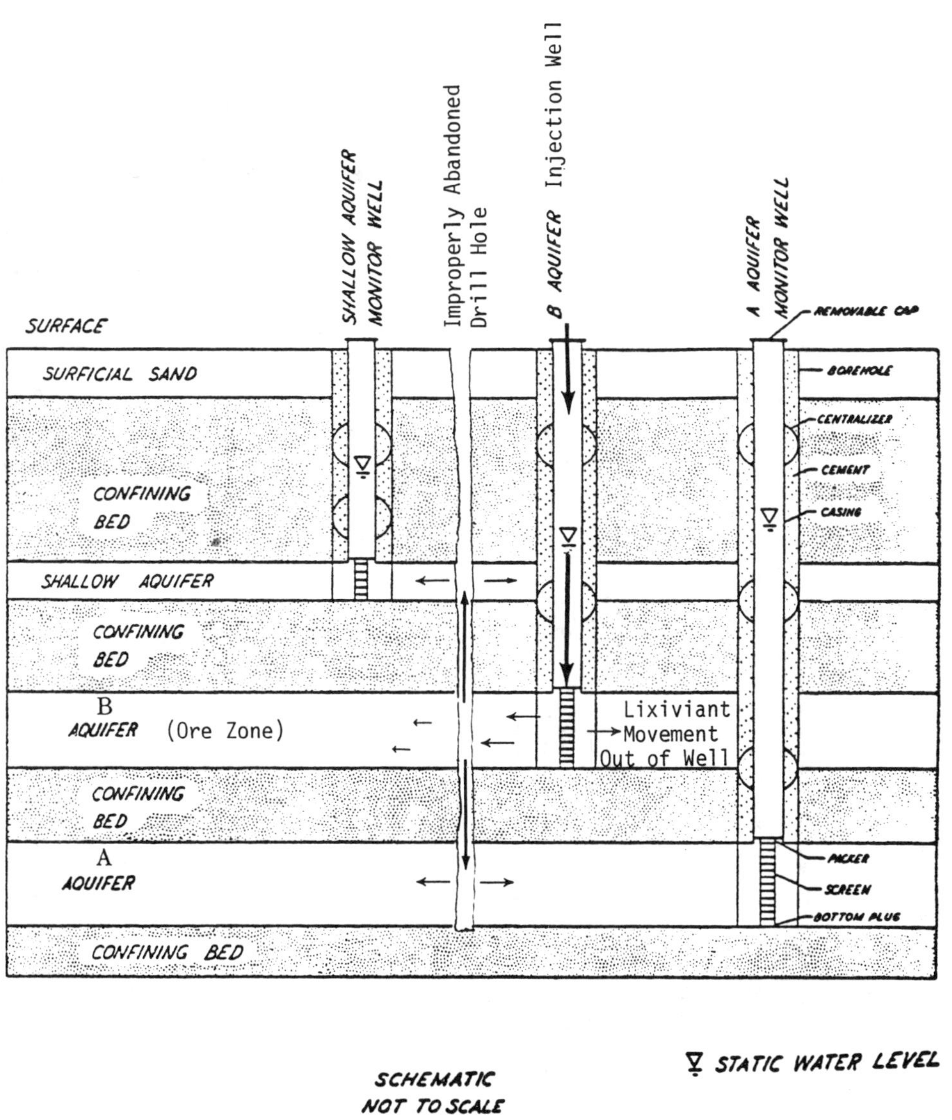

Figure 5. Schematic Of Improperly Abandoned Drill Hole (Modified from APS, June 5,1980, Application to NRC)

Table 1 Typical causes of excursions, methods used in defining the causes, and possible solutions

Cause	Method of Defining	Possible Solution(s)
Cracked Well Casings or Faulty Well Construction	1. Packer tests at various intervals. 2. Down hole TV camera. 3. Dye tracer injection.	1. Fix casing or well construction flaw. 2. Abandon well properly and drill a new well.
Improperly Abandoned Exploration Drill Holes	1. Review of maps which show exploration drill hole locations. Review drill hole plugging information records. Conduct a visual inspection in the field for drill holes near the site of the excursion. This may require the removal of the upper few feet of soil. 2. Conduct Leaky Aquifer Pumping Tests.	1. Plug, seal, and cap the drill holes properly. This may require that the hole be redrilled and cleaned out first.
Unsuitability of the Confining Units to Prevent Vertical Movement of Lixiviant Out of the Ore Zone	1. Conduct Leaky Aquifer Pumping Tests. 2. Drilling, coring and borehole geophysics in the confining units to verify their continuity, thickness, and integrity.	1. Reduce injection/recovery rates. 2. Abandon the site if it is not suitable hydrogeologically.
Undetected High Permeability Strata or Geological Faults	1. Drilling and coring. 2. Borehole geophysics. 3. Aquifer Pumping Tests.	1. Adjust mining operations in order to reduce the potential for lixiviant to migrate towards such strata. 2. Abandon the site if it is not suitable hydrogeologically. 3. Design and implement a groundwater pump-back "clean-up" system that will minimize lixiviant migration to a level that is acceptable.
Improper Balance Between Injection/ Extraction Rates	1. Conduct a Water Balance on Lixiviant Injected and Lixiviant Extracted	1. Increase extraction rates to the appropriate level needed to minimize lateral and vertical excursions.
Hydrofracing of the Ore Zone or Underlying and Overlying Strata	1. Formation pressure testing.	1. Reduce injection/recovery rates. 2. Abandon the site if it is not suitable hydrogeologically.

Hydrofracturing of the ore zone or surrounding strata can occur if the injection-recovery rates are high enough to result in failure of the beds. This could lead to the release of lixiviant out of the ore zone. If injection-recovery rates cannot be reduced in order for mining operations to be economically feasible, then the site would have to be abandoned.

Contamination to the groundwater can also result from failures on the ground surface. This might involve pipe failures from the lixiviant discharge or return lines, but more common, and failures that have already occurred, are failures of synthetic liners used for lining the evaporation ponds which receive water from the lixiviant bleed stream, spent basin wash, eluant bleed, and the residue from flushing injection wells and monitor wells.

The liner failures which have occurred resulted from improper liner installation techniques (i.e., improper sealing along the field seams, leaks around inlet pipes, no protective cover (i.e., clay) over the liner on the bottom and pond sides near the top), and chemical reactions that destroyed the liner integrity when certain reactive chemicals were discharged into the ponds.

CONCLUSIONS

In situ leach mining of uranium ore in the Powder River Basin, Wyoming, is a mining technique which can minimize environmental impacts and allow for mining of low grade uranium ore located at considerable depth that could not be mined using conventional open pit or underground mining techniques. However, the problems as outlined are a concern and need to be addressed at existing as well as at potential commercial and research and development uranium in situ leach mine sites. Although there has not as yet been any serious groundwater contamination resulting from the problems outlined, the potential for serious contamination does exist.

In order for uranium in situ leach mining to realize its full potential, contamination of groundwaters at uranium in situ leach mine sites must and can be minimized in the future by:

Obtaining a thorough understanding of the hydrogeology at the site, especially with respect to the proposed mine plan and lixiviant movement in the ore zone aquifer and potentially through the confining beds to surrounding aquifers,

Verifying that all abandoned exploration drill holes or abandoned wells within the hydrogeologic area that will be affected by mining are properly plugged, and

Designing wells and well fields properly and conducting all phases of the well field operations correctly. The mining operation must be conducted so that the occurrence of unnecessary groundwater contamination from vertical and lateral excursions from the ore zone aquifer to aquifers surrounding the ore zone aquifer is minimized.

Chapter 27

RETREATMENT AND STABILIZATION OF THE NATURITA TAILING PILE USING HEAP LEACHING TECHNIQUES

Randolf E. Scheffel
Ranchers Exploration and Development Corporation
Albuquerque, New Mexico 87197

ABSTRACT

Ranchers Exploration and Development Corp. reprocessed approximately 544 Kt (600,000 tons) of radioactive uranium-bearing tailings. Extraction efficiencies exceeded those predicted from laboratory leach columns and pilot field tests. An acid cure, heap leach system was developed which provided extractions greater that conventional processing alternatives. This was accomplished through meticulous treatment of the tailing with regard to mining, blending, reagent mixing, and material placement in leach heaps. Ultimate success was insured with the development of a counter current washing scheme. This provided an efficient washing arrangement producing a high-grade liquor for processing. Leach liquors were processed through a solvent extraction plant using a tertiary amine for extraction and separation of uranium and vanadium.

INTRODUCTION

In February, 1976, Ranchers Exploration and Development Corporation obtained an option from Foote Minerals, Inc., to evaluate reprocessing of two uranium tailing piles. Foote received ownership of the Naturita and Durango tailing piles in 1967 through the purchase of Vanadium Corporation of America.

The option was exercised in November, 1976, after initial testwork proved that a viable extraction technique existed. Preparation for production at Naturita began immediately with engineering and environmental permitting proceeding simultaneously.

Production began in December, 1977, just 22 months after acquisition of the property and was completed 17 months later.

The successful completion of this project was the culmination of Ranchers' many years of experience in heap leaching and materials handling.

Prior to Ranchers' operation, a successful heap leach was also conducted by Union Carbide Corporation near the Gas Hills area of Wyoming and Maybell, Colorado.(1) Since this time, others have recognized and benefitted from properly designed heap leaching practices.(2,3,4,5)

PROCESS DEVELOPMENT

Sample Preparation

The Naturita tailing pile was located on channel fill beyond the south bank of the San Miguel River and consisted of about 550,000 metric tons. Tailing material was deposited from a single discharge point which was moved infrequently. This resulted in a final deposition which was segregated and varied in depth from 1 meter to approximately 12 meters.

Thirty-five drill holes were driven with a percussion, reverse-circulation drill on pre-selected locations. These locations were selected to confirm more extensive drilling by Foote, and to obtain a composite sample for testing.

Moistures on these samples varied from 3 to 30% with an average of 9.8%.

A typical screen analysis and uranium-vanadium distribution is shown in Table 1.

Microprobe and x-ray diffraction examination identified the following uranium and vanadium minerals.

1. Tyuyamunite (calcium uranium vanadate)
2. Roscoelite (a vanadium muscovite)
3. A high uranium-low vanadium compound

TABLE 1

SCREEN ANALYSIS AND U_3O_8 - V_2O_5 DISTRIBUTION IN SCREEN FRACTIONS

Screen Fraction Mesh (Tyler)	Distribution wt %	Assay %		Units		% Distribution	
		U_3O_8	V_2O_5	U_3O_8	V_2O_5	U_3O_8	V_2O_5
+14	0.7	0.037	0.37	0.026	0.26	0.6	0.9
14x20	2.8	0.069	0.54	0.193	1.51	4.2	5.0
20x28	3.2	0.067	0.53	0.214	1.70	4.7	5.6
28x35	4.2	0.062	0.41	0.260	1.72	5.7	5.7
35x48	8.4	0.035	0.25	0.294	2.10	6.4	7.0
48x65	12.8	0.029	0.19	0.371	2.43	8.1	8.1
65x100	18.8	0.030	0.18	0.564	3.38	12.3	11.1
100x150	16.5	0.032	0.22	0.528	3.63	11.5	12.1
150x200	11.2	0.038	0.26	0.426	2.91	9.3	9.7
200x270	4.9	0.055	0.33	0.270	1.62	5.9	5.4
270x325	2.0	0.062	0.37	0.124	0.74	2.7	2.5
−325	14.5	0.090	0.56	1.305	8.12	28.6	26.9
Calculated heads	100.0	0.046	0.30	4.575	30.12	100.0	100.0
Assay heads		0.048	0.32				

Dissolution Tests

Naturita tailing material responded poorly to acid and alkaline agitation leaching. Recoveries were 47% U_3O_8 - 32% V_2O_5 and 26% U_3O_8 - 16% V_2O_5 respectively.

Microprobe inspection of the leached residue indicated the presence of oxidized uranium-vanadium substances surrounded by hematite. This could be expected from the previous ore treatment being an oxidizing, salt roast.

With the identification of value-bearing clays and hematite-enclosed values, a more aggressive approach was indicated. Several options, including roasting were investigated.

Extractions of 75% and 43% for U_3O_8 and V_2O_5 respectively were obtained using an acid cure technique. Moist tailing was agglomerated with 36 to 45 kg H_2SO_4 per mt (80-100 lbpst), allowed to cure overnight and percolation leached with dilute H_2SO_4.

Such encouraging results guided the testwork very quickly toward establishing an acid cure, heap leach scheme for processing the tailing material.

Small, 7.6 cm diam by 1.2 m deep, laboratory columns were used for screening tests to determine the following key process variables.

1. Effect of H_2SO_4 addition on extraction rate and pH of effluent.
2. Percolation rate.
3. Cure time.
4. Cure temperature.

Acid Addition

Concentrated sulfuric acid was added via a spray bottle to moist tailing while rotating in a polyethylene jar. Varying amounts of acid ranging from 27 kg to 73 kg H_2SO_4 per mt was added and the ore was allowed to cure. Percolation tests were then conducted. Optimum extractions occurred at addition rates of 45 kg to 68 kg H_2SO_4 per mt.

Percolation Rates

A number of parameters were tested for their effect on percolation rates. All samples were agglomerated in a similar fashion in the rotating polyethylene jar.

The effect of moisture content was discovered to be very important in the initial percolation rate tests. Moistures up to 12% gave acceptable percolation rates greater than 0.41 lpm per sq m (.01 gpmpsq ft). Moisture contents greater than 12% gave percolation rates less than 0.04 lpm per sq m. However, extremely low percolation rates did not adversely affect total recovery as similar extractions were achievable with smaller quantities of wash solution .

The addition of flocculents was investigated as a means of improving permeability. Some success was encountered with the use of Polyox WSR 301*, but it was not considered necessary.

Pilot Tests

Once the initial scoping tests were completed, the optimum conditions were applied to a 15.2 cm diam by 6 m deep column.

Figures 1 - 3 summarize the results of this test. Figures 1 and 2 shown 75% of the U_3O_8 and 43% of the V_2O_5 was extracted with the addition of 1460 liters of wash solution per metric ton of ore (350 galpst). Percolation rates varied from 0.86 lpm per sq m initially to 0.49 lpm per sq m at the termination of the leach. More importantly, the pH of the effluent returned in the range of 1.5 to 2.0. This pH was low enough to insure no precipitation problems, but was high enough to provide proper feed to an S-X or I-X recovery system. Figure 3 shows the variation in solution grades with increasing amounts of wash solution.

Although the homogeneity of the material added confidence to going to a full scale operation from 15.2 cm diam column tests, it was decided that a larger scale field test should be conducted.

It was decided to conduct this test on a Durango composite since drill hole information had indicated this material to contain a higher percentage of fines. This would be the ultimate test for the agglomeration and percolation results.

A leach column 2.44 m in diameter by 6 m deep, capable of holding 32 mt of material, was constructed for this purpose.

*Union Carbide Trademark

Figure 1

NATURITA TAILINGS : U_3O_8 Extraction rate from 6" diam. x 20ft. column

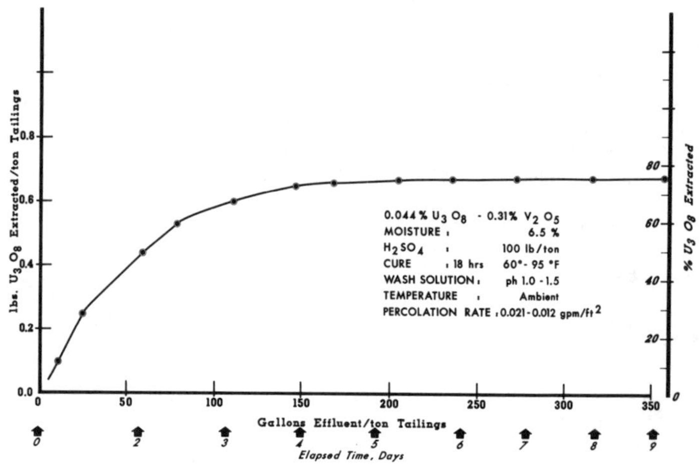

Figure 2

NATURITA TAILINGS : V_2O_5 Extraction rate from 6" diam. x 20ft. column

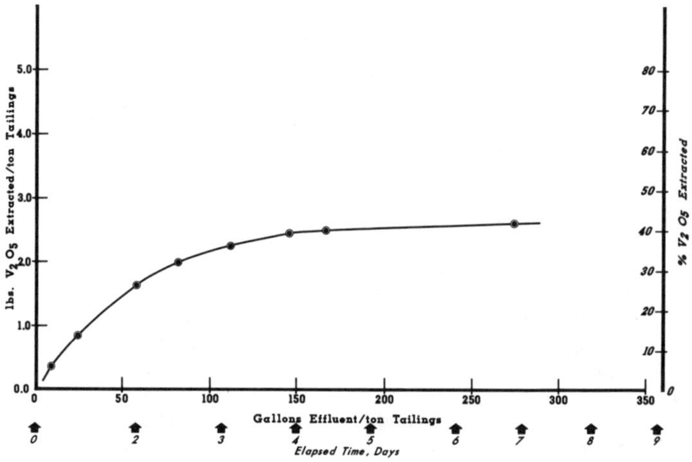

Figure 3

NATURITA TAILINGS: U_3O_8 - V_2O_5 Concentration in effluent from 6"diam. x 20 ft. column

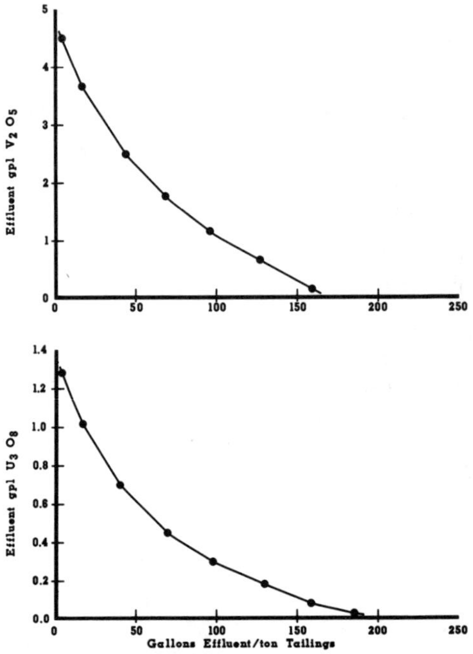

Two tests were conducted, each lasting about three weeks. Material for these tests was agglomerated in a portable cement mixer and allowed to cure for about a week. The first tests showed slower extraction rates than expected due to too little acid addition during agglomeration and improper size reduction of clay lumps.

Uranium recovery for the second test is shown in Figure 4. The washing only continued to about 584 liters effluent per metric ton (140 galpst) due to time constraints.

Figure 5 is a plot of U_3O_8 effluent grade versus gallons of effluent per ton ore. This plot is similar to that shown in Figure 3.

A similar plot for a 15.2 diam by 6 m deep column test on Durango ore is shown in Figure 6 for comparison.

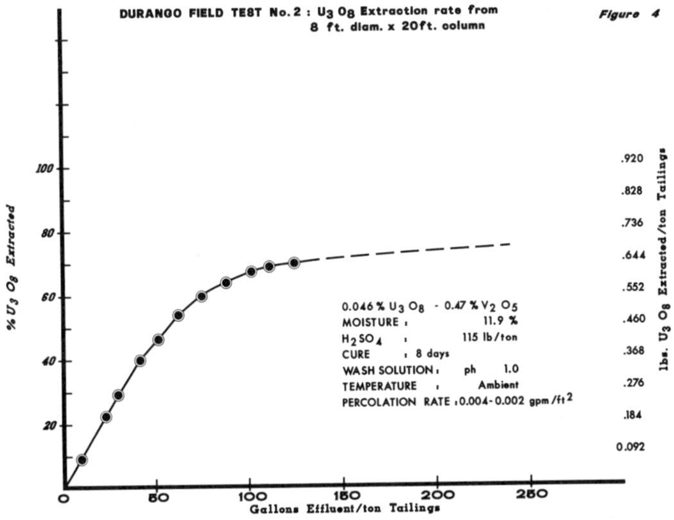

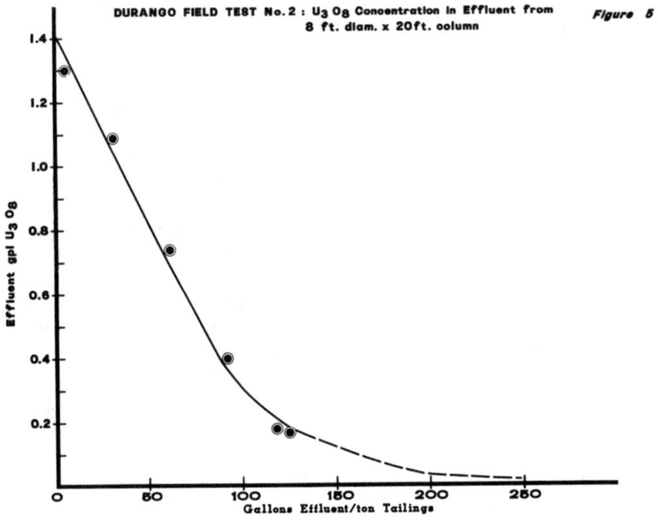

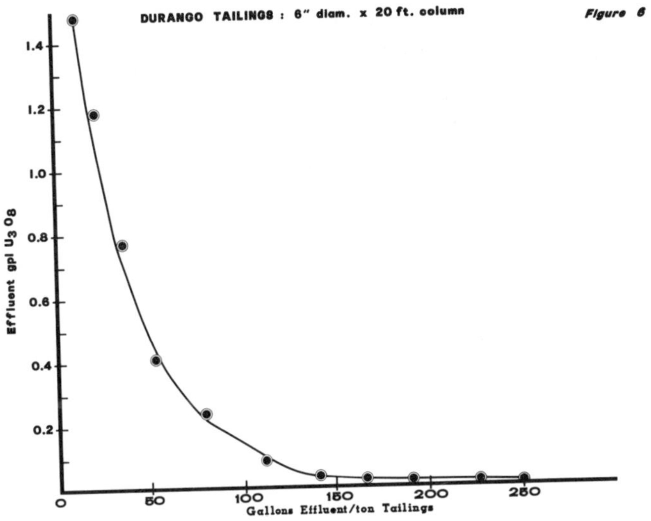

The trends appeared to be consistent, except that the recovery rate was somewhat reduced, indicating that higher wash rates might be necessary for the large scale operation. With this exception, the field tests indicated the recovery from a full scale operation could be predicted rather closely provided the material could be handled properly.

Conclusions from Leach Tests

The following observations were made from the results of the laboratory and field column tests:

- An advance indication of acid consumption was necessary to control effluent pH.
- Percolation rates were affected by initial moisture content.
- The agglomerated tailings had an initial density of 1057 kg per cubic meter.
- Saturated tailing material was extremely fluid.
- Consolidation occurred during the course of leaching.
- The majority of free solution would drain in about 10 days.
- Leached tailing had a high porosity.
- Steady effluent flow required the addition of 167 to 208 liters of leach solution per metric ton of ore.
- Impurity elements in the leach solution showed similar recovery trends to the uranium and vanadium so contaminant levels could be predicted.

The following items required particular attention to insure that maximum recovery would be obtained.

1. Tailing material should be crushed to -0.95 cm minimum.
2. Uniform mixing of acid would be essential.
3. Proper agglomerating to insure homogeneity.
4. A modified heap leach technique would be necessary to contain the leached residue and prevent unstable conditions.

5. The heaps would have to be constructed in a manner that insured uniform permeability.
6. A staged washing system could be used to minimize the volume of process liquor and enhance washing effectiveness.

Uranium and Vanadium Recovery from Leach Liquors

Solution recovery tests were initiated as sufficient leach liquor was generated from column tests.

Initial uranium and vanadium extraction was pursued in a conventional tertiary amine solvent extraction system. The vanadium was first reduced to the cationic vanadic complex with the use of iron to lower the emf to about -380 mv. The anionic uranyl trisulfate complex could then be extracted with the use of a tertiary amine. Once the uranium was extracted, the raffinate could be oxidized with sodium chlorate to produce the anionic polyvanadate species ($H_2V_6O_{17}^{-2}$). This species could then be extracted by the same tertiary amine.

Early tests showed the occurrence of uncharacteristic amounts of interfacial scum formation. Attempts to lessen the accumulation rate were not successful.

Two alternatives were developed for circumventing this problem. The first was to use activated carbon to adsorb the constituent responsible for formation of the scum. The second was to employ an amine ion exchange resin and to scrub the scum-forming constituent from the resin with a caustic solution.

Both systems were investigated extensively. The carbon adsorption/solvent extraction alternative was chosen due to the following:

1. The recovery circuit performance was more predictable.
2. Product purity was better.
3. A lower feed liquor pH could be tolerated.

Once the solvent extraction option was decided upon, additional testing was performed to establish key parameters.

Of particular importance was the rate that impurities would build up in the feed liquor and their effect on product purity. Pregnant solutions were doped with the predicted levels of the most worrisome impurities to determine the necessary extraction and stripping isotherms. Results from these tests established the need for an approximate 35% bleed stream to limit the chloride and phosphorous level of the feed liquor.

PROCESSING

With the key parameters established, engineering began on a system that addressed the items listed above.

A nominal 1352 mtpd (1500 stpd) plant was designed for continuous operation. Reclaiming, agglomeration and leach tank loading was scheduled for two ten-hour shifts five days per week. This required a nominal design rate for ore processing of 136 mtphr (150 stphr).

The ore processing flowsheet is shown in Figure 7.

Selective Mining

The first challenge was reclaiming of the tailing to insure reconstitution of the original size distribution. The key assumption was that fines content was proportional to the moisture content retained by the tailing material. This allowed for a reclaiming plan designed to give an average moisture of about 10%. This was accomplished by reclaiming specific ore blocks and selectively loading given proportions of material into the haulage trucks.

Additional blending was effected at the tailing site with the use of a Caterpillar D-6 CWT dozer. The dozer pushed to a front-end loader cutting at an acute angle to the layers of alternate slime and sand deposition.

A farm disc was also utilized to till areas that required additional air drying.

A Caterpillar 988B loader was used for loading 25-ton bottom-dump trucks for the 5.6 km haul to the processing site.

Blending

Once received at the processing site, the initial treatment of the tailing was geared toward blending to enhance uniform particle size distribution.

As seen in Figure 7, blending was effected by utilizing a radial stacker. Ore was conveyed to a Cedar Rapids 3042 Double Impeller Impact Breaker where clay chunks were reduced to -0.95 cm. The crushed product discharged on to a radial stacker that continuously traversed through a 135° arc. Consequently, material was placed in a Chevron pattern as the pile was being made. Since the stockpile contained about 8200 mt when finished, about four days worth of mined material could be reasonably blended. Two 135° arcs were capable of being constructed. As one was completed, reclaiming with a front-end loader would begin from one end of the stockpile. While reclaiming from this pile, the next one would be under construction.

Acid Consumption

Acid consumption was estimated by conducting empirical agitation tests developed in the laboratory. A short-term agitation leach test was developed by comparing terminal acid consumption in an acid-cure column test. Once a relationship was established, the short-term agitation test could be used to estimate the expected acid consumption.

These tests were conducted on a grab sample obtained while constructing a particular stockpile. The estimated amount of acid would then be used during agglomeration on that particular stockpile.

This procedure was critical to the operation as the ore was allowed to cure four to six weeks. This meant that a considerable amount of tonnage could be improperly treated and the effect would be long lasting.

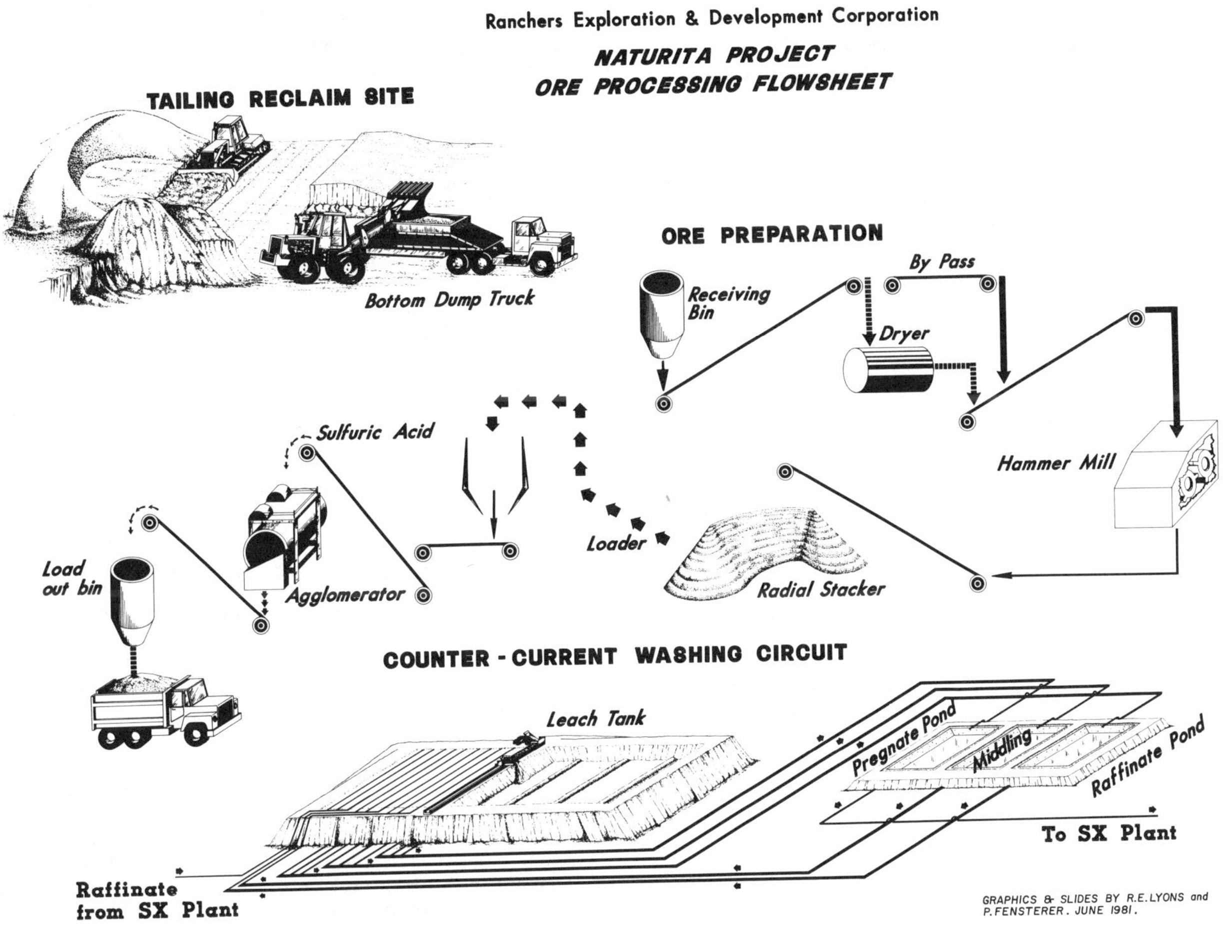

Figure 7

The technique worked extremely well as no significant over-treating or under-treating was experienced.

Agglomeration

Agglomeration was effected in a 3.7 m diam by 10.1 m long rubber-lined steel drum.

The first 3.7 m of the feed end consisted of a mixing area. Small lifters and a 15.2 cm high dam were placed in this section to enhance mixing.

Concentrated sulfuric acid was sprayed onto the ore in this section.

The drum, which weighed in excess of 18,000 kg when loaded, was suspended and driven by a patented Hurst Cable Suspension system. This system was selected for driving the drum to reduce capital and operating costs as well as to reduce delivery time for this piece of equipment. This system worked flawlessly once the proper wire ropes and guides were developed.

The slope of the drum was variable from 9.4 cm per m to 20.8 cm per m and the speed was variable from 7 - 16 rpm. Neither of these parameters were changed once the optimum conditions were established.

Leach Tanks

Containment of the leach material during washing was effected by constructing clay lined pits in which to place the material. This also addressed the requirements for proper stabilization after treatment.

Due to the geometry of the property, three leach tanks were constructed. Their approximate size was 122 m W x 274 m L x 6 m D. The inside berm slope was 1:1 and the outside slope was 2:1. The entire berms were compacted to 98% of standard proctor. A 0.3 m clay liner in the bottom of the leach tanks was compacted to 98% of standard proctor also.

The processing site was selected solely with preference toward the local Mancos Shale for purposes of obtaining this material for the leach tanks. Laboratory tests indicated permeabilities less than 1×10^{-8} cmpsec at 98% of standard proctor and optimum moisture.

The leach tanks were designed so that the cut and fill was balanced. This resulted in the bottom elevation of the leach tanks being approximately 2 m below the average grade.

A more detailed explanation of the physical plant along with the leach tanks and evaporation ponds has been presented by Zambrana and Heath.(6).

Leach Tank Loading

The necessity for uniform permeability required the development of a means for placing the agglomerated particles in the leach tanks to a depth of 6 m. This was accomplished with the development of a conveyor designed and constructed by R. R. Hensler, Tucson, Arizona.

A 122 m long, square frame was made with 15.2 cm square tubing. Within the framework, a 0.9 m conveyor belt was positioned. A tripping device was mounted on top of the frame and moved along the length of the conveyor. The ends were equipped with skids and winches for dragging along the berms. The portion of the conveyor which was inside the berms was supported by tractor pads. Winches, connected to the front of the pads via a cable and snatch block assembly, could move it when energized. A single switch started all winches which moved the conveyor laterally and in a reasonably straight line.

The tripper placed material in front of the conveyor as it traveled continuously along its length. The conveyor could be moved in 2 m increments and walked along on top of the material placed in front of it.

Conveyor feeding was from a dump truck feeding a hopper at one end.

The conveyor performed satisfactorily and no mechanical problems of significance were incurred. It accepted considerable misalignment without consequence.

Countercurrent Wash Circuit

Since testing had shown that the acid cure solubilized the majority of values, a system was developed to maximize the washing efficiency.

It was also necessary to minimize the flowrates necessary to be treated in the S-X circuit, since the plant was to be semi-portable.

A countercurrent wash system was designed whereby raffinate from the S-X facility was used as a final wash solution. By dividing the leach tanks into segments, separate washing cycles could be established. This was accomplished by placing drains along the width of the leach tanks on 7.6 m centers. These drains consisted of 10 cm perforated PVC pipe covered with -1.9 cm washed gravel. A piping system was then designed that allowed for the wash solution to be segregated into three separate storage ponds. By advancing the raffinate wash solution through these steps, a countercurrent wash was effected that resulted in an up-graded pregnant liquor. If necessary, acid could be added prior to recycling to the heaps for pH control. This scheme is illustrated in Figure 8.

The system was designed to produce 454 lpm of solution at a grade of 0.75 gpl U_3O_8 and 3 gpl V_2O_5. This allowed for a minimum wash rate of 2085 liters of wash solution per metric ton (500 galpst) of tailing treated.

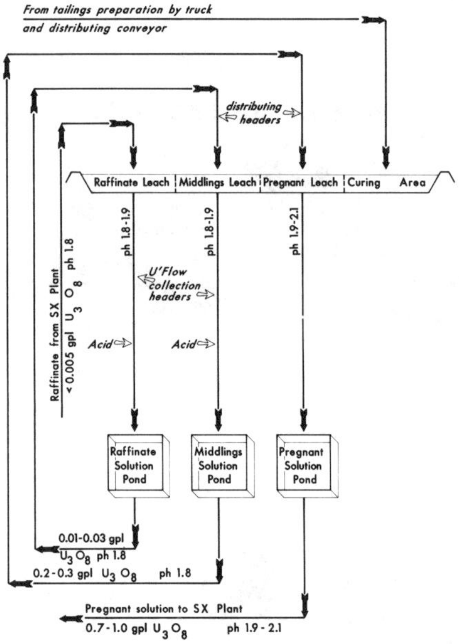

COUNTER-CURRENT WASHING CIRCUIT

Figure 8

Initially, solutions were applied to the surface through Ranchers' patented needle-valve system. However, with the high permeability and differential settling, the solution application was not uniform. To eliminate this problem Rainbird-type sprinklers were utilized. These were effective in providing uniform distribution at low application rates per square meter of area being washed. Unfortunately, these types of sprinklers are not acceptable during -28°C temperatures. A flooding system was ultimately established by leveling the surface and preparing "rice paddies". Once this system was established much improved washing displacements were effected.

As a consequence of our experimenting, about 3127 liters of wash solution per metric ton (750 galpst) of tailing was necessary in order to achieve our predicted washing effectiveness.

The pregnant solution grade was not adversely effected by this as the wash rates were increased in the final two wash stages only.

The system proved to be very flexible and effective with solution up-grading occurring as expected.

Uranium - Vanadium Recovery Plant

The flowsheet for treating the pregnant leach liquor is shown in Figure 9.

The processing plant consisted of three distinct operations:

1. Pretreatment of the pregnant liquor
2. Uranium Recovery
3. Vanadium Recovery

The emf of the pregnant liquor was adjusted to -380 mv by passing it over scrap iron in an upflow column. This was accomplished in a 1.22 m diam x 4.57 m high FRP tank. The solution was then contacted with activated carbon in 2.44 m diam x 3.66 m high FRP tanks for approximately 60 minutes. The carbon was filtered

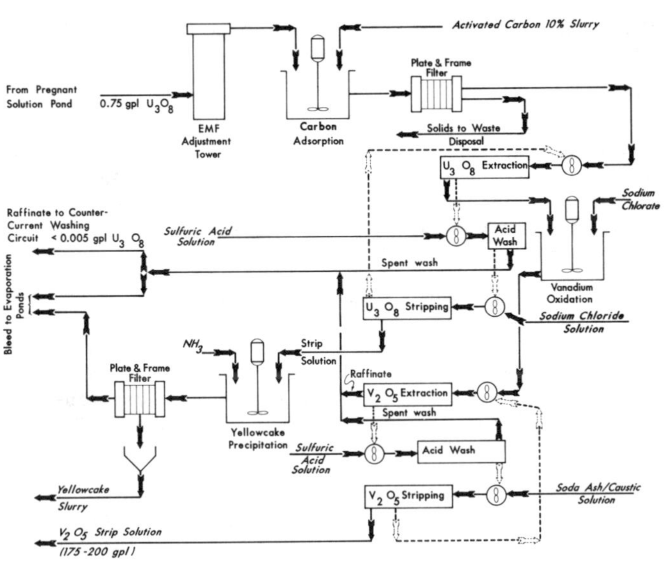

SX PROCESSING CIRCUIT
Figure 9

from the solution in a 1.22 m Sperry plate and frame press. The press contained 57 chambers each 2.54 cm in thickness.

This pretreatment provided a sparkling clean solution for solvent extraction, free of particulate matter and scum-forming organic constituents.

The treated liquor was first processed for recovery of uranium. The uranium solvent extraction circuit consisted of three extraction stages using 2.13 m x 8.53 m settlers, one acid wash stage using a 2.13 m x 8.53 m settler, and three stages of stripping using 1.83 m x 7.32 m settlers. The organic extractant was 3.5v% of Alamine 336* diluted in Kerr McGee 470B solvent with 1.5v% isodecanol as a phase modifier. Stripping was effected using 60 gpl sodium chloride solution acidified to pH 2.0 with sulfuric acid.

Recovery of yellowcake was accomplished by three-stage precipitation with ammonia. The slurry was filtered in a 1.07 m Sperry plate and frame press. The cake was washed, blown with air to 40% moisture and transferred via a screw conveyor and a Moyno pump to a 316 stainless steel tanker. This slurry was shipped to Kerr McGee's Gore, Oklahoma, conversion faciltiy for dissolution in nitric acid and off-loading. This was the first commercial shipment of yellowcake in this form.

The raffinate from the uranium extraction circuit was prepared for vanadium extraction by adjusting the emf to -680 mv. This was accomplished by adding sodium chlorate in three stages of oxidation to 3.05 m diam x 3.05 m high FRP tanks with a total residence time of 90 minutes.

The vanadium solvent extraction circuit consisted of four extraction stages using 1.83 m x 7.32 m settlers, one acid wash stage using a 0.91 m x 3.55 m settler, and two stages of stripping. One strip stage used a 0.91 m x 3.66 m settler, and the other used a 2.13 m x 8.53 m settler. The organic extractant was 2.5v% Alamine 336 diluted in Kerr McGee 627 solvent with 1.5% TBP as a modifier. Stripping was effected using 90 gpl Na_2CO_3 with NaOH being added to control the pH. Precise pH control in both stages of stripping was critical to phase disengagement.

A strip liquor assaying 11% - 15% V_2O_5 by weight was shipped to Union Carbide, Rifle, Colorado, for toll milling and conversion to V_2O_5 flake.

Auxillary Equipment

Supporting equipment for the operation consisted of the following:

- 8 - 3.66 m diam x 7.62 m acid storage tanks.
- 1 - 159 cu meter bolted construction diesel storage tank.
- 2 - 200 HP Kewanee boilers for process heat.
- 6 - Tranter plate-type heat exchangers.
- 1 - sodium chlorate bulk unloading and solution storage pad.
- 1 - soda ash bulk unloading, solution storage and dilution skid.
- 1 - sodium chloride bulk solution storage and dilution skid.
- 1 - liquid anhydrous ammonia storage tank complete with vaporizer.

A truck maintenance shop, plant maintenance shop, an office trailer, two shower trailers and an assay trailer were also provided.

Water was supplied from the San Miguel River through a 6.1 km long, 15.2 cm diam steel pipe line. Pumping had to cease in the winter when the latent heat of the water was questionable. During this time water was transported by truck.

Six evaporation ponds were also constructed in a similar manner to the leach tanks for containment of the necessary bleed streams.

RESULTS

Tables 2 and 3 show the mass balance and washing efficiencies. This information was obtained by auger sampling the leached heaps on 30 m centers and assaying the leached residues. Information on saturation of the liner and an estimate of the values associated with the liner was also obtained. This was done by cutting trenches with the D-6 CWT Cat to expose the liner for sampling.

Table 2 summarizes the mass balance and the accountability. The accountability of the uranium is better because more emphasis was placed on this metal.

It is interesting to note that the acid-cure process solubilized nearly 92% of the uranium that could be solubilized by hot sulfuric digestion.

Production credited by settlement with Kerr McGee was 73.7% of the total contained uranium in the tailing material. After accounting for all mechan-

*Henkel Trademark

TABLE 2

U_3O_8 - V_2O_5 MASS BALANCE AND ACCOUNTABILITY :

MASS BALANCE :	U_3O_8	Distribution (%)	V_2O_5	Distribution (%)
Total kg to leach	1,137,500	(100)	8,751,000	(100)
Total acid soluble kg to leach *1	977,700	(86.0)	8,485,100	(97.0)
Actual kg solubilized by acid cure *2	897,800	(78.9)	5,291,900	(60.5)
kg in residue	239,700	(21.1)	3,459,100	(39.5)
ACCOUNTABILITY :				
Production (kg)	838,100	(73.7)	4,043,400	(46.2)
Tolling Deduct (kg)	–	–	82,600	(0.9)
Soluble loss in heaps (kg)	25,300	(2.2)	608,300	(7.0)
Bleed stream losses (kg)	10,400	(0.9)	250,900	(2.9)
Carbon filter cake (kg)	900	(0.1)	–	–
Clay liner losses (kg)	9,300	(0.8)	69,600	(0.8)
U_3O_8 product losses (kg)	–	–	6,400	(0.1)
Total accountable extracted (kg)	884,000	(77.7)	5,061,200	(57.8)
Solubilized (kg) not accounted for	13,800	(1.2)	230,700	(2.6)

*1 Hot sulfuric digestion residue of 0.006 % U_3O_8 and 0.01 % V_2O_5

*2 Based on washed residue assays of 0.009 % U_3O_8 and 0.13 % V_2O_5

ical losses, an unaccountable difference of only 1.2% is indicated. This was well within the technique for sampling and assaying.

Vanadium solubilization exceeded the expected 50% as 60.5% was solubilized. However, production was only slightly in excess of the 45% predicted. This was due to the abnormally high washing and bleedstream losses due to lower than predicted V_2O_5 recovery in the S-X circuit.

Table 3 summarizes the ability to displace the solubilized values from the heaps with the counter-current wash system. As shown, the uranium washing efficiency was in excess of 97% for the three-stage washing system.

TABLE 3

EXTRACTION AND WASHING EFFICIENCIES :

	U_3O_8	V_2O_5
Total acid soluble kg to leach	977,700	8,485,100
Actual kg solubilized by acid cure	897,800	5,291,900
Soluble values left in residue	25,300	608,300
U_3O_8 Extraction Efficiency	$\left[\frac{897,800}{977,700}\right] \times 100 =$	91.8 %
V_2O_5 Extraction Efficiency	$\left[\frac{5,291,900}{8,485,100}\right] \times 100 =$	62.4 %
Actual Washing Efficiency (U_3O_8)*	$\left[\frac{897,800 - 25,300}{897,800}\right] \times 100 =$	97.1 %
Actual Washing Efficiency (V_2O_5)*	$\left[\frac{5,291,900 - 608,300}{5,291,900}\right] \times 100 =$	88.5 %

* *Based upon average wash water concentration of 0.035 gpl U_3O_8 and 0.33 gpl V_2O_5.*

The vanadium washing efficiency was not as good due to the initial washing solution grade being 0.33 gpl V_2O_5.

The high extraction efficiency was a direct consequence of the tenacity in blending the ore, providing for proper acid addition and the uniform loading of the material into the leach tanks. The high washing efficiency was a result of the designed washing scheme.

ENVIRONMENTAL AND RESTORATION

Thirty-one various licenses and permits were required by the State of Colorado and the Bureau of Land Management prior to commissioning. These consisted mainly of radioactive material licenses, emission permits and easements.

A comprehensive environmental monitoring program was required by the Colorado Department of Health. This required the following:

- Gamma-ray monitoring of the main haulage roads and the plant site.
- Air sampling at the tailing and plant sites.
- Water well and surface water monitoring.
- Alpha emission tests in the yellowcake area, both air and smears.
- A comprehensive urinanalysis program.

The entire program was monitored closely by the Colorado Department of Health. About 5 different visits to the site were performed by the Department of Health during the 18 months of operation. No major problems were detected with any portion of the monitoring program.

Restoration of the river-side tailing area consisted of removing to the plant site for burial, some 107,000 mt of contaminated gravel. Once this was completed a gamma survey of the area was conducted which indicated an average gamma radiation level of 78 μRphr. This was within the then current U. S. Surgeon General's guidelines for the Grand Junction Remedial Action Program.

Contaminated areas of the processing site were cleaned up and disposed of in the leach tanks. The leach tanks were then covered with 61.0 cm of compacted Mancos Shale and 15.2 cm of top soil and reseeded.

Random emission tests after completion of the cover were at background levels.

ACKNOWLEDGEMENTS

The author wishes to acknowledge two individuals and a corporation for their contributions toward the success of this project. First, I would like to express my sincere appreciation to the late Kenneth L. Power, not only for his technical expertise, but for his tremendous personal character expressed throughout the start-up period. Secondly, I wish to congratulate Hazen Research, Inc. for their efforts in conducting the proposed test program. A special thanks goes to Enzo Coltrinari of that organization for his brilliant work on the solvent extraction plant.

REFERENCES

1. Woolery, R.G., Hansen, D.J., Weber, J.A., and Ramachandran, S., "Heap Leaching of Uranium - A Case History," Mining Engineering, March, 1978, pp 285 - 290.
2. McClelland, G.E., and Hill, S.D., "Silver and Gold Recovery from Low-Grade Resources," Mining Congress Journal, May, 1981, pp. 17-41.
3. Potter, P.M., "Design Factors for Heap Leaching Operations," Mining Engineering, March, 1981, pp. 277 - 281.
4. Chamberlin, P.D., "Heap Leaching and Pilot Testing of Gold and Silver Ores", Mining Congress Journal, April, 1981, pp. 47 - 52.
5. Lefler, A.C., "Leaching Practices at Smoky Valley Mine", Gold and Silver--Leaching, Recovery and Economics, Schlitt, W.J., Larson, W.C., and Hiskey, J. B., eds., AIME 1981, pp. 51-55.
6. Zambrana, B., and Heath, R.A., "Reprocessing of Uranium Tailings by Heap Leaching-Ranchers Exploration Naturita Project," Colorado Mining Association 1980 Mining Yearbook, pp. 141 - 149.

Chapter 28

OXYGEN IN SITU LEACHING OF URANIUM IN A SHALLOW ACQUIFER

Velu Annamalai
Zamzow Plant
IEC Corporation
Three Rivers, Texas 78071

Lawrence M. Litz
Linde Division
Union Carbide Corporation
Tarrytown, New York 10591

ABSTRACT

Commercialization of a proprietary "froth-flow" technique is discussed. Oxygen feed concentrations of 200 - 250 parts per million by weight were successfully employed as a two-phase, gas-liquid system in a shallow aquifer. Details are reviewed for both the pilot and commercial operation in a field which was initially operated using hydrogen peroxide as the oxidant and which was subsequently fully converted to oxygen. Uranium productivity with oxygen, both in terms of head grade and daily production rate, was as high as that achieved with peroxide. Cost savings in excess of $5.00 per pound of U_3O_8 produced, were realized for the chemicals involved.

INTRODUCTION

In-situ uranium solution mining involves the injection of a lixiviant (leach solution) through appropriately placed wells into the underground ore zone. Upon contacting the uranium minerals, this solution will dissolve the uranium. The uranium containing solution is then pumped out of the wells into a processing plant where the uranium values are removed, typically in ion exchange columns, and the stripped liquor is recycled back into the field. The lixiviants may be acidic or basic. Where the ore zone mineralization contains appreciable carbonate, as in the South Texas fields, acid consumption would be excessive and the typical solution is ammonium carbonate-bicarbonate at about 5-10 g/l and a pH of 7-9.

Because the usual uranium minerals, uraninite or coffinite, contain uranium in the insoluble tetravalent state, it is necessary to add an oxidant to increase the valence to the hexavalent state where the uranium is soluble. Oxygen dissolved in the lixiviant has been shown to be quite effective for this purpose. It is also much more economical than other oxidants. Of concern is the possibility that (1) its solubility may or may not provide concentration levels adequate to give an economically attractive uranium level in the extracted liquor, and (2) the addition of oxygen in excess of the saturation level so as to increase the uranium grade in the lixiviant may cause gas blockage of the typical fine-pored ore formations.

The IEC Corporation is currently mining uranium by the in-situ leaching method at its Zamzow property near Ray Point, Texas. This field had been put into commerical operation using hydrogen peroxide as the oxidant and had satisfactorily produced uranium in this mode. However, the significant potential cost savings available by switching to oxygen as the oxidant appealed to IEC. A cooperative effort to achieve this goal was agreed upon between IEC and the Linde Division of Union Carbide. This paper reviews this effort and the successful conversion of this field to oxygen.

FIELD CHARACTERISTICS

In IEC's Zamzow field, a staggered five-spot well pattern with 41 feet between wells is used. Lixiviant pumped into the injection wells flows through the uranium bearing porous sandstone formation to the adjacent production wells. Submersible pumps are used down-hole to pump the uranium bearing solution (the pregnant liquor) out of the formation. The extraction rate is fixed by the rate at which the solution flows into the production well under the hydrostatic head of the surrounding aquifer. Environmental control requires that no excursions of the injected solution occur outside the boundaries of the approved mine area. Therefore, only as much liquid can be injected as is extractable.

A major problem with this field insofar as using oxygen is concerned is that it has a very shallow aquifer. The ore zone is at between 130 and 190 feet below grade and the natural water level is at 90-150 feet. Thus, there is only a 40 foot head of water above the ore zone. The lixiviant termperature is about 25°C. Under these conditions, a saturation O_2 level of only about 80 ppm exists.

At least 200 ppm of oxygen is believed necessary to extract the uranium at an economically viable rate. In some fields, oxygen solubility can be increased by increasing the down-hole pressure. This may be done by: (a) operating with the well filled to the top with lixiviant to increase the hydrostatic head and (b) applying hydraulic pressure by pumping the lixiviant into the well head at above atmospheric pressure. These procedures were not applicable, however, to the IEC Zamzow field because it has a relatively high permeability (5-8 Darcys). Increasing the down-hole pressure by any appreciable amount would result in injection rates much in excess of the extraction rates, an intolerable condition. Thus, to use oxygen, a technique must be used which will inject gaseous as well as dissolved oxygen into the ore zone to attain the 200 ppm minimum.

FROTH-FLOW OXYGEN INJECTION

It was decided that a technique called Froth-Flow injection might provide the best means to bring adequate quantities of oxygen into contact with the uranium in the ore zone. Froth-Flow is a procedure for employing oxygen for in-situ leaching which had a preliminary test earlier in a much deeper ore zone by Union Carbide's Metals Division at their Palanga Operations field near Benavides, Texas.[1] It involves dissolving the desired concentration of oxygen gas in the lixiviant in a surface dissolver under pressure. This pressurized lixiviant-oxygen solution is then distributed to the individual injection wells and is fed down each well via a pipe with a flow throttling device, or choke, at the bottom. It is the purpose of this choke to keep the pressure high enough in the well downcomer so that the oxygen does not come out of solution as the solution flows down the well. Without the choke the siphoning effect of the liquid in the downcomer could develop a partial vacuum in the upper sections of the pipe.

As the solution exits from the choke at the ore zone level in the well its pressure is reduced. The gas in excess of that required to maintain the saturation concentration at the pressure at this point then comes out of solution in the form of very fine bubbles. This froth flows into the ore formation as a two-phase liquid. If, for example, the solution upstream of the choke had 200 ppm of oxygen in it and the saturation concentration was 80 ppm, 120 ppm would come out of solution (ignoring any supersaturation). If the pressure were 74' w.c. absolute (32.0 psia) in the well, the gas volume would be only about 4 percent of the total liquid-gas volume.

A very small bubble size (usually a fraction of a millimeter or less) is typical of gas-liquid mixtures produced in this fashion in ionic solutions. This permits the mixture to flow readily through the bulk of the pore structure without inducing gas blockage in a formation with the pore characteristics such as exist at Zamzow. Both the oxygen in the gas phase as well as that in solution can react with the oxidizable species in the formation. As the dissolved oxygen concentration is reduced by reaction, the oxygen in the gas bubbles will go into solution, thereby keeping up the solution oxygen concentration.

OXYGEN SOLUBILITY

A key factor in the design of the Froth-Flow system is the oxygen solubility a) at the pressure in the ore zone and b) at the pressure to be selected for the surface oxygen dissolver unit. Once the desired operating oxygen concentration is selected, the operating pressure in the dissolver and in the lixiviant distribution manifolds must be high enough to insure that the required quantity of oxygen will be put into solution and will be dissolved when the solution exits from the choke at the bottom of the injection wells. Gas which is not dissolved will typically be present as fairly large bubbles (greater than a couple of millimeters). As such, they will usually not flow with the liquid into the ore zone. Instead, they will tend to rise up in the annular space between the well casing and the down-comer and be vented to the atmosphere.

The selection of the operating oxygen concentration is based both on the minimum required to economically extract the uranium as well as the maximum acceptable free gas concentration in the ore formation. The latter depends on the pore size and pore size distribution of the interconnected pores through which the gas-lixiviant froth must flow. It is selected such that significant flow reduction due to gas blockage does not occur. This maximum will normally have to be determined by testing in the field if very high free oxygen concentrations, which could result in bubble coalesence, are desired.

Oxidant concentrations of 0.025 to 0.06 oxidation equivalents per liter appear commercially desirable in most fields. This corresponds to about 0.5 to 1.0 g/l of hydrogen peroxide or about 200 to 470 parts per million (ppm) of oxygen weight of lixiviant. At the nominal salt concentrations involved and in the temperature range of 0 to 40°C and pressures between 1 and 100 atmospheres of oxygen, the oxygen solubility is given by Equation 1 as a good approximation.[2]

(1) O_2 solubility (ppm) =
2171 P, atm. (1-0.0304 lnP, atm.)/(33.5+T,°C)

Figure 1 is a plot of this relationship at 15°, 25° and 40°C with the pressure expressed in feet of water column. From these curves, it is seen that, at the respective temperatures, an oxygen saturation concentration of 200 ppm would be obtainable at a water head of about 160, 195 and 245 feet, respectively.

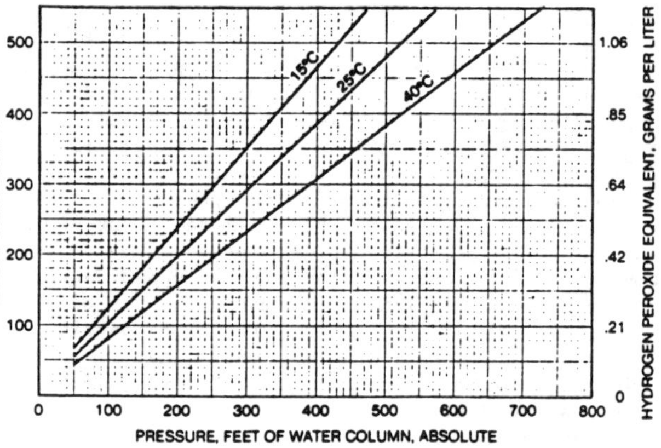

FIGURE 1. OXYGEN SOLUBILITY IN TYPICAL URANIUM LIXIVIANT

The Zamzow field with a nominal hydrostatic head of 40 feet (74 feet, absolute) at the ore zone, at the 25°C lixiviant termperature, would have a saturation oxygen concentration of about 80 ppm.

TEST SYSTEM DESIGN

The design basis for the pilot test program is given in Table I. A target oxygen concentration of 200 ppm in the ore zone approximates the 0.4-0.5 g/l hydrogen peroxide content that was being used in the commercial field operations. An extra 15 percent was to be fed on the assumption of 85 percent dissolution efficiency in the surface dissolver. When the system design was initially formulated, it was anticipated that 40 GPM of lixiviant would be distributed equally to four wells. When put into operation, the output of the available production well allowed only 25-30 gallons of lixiviant to be injected. As a result, the total flow was lower than the surface dissolver was initially designed for and the down-hole choke was also somewhat oversized, giving less pressure drop than expected. Both factors tended to reduce the dissolved oxygen level, as will be discussed below.

TABLE I
FROTH FLOW DESIGN BASIS - IEC TEST

No. of Wells	4
Lixiviant Injection, GPM/Well	10
Lixiviant Injection, Total GPM	40
Lixiviant Injection, Well-head Press, psig	100
Lixiviant Temperature, °C	25
Ore Zone Depth, Ft	170
Aquifer Surface, Ft	130
Oxygen Solubility, ppm	80
Oxygen to Ore Zone, ppm	200
Oxygen Solubility at Well-head, ppm	270
Oxygen Dissolved, % sat'n	85
Oxygen Dissolved, SCFH total	48
Oxygen Feed Rate, SCFH total	57*
Pipeline Dissolver, NPS, inches	1 1/4
Pipeline Dissolver, Length, Feet	100
Down-hole Orifice, Hole Diameter, Inches	0.217

* Assumes dissolution of 85% of gas fed.

SURFACE DISSOLVER

A variety of gas-liquid contactors, such as packed towers, gas sparged turbines, bubble towers, etc., properly designed, could serve to dissolve the oxygen in the lixiviant prior to feeding the solution to the individual injection wells. However, for the purpose of this application, it appeared most convenient and most economical to use a pipeline contactor since the barren lixiviant had to be brought by pipeline from the process plant to the well-field. For the design values of 200 ppm in 40 GPM of lixiviant, a 1 1/4" schedule 40 pipe, 100 feet long should give about 85% gas dissolution.

Figure 2 shows a flow diagram of the test arrangement. The piping sectors between the oxygen gas sparger and the gas separator served as the pipe line contactor. Initially, because 1 1/4" PVC pipe was not available, this section was made up of 120 feet each of 2" and 1" pipe in alternating 60' lengths. When, under the pressure and flow actually employed, it was determined that insufficient oxygen was being dissolved this pipe length to the separator was extended to 540 feet to achieve the desired oxygen dissolution. This was made up of 120' of 1", 120' of 1 1/2" and 300' of 2" pipe.

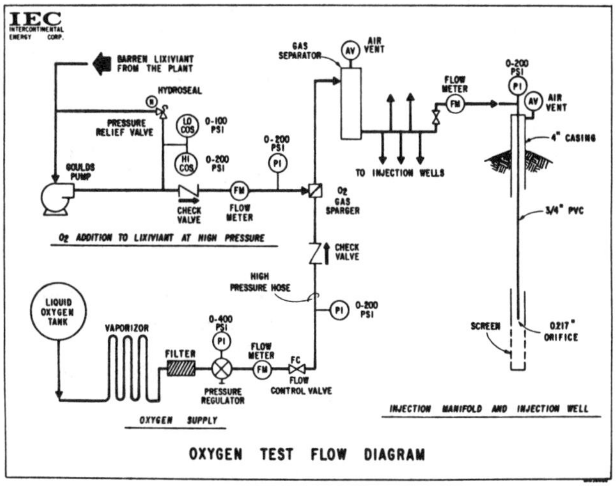

FIGURE 2. TEST SYSTEM FLOW DIAGRAM

A gas separator was included to remove undissolved gas prior to distribution of the lixiviant to the wells. The possibility of getting substantial quantities of undissolved gas into an injection well posed the problem of an inbalance of lixiviant flow. The separation unit was simply a piece of 6" pipe with the liquid fed tangentally near the top and discharged near the bottom. Low liquid velocity allowed large gas bubbles to disengage and collect in the top from which they automatically vented via the air vent valve.

LIXIVIANT SUPPLY

Barren lixiviant from the process plant was pressurized to 140-170 psig by a Goulds multi-staged centrifugal pump. A flow meter downstream of the pump indicated the total lixiviant flow to the field. A check valve in the line prior to the point of gas addition was included to block gas flow backwards into the lixiviant supply system in case the pump failed to maintain the lixiviant flow. High and low pressure cut-off switches were also installed to shut off a solenoid in the oxygen supply line if the liquid pressure dropped below 100 psig or rose above 200 psig.

After removal of free gas in the separator, the lixiviant was distributed to the injection wells from a simple manifold. The flow meter and valves in the lines from the manifold to the well heads permitted the lixiviant flow to be balanced and controlled to insure that the correct balance between injection and extraction rates was maintained.

Lixiviant flowed down the wells in 3/4" PVC downcomers which terminated at the top of the screens opening into the ore zone. A simple orifice hole drilled in a pipe cap served as the choke to maintain the pressure in the downcomer.

OXYGEN SUPPLY

Liquid oxygen from a Linde VCC-150 storage tank was delivered to the atmospheric temperature vaporizer automatically as required to maintain the desired gas pressure. The tank, vaporizer, control and safety vent systems were all pre-assembled on a small trailer unit. The gas flow volume was read from a flowmeter and manually adjusted in proportion to the total lixiviant flow to provide the specified gas/liquid ratio. Gas pressure was set to insure at least 15 psig pressure in excess of the maximum lixiviant pressure. The gas was sparged into the lixiviant via a perforated coaxial tube in the lixiviant feed line.

TEST PATTERN

A seven spot pattern was chosen for the initial oxygen test in an area which was completely isolated from the production area. The ore thickness, ore grade and the depth of these seven wells are shown as Pattern No. 1 in Fig. 3. Well 2A-828 was chosen as the production well with four of the surrounding six wells as injection wells. The ore zone exhibited a high sulfide content and a high total reducing capacity. All of these wells were cased with 4" pipe and completed in the usual manner with well screens in such a way that only the mineralized zone is contacted by the injection fluid. Packers (2") were used only when a well had to be underreamed in the ore zone.

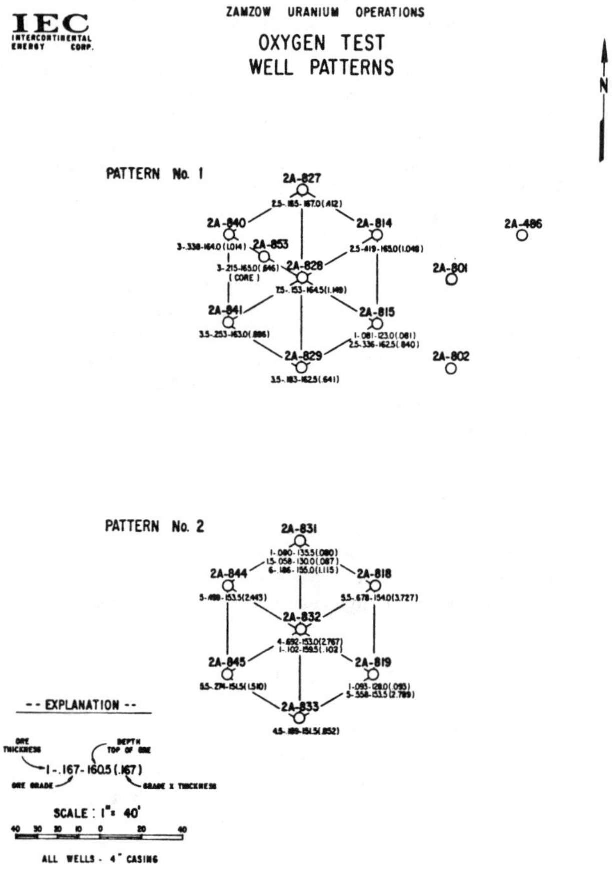

FIGURE 3. TEST WELL LOCATIONS

In selecting the test pattern, every effort was made to see that none of the commercial field's peroxide containing injection solution had come close to this pod. Water samples were taken from wells in the test pod before any external solution was put in to determine the base values of the aquifer as well as whether any hydrogen peroxide containing lixiviant had leaked into the test pod. The results tabulated below show that the test pod was very much isolated from any leaching area being used in the commercial operation.

TABLE II
ANALYSIS FROM TEST PATTERN PRIOR TO LEACHING

Well #	U_3O_8, ppm	pH
2A-814	0.6	8.19
2A-815	0.9	7.68
2A-829	2.1	8.04

This low U_3O_8 content confirmed that all the U_3O_8 response from the test pod would be entirely the contribution of the new oxidant, namely oxygen, when the pod went into test operation.

TEST OPERATION AND RESULTS

Before the test began, this pod was pretreated, as was usually done, with 20,000 gallons of 1.0M $(NH_4)_2CO_3$ solution followed by 100,000 gallons of leach solution for each well. Neither the $(NH_4)_2CO_3$ solution nor the leach solution used for the pretreat had any oxidant added to them. The purpose of pretreating any new mine area is two-fold: 1) to help displace high calcium content water from the aquifer area of the ore zone and thus minimize calcium carbonate precipitation inside pipes and in the ion exchange columns during the production period; 2) to begin leaching of naturally oxidized uranium in the ore pod to establish the initial soluble U_3O_8 level before oxidation was begun.

Before the oxygen test was begun, samples were taken from the production well (2A-828) for three days. They showed that the U_3O_8 level in the leach solution in contact with the test pod had reached 23 ppm.

Continuous injection of leach solution, without addition of oxygen, was started on 6/6/79. Oxygen injection was begun the next day. The U_3O_8 concentration in the pregnant liquor from the production well, its flow rate, the wellhead pressures, the oxygen pump delivery pressure, the amount of oxygen addition, the lixiviant injection flow rate and other required data were monitored regularly. Figure 4 shows the U_3O_8 concentration in the pregnant liquor from the production well over the first month of operation.

In the three days after the oxygen free lixiviant injection was begun, the pregnant liquor U_3O_8 concentration dropped down to about 14 ppm. Then, as the oxygen began to take effect, the U_3O_8 concentration rose to 45 ppm, and then declined to about 32 ppm. Tests of the oxygenated lixiviant showed only about 45-50 percent of oxygen saturation at the existing line pressure and temperature, indicating, as reported above, that the residence time in the surface dissolver was insufficient. After the extension of the pipeline sections, the U_3O_8 concentrations rose sharply over the next three days to peak at 80 ppm and then, as was typical, slowly declined at a rate of about one ppm U_3O_8 per day over the next month. The 3-4 day response to changes in injection variables was characteristic of subsequent operations also. Injection flow rates appeared to be unaffected by the presence of oxygen in the formation. Also, no gas locking of the production pump was observed. As expected, some small amounts of gas did escape up through the annulus of the injection well and was vented from the gas vent valve on the well head.

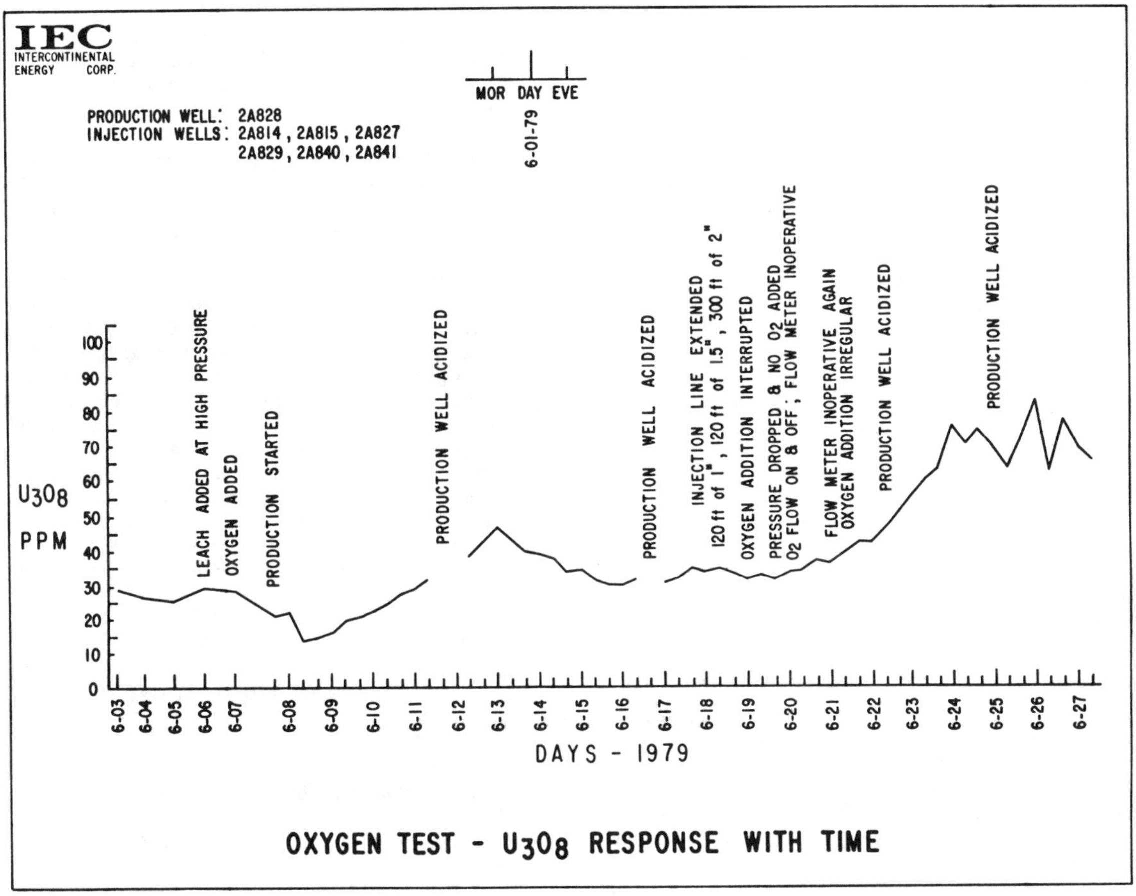

FIGURE 4. U_3O_8 CONCENTRATION IN PREGNANT LIQUOR

To further confirm that none of the hydrogen peroxide addition in the regular production area leaked into the oxygen test area and contributed one way or another to the test results, samples were taken on 6/25/69 from wells which were between the test area and the production area. The location of these wells are shown in Fig. 1. Analyses, shown on Table III, verified that the test pod is completely isolated from the hydrogen peroxide production area.

TABLE III

ANALYSES FROM WELLS BETWEEN PRODUCTION FIELD AND TEST PATTERNS

Well #	U_3O_8, ppm	pH
2A-486	0.5	7.81
2A-801	0.5	8.20
2A-802	0.5	7.71

After running the tests for a month, it was decided to extend the test to a second seven well pattern in which the orebody had a higher ore grade. The details of these wells, can also be seen in Fig. 1. This test pod was partially pre-treated with leach solution containing dissolved oxygen at higher pressure. The lixiviant line feeding test pod 1 was extended to test pod 2 and the total lixiviant flow increased to about 40 GPM. The gaseous oxygen addition was increased to about 75 SCFH (about 312 ppm) to attempt to supply sufficient oxygen to both the test pods. Test pod 2 came on line on 7/18/79.

The initial U_3O_8 concentration from production well 2A-832 was about 170 ppm. It dropped rapidly to a low level of about 40 ppm by 7/27/79. During this time, the first test pod (2A-828) also dropped to a U_3O_8 level of 22-25 ppm. It was suspected that there was not enough residence time in the surface dissolver for enough gaseous oxygen to go into solution with the oxygen addition was almost doubled. Therefore the lixiviant line between the high pressure pump and the second test pod was further extended to a length of 640 feet on 7/27/79. The U_3O_8 response from the production well 2A-832 immediately showed improvement and constantly increased to about 90 ppm from about 40 ppm over a period of 10 days. The increase in U_3O_8 response, as before, showed that a longer residence time helped the oxygen go into solution more efficiently.

While the second test pod was running satisfactorily, the U_3O_8 grade for production well 2A-828 in the first test pod gradually decreased to a low level of about 20-25 ppm. It was concluded that the first test pod did not receive a sufficient amount of injection solution due to the fixed amount of lixiviant solution available (the high pressure Goulds pump was limited in the flow it delivered at the required delivery pressure). Therefore the injection to the first test pod was totally cut off on 8/14/79 and all the available lixiviant solution was used solely in the second test pod. An immediate increase in the U_3O_8 grade from well 2A-832 was seen after this modification.

As mentioned earlier, all injection to the first test pod was cut off on 8/14/79, but the production well 2A-828 continued to operate without much of a change in U_3O_8 grade. After seeing no fluctuations in 2A-828, it was decided to inject the first test pod with hydrogen peroxide to see whether this pod responded better with hydrogen peroxide. Though there was no dramatic change in U_3O_8 concentration right away, it did increase gradually from about 23 ppm (8/21/79) to about 40 ppm (9/6/79).

A 2" Kenics 6 element static mixer was put in the lixiviant line downstream of the oxygen sparger on 9/4/79 to try to get better dissolution of the oxygen. The U_3O_8 response showed a dramatic improvement from 9/7/79. However, at the same time the total carbonate in the system also increased from about 3 gpl to about 8 gpl in about 6 days due to some changes in the chemical addition procedure in the plant. Though the increase in the total carbonate would have contributed to the higher U_3O_8 response,[3] IEC feels that the Kenics mixer made a significant contribution to the overall response of U_3O_8 in the test pod.

The second test pattern operation was continued till 10/11/79 with the U_3O_8 grade gradually falling to about 50 ppm and then continuing at that level. Anytime oxygen was cut for a few days, a lower U_3O_8 response was immediately seen in the production well assays. The first test pod, which has been run with hydrogen peroxide addition since about 8/21/79, had not shown any further improvement during this entire period.

At this time a second Goulds pump was acquired and a third test pod was selected in the commercial production mine area which had been operated for more than a year with hydrogen peroxide addition. This test pod was isolated and oxygenated lixiviant was added. The U_3O_8 response did not show any change as a consequence of the shift from peroxide to oxygen. This may have been due to the fact that this particular test pod was in an old area where most of the orebody in and around the wells would have already been leached out. By the time the hydrogen peroxide reached the location where leaching was occuring, most of the peroxide would have been decomposed to oxygen and water. Thus, oxygen and not hydrogen peroxide would be the local oxidant. Hence whether the primary feed to the well contained hydrogen peroxide or oxygen, the U_3O_8 response would not probably change to any appreciable extent.

ECONOMIC EVALUATION OF PILOT TESTS

The main reason for conducting this feasibility study of oxygen as an oxidant was the very high relative cost of hydrogen peroxide compared to oxygen. The approximate cost of the hydrogen peroxide being added to the lixiviant was about $8.00/lb of U_3O_8 produced. Savings realizable in the cost of the oxidant without lowering the daily uranium production level would justify a change from peroxide to oxygen.

The data collected during the first month in test pod 1 showed very encouraging results and these results were repeated in the second test as shown in Table IV. The cost comparison between hydrogen peroxide and oxygen (on the basis of the tests) is not extremely accurate because it compares small test pods against the entire wellfield. However, only a qualitative number was sought which would give a clear indication of the direction to move. In this cost comparison, it was assumed that all the other costs involved in the operation remained the same whether hydrogen peroxide or oxygen were used as an oxidant in the system.

TABLE IV

OXIDANT COST COMPARISONS FROM TEST RESULTS

Test Pods	1	1 & 2
Test Period	6/8-7/8/79	6/8-9/25/79
U_3O_8 Recovered, lbs.	315.13	1362.83
Costs,		
Oxygen Cost, Total	$260.80	$1063.63
Oxygen Cost/lb U_3O_8	0.80	0.78
Power Cost, Incremental/lb U_3O_8	0.50	0.50
Total Cost/lb U_3O_8	1.30	1.28
Peroxide Cost/lb U_3O_8 in Commercial Field	8.00	8.00
Savings/lb U_3O_8	6.70	6.72

It is assumed that any U_3O_8 recovered due to pre-oxidation of the aquifer (U_3O_8 recoverable without adding any oxidant) would be the same irrespective of the type of oxidant and hence no special allowances are made in this cost analysis.

While performing the oxygen test, IEC also looked for problems that might arise because of the addition of oxygen, such as 1) gas blockage in the orebody, 2) gas locking in the production well pumps, 3) any additional ionic species or any increase in the amount of the already present species in the pregnant leach solution. None of the above problems were encountered during the test period. It was therefore, at this point, on the basis of the overall performance with oxygen and its cost advantage, that a decision was made by IEC Corporation to convert the entire mine area to oxygen addition and eliminate peroxide addition in the wellfield altogether.

COMMERCIAL OPERATION

Retrofit Requirements

The previous section indicated in a pilot test that oxygen gas, introduced as a froth, could be used profitably in a uranium in-situ operation as an effective oxidant. Whether oxygen could completely replace hydrogen peroxide in this shallow aquifer could only be fully answered after oxygen was put into full commercial scale operation. IEC Corporation's Zamzow Mine was an operating mine using hydrogen peroxide as its only oxidant in about 1600 GPM of lixiviant. In order to use oxygen as the oxidant, the existing system needed modification. It was necessary to install high pressure injection pumps, the instrumentation needed to measure oxygen flow rates, the piping and fittings to inject oxygen in the lixiviant stream and certain electrical equipment. Figure 5 is a flowsheet of the entire commercial plant operation with oxygen as the oxidant.[4]

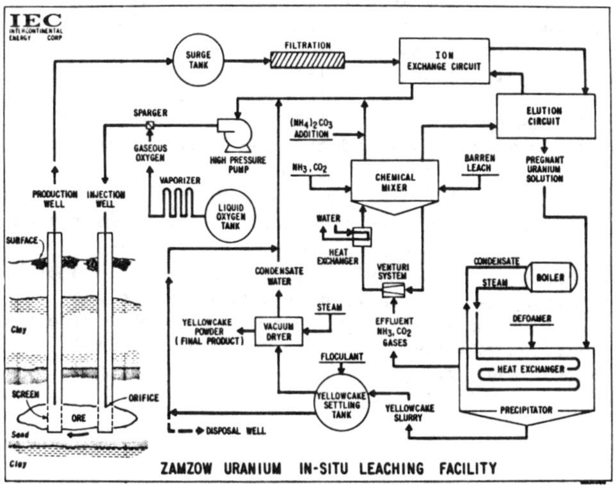

FIGURE 5. COMMERCIAL SYSTEM FLOW DIAGRAM

In the hydrogen peroxide system, the barren solution coming out of the ion exchange columns was pumped to the wellfield through 6" PVC pipes by 10 HP injection (booster type) pumps delivering about 400 GPM at 20-40 psig. Since oxygen had to be added at higher pressures, 100 HP pumps were used to deliver about 400 GPM of lixiviant at about 150 psig. These pumps are stainless steel, chemical process type with double mechanical seals and external flush. The pumps were fitted with high and low pressure sensors to shut off the oxygen input to the lixiviant if the lixiviant pressure was outside the design limits.

The instrumentation to provide the proper oxygen flow rate consists of a simple differential pressure flow meter and a needle valve to manually regulate the oxygen flow in proportion to the lixiviant flow rate. The valves, fittings, and piping used in the oxygen flow system were cleaned for oxygen service by removal of all organic and oxidizable matter. The oxygen gas is introduced into the lixiviant stream by a sparger to break up the gas flow into small bubbles. Each lixiviant line has a 4" Kenics in-line static mixer downstream of the sparger to aid the solution of the oxygen. A Linde TM-11,000 liquid oxygen tank with a 50 ton liquid capacity (about 1,265,000 ft^3 as gas), with suitable vaporizer capacity, was installed to supply the anticipated oxygen consumption of 50,000 to 60,000 ft^3/day.

It was found desireable to add a pressure gauge to each well to measure the pressure in the 4" casing. These gauges served a two-fold pupose:

1. This pressure would indicate whether the injection well is taking the solution in or not. If the pressure is high, it means that the well screen is plugged up and it needs cleaning.

2. By the state of Texas injection regulations, the operation could not exceed a pressure of about 0.43 psi per foot of well depth or about 55 to 75 psig at the top of the casing. This pressure gauge helps meet this regulation.

The old 6" barren lixiviant lines were changed to 4" lines in order to maintain the required high flow velocity to provide dissolver capacity. These lines were also run longer than needed in order to give additional residence time for oxygen to go into solution. Each injection well was fitted with a 3/4" down hole stringer capped at the bottom end with a stainless steel end cap having a 0.217" orifice drilled at its center. The down hole stringer was fixed in the well in such a way that the orifice was located just about at the top of the screen inside the well casing.

Installation and Start-Up of Oxygen System

Design, ordering and installation of the entire oxygen system only took about 6 months. Installation of the lixiviant pumps and other required modifications were completed on a phased schedule to put one-fourth of the mine area at a time on the oxygen system at an approximate flow rate of about 400-450 GPM each. Since each injection well orifice was designed for about a 10 GPM flow rate, at least 40 injection wells (most of the time 45-50 wells, since all the wells did not always take 10 GPM of solution) were needed for a lixiviant flow rate of 400 GPM. If more injection wells than specified were opened up and each of them took their full flow, the pressure would drop below the desired level of 100-120 psig at the injection well heads. If fewer injection wells were opened than required, the line pressure would rise above the design level.

Transition Performance

The first 400 GPM section was started on April 18, 1980. For the first few days no oxygen was added and only leach liquor was pumped to the well-field at elevated pressures. Quite a few leaks in the PVC injection lines were found and repaired during this time. The operators also had to be trained in the running of the high pressure pumps with all the limitations of pressures and flows. After a few days, oxygen was added to the lixiviant stream. No special problems were encountered. The other three units were added to the oxygen system one by one, and by May 24, 1980, the entire wellfield was completely under oxygen addition with a total flow rate of 1500-1600 GPM. A well head pressure of 100-120 psig was easily achieved in the entire wellfield.

Neither lixiviant flow rate reduction compared to commercial operations when peroxide was the oxidant nor extraction pump gas locking problems were experienced after the oxygen system was turned on. In fact, some gas locking problems had been encountered when the wellfield was under hydrogen peroxide addition. It was believed that this was due to a mixture of gases such as oxygen, carbon dioxide and ammonia coming out of the wellfield. No special difficulty was encountered in maintaining the pressure in the injection wells and an acceptable dissolved oxygen level in the lixiviant streams.

Oxygen is added to the lixiviant streams to provide about 235 ppm of gas at the high pressure pump with a delivery pressure of about 150 psig. Table V is provided to the operators who regulate the oxygen gas flow with respect to the lixiviant flow in each of the four pump systems:

TABLE V

OXYGEN FEED RATE VS. LIXIVIANT RATE

Lixiviant Flow Rate (GPM)	Oxygen Flow Rate (SCFM)
300	7.0
320	7.5
340	7.9
360	8.4
380	8.9
400	9.3
420	9.8
440	10.3
460	10.7
480	11.2
500	11.6

Dissolved Oxygen Measurements

A dissolved oxygen meter was obtained from Hydro Labs, Austin, Texas which is designed to read high oxygen concentrations at elevated pressure. It was found that gas dissolved in the electrolyte during the high pressure measurement and was subsequently released as gas bubbles when the probe was removed from the high pressure environment. As a result, it was necessary to change the membrane and the electrolyte after each measurement. Despite this inconvenience, this meter appears to be the best of several evaluated for these high pressure tests.

Dissolved oxygen readings were taken at the injection well heads. A set of sample readings is given in Table VI. It shows that an oxygen dissolution efficiency of about 80-90% is achieved if the correct amount of oxygen is added to the lixiviant streams. If less oxygen is added than the 235 ppm design level, the dissolution efficiency is between 85-97%. If considerable excess oxygen is added, bringing that fed in excess of the saturation level, the dissolution efficiency drops as low as 50%, depending upon how much excess oxygen is added. This may be due to the higher concentration of free gas bubbles in the lixiviant line which would promote coalescence into large size bubbles. The gas dissolution efficiency of the large bubble stream would be expected to be poorer than in a small bubble stream.

TABLE VI

DISSOLVED OXYGEN READINGS AT THE INJECTION WELL HEADS

Injection Well Head Pressure psig	Injection Well Head Pressure ft. water	Saturated Dis. Oxygen (ppm)	Lixiviant Flow Rate (GPM)	Oxygen Flow Rate cuft/min*	Oxygen Feed (ppm)	Actual Dis. Oxygen (ppm)	Oxygen Dissolution Efficiency %
118	308	295	320	7.5	233	209	89.9
125	325	315	320	7.5	233	190	81.5
99	264	260	330	8.5(+0.8)	256	225	88.0
90	243	240	360	9.1(+0.7)	251	223	88.9
105	278	270	360	8.4	232	213	92.0
110	290	280	373	11.8(+3.1)	314	185	58.9
108	285	278	360	9.8(+1.4)	270	205	75.9
125	325	315	350	10.1(+2.0)	286	198	69.1
118	308	300	350	11.5(+3.4)	326	206	63.2
105	278	270	355	5.4(-3.0)	151	142	94.0
105	278	270	365	5.4(-3.0)	147	143	97.4
95	255	250	365	11.5(+3.0)	313	170	55.4
85	232	225	365	11.5(+3.0)	313	165	52.8
92	248	242	360	5.8(-2.6)	160	151	94.4
90	243	240	365	5.5(-3.0)	150	130	86.9

* Numbers inside the parenthesis indicate cuft/min oxygen added to the lixiviant stream either more or less than actually required.

Full Scale Performance

After the entire wellfield was put under the oxygen system, the uranium productivity was as high or higher than was experienced with peroxide. Actually, for several months, a record high production was achieved. A substantial reduction in the oxidant cost for the U_3O_8 production was seen within a month after hydrogen peroxide was replaced by oxygen. Both U_3O_8 grade level ans production solution flow rate showed improvement. In addition to cutting the cost to produce a pound of U_3O_8, the fact that some high production records were reached after oxygen was added says a great deal about the oxygen injection system. The initial concern that most of the operators had for the oxygen system (that the oxygen, both liquid and gaseous, can form explosive mixtures in contact with certain organic materials) has been removed by proper education and their own experience with the system. In essence, the full scale operation of the oxygen injection system was found to be no different from the previous hydrogen peroxide system as far as operational procedues and their problems are concerned.

Comparative Economics between Oxygen and Hydrogen Peroxide at the Commercial Level

By end of May 1980 the entire wellfield had been converted to oxygen. Due to some concern that the U_3O_8 production with oxygen might be low, some peroxide was added in addition to oxygen in a new mine area which was added onto the production field at this time. The oxidant cost for the month of May 80 showed a substantial reduction even though both oxygen and peroxide were used in the wellfield. As shown in Table VII, from this time onwards the cost of oxidant in this operation was consistantly low as compared to the \$8.00 per pound of U_3O_8 in the earlier period when hydrogen peroxide was the only oxidant.

TABLE VII

COST OF OXYGEN PER POUND OF U_3O_8 PRODUCED

Month	Oxidant Cost, \$/lb U_3O_8	Month	Oxygen Cost, \$/lb U_3O_8
May 80	3.40*	Nov 80	0.16
Jun 80	1.17*	Dec 80	0.39
Jul 80	0.51	Jan 81	0.40
Aug 80	0.49	Feb 81	0.45
Sep 80	0.22	Mar 81	0.52
Oct 80	0.32	Apr 81	0.52

*During May and June 1980, in addition to oxygen, a small amount of peroxide was also added in the new mine area.

During the period July 1980 to April 1981, the average monthly cost of oxygen to recover one pound of U_3O_8 was \$0.40. The only operating cost increase due to the conversion to oxygen from hydrogen peroxide was for the additional electrical power to run the higher pressure lixiviant pumps. The cost of electricity for the total plant when peroxide was used as the only oxidant was around \$0.50 per pound of U_3O_8. All four lixiviant pumps for the oxygen system are located in the plant area. After oxygen replaced peroxide as the oxidant, the electrical power cost for the plant rose to about \$1.00 to \$1.10 per pound of U_3O_8. If this incremented electrical cost is added to the oxygen cost, the average cost to recover a pound of U_3O_8 with oxygen on a full commercial scale is only about \$0.95. Table VIII summarizes the savings achieved.

TABLE VIII

OXIDANT COST COMPARISON FROM COMMERCIAL OPERATIONS

Total average oxidant cost per pound of U_3O_8 produced

With H_2O_2, 6/78-4/80	\$ 8.00
With O_2 (incl. increm. power), 7/80-4/81	\$ 0.95
Savings (H_2O_2-O_2)	\$ 7.05

The inflationary cost increase for all the chemicals during this period was not taken into consideration. On a constant dollars basis for the period when the H_2O_2 costs were obtained, the total oxidant cost to recover a pound of U_3O_8 with oxygen would be even lower than the \$0.95. Other than the additional electrical cost referred to above, there was no other significant additional cost involved in the oxygen system.

CONCLUSIONS

1. Oxygen gas at a concentration considerably in excess of the saturation level can be successfully injected with the lixiviant for in-situ leaching of uranium in shallow ore bodies with good permeability using the Froth-Flow technique.

2. Operation in both the pilot test and the commercial field mode gave as high lixiviant flow rates through the ore zone and as high U_3O_8 concentration in the pregnant liquor with oxygen as the oxidant as with hydrogen peroxide.

3. Average oxidant cost savings of \$7.05 per pound of U_3O_8 produced were realized in full scale commercial operation with oxygen compared to the previous commercial operation with hydrogen peroxide.

ACKNOWLEDGMENT

The authors wish to express their gratitude to IEC Corporation for their support and permission to use all the data that has been included in this paper. Special thanks are due to Mr. A. D. Hutcheson, President, Mr. R. D. Gustafson, Vice President, and Mr. Jim Taylor, Mine Manager (all IEC Corporation) for their encouragement and cooperation. Our special appreciation to Mr. David Syring, IEC Maintenance Superintendent, who worked tirelessly in putting both the pilot system and the commercial system into operational form. We would also like to express particular thanks to Mr. Ivan E. Randall of the Linde Division of Union Carbide Corporation for his assistance in designing and installing the oxygen supply systems for both the pilot and commercial operations.

REFERENCES

1. Hard, Robert A., and Robert L. Ripley "Procedure for Leaching Base Elements, Such As Uranium, In-Situ", U.S. Pat. 3, 708, 206, 1/2/73.

2. Litz, Lawrence M., "In-Situ Uranium Mining With Oxygen", Paper presented at the AIME National Meeting, Las Vegas, February, 1980.

3. Hiskey, J. B., "Hydrogen Peroxide Leaching of Uranium in Carbonate Solutions", Trans. Instn. Min. Metall. (Sect. C: Mineral Process Extr. Metall.), 89, 1980, C145-52.

4. Annamalai, V., And McGarvey, F. X., "Operating Experience in the Recovery of Uranium at the Pawnee and Zamzow Sites", Paper presented at the 55th Annual Fall Technical Conference of the Society of Petroleum Engineers of AIME, SPE 9507, Dallas, TX, September, 1980.

Chapter 29

ORE SIZE, A MAJOR FACTOR AFFECTING IN SITU URANIUM ACID LEACHING

Harry Wolf
Agnew Lake Mines, Ltd.
Espanola, Ontario, Canada

ABSTRACT

The effect of ore size on uranium extraction in a sulphuric acid medium was investigated. A series of laboratory leach tests employing different size ore fractions was performed at ambient temperature. Both dilute acid and acid containing ferric sulphate were used. Thiobacillus ferro-oxidans accomplished the reoxidation of ferrous sulphate. Assays were done for uranium, thorium and iron. Upon termination of the tests mass balances were calculated. An empirical model is presented which may be used to calculate extraction-time profiles for Agnew Lake ore as a function of ore size distribution. Ore size was found to be the most significant parameter affecting overall extraction. An argument supporting radial penetration of Agnew Lake ore as a mechanism for in situ leaching is presented.

INTRODUCTION

The chemistry of the element uranium enables industry to employ either an acid or alkaline carbonate leach solution for most uranium ores. Either method may be used for conventional, in-situ, or heap leaching.

Uranium recovery in the leaching process depends both on the solubility of the uranium minerals and, on the access of the leach reagent to them. The uranium leaching rate is proportional to the liberated mineral surface area and the chemical reactions involved in the dissolution step. The dissolution of U(VI) and the oxidation of U(IV) to U(VI) are the basic chemical reactions in the leaching process. The active surface area is a function of the effective porosity and the particle size of the ore (1).

Ore pieces, with a high porosity and high permeability , may provide a large active surface area even in coarser fractions. This type of ore might be suitable for heap or in-situ leaching techniques. Nonporous and impervious ores are usually more amenable to conventional milling where the active mineral surface is exposed by grinding the ore down to the liberation size of the mineral.

MINERALOGY AND GEOLOGY

The Agnew Lake Mines property is located about 100 km east of Elliot Lake, Ontario. Mineralization occurs mainly in quartz-pebble conglomerate lenses. The rock formation is of Precambrian age. The ore is located in several zones to a vertical depth of 1 067 m. Number 3 and number 5 ore zones were actually mined. The conglomerate is similar to that of the Elliot Lake area with a lower uranium and pyrite content and a higher ratio of thorium to uranium (2).

Uranothorite, uraninite, brannerite, monazite and thorite are the main radioactive minerals which are found concentrated in the quartz-cericite matrix. The matrix also contains pyrite, pyrrhotite, rutile, feldspar, muscovite and traces of chalcopyrite, galena, zircon, magnetite and other minerals. Pyrite and pyrrhotite are the major sulphide minerals in the ore. The uranium, thorium, yttrium and rare earth elements are associated with sulphide mineralization. The ore contains 0.5 to 1 kg U_3O_8/t, 1 to 3 kg ThO_2/t and 10 to 30 kg FeS_2/t. The rare earth elements concentrations usually exceed the amount of thorium.

MINING METHOD

Ore was prepared for the in-situ leaching process by a block caving technique. Six-inch holes were drilled vertically into the ore zone and fragmentation was achieved by blasting with A.N.F.O. explosive. The swell (approximately 35%) was removed to the surface leach pile for heap leaching. Access to the ore zones is via a surface ramp and a vertical shaft.

URANIUM LEACHING AT AGNEW LAKE MINES

The ore body was first discovered in the early fifties, and a development program was instituted from 1967 to 1970 to establish a conventional underground mine. The low price of uranium forced the curtailment of this program in 1971. In the early seventies a research program dealing with the leaching of Agnew Lake ore was instituted. The research was focussed mainly on the in-situ leaching of -203 mm (-8 in.) ore. A sulphuric acid-ferric sulphate-Thiobacillus ferrooxidans bacteria type lixiviant was employed. A series of pilot trickle leaching tests using this solution provided promising results on the -203 mm, run-of-mine ore (5-8). About 70% uranium extraction was attained within one year. During this period studies on the permeability, porosity and mineralogy of the ore were carried out. The ore was found to be porous and could be permeated by a surface active oil (3-5,9).

The results of these tests predicated the heap and in-situ route and the necessary development commenced in March 1976. By May 1980, eight underground stopes and one surface pile had been completed. The underground stopes contain approximately 2×10^6 t of material with an average grade of 0.44 kg U_3O_8/t and the surface pile of 2.18×10^6 t had an average grade of 0.23 kg U_3O_8/t. Trickle leaching was originally proposed for both the surface pile and the underground stopes, however, in 1978 the underground stopes were changed to submerged (flood) leaching.

It was not possible to duplicate the results predicted by the testwork in the mine and when the price of uranium dropped considerably the company announced the cessation of further development. However, the mine would continue leaching and uranium separation until this proved uneconomical.

Uranium Extraction and Recovery

The lixiviant (comprising of a mixture of barren solution coming from ion exchange columns, raffinate emerging from solvent extraction, water, and sulphuric acid) is sprayed on the surface pile. It percolates the pile, dissolving uranium and other components present in the ore, and allowing most of the available Fe(II) to be oxidized to Fe(III) by the bacteria Thiobacillus ferrooxidans. The effluent from the pile is collected in a holding pond, part of the solution is recirculated back to the pile and the remainder is sent underground to flow through the stopes. Uranium is extracted from the solution recovered from the last stope. This is achieved in Himsley moving-bed ion exchange columns operating with Diamond Shamrock Duolite ES 131 anionic resin. The loaded resin is eluted with 10 to 13% sulphuric acid. This is further purified and concentrated by solvent extraction using a mixture of tertiary amine "Alamine 336", isodecanol and kerosene. The uranium is finally precipitated with ammonia forming insoluble ammonium diuranate (yellowcake). The product is thickened, washed, dried and then packed in drums. Only uranium is recovered. A portion of the barren solution is bled off and treated with lime and barium chloride (to coprecipitate radium), and is pumped as a slurry to the waste management area.

NEW RESEARCH

In 1978, a further research program was initiated to investigate the in-situ and heap leaching of broken ore, and to obtain additional basic data on the physical and chemical properties of Agnew Lake ore.

MEASUREMENTS OF PHYSICAL PROPERTIES OF THE ORE

It was found that the absolute permeability of the ore was extremely low, thus inferring either that the pores were extremely small or that there were few of them. This meant that the flow of lixiviant into the interior of the ore would be restricted. The average effective porosity was found to be approximately 5% and the average absolute permeability was measured at about 0.05 millidarcies. The absolute permeability was higher when measured parallel to the ore schistosity. The bulk density of the ore was found to be in a range of 2.60 to 2.70 g/cm^3, and the grain density varied between 2.80 and 2.90 g/cm^3.

The porosity of the ore was measured using a gas (helium) expansion porosimeter and the absolute permeability of the ore was measured with a gas permeameter also using helium. The water uptake of the ore was also investigated. The measurements were performed both on cylindrical core plugs (2.54 cm in diameter, and 5 cm long), and on pieces of broken ore. The measurements showed that Agnew Lake ore absorbs, on average, 0.55% water by volume.

SULPHURIC ACID LEACHING TESTWORK

The objective of this testwork was to determine the extraction rate of uranium and thorium as a function of ore size under flood leaching conditions. This was done using dilute sulphuric acid at an ambient temperature of 22 ± 5° C. The tests were arranged so that the uranium and thorium extractions of each ore size fraction could be studied individually.

Leach tests employing six different sizes of 3-zone ore and ten different sizes of 5-zone ore were performed at Agnew Lake Mines.

ORE SAMPLE PREPARATION

The 3-zone ore was crushed to minus 25.4 mm (1 in.) in a jaw crusher and screened. A portion of coarser ore was further crushed in a rolls crusher and six different size fractions were separated from the combined material for the first series of leaching tests.

Ore from the 5-zone area was used for the second series of leaching tests. Ten different size fractions in a range of 0.36 to 152.4 mm (0.0139 to 6 in.) were prepared for this investigation. The ore was first separated from the gangue by means of a gamma counter. Larger size fractions were selected by hand and the smaller size fractions were prepared by crushing and screening. The larger pieces were carefully examined for visible cracks caused by blasting or crushing. Pieces of ore with evident cracks were excluded from the testwork. Dust adhering to the coarse particles was removed with compressed air.

Preparation of Solid Samples for Assay

Head Samples. Since the Agnew Lake ore is non-homogenous, analytical head samples were prepared only for the leaching tests on fractions smaller than 25 mm. These head samples were prepared by crushing and quartering until an analytical sample was obtained. Head values for the larger ore fractions were obtained by the calculation of metallurgical balances after the end of the leaching stage.

Tailing Samples. At the end of the leaching stage, the ore in the columns was washed first with 0.5% sulphuric acid and finally with distilled water. It was then dried for 16 hours at 110° C. The entire mass of the ore employed in each test was then used to prepare analytical samples.

ANALYTICAL METHODS AND MEASUREMENTS

Uranium and thorium were measured colorimetrically as U(VI) and Th(IV) using the colour forming reagent arsenazo III after separation of uranium on chromatographic columns of silica gel (10,11). Corrections for the thorium interference on uranium measurements were applied. Iron was measured volumetrically using potassium dichromate and colorimetrically with ammonium thiocyanate.

The pH of solutions was measured with a combination glass electrode and redox-potential (EMF) with a combination platinum-silver/silver chloride electrode.

PRELIMINARY LEACHING OF 3-ZONE ORE

The six leach tests were carried out in the plexiglass columns 40 cm high with an inside diameter of 9.5 cm. Each column was charged with 2 kg of ore and flooded from the bottom with 2L of 1% sulphuric acid. The leach solution in the columns was constantly recycled at 150 to 200 ml/min, using air pumps. The tests ran for 105 days, the pH varying between 1.2 and 1.42.

The solid head and tailings samples, as well as all solution samples, were analysed for uranium, thorium and iron. The head analysis of each size fraction is presented in Table 1.

TABLE 1 Size Fractions Assays of 3-Zone Ore

Test No.	Ore Size (mm)	(kg/t)			
		U	U*	Th*	Fe
1	13.50 - 25.40	0.230	0.222	0.926	6.8
2	6.35 - 13.50	0.175	0.178	0.740	6.3
3	3.35 - 6.35	0.209	0.205	0.807	7.0
4	1.70 - 3.35	0.189	0.197	0.788	7.6
5	0.85 - 1.70	0.138	0.142	0.550	8.3
6	0.36 - 0.85	0.146	0.145	0.543	6.7

* Calculated heads from metallurgical balances.

From the analyses of the solid tailings and solution samples, a metallurgical balance was performed to determine the head grade for each test. Uranium and thorium extractions were calculated for each test at discrete time intervals. Data for uranium and thorium extraction vs. particle size on 3-zone ore are presented in Table 7.

The terminal concentration of dissolved salts in the leach solution was relatively low due to the low reactivity of these ore samples with dilute sulphuric acid.

The only visible precipitate which formed during leaching of 3-zone ore was identified as mold.

Bacteria cultures of the genus Thiobacillus ferrooxidans developed after 21 days of leaching. The redox-potential EMF changed rapidly from -400 mV to -700 mV at this time and remained at this level for the rest of the leaching period.

The total sulphuric acid consumption ranged from 4.5 to 6 kg/t. The initial amount of sulphuric acid introduced with the lixiviant was 10 kg/t of ore.

The testwork indicates that uranium and thorium extraction depends primarily on the size of ore. As the size of ore increases, the extractions drop off considerably. To verify this information a new series of tests was initiated to investigate 5-zone ore utilizing size fractions up to 152.4 mm (6 in.).

SULPHURIC ACID LEACHING OF 5-ZONE ORE

Ten laboratory scale sulphuric acid flood leaching tests on different size fractions of 5-zone Agnew Lake ore were carried out as above. Each fraction was carefully placed into a column and flooded from the bottom with 1% sulphuric acid prepared from distilled water and reagent grade acid. The initial pH and EMF of this leachant was 1.0 and -545 mV, respectively. The ore weight of each sample ranged from 2 kg for fine fractions and up to 20 kg for the test with the largest ore pieces. The ore size fractions and the weights of samples used for each test are presented in Table 2. The weight of ore to the leachant ratio, was 1:1 for all tests.

Neither during the initial flooding of either ore, nor at any other time during the tests was the formation of hydrogen sulphide observed.

TABLE 2 5-Zone Ore Size Fractions and Weight

Test No.	Ore Size (mm)	Weight (kg)	Test No.	Ore Size (mm)	Weight (kg)
1	101.6-152.4	20.00	6	6.35-13.5	2.0
2	76.2-101.6	7.25	7	3.35- 6.35	2.0
3	50.8- 76.2	6.31	8	1.70- 3.35	2.0
4	25.4- 50.8	6.50	9	0.85- 1.70	2.0
5	13.5- 25.4	2.0	10	0.36- 0.85	2.0

The leach solution in the columns was constantly recycled by means of air pumps. This ensured that the solution was continuously saturated with the oxygen required for the bacterial oxidation of Fe(II) to Fe(III).

This ore had a higher uranium and thorium content and was much more reactive with sulphuric acid than was the ore from 3-zone. The mineral contents of 5-zone ore fractions can be found in Table 3. These values were obtained from metallurgical balances. Some fractions were also analysed for uranium head values.

Leach solution samples were taken periodically and analysed for uranium, thorium and iron. The pH and EMF were also measured at the time of sampling. The sample aliquots were replaced with 1% sulphuric acid. Distilled water was used to compensate for evaporation.

Upon termination of the tests, the solution in each column was replaced with wash solution (0.3% sulphuric acid), and this was also recycled using air pumps. Tests 1 to 6 were washed twice while tests 7 to 10 were washed three times. The last wash in each case was performed with distilled water. These solutions were combined and the total volume measured for each test.

Solutions recovered at the termination of the tests were analysed for uranium, thorium and iron. The results were then used to complete the metallurgical balances.

TABLE 3 Composition of Each Fraction of 5-Zone Ore

Test No.	Ore Size (mm)	(kg/t)			
		U[1]	U[2]	Th[2]	Fe[2]
1	101.6 -152.4	---	0.784	1.87	24.5
2	76.2 -101.6	---	0.953	1.78	23.5
3	50.8 - 76.2	---	0.686	1.57	22.1
4	25.4 - 50.8	---	0.876	1.92	24.8
5	13.5 - 25.4	1.085	1.049	2.45	24.6
6	6.35- 13.5	1.060	1.082	2.44	25.0
7	3.35- 6.35	1.111	1.202	2.64	29.7
8	1.70- 3.35	1.238	1.296	2.76	28.7
9	0.85- 1.70	1.187	1.251	2.52	27.3
10	0.36- 0.85	1.136	1.250	2.49	23.9

1) Spectrophotometric determination

2) Calculated values

As previously discussed, analytical head samples were obtained in the tests employing finer materials, and the head values for tests with coarser materials were obtained from the metallurgical balances. The head values and solution concentrations were used to calculate the extractions of uranium, thorium and iron. The uranium, thorium and iron extractions at discrete time intervals, and the solution concentrations for uranium, thorium and iron as well as pH and EMF values, are presented in Table 8. The extraction of uranium for different size fractions vs. leaching time is depicted in Figure 1.

The leaching tests employing 5-zone ore were performed for various durations ranging from 89 to 185 days. After 89 days of leaching, the smallest ore fraction (0.36 to 0.85 mm) yielded extractions of 93.1% U, 87.6% Th and 49.6% Fe. The corresponding leach solution contained 1.16 g U/L, 2.18 g Th/L and 11.8 g Fe/L. In contrast, leaching of the largest ore fraction (101.6 to 152.4 mm) yielded extractions of 14.9% U, 14.7% Th and 4.2% Fe. The corresponding leach solution contained 0.117 g U/L, 0.273 g Th/L, and 1.02 g Fe/L.

After 185 days of leaching of the largest ore fraction, the extractions increased to 18.0% U, 18.3% Th and 6.0% Fe. The solution then contained 0.14 g U/L, 0.34 g Th/L and 1.46 g Fe/L.

A substantial amount of the 5-zone ore was soluble in sulphuric acid. The ore weight loss for each fraction was measured at the end of each test. The weight loss after 89 days of leaching is presented in Table 4.

TABLE 4 Dissolved Amounts of 5-Zone Ore Vs. Particle Size After 89 Days of Leaching

Test No.	Ore Size (mm)	Weight Loss (%)	Test No.	Ore Size (mm)	Weight Loss (%)
1	101.6-152.4	0.40	6	6.35-13.5	2.65
2	76.2-101.6	0.67	7	3.35- 6.35	2.90
3	50.8- 76.2	0.69	8	1.70- 3.35	3.37
4	25.4- 50.8	0.92	9	0.85- 1.70	3.61
5	13.5- 25.4	1.93	10	0.36- 0.85	3.56

The Mechanism of Uranium Leaching

Uranium leaching with sulphuric acid is a complex process governed by chemical dissolution, and the oxidation of uranium. It is controlled by surface etching, diffusion, and mass transfer within the ore. The physical parameters are in turn controlled by the physical properties of ore, and the ore size.

The leaching process may be conceptually divided into four basic phases. During the first phase of leaching (first 5 to 10 days), the accessible, soluble uranium minerals are solubilized by dilute sulphuric acid. This phase is generally controlled by the chemical reactions involving the ore minerals and sulphuric acid. The bacteria are not yet active and under this weak oxidizing condition (EMF around -400 mV), 70% uranium and 73% thorium are dissolved from an ore fraction of 0.36 to 0.85 mm after only 5 days of leaching. For more details refer to Table 8.

During the second phase, from the 5th to the 30th day (time could vary according to the leaching conditions), the process is more controlled by diffusion and the leaching rate decreases. The oxidation is still weak. At the end of the second phase the bacteria start to proliferate and the system proceeds to the third phase.

Once the bacteria proliferate, more Fe(II) is oxidized to Fe(III), the pH increases, and EMF changes quickly from approximately -400 mV to -700 mV. U(IV) is oxidized to U(VI) and simultaneously solubilized and the leaching rate improves. This stage lasts approximately until the 45th day of leaching.

Once the accessible U(IV) is oxidized, the process is again controlled by etching and diffusion within ore. The leaching rate gradually decreases. This decrease in the extraction rate during the last stage of leaching might also be attributed to the depletion of uranium in the ore.

If the leaching is performed with a solution containing the active iron oxidizing bacteria and sufficient Fe(III) ion for the oxidation of U(IV) to U(VI), the first three phases become difficult to separate. The uranium extraction rate will be higher, initially. However, after sufficient time, the overall uranium recovery will be similar to the phase 4 case.

The Mechanism of Thorium Leaching

Agnew Lake ore contains about three times as much thorium as uranium. The testwork shows that thorium is extracted almost to the same extent as the uranium. After two days of leaching about 50% Th is leached from 0.36 to 0.85 mm ore. Thereafter, the content of thorium in solution starts to decline. Later as the tests progress the concentration of thorium in solution again begins to increase steadily. This effect is more noticeable on smaller sized ore.

Precipitate Formation

Whitish flakes are formed in the columns during the early stages of leaching. This precipitate has been identified as mold. As leaching tests progress a second type of a yellow voluminous precipitate forms and is deposited on the surface of the ore particles and on the surface of the columns. This precipitate is insoluble in 1% sulphuric acid and its composition was not investigated. The formation of yellow precipitate is first observed in the columns with the finest ore, these also having the highest concentration of dissolved salts. The salts in the leach solution contained mainly iron, thorium, aluminum, calcium, yttrium and rare earth elements. The polymerization of sulphate complexes is enhanced by the aging of the leached solution and is thought to be the source of the yellow precipitate. Uranium precipitation during both leaching tests was not observed. To prevent or decrease the formation of yellow precipitate, leach solutions with high salt contents might be diluted or replaced with fresh lixiviant.

Bacterial Activity

At Agnew Lake, the bacteria become well established after about three weeks. This activity is accompanied by a rapid decrease in the EMF (-380 mV to -700 mV). It was found that during the testwork bacteria became active at various times depending on the size of ore. A high content of dissolved salts in the leach solution and frequent pH changes due to additions of sulphuric acid (primarily for the tests with smaller size of ore) tended to inhibit the growth of bacteria (12).

Bacteria started to become active after about 4 weeks in tests 1 to 4, and were well established with EMF at -700 mV by 6 weeks. The bacteria did not develop as quickly in tests 5 to 10 and these tests were inoculated with stock bacteria on the 33rd and 49th day of leaching. After 7 to 8 weeks the bacteria became more active and the EMF levelled in a range of -500 to -600 mV. For more details refer to Table 8. The bacteria activity was rapidly improved by diluting the leach solution with 0.5% sulphuric acid, the EMF values then stabilized in a range of -700 to -720 mV.

Sulphuric Acid Consumption

Concentrated sulphuric acid was added during the leaching testwork in order to maintain the pH below 1.9 Test 1 (largest ore) did not require any addition of acid for pH adjustments.

The acid consumption increased with decreasing the size of ore, particularly in the early stages of leaching. Tests employing the smaller size fractions required frequent additions of sulphuric acid. The required additions of sulphuric acid (as 100%), in excess of the 10 kg/t initially introduced to the leaching system for the leaching of 5-zone ore samples, are presented in Table 5.

TABLE 5 Sulphuric Acid Addition

Test No.	Ore Size (mm)	Leaching (days)	H_2SO_4 (kg/t)
1	101.6 -152.4	185	0.0
2	76.2 -101.6	185	1.5
3	50.8 - 76.2	185	1.7
4	25.4 - 50.8	185	6.7
5	13.5 - 25.4	132	16.3
6	6.35- 13.5	132	20.7
7	3.35- 6.35	89	29.8
8	1.70- 3.35	89	33.5
9	0.85- 1.70	89	33.9
10	0.36- 0.85	89	35.3

URANIUM EXTRACTION AS A FUNCTION OF TIME AND SIZE

It would be useful to have a simple mathematical expression to represent uranium extraction vs. time. Earlier works (6,7) reported the uranium extraction of Agnew Lake ore as a linear function of time in the log-log system. This was obtained by linear regression analysis. However, the current research carried out on Agnew Lake ore indicates that the uranium extraction is a complicated system and generally does not follow this linear relationship, even for separate ore size fractions. The equations obtained by linear regression represent uranium extractions reasonably well for large pieces of broken ore, and for smaller pieces of ore over shorter periods. Calculated extractions have positive errors for the smaller size of ore with longer leaching periods. For example, an extraction of 106.7% is calculated by linear regression for the ore 0.36 to 0.85 mm in size after 89 days of leaching compared to the 93.1% actually measured.

Polynomial regression of the data from the Test 1 (101.6 to 152.4 mm ore) yielded the following relationship:

$$E = 2.457 + 2.026 \log t + 2.135 (\log t)^2 \qquad (1)$$

E = Uranium extraction (%)

t = Leaching time (days)

This equation has been used to predict uranium extractions for the largest size ore for periods of 1 and 2 years. The calculated data is presented as a graph of uranium extractions as a function of ore size and time in Figure 2.

Overall uranium extraction, at discrete times for ore with a known grading analysis may be calculated using equation (2).

$$E_t = 10^{-2} \sum_{i=1}^{n} w_i e_i \qquad (2)$$

E_t = Overall extraction (%) at time (t)

n = Number of size fractions

w_i = Weight percentage of ith fraction

e_i = Extraction of ith fraction

A graphical expression of uranium extractions presented in Figure 2 and equation (2) may be used to calculate the overall extraction of uranium for any given size distribution. This has been done for two independent tests performed at Agnew Lake Mines in 1978.

Some 78.5% uranium extraction was calculated after 6 months compared to the 79.0% measured in the first test. In the second test, 78.4% uranium extraction was calculated compared to 80.0% measured. Other laboratory tests and the plant extraction also compare well. Uranium leaching kinetics, for either trickle or the submerge technique, were found to be similar.

Sulphuric acid uranium leaching of Agnew Lake ore is found to be mainly dependent on ore size. With increasing ore size, the uranium (and thorium) extraction rate decreased. Measured and calculated uranium extraction data at various leaching times is presented in Table 6.

TABLE 6 Estimated or Measured Uranium Extractions Vs. Ore Size

Leaching Time (months)	Uranium Extracted (%), for Ore Sizes (mm)						
	-2.54	12.7	25.4	38.1	76.2	152.4	254.0
3	92	67	48	36	23	12	7
6	93	73	55	43	28	15	9
12	95	78	61	48	31	18	11
24	95	82	67	55	35	21	13

PROPOSED ETCHING MODEL

A Single Piece of Ore Exposed to Lixiviant

Suppose a single ore particle is leached with a sulphuric acid lixiviant. After some time, the ore is penetrated by the lixiviant to a specific depth. The degree of penetration is dependent on the kinetics of the leaching process and the ore permeability. Under ideal conditions, a wetted portion of Agnew Lake ore will yield 95% of its uranium. The total recovery of uranium can thus be expressed by equation (3).

$$\% U_{rec} = \frac{100 (V_{tot} - V_{un}) f}{V_{tot}} \qquad (3)$$

V_{tot} = Total volume of ore piece

V_{un} = Unleached volume of ore piece

f = Recovery co-efficient equivalent to 0.95

To simplify the problem, one can assume that the particle is spherical or cubic having a diameter (R) or side length (A). The ore will be penetrated to a certain depth. The unleached portion of round ore has a diameter (r) and the cubic shape has side length (a). In this case, the % U recovery can be expressed by equation (4) for a spherical particle and equation (5) for a cubic particle.

$$\% U_{rec} = 95 \left(1 - \frac{r^3}{R^3}\right) \qquad (4)$$

$$\%U_{rec} = 95\left(1 - \frac{a^3}{A^3}\right) \qquad (5)$$

Thus, it is found that the percentage of ore penetrated does not depend on the shape of the ore particle if the particle is equilateral. These formulae may then be applied to broken ore if the penetration depth is known.

This model was used to calculate uranium recovery for various sizes of ore using a penetration depth of 3 and 5 mm and compared with the data which was obtained from tests with 5-zone ore for ½ and 1 year leaching. The results are depicted in Figure 3. From graph 3 it is seen that the uranium leaching follows the proposed ore penetration -- etching depth model for a penetration of 3 to 5 mm for ½ and 1 year of leaching.

CONCLUSIONS

Uranium and thorium minerals present in the Agnew Lake quartz pebble conglomerate ore are easily dissolved in dilute sulphuric acid under weak oxidizing conditions (EMF less than -400 mV). Under these conditions about 70% of the uranium and thorium is leached from 0.36 to 0.85 mm (-20 to 42 mesh) particles of ore after 5 days of leaching at ambient temperature (22 ± 5° C).

Independent of the particle size, the uranium leaching rates in the early stages of leaching are increased when initial oxidizing conditions in the leach solution are increased. A high ratio of Fe(III)/Fe(II) promotes uranium dissolution. Concentrations higher than 3 g Fe(III)/L do not have a significant effect.

Thiobacillus ferrooxidans is inhibited by high concentrations of dissolved salts in the leach solutions. This results in lower negative EMF values and also in a lower Fe(III)/Fe(II) ratio.

The formation of molds is common in the leach environment.

High concentrations of dissolved impurities (Fe, Th, Y, Al, Ca, rare earth elements and other components) in the leach solution are associated with the formation of a yellow precipitate. This is probably caused by the polymerization of hydrated sulphate complexes. Aging, higher temperature, and higher pH values are factors which favour precipitate formation.

Uranium precipitation is not observed in the leaching systems if the negative value of EMF is higher than -380 mV.

The Agnew Lake ore is found to be compact, displaying approximately 5% effective porosity and very low absolute permeability (average of 0.05 millidarcy).

The fragmentation of Agnew Lake ore is found to be the most important factor affecting uranium leaching rates. The uranium leaching rates decreased rapidly when the size of ore increased.

The analysis tends to suggest that Agnew Lake ore is radially penetrated by solutions during leaching.

ACKNOWLEDGEMENT

The author wishes to thank Kerr Addison Mines Limited, Toronto, for permission to publish this paper.

REFERENCES

1. Celeda, J., and J.G. Lara, "The Kinetics and Mechanism of the Dissolution of Uranous-Uranic Oxide in Sulphuric Acid in the Presence of Ferric Ion". J. Inorg, Nucl., Chem., Vol. 27, 2561-72, 1965.

2. Card, K.D., "Geology of the Sudbury-Manitoulin Area. Districts of Sudbury and Manitoulin". Ontario Geological Survey Report 166, 210-11, 1978.

3. Hughson, M.R., and S. Kaiman, "Mineralogical Report on Drill-Core Samples from Agnew Lake Property of Kerr Addison Mines Ltd.". Canmet-Ottawa, Mines Branch Investigation Report IR 67-41. June 22, 1967.

4. Kaiman, S., "Mineralogical Study of Leach Residue of Uranium Ore from Agnew Lake Area, Ontario". Canmet-Ottawa, Mines Branch Investigation Report IR 67-51. August 1, 1967.

5. Kaiman, S., "Leaching of Agnew Lake Minerals in Polished Briquettes". Canmet-Ottawa, Extractive Metallurgy Division, Internal Report EMT 72-5, July 7, 1972.

6. Willmann, A., "Underground Bacterial Leaching Experiment". Agnew Lake Mines Ltd. Internal Report. April 7, 1971.

7. McCreedy, H.H., F.C. Lendrum and W.A. Gow, "Stope Leaching of Uranium". Presented at Conference of Metallurgists, Quebec City, August 26-29, 1973.

8. Manchee, R.J., and K.H. Garrett, "Leaching of Individual Half-Inch Lumps of Uranium Ore from Agnew Lake, Ontario". Trans. Inst. of Min. Metall., Sect. C. 83 (811), C115-17, 1974.

9. Kaiman, S., "The Use of a Penetrant Dye to Study Ore Permeability". Canmet-Ottawa, Mines Branch Technical Bulletin TB 165. January 1973.

10. Volf, J., and Z. Sulcek, "Contribution to the Determination of Low Amounts of Uranium in Ores". Acta Geologica et Geographica Universitatis Comenianae, Bratislava, Czechoslovakia, 15, 153-64, 1968 (in German).

11. Savvin, S.B., "Analytical Use of Arsenazo III; Determination of Thorium, Zirconium, Uranium and Rare Earth Elements". Talanta, Vol. 8, 673-85, 1961.

12. Silver, M., Private Communication, 1981.

TABLE 7 Leaching of Different 3-Zone Ore Size Fractions at Discrete Time Intervals*

Time (days)	U Content in Leach Solution (g/L)						U Leached (%)					
	1	2	3	4	5	6	1	2	3	4	5	6
0.8	0.040	0.052	0.080	0.091	0.074	0.082	18.0	29.1	38.9	45.9	52.1	56.6
1.9	0.053	0.065	0.095	0.100	0.079	0.080	23.7	36.8	46.3	50.6	55.5	55.1
2.8	0.060	0.074	0.104	0.105	0.085	0.082	27.2	41.3	50.9	53.3	59.6	56.6
5	0.070	0.084	0.112	0.107	0.084	0.083	31.7	46.9	54.4	54.2	58.9	57.3
8	0.082	0.089	0.116	0.111	0.085	0.083	37.1	49.9	56.8	56.1	59.8	57.8
14	0.092	0.100	0.122	0.116	0.088	0.089	41.4	55.9	59.6	59.0	61.8	61.3
20	0.094	0.104	0.124	0.122	0.090	0.090	42.4	58.6	60.5	61.9	63.2	62.2
27	0.102	0.111	0.129	0.136	0.104	0.112	46.1	62.2	62.6	68.5	73.6	77.0
36	0.108	0.114	0.135	0.140	0.110	0.117	48.6	64.0	65.7	70.8	77.6	81.0
45	0.114	0.119	0.144	0.145	0.119	0.125	51.2	66.8	70.3	73.5	83.4	86.2
64	0.122	0.125	0.147	0.151	0.120	0.127	54.9	70.1	71.4	76.6	84.5	87.4
98	0.123	0.124	0.149	0.154	0.122	0.123	55.9	69.1	72.7	78.0	85.6	84.7
105	0.122	0.126	0.149	0.154	0.120	0.125	55.2	70.8	72.7	78.0	84.8	86.2
	Th Content in Leach Solution (g/L)						Th Leached (%)					
0.8	0.146	0.189	0.302	0.359	0.318	0.294	15.7	25.5	37.4	45.5	57.8	54.2
1.9	0.189	0.237	0.353	0.391	0.324	0.285	20.4	32.0	43.8	49.6	58.9	52.5
2.8	0.214	0.277	0.385	0.404	0.310	0.276	23.1	37.4	47.8	51.3	56.4	50.8
5	0.241	0.310	0.394	0.383	0.293	0.264	26.1	41.8	48.8	48.6	53.3	48.6
8	0.280	0.319	0.400	0.354	0.282	0.252	30.2	43.1	49.6	44.9	51.3	46.5
14	0.309	0.348	0.356	0.304	0.264	0.241	33.3	47.1	44.1	38.5	48.0	44.4
20	0.318	0.363	0.346	0.323	0.267	0.249	34.4	49.1	42.9	41.0	48.6	45.9
27	0.332	0.368	0.365	0.348	0.293	0.288	35.9	49.8	45.3	44.1	53.2	53.0
36	0.347	0.380	0.376	0.362	0.312	0.317	37.5	51.4	46.6	45.9	56.7	58.3
45	0.357	0.390	0.386	0.382	0.331	0.330	38.5	52.8	47.8	48.5	60.1	60.8
64	0.382	0.401	0.391	0.402	0.345	0.350	41.2	54.1	48.5	51.0	62.7	64.5
98	0.392	0.401	0.406	0.423	0.364	0.360	42.3	54.2	50.4	53.6	66.3	66.3
105	0.399	0.403	0.408	0.424	0.361	0.367	43.0	54.4	50.6	53.8	65.6	67.5

* For ore size refer to Table 1.

TABLE 8 Leaching of Different 5-Zone Ore Size Fractions at Discrete Time Intervals

Test No. Ore Size (mm)	Leaching Time (days)	Solution Concentration (g/L)			pH	EMF (-mV)	Extraction (%)		
		U	Th	Fe			U	Th	Fe
1	0.8	0.017	0.044	0.11	1.02	395	2.2	2.4	0.5
101.6-152.4	5	0.040	0.100	0.29	1.09	395	5.1	5.4	1.2
	11	0.055	0.136	0.41	1.08	395	7.1	7.3	1.7
	49	0.094	0.223	0.80	1.22	590	12.1	11.9	3.3
	89	0.117	0.273	1.02	1.34	705	14.9	14.7	4.2
	185	0.140	0.341	1.46	1.60	685	18.0	18.3	6.0
2	0.8	0.032	0.086	0.16	1.02	385	3.4	4.8	0.7
76.2-101.6	5	0.075	0.184	0.50	1.10	400	7.9	10.4	2.2
	11	0.100	0.234	0.84	1.18	395	10.5	13.2	3.6
	49	0.165	0.361	1.31	1.48	705	17.4	20.3	5.6
	89	0.195	0.422	1.67	1.72	720	20.3	23.7	7.1
	185	0.238	0.544	2.24	1.83	700	25.0	30.6	9.5
3	0.8	0.037	0.106	0.20	1.04	395	5.4	6.8	0.9
50.8-76.2	5	0.080	0.205	0.56	1.12	405	11.7	13.1	2.5
	11	0.107	0.250	0.80	1.21	405	15.6	16.0	3.7
	49	0.161	0.378	1.25	1.53	715	23.5	24.1	5.7
	89	0.185	0.439	1.54	1.78	715	26.9	28.0	7.0
	185	0.222	0.549	2.15	1.79	700	32.4	35.0	9.8

cont'd......

TABLE 8 (cont'd)

Test No. Ore Size (mm)	Leaching Time (days)	Solution Concentration (g/L)			pH	EMF (-mV)	Extraction (%)		
		U	Th	Fe			U	Th	Fe
	0.8	0.050	0.132	0.23	1.05	395	5.7	6.9	0.9
4	5	0.113	0.274	0.64	1.13	395	12.9	14.3	2.6
	11	0.146	0.326	0.95	1.25	390	16.7	17.1	3.9
25.4-50.8	49	0.226	0.484	1.82	1.50	720	25.9	25.3	7.4
	89	0.277	0.621	2.56	1.84	725	31.7	32.5	10.4
	185	0.365	0.813	3.56	1.81	715	41.7	42.5	14.4
	0.8	0.109	0.280	0.34	1.08	390	10.4	11.4	1.4
5	5	0.213	0.526	1.05	1.32	390	20.4	21.5	4.3
	11	0.263	0.577	1.36	1.65	395	25.1	23.6	5.5
13.5-25.4	49	0.483	1.09	3.78	1.58	570	46.2	44.7	15.4
	89	0.597	1.33	5.06	1.92	590	57.0	54.1	20.6
	185	0.719	1.59	7.10	1.82	-	68.2	64.4	28.6
	0.8	0.157	0.412	0.54	1.12	385	14.5	16.9	2.6
6	5	0.315	0.712	1.21	1.49	385	29.1	29.2	4.9
	11	0.382	0.816	1.92	1.52	390	35.3	33.5	7.7
6.35-13.5	49	0.646	1.42	4.64	1.55	500	59.8	58.4	18.6
	89	0.791	1.71	6.57	1.88	570	73.2	70.2	26.3
	185	0.916	1.98	9.13	1.73	-	83.0	79.4	35.7
	0.8	0.204	0.675	0.87	1.30	395	16.9	25.6	2.9
7	5	0.526	1.24	2.03	1.29	385	43.8	47.1	6.8
	11	0.670	1.44	2.82	1.78	390	55.8	54.7	9.5
3.35-6.35	49	0.873	1.92	5.37	1.58	435	72.6	72.8	18.1
	89	1.05	2.20	9.30	1.75	560	87.8	83.6	31.3
	0.8	0.346	0.767	1.13	1.52	390	26.7	27.8	3.9
8	5	0.761	1.74	3.32	1.43	395	58.7	63.3	11.6
	11	0.882	1.82	3.99	1.69	400	68.3	66.5	13.9
1.70-3.35	49	1.04	2.16	6.38	1.55	420	80.4	78.5	22.1
	89	1.19	2.40	10.4	1.64	530	91.5	87.0	36.1
	0.8	0.428	0.767	1.44	2.12	370	34.3	30.4	5.3
9	5	0.811	1.80	4.86	1.49	395	64.9	71.4	17.8
	11	0.907	1.79	4.83	1.60	410	72.7	71.0	17.7
0.85-1.70	49	1.06	2.07	7.54	1.50	430	85.0	82.0	27.6
	89	1.15	2.19	11.1	1.59	535	92.6	86.9	40.7
	0.8	0.349	0.433	2.12	3.82	75	28.0	17.4	8.9
10	5	0.882	1.83	5.80	1.46	400	70.7	73.6	24.3
	11	0.992	1.58	6.39	1.57	415	79.3	63.7	26.8
0.36-0.85	49	1.10	2.12	9.13	1.50	435	88.4	85.2	38.3
	89	1.16	2.18	11.8	1.63	525	93.1	87.6	49.6

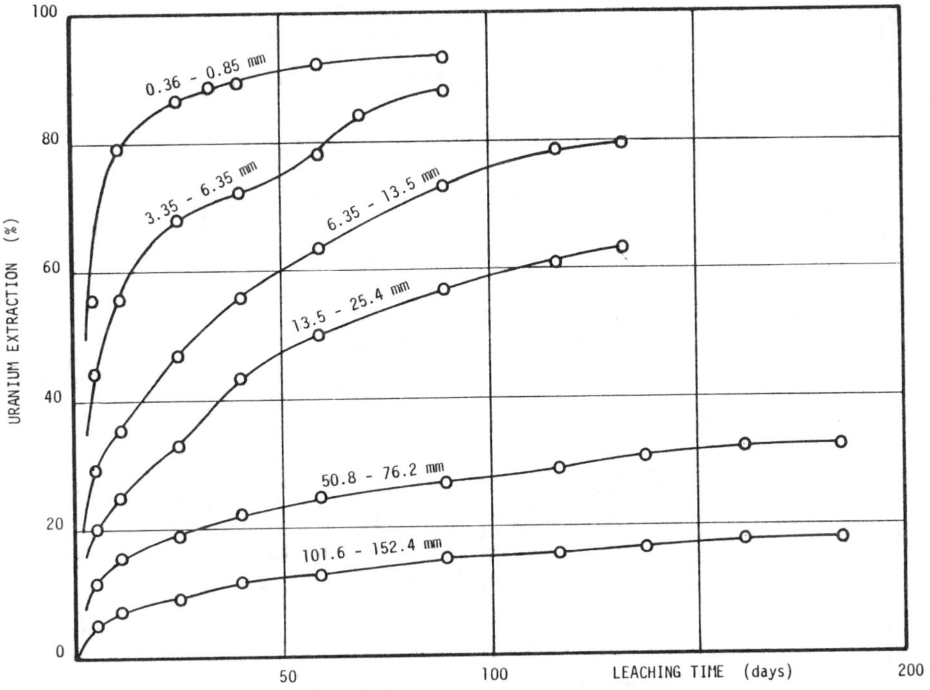

FIGURE 1. URANIUM EXTRACTION OF 5-ZONE ORE SIZE FRACTIONS VS. LEACHING TIME

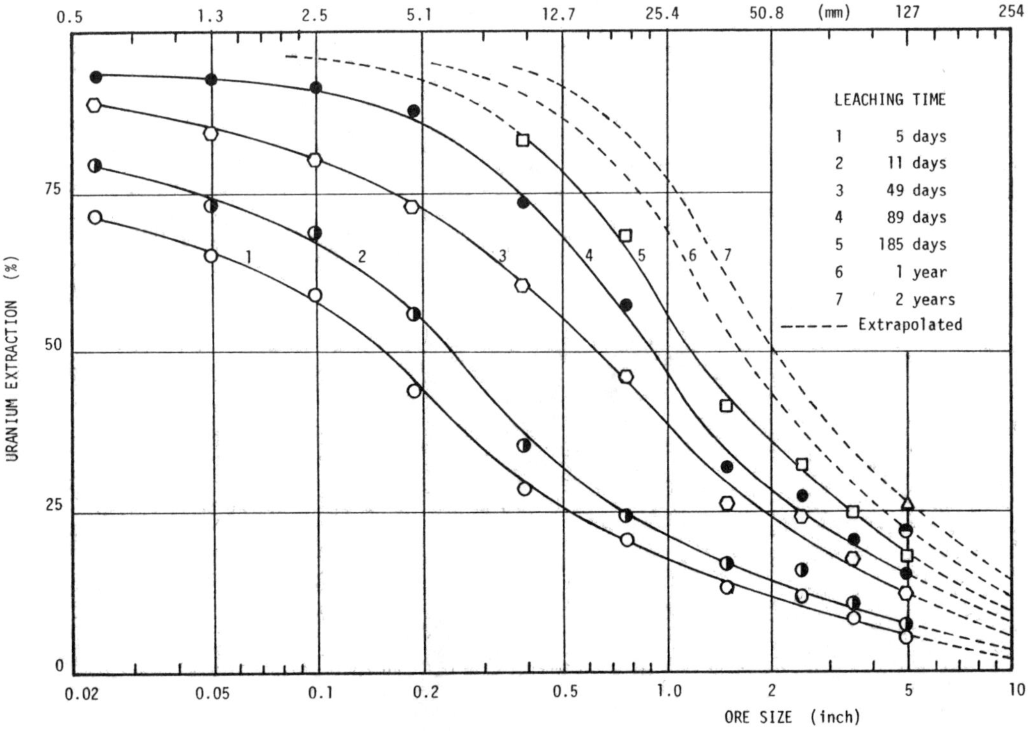

FIGURE 2. URANIUM EXTRACTION AT DISCRETE TIME INTERVALS VS. SIZE OF 5-ZONE ORE

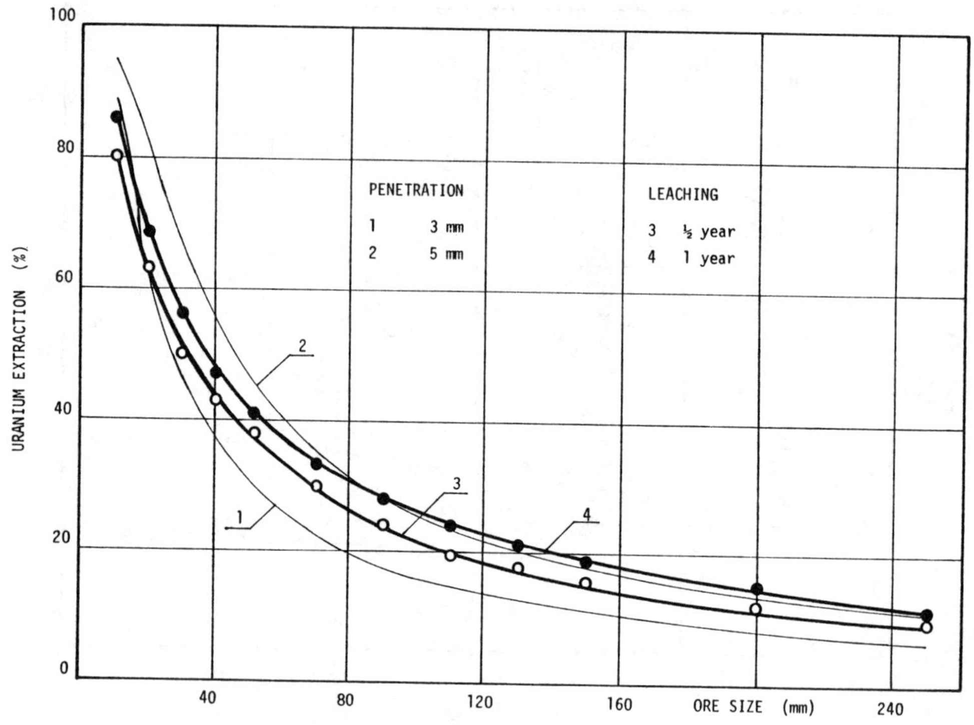

FIGURE 3. ACTUAL URANIUM EXTRACTION RATES OF 5-ZONE ORE COMPARED TO CALCULATED EXTRACTION RATES FOR VARIOUS RADIAL PENETRATION THICKNESSES

Chapter 30

DEVELOPMENT AND APPLICATION OF THE KREBS SOLVENT EXTRACTION SYSTEM

Alain A. Sonntag
David A. Goodman
Krebs and Company
Paris, France

ABSTRACT

The KREBS Solvent Extraction System utilizes a uniquely designed mixer-settler having very high capacity per unit area and great flexibility. The development of the mixer-settler is traced through its various stages wherein the independence of mixing and pumping, reduction in settler area requirements and simplicity of plant layout were achieved. The laboratory, pilot plant and full scale tests contributing to the development of the machine are described. The existing KREBS SX plants in Europe and Africa are described as well as the plants presently under design. Results of recent testwork are also given.

INTRODUCTION

KREBS has been involved in solvent extraction for several years, starting initially in the treatment of basic materials for the fertilizer industry and later moving into the metallurgical field. In the early years KREBS was faced with problems of materials of construction and in the excessive size of settlers. It was to solve some of these problems that the design of a new mixer-settler was initiated, particularly as the requirement for increased capacities of processing plants became evident.

As a result KREBS has developed a mixer-settler system with substantial improvements in design and performance over the conventional system. The heart of the system is the low-speed conical pump which adjusts automatically to flow variations. It introduces very little turbulence, and the centrifugal forces developed by the pump induce initial phase separation and coalescence while pumping the mixed phases. Laboratory and industrial experience has shown that the combination of the conical pump and superposed launder allows a reduction in settler area of 50-75% of that normally required.

Development of the KREBS System began in 1968 on a bench scale, and within a period of just over 10 years pumps of up to 2,000 m^3/h capacity have been constructed, confirming the scale-up parameters and the feasibility of large units.

Application of the KREBS System to different hydrometallurgical processes has been tested in pilot scale units. The results from these tests have proved to be reliable for purposes of scale-up. Full-scale plants using the KREBS Solvent Extraction System have now been in operation for over six years in the field of nickel-cobalt separation and uranium extraction. Full operating and maintenance history is available.

The advantages of the KREBS System over the conventional system can be stated as follows:

- Reduction in equipment size and, hence, plant size resulting in lower capital requirements and lower operating costs.
- Independence of agitation from pumping allowing optimum conditions of mixing.
- Simplified layout due to the "head-to-head" arrangement of the mixer-settler units and, hence, easier maintenance.
- Reduction in piping and instrumentation and easy bypassing of any mixer-settler.
- Flexibility of the mixing and pumping unit and, hence, easier operation.

DESCRIPTION OF THE KREBS MIXER-SETTLER SYSTEM

The KREBS Mixer-Settler, as shown in Fig. 1, includes the following features:

- A MIXING IMPELLER selected to best satisfy dispersion requirements.
- A low turbulence CONICAL PUMP, built around the impeller shaft, to transfer the dispersion from the mixer to the settler and induce phase separation.
- A SUPERPOSED LAUNDER to provide further phase separation and simplify the layout of the equipment.
- A SETTLER of simple design having a short angled plate distributor which directs the phases into the main part of the settler without turbulence and contributes to the phase disengagement. The settler itself is free of internals and provided with removable inspection covers for easy maintenance and cleaning.

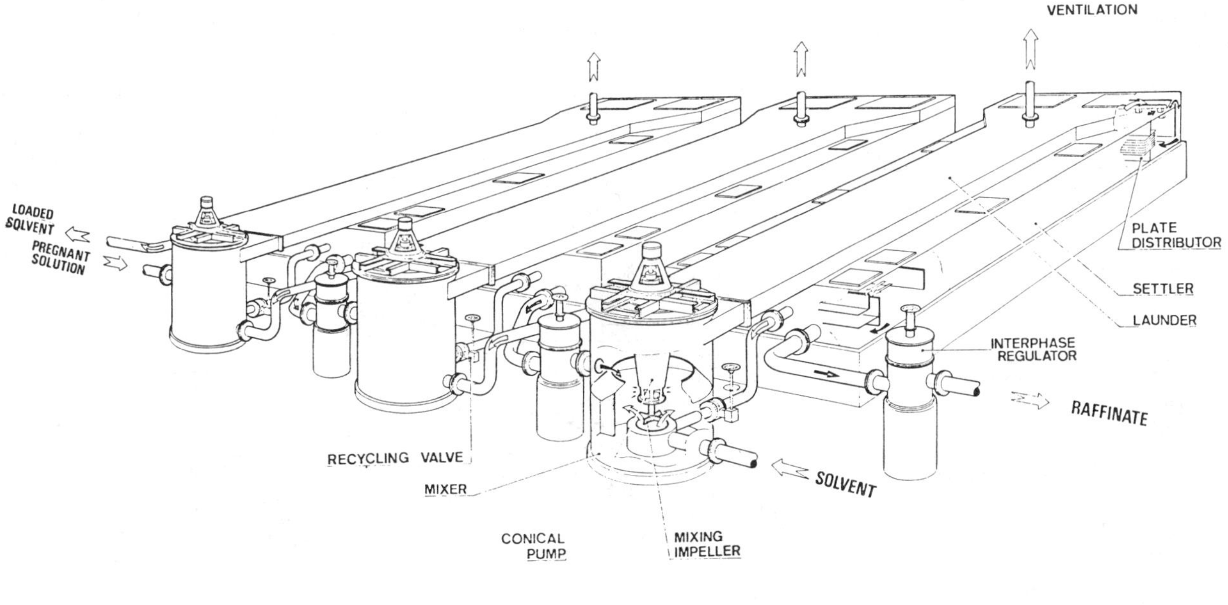

Fig : 1 _ GENERAL ARRANGEMENT OF KREBS MIXER SETTLERS

- An INTERFACE REGULATOR used to adjust the height of the liquid in the settler in case of plant shutdown or bypass of a particular unit.

The Mixer

The KREBBS MIXER, as shown in Fig. 2, is an assembly of three main components:

- the mixing chamber with its internals;
- the rotor assembly composed of mixing impeller, conical pump rotor, and drive mechanism;
- the conical pump stator and the discharge deflector.

The Mixing Chamber. The mixing chamber is designed for highest stage efficiency. Its volume and dimensions are defined to best satisfy process requirements such as phase flow rate and mass transfer rate. The chamber may be of cylindrical, rectangular, or any intermediate shape.

The mixing chamber is fitted with antivortex baffles and the arrangement of the inlets is such that short-circuiting is avoided and stability of the continuous phase is assured. Each mixing chamber is normally equipped with nozzles for sampling and instrumentation as required.

The Rotor Assembly. The rotor assembly is designed and constructed according to the mixing and pumping requirements of each individual system and is mounted on the vertical shaft.

The Mixing Impeller. The shape, dimensions and rotation speed of the mixing impeller are selected to obtain the optimum droplet size in the dispersion of one phase into the other. Wherever possible, the characteristics of the impeller are defined by testing in a pilot unit and scaled up according to well established principles.

The Conical Pump Rotor. The conical pump rotor comprises the vertical shaft with radial trapezoidal blades attached to it. The size and angle of the rotor blades are selected to achieve pumping of the total flow rate of the phases within the range of speeds chosen for the mixing impeller.

The rotor induces very little turbulence in the flow of the mixture and, due to the centrifugal forces developed, phase separation is induced in the conical pump itself. Moreover, the pumping curve is relatively flat and variations in the total flow rate of the feed solutions produce only slight changes in the height of the liquid level in the mixing chamber. The conical pump adjusts itself to the required "impeller diameter" for a fixed speed of rotation (Fig. 3).

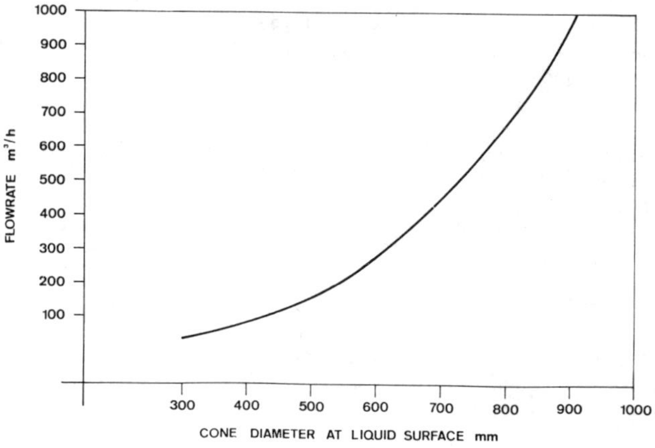

Fig: 3 – CONICAL PUMP CHARACTERISTICS
Flowrate versus cone diameter at liquid surface for 150 RPM

Similarly, if the shaft speed is regulated to improve mixing or to change capacity, the level adjusts itself to the corresponding "immersed diameter" (Fig. 4).

The clearance between the rotor blades and the frustum conical stator is kept large enough to avoid expensive tooling of the pump components.

The Drive Mechanism. The drive mechanism supports the rotor assembly and provides the required power for mixing and pumping. The KREBBS Mixer can be equipped for infinite variations of impeller speed via hydraulic drive mounted on the supporting bearing box or for fixed impeller speed via fixed speed electrical motor with appropriate reducer.

The Pump Stator. The pump stator is an upwards flared, stationary, hollow cone which is at a fixed location in the mixing chamber. The mixture enters at the bottom of the cone and is "centrifuged" by the pump blades toward the top of the stator. During its rise, the

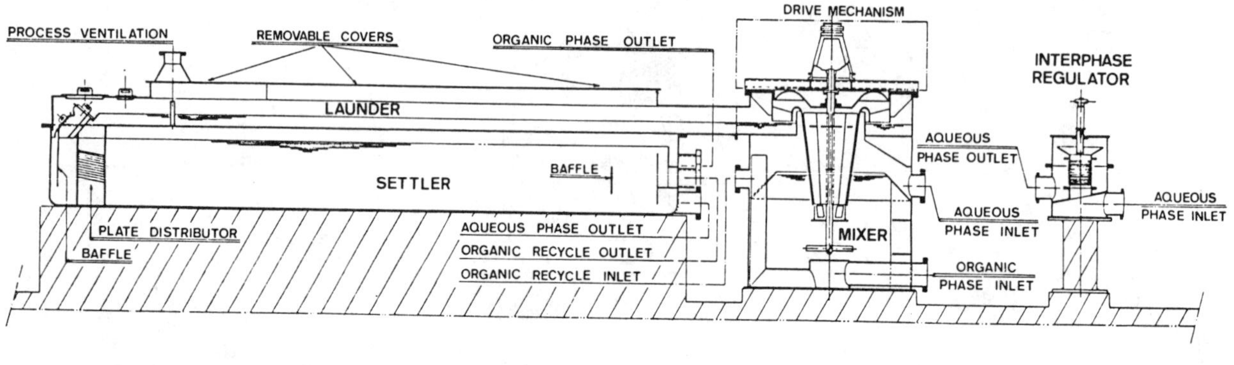

Fig: 2 – SECTION VIEW OF THE KREBS MIXER–SETTLER

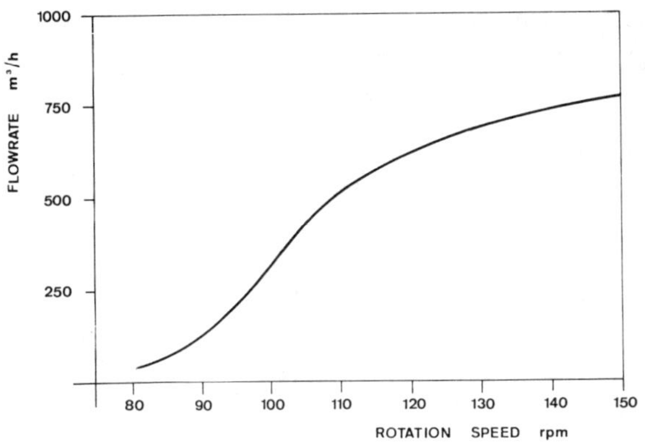

Fig: 4 – CONICAL PUMP CHARACTERISTICS
Flowrate versus rotation speed
for 800mm cone diameter at liquid surface

centrifugal forces induce the phase separation which is continued in the launder connecting the mixer to the settler.

The upper portion of the stator is curved to allow smooth flow of the solution into the launder. This discharge deflector prevents possible spraying. It covers the entire stator and rotor of the conical pump.

An entry deflector is fixed at the bottom of the conical stator, thus preventing direct flow of the mixture from the impeller to the pump stator and forces radial entry of the mixture.

The Settler

The partially separated phases leaving the conical pump are fed into a wide launder which directs the pump overflow toward the opposite end of the settler, and further enhances phase separation.

At the end of the launder the light phase, which is mainly organic, is fed into the upper part of the settler by means of an overflow weir provided with downpipes. The heavy phase, which is mainly aqueous, is fed into the lower part of the settler by means of an overflow weir. The height of both overflow weirs can be adjusted to suit a wide range of flow rates.

The light and the heavy phases enter the settling tank through a short-angled plate distributor which eliminates turbulence and helps to stabilize the interface throughout the settler. The distributor is made of parallel plates which cover the entire width and about one-third of the height of the tank.

The phases are progressively separated in conventional manner but with much greater efficiency owing to the preseparation which has occurred in the conical pump and the launder. A dispersion band is formed in the middle zone of the settling tank. In order to hold back this band and to reduce entrainment of small droplets with the separated phases, a baffle plate is provided at the outlet end of the tank.

Three separate takeoffs are provided at the end of the settler:

- The organic phase is fed to the next up-stream mixer.
- The aqueous phase is fed through the interface regulator to the next down-stream mixer.
- A recycle can be fed back to the mixer to ensure correct phase ratio in that stage, as required.

The settler as well as the superposed launder is provided with removable inspection covers to provide access for checking and cleaning.

The Interphase Regulator

The KREBS INTERFACE REGULATOR, as shown in Fig. 2, is used to adjust the height of the interface by increasing or decreasing the overflow level of the aqueous phase. The regulator is fluid-tight and maintains the liquid level in the settler in case of pump failure or plant shutdown.

The regulator is of a very simple design and involves a large cylinder in which a bellows is moved upwards or downwards by means of a mechanical device. The position of the inlet and outlet flange is defined according to the piping arrangement of the mixer-settler unit.

Materials of Construction

Different extractants and different chemical systems have required equipment fabrication in a variety of materials. Preferred materials of construction are free standing FRP or stainless steel supported by a steel frame. PVC, PVC reinforced with FRP, and concrete tanks lined with stainless steel can also be used depending on the local economic conditions; for highly corrosive liquors, special alloys and titanium are used.

Plant Layout

The KREBS Mixer-Settler System offers maximum flexibility in equipment installation and substantial reduction in overall plant area. As shown in Fig. 1, all the mixers are on the same side, this arrangement being possible thanks to the top launder. This offers the advantage of very short connections for pipes and cables and easy installation of bypass pipes around the mixer if required. All pipes and valves are arranged around the mixer which facilitates supervision and maintenance. Moreover, the area requiring protection with expensive antiacid coatings is very small.

THE KREBS MIXER-SETTLER DEVELOPMENT

Development of the KREBS System began in 1968 on a bench scale, and by 1979 a 2,000 m^3/h conical pump had been run to confirm the scale-up parameters and the feasibility of large unit operation.

Chronological Steps of the Development

The chronological development of the KREBS Mixer-Settler System from the first small pump to the full-scale plant can be summarized as follows:

YEAR	STEP	TOTAL FLOWRATE m^3/h
1968-1969	Experimental work with the first conical pump	2
1973	Studies for a nickel project	15 - 30
1975	Commissioning of a solvent extraction plant for nickel solution purification	10 - 15
1976	Construction of a pilot mixer-settler and experimental work in an uranium mill	15
1976	Construction and experimental work of a conical pump for extrapolation	800 - 1000
1977	Construction of the solvent extraction plant of the LANGOGNE uranium mill (FRANCE)	120
1977	Completion and start-up of the solvent extraction plant of the MOUNANA uranium mill (GABON)	400
1976-1978	Extension of LE HAVRE nickel solvent extraction plant (FRANCE)	20
1979	Construction of a large size conical pump and experimental work	2000
1980-1981	Construction of the solvent extraction plant of the VAAL REEFS uranium mill (REP. SOUTH AFRICA)	575
1981	Engineering of the solvent extraction plant of the KEY LAKE project (CANADA)	475

The Mixer-Settler Development

Initial Experimental Work on the Conical Pump. The design of a solvent extraction system requires that, as well as mixing two phases, relatively large volumes be pumped from stage to stage against low heads. Conventionally, these two functions are achieved by using pumper-mixer impellers. The design of such impellers is necessarily a compromise between a mixing impeller and a pumping impeller.

In 1968 KREBS was studying a particular system for which the compromise was found not to be entirely satisfactory. Therefore, a program was initiated to develop a new type of pump which would more efficiently pump large volumes against low heads.

Among the different pumps studied was the conical pump used by KREBS to transfer mercury in salt electrolysis cells. These are rather small pumps with a rotating cone impeller. In 1968 a model pump was built in which the cone was the revolving part. Flows obtained were in the range of 1 m^3/h but this model was abandoned due to poor power efficiency.

In 1969 a conical pump with a fixed cone and revolving blades was designed and tested. The results were very promising and complete hydraulic tests were undertaken and basic parameters studied. A total flow of 2 to 2.5 m^3/h was obtained with pump efficiency in the range of 30 to 35% for a head of 100 mm.

Nickel Pilot Plant. As a next step in the development, conical pumps were integrated into three mixer-settlers of small industrial scale to study the purification of nickel chloride solution in a pilot plant. The successful operation of the pumps confirmed the expected performances. In addition, the test work provided useful information on the performance of a settler with a conventional design, the mixing conditions, and the chemistry of the process. The pump flow rates ranged from 12 m^3/h to 32 m^3/h for heads of 300 to 400 mm.

Based on these results, KREBS built a full-size plant to treat nickel solutions and produce cobalt chloride, ferric chloride, and nickel carbonate.

Uranium Pilot Plant. Shortly afterwards, KREBS became involved in uranium hydrometallurgy, proceeding with a full program of solvent extraction tests using a new type of pilot plant mixer-settler. The design of this unit included for the first time the launder on the top of the settler. The unit had the following characteristics:

Mixer:	Volume.........0.3 m^3
Conical Pump:	Flow Rate......10 - 12.5 m^3/h
Settler:	Area...........0.75 m^2

The mixer-settler was installed in an operating uranium mill in FRANCE (LANGOGNE plant) and was operated for several months in parallel to the commercial solvent extraction circuit. As a matter of interest, the tests gave good settling rate figures for both organic continuous and aqueous continuous operating conditions. Extrapolation factors were calculated for the mixer size and the settler area. Coalescence aids such as "knit mesh" and picket fences were tested but the benefits gained from their use were minimal and the problem of cleaning when crud was present further obviated their use.

The results of this testing program were used in 1977 for the design of the LANGOGNE and MOUNANA KREBS SX plants.

Testing of a Large Size Pump. To scale-up the conical pump to the needs of the MOUNANA plant and other plants which required a range of 800-1,000 m^3/h, a prototype was built in 1977. This pump was tested on a water tank and the hydraulic results gave the design parameters for the commercial units subsequently installed.

A later scale-up of the conical pump was accomplished in 1979 with the construction of a 2000 m^3/h pump. It was tested with water and the various parameters of flow rate, power consumption, head, RMP, geometry, and pump efficiency were studied and determined.

Summary. As a result of 13 years of mixer-settler development, KREBS has now in its hand a very useful and efficient piece of equipment offering:

- A wide range of steady flow rates.
- Efficient pumping at low rpm.
- Simple construction and low cost.

The Industrial Development

The nickel treatment plant in LE HAVRE (FRANCE), built in 1974, was the first commercial plant using the KREBS Mixer-Settler as the solvent extraction units. The mixer design included the conical pump but the settler was conventional and the plant layout used head-to-foot arrangement.

In 1977 KREBS built two uranium solvent extraction plants using the new mixer-settler with the conical pump and the launder. During designing of these plants basic engineering was extended and resulted in the new layout which is now part of the KREBS Mixer-Settler System, as described above and shown in Fig. 1.

In 1978, for the extension of the nickel plant in LE HAVRE, 18 mixer-settlers were manufactured and installed. The design of this equipment did not include the launder to maintain homogeneity with the first unit built in 1974. In 1979 and 1981 KREBS was awarded two more contracts for solvent extraction plants in two uranium mills: one in SOUTH AFRICA and one in CANADA.

More details for the operating plants are given below:

COMMERCIAL KREBS SX PLANTS

Full-scale plants using the KREBS solvent extraction systems have now been in operation for over six years and full operating and maintenance history is available.

The first SX plant designed and built by KREBS was a nickel and cobalt separation plant at LE HAVRE (FRANCE). Later KREBS designed several SX plants for uranium recovery.

Le Havre - France

In 1974-1975 KREBS designed and built a plant for the recovery of nickel, cobalt, and iron chloride from nickel slag for SOCIETE LE NICKEL (SLN) at LE HAVRE (FRANCE). After two years of continuous and entirely satisfactory operation, SLN awarded KREBS another contract for the expansion and modification of the original plant in order to treat a nickel matte and produce 16,000 metric tons of high purity nickel and 600 metric tons of cobalt per year.

The nickel refinery process includes:

- Leaching of ground nickel matte with a chloride solution.
- Solvent extraction and nickel solution purification.
- Electrolysis of a pure nickel chloride solution.
- Cutting and packaging of nickel cathodes.

The SX process includes iron removal and cobalt removal sections. Iron is first extracted with tributyl phosphate (TBP) and cobalt is then extracted with tri-iso octylamine (TiOA). Water is used for stripping in both circuits. The pregnant liquor flow to the extraction units is about 10 m^3/h.

The iron solvent extraction circuit consists of five extraction, four stripping, and four regeneration/wash mixer-settlers. The cobalt solvent extraction circuit consists of two parallel trains, each composed of four extraction, two wash, five stripping, and three regeneration/wash mixer-settlers.

There are a total of 41 mixer-settlers of various sizes. The area of the solvent extraction settlers of the iron removal section is 5.9 m^2. The mixer-settler design includes a conical pump and a standard settler without a launder. All units are installed "head-to-tail" on the same level. Fixed speed electric motor drives are used.

Mounana - Gabon

MOUNANA is the first complete KREBS uranium SX plant and was commissioned three years ago for COMUF. The plant is operated by MOKTA and is part of the ROTHSCHILD business interests. The plant has been operating since 1961, although solvent extraction has been in operation only since 1977. Prior to that time uranium was precipitated as a 35% U product directly from leach solution.

The plant presently operates at about 1,000 tpd of ore containing 0.3-0.5% U. With construction of the new facilities, tonnage will increase to 2,000 tpd by the end of this year.

The process flow sheet includes autogenous grinding, acid leaching, sands washing in rake classifiers, slimes washing in CCD thickeners, clarification, solvent extraction with salt stripping, and magnesia precipitation of yellow cake.

The SX plant consists of four extraction, four stripping, one wash, and two regeneration/wash mixer-settlers. The area of the settlers is 31.8 m^2 for the extraction and 10 m^2 for the stripping section. A raffinate after-settler is located adjacent to the No. 4 solvent extraction settler. The settlers are supported by steel beams on top of small concrete piers and each pair of settlers is separated by a walkway. The barren solvent tank is located adjacent to the No. 4 stripping settler.

The layout of the MOUNANA SX plant is functional due mainly to the fact that the mixers are all located at the same end of the settlers and are aligned in a straight row. This allows for simplified piping and access platform arrangements. A centrifuge for crud treatment is located between the extraction and stripping sections. The plant is designed for a pregnant solution flow of 180 m^3/h corresponding to a specific flow of 12.5 $m^3/h.m^2$.

Langogne - France

The CFMU plant at LANGOGNE in FRANCE is operated by MOKTA and is partly owned by COGEMA which is a subsidiary of CEA (Automatic Energy Commission). The plant is located near LANGOGNE in central FRANCE which is about two hours drive from LYON.

The mine has been operating since 1958 at which time the ore was shipped to a custom mill. The original SX plant has been operating since 1970. In 1977 the old extraction settlers were replaced by the KREBS Mixer-Settlers and the old circular strip cells with external mixers were retained.

Presently the mill processes 500-845 tpd of ore containing an average of 0.15% U. The ore comes from both open-pit and underground operations. The mill flow sheet includes primary and secondary crushing, grinding consisting of a rod mill in open circuit with a ball mill, acid leaching, filtration, clarification, solvent extraction with salt stripping, and magnesia precipitation of yellow cake. In addition, solution is supplied to an ion exchange plant from a heap leaching operation, the eluate from which is mixed with pregnant solution for processing in the SX plant.

The SX plant is located indoors and consists of the four KREBS extraction mixer-settlers, five circular stripping mixer-settlers, and one regeneration cell. A raffinate settler is located transversely at the end of the extraction cells. The area of the KREBS Mixer-Settlers and raffinate settler is 10 m^2. The extraction pump mixers are powered by fixed-speed electric motors and are serviced by a crawl beam and hoist. The settlers are supported on steel beams on concrete. Alongside each extraction mixer is located the jack-leg for adjustment of the interface levels. A centrifuge similar to that at COMUF in GABON was supplied for crud treatment.

Specific Flow. The aqueous flow to the extraction units is about 50 m^3/h giving a specific flow of 5 $m^3/h.m^2$. At a O/A ratio of 1.5, the total specific flow is about 12.5 $m^3/h.m^2$. Both the aqueous and total specific flows are about 3 times greater than for conventional systems or, conversely, for any given flow, requirements are about 1/3 of the settler area of a conventional settler.

Vaal Reefs - Republic of South Africa

The AFRIKANDER LEASE plant is to be operated by the ANGLO-AMERICAN CORPORATION (AAC) as part of the VAAL REEFS complex and is located about 15 km from KLERKSDORP.

The mill process includes SAG milling, autoclave acid leaching, CCD thickening, clarification with a hopper clarifier and leaf clarifiers, and SX followed by ammonia precipitation of yellow cake. The plant is now under construction.

The SX plant consists of three extraction, four scrubbing, three stripping, and one regeneration mixer-settler. They are located outdoors. The extraction settlers are separated from the remaining settlers by the fresh solvent and loaded solvent tanks, which are elevated on a concrete platform under which is located the Sperry Vickers hydraulic drive unit for the mixers.

The area of the settlers is 53 m^2 for extraction and 8 m^2 for stripping. Each extraction settler is equipped with a vertical, movable, stainless steel baffle located near the discharge end of the settler. External jack-legs are provided on each settler for adjustment of interface levels.

Specific Flow. The design of the plant was based on the results of extensive test work conducted by Anglo-American on a KREBS pilot scale unit set up at VAAL REEFS EAST. The design of the settlers includes a 25% over design provision to allow expansion of throughput. The design specific flow for aqueous (dispersed phase) is 4.3 $m^3/h.m^2$ whereas the total specific flow is 10.8 $m^3/h.m^2$.

Key Lake - Canada

At the beginning of this year the KEY LAKE MINING COPRORATION and WRIGHT/COMMONWEALTH selected the KREBS Solvent Extraction System for the KEY LAKE project.

The SX plant consists of four extraction, three scrubbing, four stripping, one ammonia scrubbing, and one regeneration mixer-settlers.

Specific Flow. At the nominal operating rate the specific aqueous flow will be 3.5 $m^3/h.m^2$ and the total specific flow will be 8.8 $m^3/h.m^2$.

APPLICATION OF THE KREBS MIXER-SETTLER TO DIFFERENT HYDROMETALLURGICAL PROCESSES

From the very first development stages of the conical pump, the subsequent detailed development of the launder and settler, and the construction of the first commercial plants, KREBS has steadily accumulated data on scale-up factors for the system. In dealing with ores and the metallurgical products derived from them, solutions are never identical. The differences affect the treatment processes including solvent extraction. The policy of KREBS when undertaking a new project is to carry out the experimental work aimed at:

- Determining the specific settling rate.
- Checking the extraction efficiencies where possible

The specific settling rate is experimentally determined with actual solutions using a KREBS Mixer-Settler pilot unit specially designed for scale-up purposes. Dispersion band height in the settler is measured for various flow rates, O/A ratios, and temperatures. Specific flow rate in $m^3/h.m^2$ of settling area is calculated typically for dispersion band heights of about 200 mm or less.

The KREBS Mixer-Settler Pilot Unit

The specifications of the KREBS pilot unit are as follows (See Fig. 5 for drawing of this unit):

- Mixer	diameter	400 mm
	total height	580 mm
	working volume	40 l
- Impeller	diameter	120-150 mm
	four pitched blades	
- Conical Pump	cone diameter bottom	60 mm
	top	190 mm
	rpm	100-400
- Launder	length	1350 mm
	width	130 mm
	height	100 mm
- Settler	length	1060 mm
	width	180 mm
	height	860 mm
	working volume	152 l
	working area	0.173 m^2

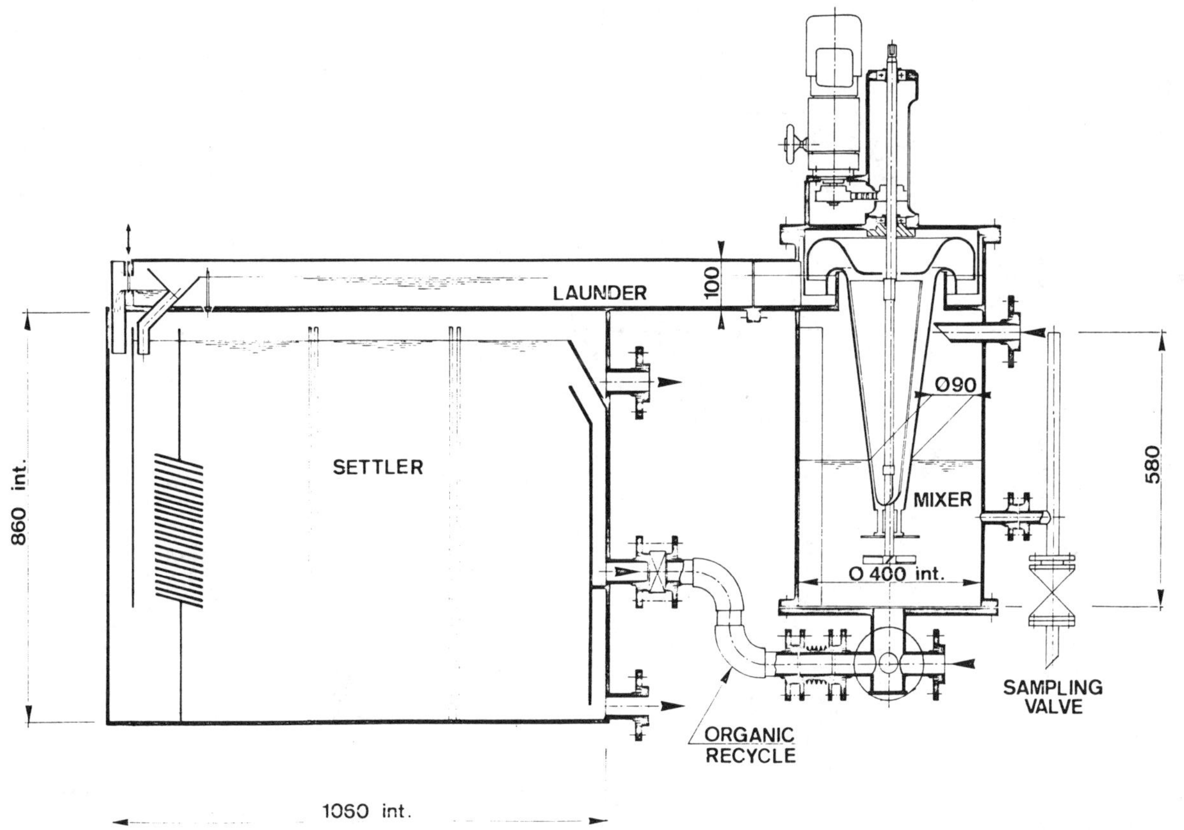

Fig : 5 _ KREBS MIXER SETTLER PILOT UNIT
SECTION VIEW

The pump can handle total flow rate up to 4.5 m^3/h. One or two baffles can be inserted in the settler at the dispersion band level.

Uranium Solvent Extraction

Uranium from H_2SO_4 Leaching of Ore. Solvent extraction of uranium from sulfuric acid leaching solution using tertiary amine as extractant is a widely used technique. Over the last five years many test programs have been undertaken by KREBS with different clients. Some of these programs were run at client's pilot plants where available solution was limited in quantity, while other programs were run in commercial plants when extension or revamping of existing plant was being considered. All the tests were run under similar conditions:

- Aqueous and organic solutions were fed to the unit at the advanced O/A designed for the project or used in the plant.
- Organic recycle was used when needed.
- Settling conditions: i.e., dispersion band height was measured as a function of the variation of aqueous flow, total flow, and pump rpm.
- Stage efficiency was checked whenever possible by sampling the two phases from the settler and performing shake-out tests. Assays on organic and aqueous give the stage efficiency.
- Specific flow rate per unit area was plotted against dispersion band height.
- Organic in aqueous and aqueous in organic entrainments were determined by assay.

Results of these tests show that better specific settling flow rates are achieved using the KREBS Mixer-Settler than any other system to which it has been compared. Results vary over a wide range due to different pregnant solution characteristics and operating temperatures which differ for each plant.

Under similar testing conditions, i.e., for agitator and conical pump rotation speed in the range of 300-350 rpm and for O/A ratios of 1.0 to 1.5, the specific flow rates achieved ranged from 3.9 to 10.6 $m^3/h.m^2$ of aqueous feed or 8.0 to 26.0 $m^3/h.m^2$ of combined flow.

Table 1 gives the results of tests carried out in different operating plants in the USA, CANADA, and SOUTH AFRICA.

Figure 6 plots the dispersion band height versus dispersed phase flow rate. Narrow dispersion bands were never obtained in the WYOMING uranium plant due to long disengagement times and low temperatures.

TABLE 1

Determination of the Specific Settling Flow Rate
(Tests in Operating Plants)

URANIUM MILL LOCATION	TEST N°	FLOWRATES AQUEOUS m^3/h	FLOWRATES ORGANIC m^3/h	O/A RATIO	DISPERSION BAND mm	SPECIFIC FLOWRATE DISPERSED PHASE $m^3/h.m^2$	SPECIFIC FLOWRATE COMBINED FLOWS $m^3/h.m^2$	ORGANIC ENTRAINMENT ppm
SOUTH AFRICA	5	0.71	1.02	1.44	4	4.1	10.0	16 (1)
	7	0.98	1.09	1.11	7	5.6	11.9	13 (1)
	9	1.09	1.35	1.24	15	6.3	14.1	7 (1)
	11	1.35	1.51	1.12	40	7.8	16.5	-
NEW MEXICO U S A	1	0.7	0.24	1.2	15	4.0	8.9	39
	2	0.75	0.36	1.46	45	4.3	10.7	37
	3	1.04	0.27	0.8	40	6.0	10.8	37
	7	0.908	0.30	1.2	50	5.2	11.5	200
	8	1.19	0.45	1.09	70	6.9	14.4	92
WYOMING U S A	9	0.77	0.26	1.31	362	4.4	10.3	-
	10	0.68	0.24	1.05	362	3.9	8.0	-
	11	0.66	0.24	1.04	254	3.8	7.8	-
	12	0.68	0.24	1.15	317	3.9	8.4	-
SASKATCHEWAN CANADA	15	0.80	0.57	1.9	13	4.6	13.4	-
	17	1.39	0.52	1.3	89	8.0	18.5	-
	24	1.83	0.52	1.1	0	10.6	22.2	81
	27	1.8	0.52	1.5	25	10.4	26.0	30

(1): Amine content only

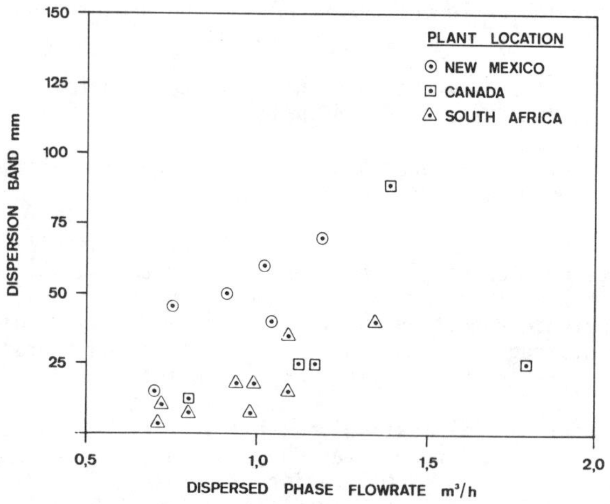

Fig:6 _ DISPERSION BAND VERSUS DISPERSED PHASE FLOWRATE

Stage efficiency was calculated during the pilot campaigns in NEW MEXICO and SOUTH AFRICA. The method consisted of assaying uranium in samples of raffinate and organic from the mixer before and after shake out with O/A = 1/1. Stage efficiency achieved was 99-100% with some low figures of 95-96%.

In the case of new projects or studies where large amounts of solution were not available, tests were done using the clients' pilot plant facilities. In this case, aqueous solution was recycled to the mixer-settler and hydraulic measurements (dispersion band, flow rate) only were taken. Stage efficiency and organic entrainment were not determined. Generally, results are consistent and reliable for scaling-up.

Uranium from Phosphoric Acid Solution. Phosphoric acid solution obtained by processing phosphate rocks with concentrated sulfuric acid is a large potential source of uranium. Phosphate rocks from FLORIDA and MOROCCO contain valuable amounts of uranium.

Two series of tests were performed in 1977 and 1979 with phosphoric acid ex-FLORIDA and ex-MOROCCO. Results are summarized in Table 2.

TABLE 2

Extraction of Uranium from Phosphoric Acid

	H_3PO_4 30 % P_2O_5 Ex FLORIDA	H_3PO_4 27/30% P_2O_5 Ex MOROCCO
Extraction Solvent	O. P. P. A.	DEPA - TOPO
Acid Treatment	Settled and filtered	Stored and Settled
Temperature	30 °C	35 - 40 °C
O/A ratio	1.1	1.1
Mixer-Settler		
Mixer retention time	1 min	3 min
Settler area	0.125 m^2	0.4 m^2
RESULTS		
Aqueous specific flowrate	4,3 $m^3/h.m^2$	3.75 $m^3/h.m^2$
Combined specific flowrate	9.3 $m^3/h.m^2$	7.9 $m^3/h.m^2$
Solvent entrainment	70 - 200 ppm	100-150 ppm
KREBS Specific Flowrate / Conventional Specific Flowrate	1.0/0.36	-

Vanadium Solvent Extraction

Solvent extraction of vanadium has been tested with a KREBS Mixer-Settler pilot unit. The company which performed the tests treats several different vanadium-bearing materials from various sources, spent catalysts, and fly ash from oil-burning facilities.

Tests were done with solutions coming from six different products. The solvent is a mixture of organophosphoric compounds in kerosene. Extraction was operated at an O/A ratio of 1/1. The pilot unit was run both aqueous continuous and organic continuous. In both cases the machine performed very well. For five solutions out of six the specific settling rate obtained for dispersion bands of less than 200 mm was high and varied from 6.5 $m^3/h.m^2$ up to 14.0 $m^3/h.m^2$ of aqueous, depending on the solutions. These figures correspond to specific total flows of 10 to 18 $m^3/h.m^2$. So as to take into account the behavior of the worst solution, a specific flow rate of 5.3 $m^3/h.m^2$ of aqueous phase was chosen for plant design purposes.

Copper Solvent Extraction

Recently, some work has been done in copper solvent extraction for a large copper company. The hydrodynamics of the copper sulfide leach solution and LIX 64N in kerosene were studied over a rather limited period and no attempt at optimization was made. The phase characteristics were:

Copper solution	pH	2.4
Solvent	LIX 64N	10% vol
	Kerosene	90% vol

During the tests flow rats were varied from 0.61 to 1.56 m^3/h of aqueous feed and from 0.75 to 1.77 m^3/h of organic feed. O/A ratios were kept in the range of 1.1 to 1.5. Dispersion bands were not noticed for any flow. Raffinate was always clear and free of visible entrainment. Rotation speed of the agitator and conical pump varied from 275 to 350 rpm, having little effect on the disengagement time which ranged from 65 to 100 sec.

Results show that high specific flow rates may be achieved with reasonable organic losses. These losses may be improved by optimization of the impeller size and speed. Average results for different operating conditions are given in Table 3.

Miscellaneous

The KREBS Mixer-Settler pilot unit has also been used for the solvent extraction of several high-value metals. Plants producing these metals are generally of small capacities.

Advantages of the KREBS Mixer-Settler are in space savings requirements where dozens of mixer-settlers are needed and also in systems where conventional equipment, with the tendency to overmix, is not suitable.

Rare Earths Separation. A series of tests were performed by a French producer of rare earths. Rare earths analyses and solvent used in the extraction are confidential. The extraction was run aqueous continuous. Table 4 gives a specific flow rate obtained at various O/A ratios for a dispersion band of 150 mm.

TABLE 3

Solvent Extraction of Copper from Heap Leaching Solution
Using LIX 64N 10% Vol Solvent

FLOWRATES		O/A	SPECIFIC FLOWRATE		ORGANIC
AQUEOUS m^3/h	ORGANIC m^3/h	RATIO	DISPERSED PHASE $m^3/h.m^2$	COMBINED PHASES $m^3/h.m^2$	ENTRAINMENT ppm
0.61	0.79	1.3	3.5	8.1	20
0.59	0.84	1.4	3.4	8.2	49
1.05	1.34	1.3	6.1	13.8	54
1.57	1.77	1.13	9.1	19.3	60
1.57	2.35	1.5	9.1	22.6	61

TABLE 4

Solvent Extraction of Rare Earths

O/A	SPECIFIC FLOWRATE	
	DISPERSED PHASE $m^3/h.m^2$	COMBINED PHASES $m^3/h.m^2$
1.67	3.5	9.3
1.25	5.8	13.0
0.83	5.6	10.2

Rhenium-Molybdenum Separation. Solvent separation of rhenium from solutions of Re-Mo fly ash was carried out by a US company in their own plant. Solvent and solution characteristics are confidential. The KREBS Mixer-Settler was compared to the existing conventional circuit. Results are given in Table 5.

TABLE 5

Re-Mo Separation by Solvent Extraction
Comparison of Specific Flow

	CONVENTIONAL SYSTEM TOTAL FLOW $m^3/h.m^2$	KREBS SYSTEM TOTAL FLOW $m^3/h.m^2$	% INCREASE
SOLUTION 1	6.8 - 7.3	10.2	45
SOLUTION 2	3.9	13.2	240

Solvent Extraction of Niobium. Solvent extraction of niobium from solutions derived from sulfuric acid leaching of niobium concentrates was carried out. The solvent used, tributylphosphate (TBP), is very viscous and tests done with the KREBS Mixer-Settler gave a specific flow rate of 4.6 $m^3/h.m^2$ of dispersed phase when operating at O/A of 1 and with a dispersion band of 100 to 150 mm. Even this low rate is some six times better than rates with conventional mixer-settlers. Rotation speed higher than 320 rpm was found to be detrimental to phase separation.

ACKNOWLEDGMENTS

KREBS acknowledges all the companies and organizations which have welcomed our people and our equipment, and given us the valuable opportunities of working with them. We are proud of the confidence they have shown in our technology.

INDEX

A

B

C

F

G

H

I

J

K

L

M

Q

R

T

U